DATE DUE
F.61
APR 1 6 1991
991

Physical Principles of Pesticide Behaviour

Physical Principles of Pesticide Behaviour

The dynamics of applied pesticides in the local environment in relation to biological response

VOLUME 2

G. S. Hartley

36, Surrey Crescent,
Palmerston North,
New Zealand

I. J. Graham-Bryce

East Malling Research Station,
East Malling, Maidstone,
Kent, England

1980

ACADEMIC PRESS

A Subsidiary of Harcourt Brace Jovanovich, Publishers

London · New York · Toronto · Sydney · San Francisco

ACADEMIC PRESS INC. (LONDON) LTD
24–28 Oval Road,
London NW1

U.S. Edition published by
ACADEMIC PRESS INC.
111 Fifth Avenue,
New York, New York 10003

British Library Cataloguing in Publication Data

Hartley, Gilbert Spencer
Physical principles of pesticide behaviour.
Vol. 2
1. Pesticides
I. Title II. Graham-Bryce, I J
632′.95 SB951 80-40082

ISBN 0-12-328402-3

Filmset by The Universities Press (Belfast) Ltd and printed by John Wright & Sons Ltd, Bristol

Contents

Contents of Volume 1 x
Note on Internal References xi
List of Symbols xii

10. Relation of Permeation to Toxicity

I. Introduction 519
II. The "Hansch" approach to internal access factors 520
III. Optimal partition in the crescent state 522
IV. Optimum partition for eventual transfer 529
A. Finite applied dose 529
B. Different partition effects on uptake and loss 531
C. Interaction of permeation and decay 533
V. Optimum partition for efficient toxic reaction 536
VI. Optima correlated with partition 538
VII. Influence of solubility 541
VIII. Partition into other phases 542

11. Penetration of Pesticides into Higher Plants

I. Introduction 545
II. External morphology 547
A. The cuticle 547
B. Epicuticular wax 549
C. The cutin layer 550
D. Stomata, trichomes and grooves 553
E. Roots 555
III. Internal morphology 556
A. The translocation systems 556
B. Apoplast, symplast and free space 557
IV. Entry through stomata 559

V. Permeation of leaf cuticle . . . 561
A. Introduction . . . 561
B. Cuticular waxes . . . 561
C. Permeation of detached cuticle by water . . . 563
D. Swelling of the cuticle . . . 566
E. Permeation of the detached cuticle by other substances . . . 569
VI. Penetration into aerial parts . . . 579
A. Relevance of detached cuticle studies . . . 579
B. Measurement of uptake . . . 582
C. Reversibility of uptake . . . 587
D. Washing out by rain . . . 595
VII. Uptake into roots and xylem . . . 596
A. Entry into roots . . . 596
B. Translocation from roots to shoots . . . 600
C. Mechanisms of passage through selective barriers in roots . . . 606
D. Transport in the xylem . . . 610
E. Summary of the properties required for translocation in the apoplast . . . 611
VIII. Transport in the symplast . . . 612
A. Introduction . . . 612
B. Mechanism of phloem transport . . . 616
C. Active and passive transport . . . 619
D. Sources and sinks . . . 622
E. Transport in counter flow . . . 623
F. Selective entry into transport systems . . . 631
IX. Environmental effects . . . 633
A. Effect of light . . . 633
B. Effect of temperature on foliar uptake . . . 637
C. Effects of atmospheric humidity on foliar uptake . . . 639
D. Effects of temperature and humidity on uptake by roots . . . 642
X. Effect of additives on penetration . . . 646
A. Introduction . . . 646
B. Humectants . . . 647
C. Other auxiliary solvents . . . 649
D. Surfactants . . . 652
E. Nutrient and salt additives . . . 654

12. Penetration of Pesticides into Insects and Fungi

I. Introduction . . . 658
II. Penetration into insects . . . 659
A. Composition of the insect cuticle in relation to penetration . . . 659
B. General features of the penetration process . . . 662
C. Effects of solvent . . . 669
D. Effects of penetrant properties . . . 672
E. Uptake of insecticide vapour . . . 676
III. Penetration into fungi . . . 679
A. Barriers to penetration . . . 679
B. Routes to absorbing surfaces in fungi . . . 681

C. Uptake of fungicides by fungi . . . 684
D. The detailed pattern of fungicide accumulation . . . 686

13. Effects of Growth and Movement of Organisms on Interception of Pesticides

I. Introduction . . . 698
II. Control of soil-inhabiting pests and pathogens . . . 699
A. Introduction . . . 699
B. Action of seed treatments . . . 701
C. Interaction between biological factors and toxicant supply from seed treatments . . . 704
III. Control of insects and pathogens attacking aerial parts of plants . . 710
A. Introduction . . . 710
B. Pick-up of pesticide by insects from residual deposits . . . 713
C. Effects of humidity on pick-up from deposits . . . 720
D. Sites of uptake on insects walking on residual deposits . . . 722
E. Uptake of pesticides from residual deposits by crawling insects . 725
F. Selectivity between different insect species . . . 727
G. Interception of pesticide sprays by flying insects . . . 730
IV. Movement of fungal spores in relation to contact with fungicides . . 735
V. Attractants, repellents and related behaviour-controlling chemicals . . . 740
A. Introduction . . . 740
B. Directional response to behaviour-controlling chemicals . . . 742
C. Sensitivity of response to behaviour-controlling chemicals . . . 748
D. Practical use of behaviour-controlling chemicals in pest control . 754
VI. Growth of plants in relation to uptake of soil-applied pesticides . . 757

14. Application and Formulation

I. Introduction . . . 764
II. Application methods . . . 765
A. Dispersion and localization of spray . . . 765
B. Formulation for spray . . . 766
C. Dusts . . . 768
D. Fumigation . . . 768
E. Smokes . . . 769
F. Granules . . . 770
G. Foam . . . 773
H. Baits . . . 774
III. Limiting factors in formulations . . . 774
A. Introduction . . . 774
B. Melting point and solubility limits . . . 775
C. Chemical stability . . . 777
D. Corrosion and compatibility . . . 778

IV. Uniformity and control of applicance 781
A. Metering 781
B. Uniformity across the swath 784
C. Uniformity along the swath 790
D. Uniformity of penetrated spray 791
V. Retention 792
A. Introduction 792
B. Reflection of liquid drops 793
C. Retention of solid particles 796
D. Electrostatic attraction 798
E. Adhesion of particles 800
F. Weathering of deposits 803
G. Selective adhesion 806
H. Adhesive formulations 809
J. Selective availability 811
VI. Control of drift 812
A. General 812
B. Avoidance of small drops 814
C. Spray thickeners 815
D. Deposit stickers 818
E. Reduction of vaporization 818
F. Reduction of reception 818
G. Drift of dust and micro-granules 818
VII. Application of pesticides to soil 820
A. Introduction 820
B. Release from initial source and its significance 821
C. Patterns of distribution 823
D. Diffusion from sources distributed in limited volume 824
E. Effect of chemical decay 831
F. Leaching from discrete sources 835
G. Accelerated solution of active ingredient 837
H. Granule–soil-water contact 840
J. Internal and external delay 841
K. Loss from surface application 846
VIII. Seed treatment 849
A. Introduction 849
B. Methods of applying pesticides to seeds 851
C. Evaluation of seed treatment processes 853
D. Effects of seed treatment formulation on biological activity 854
E. Pelleting and coating 856
IX. Controlled release 857
A. The advantages of controlled release 857
B. Retardation of release—physical 860
C. Retardation of release—chemical 864
D. Delayed release 866
E. Encapsulation 868

Appendix 1. Coefficients of molecular diffusion 871
Appendix 2. Vapour pressure—general 883
Appendix 3. Ionization constants—general 887
Appendix 4. Physical properties of pesticides 896
Appendix 5. Meteorological factors influencing pesticide behaviour 926
Appendix 6. Drop geometry and population density 932
Appendix 7. Interception of non-vertical spray by non-horizontal leaves 937
Appendix 8. Some dimensions of organs and pests 940
Appendix 9. Structural formulae of pesticides 948
References 975
Index 1009

Contents of Volume 1

Preface v
Acknowledgements xiii
Contents xv
Contents of Volume 2 xx
Note on internal references xxi
Lists of Symbols xxii

1. Dosage and Response 1
2. Solubility, Partition and Adsorption 10
3. Principles of Diffusion and Flow 110
4. Chemical Transformation 204
5. Behaviour in Soil and Other Porous Granular Materials . . 236
6. Behaviour of Pesticides in Air 332
7. Pesticides in Aquatic Situations 386
8. Coverage and Spreading 402
9. Permeation 470

Note on Internal References

There is much cross-referencing in this book to "other" aspects of a particular description or argument. To facilitate location of these, all pages carry the chapter number (arabic), section number (roman) and sub-section (capital letter) in the running headline—e.g. 3.VIII.B. Not all sections are sub-divided, so the final letter may be missing—e.g. 6.V.

References to other parts of the text are made by similar abbreviation in brackets—e.g. (see Section 5.VIII.B). In a few cases, where the point referred to is very specific or rather outside the main subject matter of the sub-section, a *page* reference is given.

Tables, equations and figures (including photographs) are numbered consecutively within chapters, each in its own series. When reference is made to these locally, this number only is given, but where the equation or other item is more remote, the number is followed by the standard code above—e.g. (eq. (63), Section 3.VIII.B). We hope this will save the reader the frustration of flicking through many irrelevant pages.

List of Symbols

Latins

Symbol	Meaning
A	Molecular species.
A	Equation coefficient; varies with context.
$\mathbf{A}$	Area.
a	Radius of drop or particle.
$\mathscr{A}$	Applicance.
B	Molecular species.
B	Equation coefficient; varies with context.
b	Radius (outer limit).
C	Carbon.
C, c	Concentration in solution or vapour.
C_{d}	Drag coefficient.
D	Dielectric constant.
D	Diffusion coefficient.
d, ∂	Differential operator, complete and partial.
E	Energy of adhesion.
e	Base of natural logarithms.
F	Fluorine.
F	Retention function.
$\mathbf{F}$	Total flux (of pesticide unless otherwise specified by suffix).
f	Function, general. Varies with context.
f	Fraction free (not adsorbed).
g	Acceleration of gravity.
H	Hydrogen.
H	Relative humidity.
$\mathbf{H}$	Heat content.
h	Height of drop or water column.
I	Availance.
I	Light intensity.
K	Potassium.

Symbol	Meaning
$\mathbf{K}$	Equilibrium constant.
K_{d}	Adsorption coefficient.
K_{θ}	Drop geometry function.
K_{R}	Retention factor.
K	Convenient local constant, depends on context.
k	(Reaction) velocity constant.
$\mathbf{k}$	Boltzmann constant.
$\mathcal{K}$	Transfer coefficient.
L	Length in dimension equations.
ℓ	Thickness or distance of separation (n.b. ℓ. = litres).
M	Cation (metal) species.
M	Mass in dimension equations.
$\mathbf{M}$	Mass applied or transferred, dose.
m	Population or convenient number.
N	Nitrogen.
N	Mol fraction.
n	Freundlich index.
n	Population or convenient number.
O	Oxygen.
P	Phosphorus.
P	Permeance.
p	Permeability.
p	Convenient index coefficient.
$\mathbf{P}$	Hydrostatic pressure.
$\mathbf{p}$	Vapour pressure.
$\mathcal{P}$	Partition coefficient.
Q	Amount adsorbed.
Q	Quantity of electricity.
q	Convenient index coefficient.
$\mathbf{q}$	Concertina factor.
R	Anion species.
R	Radius of curvature.
$\mathbf{R}$	Gas constant.
$\mathcal{R}$	Resistance.
r	Radius in general or of contact circle.
r	Number in a series.
S	Sulphur.
s	Porosity, i.e. fractional volume, not solid.
T	Time in dimensional equations.
$\mathbf{T}$	Temperature, absolute.
t	Time elapsed.

u_* Frictional velocity.
u Linear velocity.
$\mathbf{u}$ Ionic conductivity.
V Specific or mol volume.
$\mathbf{V}$ Volume in system.
v Fractional volume.
v Linear velocity.
$\mathcal{V}_R$ Rainfall rate.
w Width.
X Definite distance.
x Variable distance.
y Variable distance, *or* convenient variable.
z Variable distance, *or* convenient variable.
Z Definite distance, *or*, with ± index, ionic valence.

Greeks

α Angle to horizontal.
$\boldsymbol{\alpha}$ $\sqrt{k/D}$ [(reaction velo.)$^{\frac{1}{2}}$ (diffusion coeff.)$^{-\frac{1}{2}}$].
β Arbitrary angle *or* roughness factor.
$\boldsymbol{\beta}$ Bowen ratio.
γ Surface tension.
$\boldsymbol{\gamma}$ Shape factor.
δx Small increment of x.
δ Small thickness.
ε Electronic charge.
ζ Electrokinetic potential.
η Viscosity *or* convenient variable.
θ Angle in general *or* contact angle *or* convenience index.
$\boldsymbol{\theta}$ Ratio of radii.
κ Reciprocal ionic atmosphere radius *or* thermal conductivity.
λ Latent heat.
Λ Shape factor.
μ Chemical potential.
$\boldsymbol{\mu}$ Shape index.
ν Kinematic viscosity.
ξ Permeation function.
π Circumference/diameter.
π $\mathrm{Log}_{10}\,\mathcal{P}$.
Π Osmotic pressure.

ρ	Density.
σ	Specific heat.
$\boldsymbol{\sigma}$	Charge density.
τ	Tortuosity factor.
$\boldsymbol{\tau}$	Compound time measure.
ϕ	Arbitrary angle *or* water potential.
$\boldsymbol{\phi}$	Friction coefficient.
Φ	Volume velocity of solvent or solution.
χ	Brenner number.
ψ	Electric potential.

Suffices

Where the suffix is not self-explanatory (e.g. liq = liquid, aq = water don = donor, rec = receiver etc.) the following are used:

A	Advancing (contact angle).
B	Bulk (concentration in whole solution, or bulk density).
G	Gas.
x	Value at distance x etc.
0	Value at $x=0$ or $t=0$ according to context.
O	Oil, in contrast to W = Water.
R	Radiation (of flux) *or* receding (of contact angle).
M	Momentum.
S	Solid (in contrast to L = Liquid) *or* surface (in contrast to B = Bulk).

Other suffices are specified locally.

Mathematical Functions

$\sinh x =$ hyperbolic sine of $x = \frac{1}{2}(e^x - e^{-x})$

$\cosh x =$ hyperbolic cosine of $x = \frac{1}{2}(e^x + e^{-x})$

$\cosh^2 x - \sinh^2 x = 1$

$$\frac{d \sinh x}{dx} = \cosh x \qquad \frac{d \cosh x}{dx} = \sinh x$$

Other relations as for trigonometric functions, e.g. $\tanh x = \sinh x/\cosh x$.

$$\text{erf } z = \text{error function of } z = \frac{2}{\sqrt{\pi}}\int_0^z e^{-\eta^2}\, d\eta$$

$$\text{erfc } z = 1 - \text{erf } z$$

$$\text{ierfc } z = \int_0^z \text{erfc } \eta \, d\eta$$

Notes Added in Proof

Corrections to Volume 1

p. 173, line 9 *up*: for "*K*-independent" read "*K*-dependent".
p. 182, line 3: for "above" read "below".
p. 193, eq. (108): $\int$ in denominator should be $\sqrt{\ }$.
p. 206, "Diffusion" in last paragrah of section refers to simple liquids only.
p. 430, line 7 *up*: for "Phosphoric" read "phosphine".
p. 465, text reference above Fig. 18 should be to Figs 17 and 18.
p. 493, line 5: for "50" and "60" read "5" and "6".

Addenda

1. Rapid penetration of massive local application of *liquid* ("drenching") into soil is an extreme form of the process described in Section 5.III.D. Reluctant wetting of dry soil surface can be overcome by wetting agent but the soil must have coarse crumb structure below to enable rapidly channelling water to carry down dissolved or finely-dispersed pesticide with little restraint by adsorption or adhesion. T. E. T. Trought (*N.Z. J. Agric.* 1971, **123,** 20; see also R. N. Watson and N. R. Wrenn, *Proc. 23rd N.Z. Weed Pest Cont. Conf.* 1980, p. 151) applied water-based insecticide to pasture in non-spreading jets several cm apart, so reducing evaporative loss (cf. p. 848) and reaching mobile root-feeding grubs with minimal effect on non-mobile microorganisms. Although referred to as the "jet-squirt" or "solid-stream" method, the essential feature is localization which might also be achieved with large (*c.* 5 mm) unbroken drops (pp. 704, 822 are also relevant).

2. Further to p. 759, protection of crop seedlings from soil-applied herbicides has been made more practicable by mixing seed (tomato) and active charcoal into potting compost with sufficient water to permit

plug-flow under pressure. Cylinders of mixture are forced into the treated soil (G. H. Friesen and P. B. Marriage, *Br. Crop Prot. Conf.-Weeds* 1978, p. 503; D. J. Swain, *Proc. 23rd N.Z. Weed Pest Cont. Conf.*, 1980, p. 186) and excess healthy seedlings later removed. The method could be made more efficient if mechanically elaborated to confine the seed to a central core.

10. Relation of Permeation to Toxicity

I. Introduction 519
II. The "Hansch approach" to internal access factors 520
III. Optimal partition in the crescent state 522
IV. Optimum partition for eventual transfer 529
A. Finite applied dose 529
B. Different partition effects on uptake and loss 531
C. Interaction of permeation and decay 533
V. Optimum partition for efficient toxic reaction 536
VI. Optima correlated with partition 538
VII. Influence of solubility 541
VIII. Partition into other phases 542

I. INTRODUCTION

The biochemical processes responsible for the toxic effects of pesticides lie outside the scope of this book. However, penetration of chemicals from the outside of organisms is inseparably linked with their fate within, and physical chemistry has become thoroughly enmeshed with biochemistry in attempts to unravel this fate, so that some incursion into behaviour within the organism is essential.

It is a truism to state that the effectiveness of a toxicant applied to the external surface of an organism depends not only on its intrinsic activity at the site of action, but also on the efficiency of its transfer to that site which in turn depends on the relative rates of transport, chemical decay, metabolism and excretion. There has therefore been considerable interest in the molecular properties influencing this efficiency of transfer. It has long been recognized, as indicated in other sections of this book (Section 2.X.C, 3.VI), that transfer across lipid barriers, including cell membranes, is largely determined by the oil/water partition ratio of the penetrating molecules. Cell membranes, despite their apparently delicate structure and specialized permeability mechanisms for ions, behave in some respects more simply with regard to unionized foreign molecules than do external cuticles. They are much thinner and more fluid than the macromolecular cuticles and usually encounter molecules of toxicant only at great dilution, whereas the active compound is often applied to the outside of a cuticle in a concentrated non-aqueous solution. What is not so generally recognized is the way in which activity depends on how the kinetics of the toxic reaction relate to thôse of penetration, for example whether uptake corresponds to a crescent or steady state. Not only absolute toxicities, but also relationships between the toxicities and molecular properties of different compounds can vary greatly with the way in which the tests are conducted, for example they may depend on the solvents in which the chemical is administered, and on whether it is applied as a discrete dose (as with topical application to insects) or as a sustained supply (as when fungal spores are treated in aqueous suspension or in agar incorporating the toxicant). It is worth stressing strongly that structure–activity relationships established in laboratory tests under conditions far-removed from those applying in field treatments could lead to the selection for practical development of quite the wrong compound from a series.

The object of this chapter is to examine some general features of relationships between permeation and toxicity, to identify some of the important factors and establish how they can exert their influence. A useful starting point for this examination is the widely held view that there is an optimum efficiency of dosage transfer to the active biochemical site at a finite value of the partition coefficient. The concept that there is a convex-upwards parabolic relationship between efficiency and the logarithm of the partition coefficient has been particularly emphasized by Hansch and co-workers. The immense number of chemical analogues which can be made once some new physiologically active group has been discovered has naturally led to efforts to codify and predict the effects of variously placed substituents. Of these the "Hansch approach" has been most prolific of further research. It was originated by Hansch and Fujita (1964) and has subsequently formed the subject of a multi-author review (ACS, 1970) where its application in various fields is discussed and it is compared with the approach of Free and Wilson (1964). Although we are mainly concerned with the permeability aspect, it will be helpful to review the Hansch treatment more generally first.

II. THE "HANSCH APPROACH" TO INTERNAL ACCESS FACTORS

The Hansch approach is most successful for examining the structure-activity patterns of chemicals administered in solution in systems which are physically rather simple. On the basis of tests with a wide range of analogues of a parent compound, an empirical relationship is worked out between the logarithm of the molar concentration causing a standard response and various parameters which reflect the partitioning, electronic and steric properties governing penetration to the site of action and reactivity at this site. All these have characteristic contributions from each substituent. These relationships can then be used for predicting the potency of unmade compounds from the same series. One does not expect the empirical relationship to apply when a substituent can alter the nature of the toxic reaction itself; a separate relationship must be worked out for each different type of toxic reaction.

The substituent constants used in the Hansch relationships are π, a free energy-related constant defined as $\log \mathscr{P}_X - \log \mathscr{P}_H$ where $\mathscr{P}_X$ is the partition coefficient (in favour of octanol from water) of derivative X and $\mathscr{P}_H$ is that of the parent molecule; the Hammett (1970) polarity constant, σ, governing the electronic displacement in the molecule and hence the log (dissociation constant) for an acidic group, and the Taft (1956) steric constant, $\mathscr{E}_S$, governing the interfering effect on the hydrolysis velocity of esters of groups introduced on C atoms vicinal to the –COO– group. Within limits (see Section 2.X.C) partition coefficients for other systems can be related to those between octanol and water using the relationship (Collander, 1951 and Section 2.X.B) $\log \mathscr{P}_1 = (\text{const})_1 \times \log \mathscr{P}_2 + (\text{const})_2$ or to mobility on reverse-phase thin layer chromatography (Boyce and Milborrow, 1965) using the relationship $\mathscr{P} = (\text{const})_3\ (1/R_F - 1)$. Revised values of the Taft constant according to Hancock *et al.* (1961) making allowance for hydrogen bonding effects are sometimes used. Some authors prefer to use log (dissociation constant) itself, when applicable, in place of the Hammett constant and molecular volume in place of the Taft constant. The latter are by no means equivalent and the total molecular volume seems certainly more relevant to permeation rates. The Taft and Hammett constants are introduced to allow for non-specific reactivity effects.

All these constants are, to a first approximation, additive properties of the molecular groups comprising the molecule and, provided extrapolation is limited to a small range and care is taken where groups are in a position to interfere (e.g. –OH and $-NO_2$ groups in a benzene ring can be regarded as reasonably additive in effect except where adjacent) the calculation of partition coefficients is acceptable. The partition coefficient is the major factor in activity changes due to permeation effects when compounds of closely similar chemical function are compared.

The approach of Free and Wilson (1964) is more completely empirical in that no attempt is made to relate the effect of substitutions on other properties than the drug action under examination. It assumes that the effects of various substitutions are additive except that in some locations a special value for a substituent may be assumed. If there are, say, six different substitutable H atoms in the parent molecule and each may be left unsubstituted or carry one of four substituent groups, the number of possible compounds is $5^6 \simeq 15\,000$. As few as 30 well-chosen compounds could provide a reasonable prediction of the behaviour of the rest. The success of such a purely empirical approach should make us cautious of reading too much into the Hansch equations. An equation based on almost any three physical properties could provide a reasonable fit with another set of data. This is in no way to decry the value of the approach but only to warn against unjustifiably definite conclusions.

III. OPTIMAL PARTITION IN THE CRESCENT STATE

Many simple toxicity tests—reaction of isolated organelles or enzymes to dissolved drugs, shake-culture tests against fungi or bacteria—can be correlated to structure by equations involving one or two of the constants mentioned, usually including $\pi(=\log \mathscr{P})$ and usually linear equations only. Many such are listed by Fujita (1972) and Hansch (1972), but Hansch considers that a correlation over a wide range of partition values cannot be generally successful without a π^2 term also. This term is negative, the equation being of the form $\log(1/C) = K_1\pi - K_2\pi^2 + K_3\sigma + K_4\mathscr{E}_S + K_5$. There is therefore a general tendency for there to be an optimum value of π or $\log \mathscr{P}$. The evidence for an optimum value, particularly when the tests are more complex, seems clear.

A theoretical justification for the π^2 term is given by Penniston *et al.* (1969) but applies only when the toxic reaction occurs before the steady state is reached. It cannot be true of the steady state. The toxicant must permeate structures having both aqueous and lipid elements. It is considered that if the partition in favour of the lipid phase is low, permeation will be low since the chemical can only slowly cross the lipid layers. If the partition is very high, the chemical will be held up in the lipid layers. The latter statement is only true when a sufficiently early stage of the permeation process is considered, as is implicit in the subtitle of Penniston's paper—"A non-steady state theory"—but the restriction and its further implications are not discussed in the text.

In the model used by Penniston *et al.* for computation, transfer is considered to occur across a series of alternate aqueous and lipid layers.

The rate is considered to be controlled by processes at the interfaces, there being no resistance within each layer, so that concentration within each layer may vary with time but is uniform from one interface to the next. At each interface the flux of penetrant substance from aqueous to lipid $= k_1 \times$ concentration in aqueous and from lipid to aqueous $= k_2 \times$ concentration in lipid (we use our own symbols to avoid confusion in this text). Zero net transfer will result when

$$k_1 C_{\mathrm{aq}} = k_2 C_{\mathrm{lip}} \tag{1}$$

and, since this means that the layers are in equilibrium, it follows that

$$k_1/k_2 = \text{partition coeff. } \mathscr{P}_{\mathrm{aq}}^{\mathrm{lip}} \tag{2}$$

We have elsewhere (Section 3.VI) reviewed the evidence for an interfacial barrier to diffusion and concluded that it is certainly far less than the resistance of the thinnest manipulable liquid layers. The model therefore has an unsatisfactory physical basis but it may nevertheless indicate a valid mathematical trend.

It is assumed that there are a large number, *n*, of layers of each kind and it can be shown that the model leads to the expression

$$\frac{\mathbf{F}}{\mathbf{A}} = \frac{k_1}{2n} C_{\mathrm{aq},0} \tag{3}$$

for the flux per unit area at the steady state from $C_{\mathrm{aq},0}$ in water on one face to $C = 0$ on the other. If the supply concentration is measured in the lipid phase

$$\frac{\mathbf{F}}{\mathbf{A}} = \frac{k_2}{2n} C_{\mathrm{lip},0} \tag{4}$$

Substitution of eq. (2) makes (3) and (4) identical.

For convenience of computation, Penniston *et al.* make the arbitrary assumption that $k_1 \times k_2 = 1$. Combining with (2) and (3) this gives, for the steady-state flux across the multilaminate from a constant aqueous reservoir to a sink,

$$\frac{\mathbf{F}}{\mathbf{A}} = \frac{1}{2n} \sqrt{\mathscr{P}_{\mathrm{aq}}^{\mathrm{lip}}} \cdot C_{\mathrm{aq},0} \tag{5}$$

indicating a monotonic increase of **F** with $\mathscr{P}$. It is difficult to give a physical interpretation of the $k_1 \times k_2 = 1$ assumption in a model the physical basis of which is unsatisfactory. The authors themselves say that it "results in a choice of time units", but one needs assurance that this choice is independent of the variable $\mathcal{P}$, the significance of which one wants to assess. We shall see below that a more realistic model gives the same qualitative conclusion—that, although the steady state flux is always a monotonic function of $\mathcal{P}$ (if this is the only variable) the amount or rate permeating at a given time well before the steady state is reached is maximal at some finite optimum value of $\mathcal{P}$.

The alternative diffusion treatment of transfer across a multilaminate septum assumes that, at each interface, the concentration on the immediate lipid side is always $\mathcal{P}$ times that on the immediate aqueous side. Across each lamina the normal diffusion equation applies. We will consider the simple system where all the laminae have the same small thickness, δ. In the treatment below we will denote by C_r the concentration at the centre of the rth aqueous layer, and by c_1, c_2 the concentrations in the aqueous layers immediately adjacent to the lipid layer between the rth and (r+1)th aqueous layers. At the steady state we can write

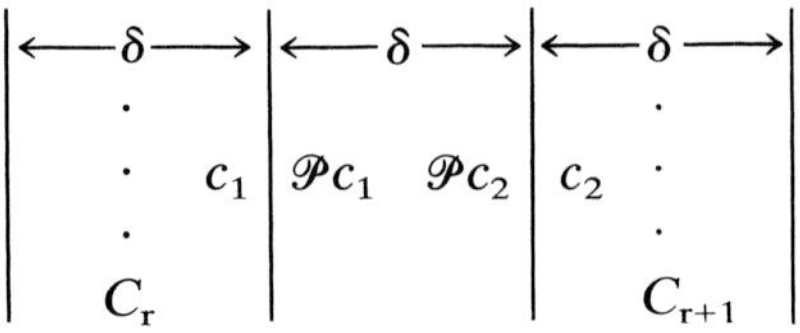

$\leftarrow\delta\rightarrow$	$\leftarrow\delta\rightarrow$	$\leftarrow\delta\rightarrow$
c_1	$\mathcal{P}c_1$ $\quad$ $\mathcal{P}c_2$	c_2
C_r		C_{r+1}

three values for the constant flux,

$$\frac{\mathbf{F}}{\mathbf{A}}=\frac{2D}{\delta}(C_r-c_1)=\frac{D}{\delta}\mathcal{P}(c_1-c_2)=\frac{2D}{\delta}(c_2-C_{r+1}) \tag{6}$$

Re-forming the equations with one side containing C and c only, adding and re-arranging, we obtain

$$\frac{\mathbf{F}}{\mathbf{A}}=\frac{D}{2\delta}\cdot\frac{2\mathcal{P}}{1+\mathcal{P}}(C_r-C_{r+1}) \tag{7}$$

D is a diffusion coefficient. This is assumed the same in each of the layers. The equations are not altered in any qualitative respect if worked out on the more realistic assumptions of different thicknesses and different diffusion coefficients. The form of the variation of fluxes with $\mathcal{P}$ is not altered, only absolute values, including that of the optimum value (if one exists) of $\mathcal{P}$. The assumptions of equal thickness and equal diffusion coefficients are therefore made for the sake of simplicity.

Since we have a large number of equal laminae and a linear gradient, (7) can be rewritten

$$\frac{\mathbf{F}}{\mathbf{A}}=\frac{D}{X}\cdot\frac{2\mathcal{P}}{1+\mathcal{P}}C_0 \tag{8}$$

where X is the thickness of the whole septum ($=2\delta n$), C_0 is the constant concentration in an aqueous reservoir on one side and the other side is a perfect sink.

The steady-state flux is, as always, a monotonic function of $\mathcal{P}$ where this is the only variable, but a different function from that obtained from the Penniston *et al.* model. While zero for $\mathcal{P}=0$ on both models, consistent with the lipid layers becoming impermeable, it has a ceiling limit of 2 from eq. (8) for $\mathcal{P}\rightarrow\infty$ consistent with the resistance of half the thickness approaching zero. The permeance, unrealistically, increases indefinitely with $\mathcal{P}$ on the Penniston *et al.* model.

To evaluate the crescent state within the multilaminate septum it is convenient to assume that δ is so small and the layers so numerous that every part of it has an equal content of aqueous and lipid volumes and its concentration can be described by a mean value $\bar{C}$ and the flux related to this by a mean diffusion coefficient, $\bar{D}$. $\bar{C}=((1+\mathcal{P})/2)\times C_{aq}$. $\bar{D}$ can be obtained from the formal equation for the steady state

$$\frac{\mathbf{F}}{\mathbf{A}}=\frac{\bar{D}}{X}\cdot\bar{C}_0=\frac{\bar{D}}{X}\cdot\frac{1+\mathcal{P}}{2}\cdot C_0 \tag{9}$$

and the identification of this with (8), yielding

$$\bar{D}=\frac{4\mathcal{P}}{(1+\mathcal{P})^2}D \tag{10}$$

This quantity does indeed go through a maximum, at $\mathcal{P}=1$, tending to zero both for $\mathcal{P}\rightarrow 0$ and $\mathcal{P}\rightarrow\infty$. This is, however, relevant only to the crescent state, because, if a permanent external source is offered it must be in dilute solution in a solvent. The maintained concentration in this solvent is an (inverse) measure of toxicity when the exposures result in similar response. If this solvent is water, the flux, from eq. (8), is proportional to $\mathcal{P}/1+\mathcal{P}$, a monotonic function increasing with $\mathcal{P}$. If it is an oil having the same value of $\mathcal{P}$ as the lipid layers of the septum, the flux is proportional to $1/1+\mathcal{P}$, also a monotonic function but decreasing with $\mathcal{P}$. Application of a maintained concentration in some solvent of intermediate polarity, for which, for example, the partition coefficient from water might have the value $\sqrt{\mathcal{P}}$, could yield a maximum steady transfer at $\mathcal{P}=1$, but this is not the kind of highly artificial exposure in which an optimum is found: indeed few organisms could survive such solvent treatment even in the absence of the toxicant. If a discrete dose of toxicant is applied, soluble in the composite cuticle so that $\bar{D}$ from eq. (10) is applicable throughout, there will not be a steady state and the factors specifically relevant to the crescent state become more important. If the discrete dose is not wholly soluble, solubility, influenced by other factors than $\mathcal{P}$, becomes important, as described in Section 10.VII.

The time course of the flux into a sink after the sudden application of a (thereafter constant) concentration to the other side of an initially empty septum is given by Crank (1956, p. 48). In differential form and using our symbols, Crank's equation (4.25) becomes

$$\frac{1}{X}\frac{\mathrm{d}y}{\mathrm{d}\boldsymbol{\tau}} = 1 + 2\sum_{1}^{\infty}(-1)^n \exp(-n^2\pi^2\boldsymbol{\tau}) \tag{11}$$

where

$$y = \frac{\mathbf{M}}{\bar{C}_0 X\mathbf{A}} \quad \text{and} \quad \boldsymbol{\tau} = \frac{\bar{D}t}{X^2}$$

M being the amount passed per unit area up to time t, i.e. $\mathbf{F} = \mathrm{d}\mathbf{M}/\mathrm{d}t$. If $D = 1$, $\mathcal{P} = 1$ whence $\bar{D} = 1$ and $X = 1$, $x =$ time (in arbitrary units) and $\bar{C}_0 = 1$, then $1/X/(\mathrm{d}y/\mathrm{d}\boldsymbol{\tau})$ = flux *out* in arbitrary units. With these assumptions eq. (11) gives the basic curve, labelled $\mathcal{P} = 1$, in Fig. 1. If the only effect of

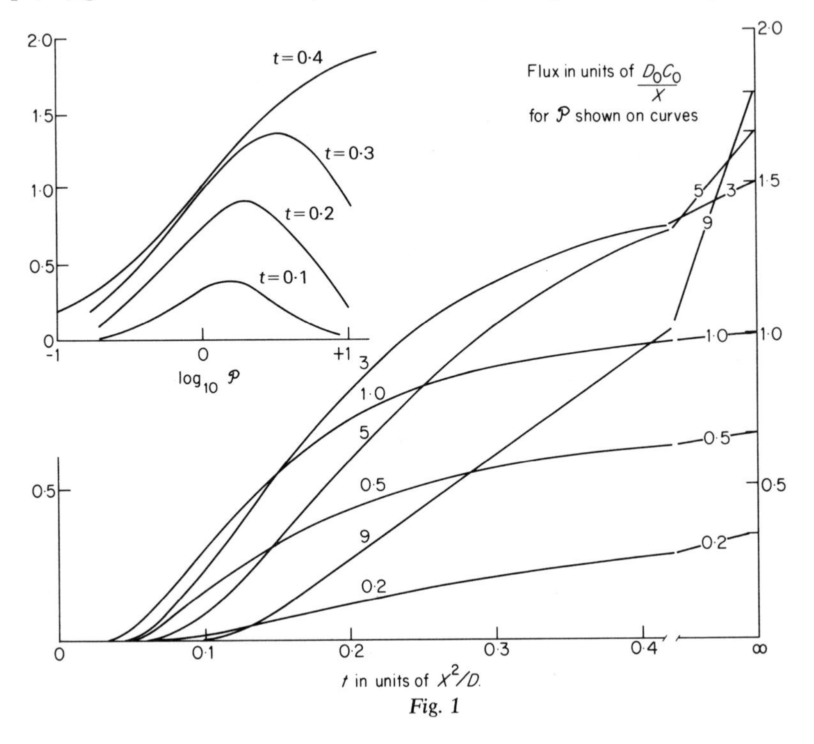

Fig. 1

change of $\mathscr{P}$ were on $\bar{D}$, according to eq. (10) we should find that $\mathscr{P}=1$ gave maximum rate at any one value of t because all the curves would be stretched on the t axis as $(1+\mathscr{P})^2/4\mathscr{P}$ increases from 1. This it does as $\mathscr{P}$ moves from 1 in either direction, $(1+\mathscr{P})^2/4\mathscr{P}$ being a symmetrical function, having the same value if $1/\mathscr{P}$ is substituted for $\mathscr{P}$.

However, the effect on $\bar{D}$ is not the only effect of $\mathscr{P}$ in any real situation. If a concentration can be suddenly presented it must be in the local environment, the ambient air or water. If the latter, the surface concentration, $\bar{C}_0$, in the composite septum is $1+\mathscr{P}/2\times$ the ambient water concentration, so that, to adjust the basic $\mathbf{F}-t$ curve to allow for change of $\mathscr{P}$ only, the t values must be increased by $(1+\mathscr{P})^2/4\mathscr{P}$ and the $\mathbf{F}$ values by $(1+\mathscr{P})/2/(1+\mathscr{P})^2/4\mathscr{P}=2\mathscr{P}/(1+\mathscr{P})$. This has been done on the figure for several values of $\mathscr{P}$. It will be seen, and is exemplified on the inset graph, that the eventual steady flux is now a monotonic function of $\mathscr{P}$ but that, at some early time during the crescent stage a maximum flux is found at a finite value of $\mathscr{P}$, this value increasing with increasing time. Correspondingly, if a critical flux must be achieved in a given time subsequent to the start of exposure, there would be an optimum value of $\mathscr{P}$ for which the necessary external concentration would be minimal.

Even more critically dependent on D, and therefore on $\mathscr{P}$, is the total amount entering the sink from the septum. Thirty times as much will have entered for $Dt/X^2=0{\cdot}1$ as for $Dt/X^2=0{\cdot}05$ where t is the time since the sudden application of concentration to the outside of the septum. The 30-fold increase would result from doubling either D or t, but it must be noted that this extreme dependence is confined to the very early stage of entry. At $Dt/X^2=0{\cdot}1$, the amount in the sink is only 1·6% of the eventual steady content of the septum itself. Even more extreme behaviour is found if a flow process is introduced into a partitioning system, as is exploited in partition chromatography, but all such early-stage differences are not likely to be important for eventual toxicity. They might explain an optimal effect of $\mathscr{P}$ on the rate of symptom development rather than on ultimate severity of symptoms. A crescent-state explanation might be acceptable where the duration of the major part of the internal pulse arising from an external application exceeded the normal life-span of the organism. This is unlikely to apply in the majority of cases examined by Hansch and followers.

So far we have considered only the multilaminate model and diffusion across the successive laminae. Permeation has a finite limit even for $\mathscr{P}\rightarrow\infty$ because the resistance of the (in our case) aqueous phase always operates. The opposite extreme would be a series of parallel aqueous and lipid paths. In this case permeance and capacity will both increase proportionally with $\mathscr{P}$ without limit. A cellular structure or a dispersion of lipid globules will behave in an intermediate manner. While the effect on capacity is proportional (in our case) to the lipid content, the effect on permeance shows steepening increase. Barrer (1968) has reviewed the various theories, models and experimental data for composite septa although his main interest was in the effect of an obstructive dispersed phase. It would be theoretically possible for the dispersed phase to contribute to capacity only, or even to reduce permeability at the same time, if it were enclosed in impervious capsules, each with only a single small entrance (as in the "ink-bottle" form of a porous structure, Section 3.IV.C). There is, on the other hand, no mechanism for the permeability factor to be increased by increase of $\mathscr{P}$ without simultaneous increase of the capacity factor.

The existence of an optimum $\mathscr{P}$ for early transfer through a composite septum depends upon the permeability and capacity factors being differently influenced. In the early stages, increased delay due to increased capacity (Section 9.II.D), has the major effect but must eventually give way to the increased steady transfer due to increased permeance. Calculation for a large number of geometrically different models has convinced us that steady-state permeation of any one septum from a constant source will always vary monotonically with $\mathscr{P}$. The form of variation and whether there is an upper finite limit can vary.

The Penniston *et al.* model has recently been the subject of a computer analysis (Dearden and Townend, 1976) which arrives at similar conclusions. The authors emphasize that "the wide variation with $\mathscr{P}$ of time to maximal concentration must arouse concern at the widespread practice of measuring biological effects of a series of congeners a fixed time interval after dosage". They adopt the same $k_1 \times k_2 = 1$ convention as Penniston *et al.* without pointing out its significant arbitrariness. It seems useful therefore to comment further on the relationship of the two models of diffusion through a multilaminate septum.

Equating the two expressions for flux, eq. (7) on the diffusion model and eqs (1), (3), (4) on the "interface jump" model without the arbitrary $k_1 k_2 = 1$ assumption, we obtain

$$k_1 = 2\frac{D}{\delta} \cdot \frac{1}{1+\mathscr{P}} \quad \text{and} \quad k_2 = 2\frac{D}{\delta} \cdot \frac{\mathscr{P}}{1+\mathscr{P}} \tag{12}$$

so that

$$k_1 k_2 = \left(2\frac{D}{\delta}\right)^2 \cdot \frac{\mathscr{P}}{(1+\mathscr{P})^2} \tag{13}$$

The assumption that D and δ are constant is made no more arbitrary, as far as the course of variation of permeance with $\mathscr{P}$ is concerned, by putting $D/2\delta = 1$. This is equivalent to assuming $k_1k_2 = \mathscr{P}(1+\mathscr{P})^2$. The Penniston *et al.* assumption of $k_1k_2 = 1$ is equivalent to putting $\mathscr{P}/(1+\mathscr{P})^2 = 1$ into the diffusion equation. This is more than just arbitrary when the variation of permeance with $\mathscr{P}$ is the subject of enquiry. The assumptions are incompatible physically. We have already shown (Section 3.VI) that there is no evidence of resistance in an interface, as compared with the phases of small but finite depth on either side of it, and therefore prefer the classical diffusion treatment.

IV. OPTIMUM PARTITION FOR EVENTUAL TRANSFER

If the optimum partition coefficient is both a real and rather general phenomenon, an explanation restricted to a very early stage of the crescent state is unsatisfactory because it is likely to apply only to rather artificially limited conditions of experiment. One needs an explanation not critically dependent on time factors. The permeability must exert a more complex sustained effect which it can do by interaction with *other rates.* Four broad possibilities present themselves: (1) limitation of external supply—applicable when response is to a discrete dose rather than a maintained exposure; (2) a sequence of differently affected permeances (modelling uptake and loss, intake and excretion); (3) interaction with chemical decay; (4) rate processes important in the actual toxic reaction.

A. Finite Applied Dose

The first mechanism is most simply exemplified by change in the boundary conditions only. We return to the multilaminate septum but now consider this to model the complete organism with a discrete dose applied on one face ($x = 0$) and complete elimination of the toxicant at the other face ($C = 0$ for $x = X$). A vulnerable site is considered to lie between, for convenience at $x = X/2$. After the application of a wholly dissolved dose to the $x = 0$ surface at time $t = 0$, the concentration at depth x in a semi-infinite continuum is given by

$$C = \frac{\mathbf{M}}{\mathbf{A}(\pi Dt)^{\frac{1}{2}}} \exp\left(-\frac{x^2}{4Dt}\right) \tag{14}$$

To obtain the value for a system where there is no escape at $x = 0$ and complete extraction at $x = X$, we repeatedly reflect this curve at X, $2X$,

$3X$ etc. without change of sign at the even numbers and with reversal at the odd numbers. The value at $X/2$ becomes

$$C_{X/2} = \frac{\mathbf{M}}{\mathbf{A}X} \cdot \frac{4}{\sqrt{\pi}} \cdot y \tag{15}$$

where

$$y = Z^{-\frac{1}{2}} \sum_{n=1}^{n=\infty} (-1)^{n+1} \exp((2n-1)^2/Z)$$

where

$$Z = \frac{16Dt}{X^2}$$

The pulse form of y against Z is plotted in Fig. 2. It represents $C_{X/2}$ against t when $D = 1$, $X = 4$ and $\mathbf{M}/\mathbf{A} = \sqrt{\pi}/16$. For comparison the curve for C_0 (at the site of application) is also shown. C_0 descends continuously at first very steeply from an artificial infinite $t = 0$ value due to assuming the applied surface dose to be dissolved in zero surface volume. The pulse

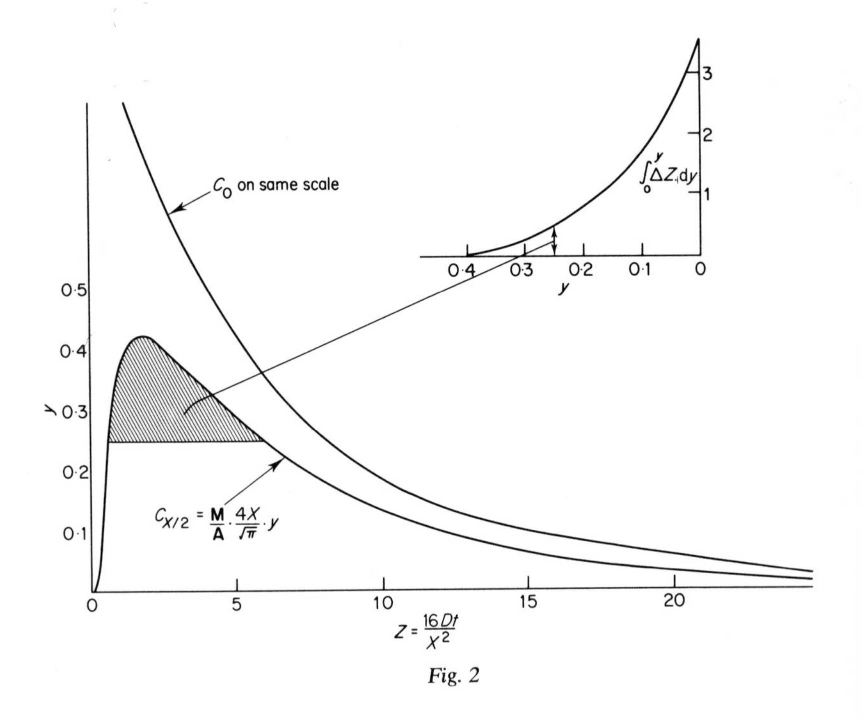

Fig. 2

form for $C_{X/2}$ is a very general one in such models and in practice. The rise is always steeper than the descent for the general reason that one starts from a sudden local application of high concentration with subsequent loss only after dispersal. In many cases the "tailing" is even more gradual than in this model.

One expects the area under the pulse curve, i.e. $\int_0^\infty C\,dt$, or the "availance", to be the quantity exerting major control over the amount of chemical damage resulting at some "micro" site at level $x = X/2$ (i.e. a site which does not extract sufficient of the toxicant to disturb the main time course of concentration). Denoting the availance by I, we find

$$I_{X/2} = \int_0^\infty C_{X/2}\,dt = \frac{\mathbf{M}}{\mathbf{A}X}\frac{4}{\sqrt{\pi}}\int_0^\infty y\,dt = \frac{\mathbf{M}X}{\mathbf{A}}\frac{1}{4\sqrt{\pi}D}\int_0^\infty y\,dZ \qquad (16)$$

The area $\int_0^\infty y\,dZ$ (Fig. 2) is found to be $2\sqrt{\pi}$ so that $I_{X/2} = \mathbf{M}X/2\mathbf{A}D$, a result which follows much more simply from the steady-state analogy developed by Hartley (1963, see Section 3.V.C). The ratio of availance at $x = X/2$ to dose entering at $x = 0$ is that of concentration at $x = X/2$ to flux at $x = 0$ when a steady concentration is maintained indefinitely at $x = 0$ (and $C = 0$ at $x = X$). Let this concentration be C_0. The septum is uniform so that $C_{X/2} = \frac{1}{2}C_0$. The uniform steady flux $= C_0 \,.\, (D/X)$, whence, at once

$$I_{X/2} = \frac{\mathbf{M}}{\mathbf{A}} \cdot \frac{X}{2D} \qquad (17)$$

We have assumed the dose to be applied directly into the composite substance of the septum itself. When concentrations, $C = ((1 + \mathscr{P})/2) \,.\, C_{aq}$ (p. 525), are measured in the whole composite, the corresponding $\bar{D}$ is that given by eq. (10). D therefore passes through a maximum at $\mathscr{P} = 1$ and therefore the availance for a given dose goes through a *minimum*. A corresponding "pessimum" value of $\mathscr{P}$ (giving least transfer at a finite value) has never been reported in the structure-activity literature. For the aqueous layers the concentration will be $2/(1 + \mathscr{P})$ times the composite value, for the lipid layers $2\mathscr{P}/(1 + \mathscr{P})$ times the composite value. The availances in the separate phases of the composite, a better measure of what can be extracted by a reacting organelle, vary therefore as $(1 + \mathscr{P})/2\mathscr{P}$ and $(1 + \mathscr{P})$ respectively for water and lipid, which are monotonic functions of opposite sign.

B. Different Partition Effects on Uptake and Loss

This "false start" has been included to illustrate an easy pitfall in this sort of modelling and because we shall need the pulse curve for $C_{X/2}$ for other illustrations. We need a bigger difference between the uptake and loss processes to obtain a clear optimum $\mathscr{P}$ effect. Consider a general model. Let some readily permeable (or, by internal circulation, effectively agitated) tissue be separated by one septum from a source and by another from a sink. Let the permeances of the two septa be different functions, $f_1(\mathscr{P})$, $f_2(\mathscr{P})$ of the partition coefficient. A steady external concentration, C_0, is maintained outside the first septum and zero concentration outside the second. Let C be the constant concentration in the intermediate tissue

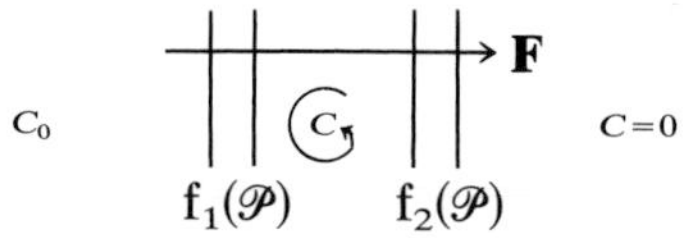

C_0 | C | $C=0$ → **F**

$f_1(\mathscr{P})$ $f_2(\mathscr{P})$

For the constant flux, **F**, we have $(C_0 - C)f_1(\mathscr{P}) = C\,.\,f_2(\mathscr{P})$ whence

$$C = C_0 \frac{f_1(\mathscr{P})}{f_1(\mathscr{P}) + f_2(\mathscr{P})} \tag{18}$$

If both functions increase monotonically with $\mathscr{P}$ but f_1 initially faster and f_2 finally to a higher value, C will go through a maximum. Two simple septum models already considered would show this effect, i.e. the first septum a multilaminate with permeance $2\mathscr{P}/(1+\mathscr{P})$ and the second a group of equal parallel aqueous and lipid paths, permeance $(1+\mathscr{P})/2$ giving

$$C = C_0 \frac{4\mathscr{P}}{1 + 6\mathscr{P} + \mathscr{P}^2} \tag{19}$$

which predicts a maximum value of C (for constant C_0) at $\mathscr{P} = 1$. This relationship has the same value for $\mathscr{P}$ and $1/\mathscr{P}$ and is thus a symmetrical hump (but not a parabola) when plotted against $\log \mathscr{P}$. If the permeabilities were reversed, C would go through a minimum, but the range would be from 1 ($\mathscr{P} = 0$) through 0·5 ($\mathscr{P} = 1$) back to 1 ($\mathscr{P} = \infty$) instead of the complements, 0–0·5–0 which result from the first disposition. A pessimum value of $\mathscr{P}$ would therefore be less noticeable and less distinguishable from other variations, than an optimum value.

Hartley's steady-state analogy indicates that the ratio of internal availance to a finite external exposure is the ratio of concentrations at the steady state, given, for this model, by eq. (18). The relationship between internal availance and consumed dose (= applied dose in this model) is that between steady internal concentration and steady flux. In the absence of chemical decay and loss from the application site, this is simply the function $f_2(\mathscr{P})$, independent of the permeance on the entry side and, in the example given, is a monotonic function. It should be noted that most experimental data would not reveal the differences in optimum values of $\mathscr{P}$ because the toxicant is presented in only one standardized way. It could well be that, in plotting toxic response for a series of compounds against $\log \mathscr{P}$, the optimum for LC_{50} (indefinite exposure), for optimum $L(Ct)_{50}$, for LC_{50} for a standard exposure time and for LD_{50} for a discrete dose would lie at different $\log \mathscr{P}$ values or, in one or more cases, not be evident. Inadequate attention is paid to this possibility in routine screening.

C. Interaction of Permeation and Decay

The third possibility which could introduce an optimum in efficiency of total transfer is the interaction of effect on permeation with that on chemical decay. This can be illustrated by return to the pulse of concentration against time (Fig. 2) following application of a discrete dose to one side of a multilaminate septum while a perfect sink keeps the other side at zero concentration. The heavy-line curve of y against Z (Fig. 3a), replotted from Fig. 2, is correct for $\bar{C}_{X/2}$ against t when $D=1$, $X=4$, $\mathbf{M}/\mathbf{A}=\sqrt{\pi}/16$, $\mathscr{P}=1$. It is converted to represent $\bar{C}_{X/2}$ against t for other $\mathscr{P}$ values by stretching the curve along the axis of Z by the factor $D/\bar{D}=(1+\mathscr{P})^2/4\mathscr{P}$. Stretching on the axis of y by $2/(1+\mathscr{P})$ converts the curves to indicate $C_{aq.X/2}$. This has been done in Fig. 3a for the two faint

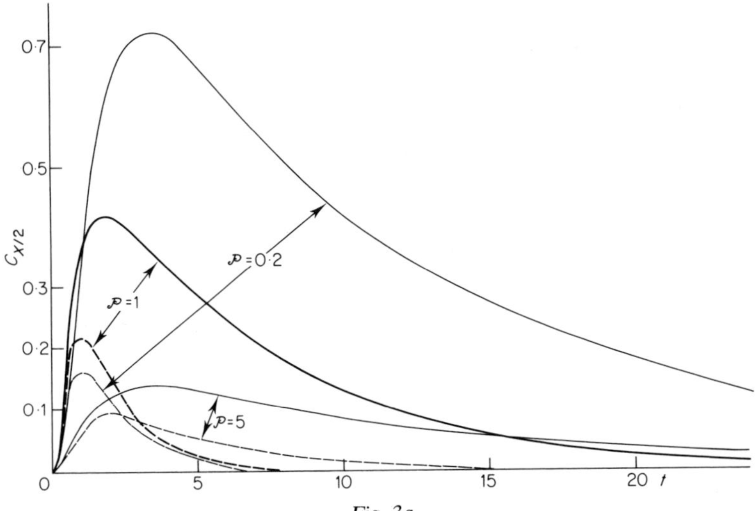

Fig. 3a

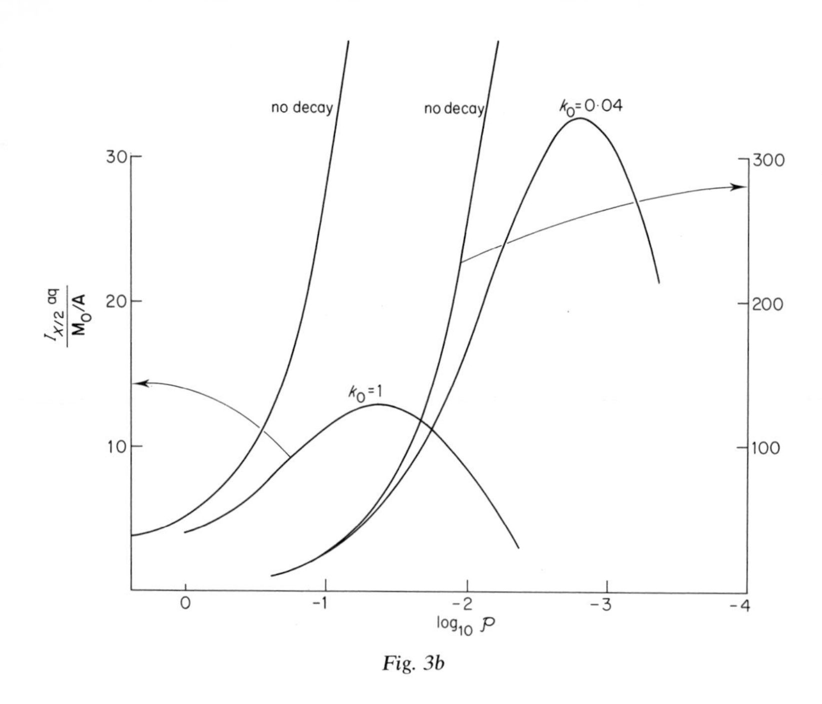

Fig. 3b

lines for $\mathcal{P} = 5$ and 1/5. D, X, **M/A** are assumed to remain constant. As pointed out before, $\int_0^\infty C_{aq.x/2} \, . \, dt$ is a monotonic function of $\mathcal{P}$.

Reaction will normally occur in the aqueous phase only. A constant, k_0, for first order decay in the aqueous phase becomes $k_0/(1+\mathcal{P})$ when it has to multiply the greater concentration in the composite. In the composite as a whole therefore, we have

$$\bar{D} = D\frac{4\mathcal{P}}{(1+\mathcal{P})^2}, \qquad \bar{C} = C_{aq} \, . \, \frac{1+\mathcal{P}}{2}, \qquad k = k_0/(1+\mathcal{P}).$$

To allow for chemical decay, any C value on a curve must be multiplied by $\exp(-k_0 t/(1+\mathcal{P}))$ for the appropriate t and $\mathcal{P}$ values. This is done, for $k_0 = 1$ in reciprocal time units of the arbitrary scale, in the broken curves. Estimating the effect of this on the total availance (area under the curves) would be tedious, but, once again, the steady-state analogy simplifies the problem. Using equations (6.3) with $x = X/2$ and (6.5) from Hartley (1963) we obtain

$$\frac{I_{X/2}}{I_0} = \frac{\sinh(\alpha X/2)}{\sinh(\alpha X)} = \frac{1}{2}\operatorname{sech}\left(\frac{\alpha X}{2}\right) \tag{20}$$

and

$$\frac{I_{X/2}}{\mathbf{M}_0/\mathbf{A}} = \frac{\sinh(\alpha X/2)}{\bar{D}\alpha \cosh(\alpha X)} \tag{21}$$

where

$$I = \int_0^\infty C\,\mathrm{d}t \quad \text{and} \quad \alpha^2 = k/\bar{D} = \frac{k_0}{D} \cdot \frac{1+\mathscr{P}}{4\mathscr{P}} = \alpha_0{}^2 \frac{1+\mathscr{P}}{4\mathscr{P}} \tag{22}$$

In eq. (20) it does not matter whether I refers to concentration in the whole composite, $\bar{C}$, or concentration in the aqueous phase, C_{aq}, provided the choice is consistent. α shows a monotonic decrease with increase of $\mathscr{P}$, other quantities constant, from ∞ at $\mathscr{P}=0$ to $\alpha_0/2$ at $\mathscr{P}=\infty$, so that eq. (20) shows monotonic increase of $I_{x/2}/I_0$ from 0 at $\mathscr{P}=0$ to $\frac{1}{2}\operatorname{sech}(\alpha_0 X/4)$ at $\mathscr{P}=\infty$.

In eq. (21), $I_{x/2}$ refers to mean concentration in the composite medium. To obtain the value for the aqueous phase we must divide by $(1+\mathscr{P})$. Making other substitutions listed above, this equation becomes

$$\frac{I_{x/2,\mathrm{aq}}}{\mathbf{M}_0/\mathbf{A}} = \frac{4}{k_0 X} \frac{(\alpha X/2)\sinh(\alpha X/2)}{\cosh(\alpha X)} \tag{23}$$

The function of (αX) goes through a maximum of 0·326 at $\alpha X/2 = 1{\cdot}25$ on its way from zero at $\alpha X = 0$ to zero at $\alpha X = \infty$. Since αX is a monotonic function of $\mathscr{P}$ this model appears to have the potential to explain the type of optimum found, but some qualifications must be noted.

$$x/2 = \alpha_0 X/4\sqrt{(1+\mathscr{P})/\mathscr{P}} \tag{24}$$

and has therefore a minimum value of $\alpha_0 X/4$ (for $\mathscr{P}=\infty$). If this minimum is greater than the value of $\alpha X/2 = 1{\cdot}25$ giving the maximum value of the (αX) function in eq. (23) the optimum cannot be realized. The efficiency of availance transfer (eq. 20)) is 0·26 at the optimum and therefore lies in a realistic range. In a non-reactive system $\alpha = 0$ and the limiting form of eq. (23), from eq. (17), is

$$\frac{I_{x/2,\mathrm{aq}}}{\mathbf{M}_0/\mathbf{A}} = \frac{1+\mathscr{P}}{4\mathscr{P}} \times \frac{X}{D} \tag{25}$$

In Fig. 3b we have plotted against log $\mathscr{P}$ values of $I_{X/2,\mathrm{aq}}/\mathbf{M}_0/\mathbf{A}$ for two reactive systems according to eq. (23) and for the non-reactive system, eq. (25) taking the arbitrary values $D=1$, $X=1$ throughout and $k_0 = 0{\cdot}04$, $\alpha_0 = 0{\cdot}2$ for one curve and $k_0 = 1$, $\alpha_0 = 1$ for the other, both in the range where the optimum is realized. A scale adjustment is necessary. It will be seen that in the more reactive system, the eq. (23) curve "peels away" from the no reaction (eq. (25)) curve at an earlier stage as $\mathscr{P}$ decreases. The less reactive system follows the unreactive more closely and then peels away more sharply. A much more reactive system than the first would show no maximum.

Very low $\mathscr{P}$ values (water favourable) are of interest in this graph, but, of course, the parameters other than $\mathscr{P}$ are arbitrary. We have assumed that the aqueous and lipid layers have equal thickness and that the diffusion coefficients within the two layers are the same. The position of the optimum would of course be different for other arbitrary assumptions.

V. OPTIMUM PARTITION FOR EFFICIENT TOXIC REACTION

The fourth mechanism envisaged to explain an optimum partition coefficient involves mechanisms less easily illustrated by models but is probably much the most important. The optimum is not of toxicant transferred but of the effectiveness of the toxic reaction.

We have seen that the efficiency of steady-state transfer is probably always a monotonic function of $\mathscr{P}$ unless other rate-processes are involved. By introducing other processes models can be designed in which the availance—i.e. the area under a curve of C against t—can go through a maximum. Far greater differences, however, can be created in the *form* of the C–t curve than in its total area. Since continuous lipid paths from site of application to site of action do not exist, increase of permeance by increase of $\mathscr{P}$ will always be limited, but capacity will always increase with $\mathscr{P}$ indefinitely. At high $\mathscr{P}$ values, pulse forms at an internal site after application of a discrete external one will always be stretched by increase of $\mathscr{P}$.

Although it is usually considered that response to an external exposure is mainly governed by its total value—i.e. by the Ct product, or more generally $\int_0^\infty C\,dt$—it is also widely recognized that there is usually some finite lower limit of concentration below which no response is evoked no matter how long the exposure. Nature does not jump, and there must be a region of low C, long t where the value of $\int_0^\infty C\,dt$ necessary to evoke a standard response, increases with decrease of C. In work with fumigants, where control of exposure is easy, it is usually found that the external Ct product necessary to evoke the standard response goes through a rather broad minimum, tending to increase indefinitely with decrease of C beyond the broad optimum range and sometimes also to increase for very short exposures to very high concentrations. The range is very dependent on the nature of the toxic mechanism.

An optimum compactness of the pulse of C plotted against t is entirely to be expected on a very general kinetic argument. $\int_0^\infty C\,dt$, for actual values at a local site, would be a valid measure of the total amount of some chemical reaction if the rate were first order with regard to the toxicant. First order reactions are rather general for reactions of a compound present in very low concentrations when the other reactant is present in great excess. Thus hydrolysis of the active compound, by water present in enormous excess and catalysed by H^+ or OH^- ions held at steady level in a highly buffered system, is very likely to be an apparent first order reaction. However, the process leading to the biochemical damage which itself leads to the overt symptoms is in general unlikely to be first order, because the other reactant (usually an enzyme) is not present in great excess. If it were, no damage would result. It is only because some essential biochemical function is seriously reduced that damage follows. With excess toxicant, the reaction will be limited by consumption of the reserve of native reagent or by its rate of replacement being exceeded. If toxicant arrives at the site too slowly, replacement and repair reactions are adequate to ensure no irreversible damage. If toxicant arrives too quickly, replacement is for a short time exceeded but the excess toxicant may have decayed by some slower reaction before the temporary disturbance of the essential process has led to other, irreversible effects. In some cases (mustard gas is a good example) the alternative decay reaction may proceed at the same speed as the toxic one because the rate of the (alkylating) reaction is set by the toxicant itself (by the "SNI" mechanism) independently of the reagent being alkylated. Depending on the kinetic mechanisms involved there could be a wide range of variations of Ct_{50} with C and t separately, but always it would be of a kind showing an optimum pulse form and probably, therefore, in a series of related compounds, an optimum value of $\mathscr{P}$.

This tendency can be illustrated if we take the rather typical pulse form of the heavy line in Figs 2 and 3a and repeat it (Fig. 4) as a plot of C against t in arbitrary units. We will assume that concentrations below 0·1 are ineffective and that any concentration above 0·2 saturates temporarily the replacement reaction and is diverted into other reactions. The amount of irreversible damage is thus proportional to the area bounded on the sides by the curve between the broken horizontal lines. The dose necessary for critical damage is inversely proportional to this "effective" area. The area is calculated from the inset curve in Fig. 2. This is, of course, a staccato caricature of a real situation and is intended only to illustrate

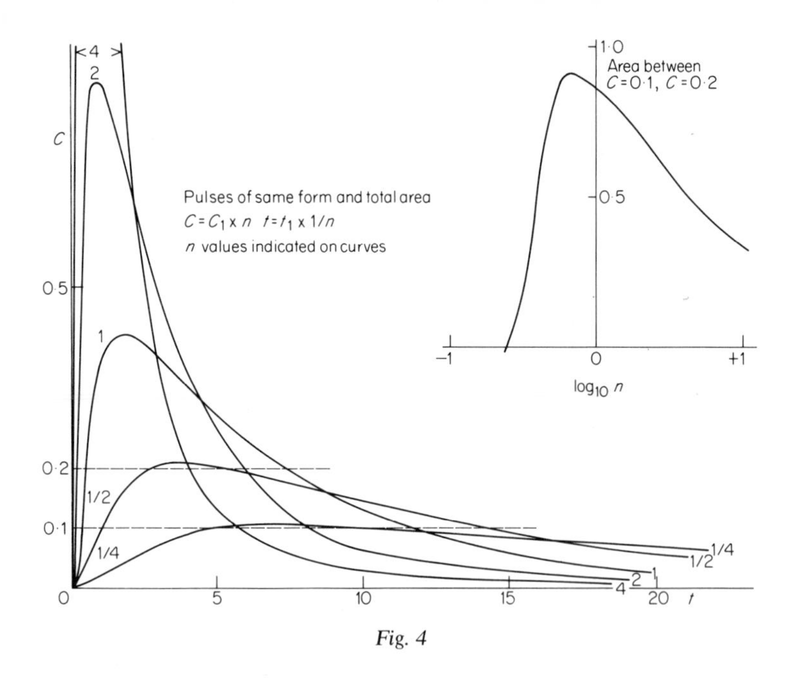

Fig. 4

the trend. We will assume that change of partition coefficient produces a factorial increase of C and corresponding decrease of t, so that the total area under the curves remains constant. Curves are drawn for values of this factor (n) of 4, 2, $\frac{1}{2}$, $\frac{1}{4}$. It will be seen that increase of n decreases the effective area while, initially, for decrease below 1, the stretch and compression are nearly compensating. Further decrease leads to a sharp fall and for $n \lesssim 1/4{\cdot}2$ the peak fails to reach the lower limit. Inset is plotted the area against log n, showing a maximum about $n = 0{\cdot}7$.

VI. OPTIMA CORRELATED WITH PARTITION

Summarizing this excursion into models we can conclude that a fraction of an applied dose or exposure having a maximum at a finite value of $\mathscr{P}$ may have penetrated to some internal site in a time very far short of that necessary for attainment of the steady state, but we do not think this is of practical importance. A maximum of internal availance ($\int_0^\infty C\,dt$) at some optimum value of $\mathscr{P}$ will only be realized when the effect of $\mathscr{P}$ on the permeance in the uptake process is in competition with chemical decay or with a differently determined permeance in the excretion process. We consider that a more universal mechanism for a more clearly-marked optimum of $\mathscr{P}$ is to be found not in optimum efficiency of transfer but in optimum pulse form so that *the effect of* what is transferred is optimal. Too long a pulse at low maximum concentration is ineffective because rate of replacement or repair is too fast for damage to build up: too short a pulse of high concentration is ineffective because excess toxicant is dispersed or destroyed by other mechanisms.

We have given possible mechanisms more attention than we would feel them to deserve had the optimum partition not achieved great prominence in the literature. We have also, in our models, considered the permeability to be affected by $\mathscr{P}$ only, because the Hansch approach has given this quantity the exclusive right to determine that there should be a maximum effect in any series of compounds, other quantities being introduced into the equations in linear forms only.

In discussion of a recent extensive computation to fit activity (two measures) to physical properties for two series of fungicides, Brown and Woodcock (1975) give a healthy warning against too definite conclusions about causal factors. They have, however, in putting much computer effort into determining the best fit among equations in $\log \mathscr{P}$ and $(\log \mathscr{P})^2$ accepted that this factor has a dominant role by mechanisms only vaguely expressed. That the best fits are still not good suggests that alternative controlling factors should be considered. The purely empirical approach of Free and Wilson (1964) has the advantage in this respect of not making any assumption about mechanism.

In fatty monolayers (Section 9.III.D) and in macromolecular networks (Section 9.II.F, App. 1.VI), the dependence of D on molecular size (and shape) is very much greater than in simple liquids. Increase of $\mathscr{P}$ in a series of compounds usually results from increase of non-polar volume. A probable mechanism is that penetration of small molecules is at first favoured by increased lipid partition but that with large molecules the decrease of diffusion coefficient assumes overriding importance.

Hussain *et al.* (1974) measured the rate of transfer of some organophosphorus insecticides from the outside of the petiole to the inside of detached cotton leaves maintained turgid with the cut end in water. One homologous series of *n*-alkyl compounds fits a Hansch parabola but when *n*-butyl (near optimal, the *n*-propyl being slightly more mobile) is replaced by *sec*-, iso- and *tert*-butyl there was a marked decrease in that order. Although the authors consider, in the text, that this could be due to increased hydrophobic behaviour their tabulated measured partition coefficients show no appreciable change. That of the *tert*-butyl compound is 3% more oil favourable but even the *n*-octyl, four times more oil favourable, is more mobile. An effect on diffusion in the cutin (p. 610) is much more probable.

In our models we assumed D independent of $\mathcal{P}$ in an attempt to justify the $(\log \mathcal{P})^2$ term. In no model based on change of partition only does the final transfer-determining function of $\mathcal{P}$ vary as much as $\mathcal{P}$ itself. The ratio of log (concentration for standard response) to ratio of $\log \mathcal{P}$ between any two compounds would therefore be expected to be numerically less than 1. Values in excess of 1 would indicate some additional mechanism.

The clearest example of such excess is in a series of alkyl tritylamines supplied in the ambient water to the snail (*Australorbis glabratus*; Boyce and Milborrow, 1965). The C_4 compound was most toxic. From C_1 to C_2 and from C_6 to C_7, $\log_{10} (LC_{50})$ moves through more than 1 unit. For the most extreme case (paraffin/water) the well established value for $\delta \log_{10} \mathcal{P}$ for 1 CH_2 group is only 0·5. Where the parabolic element is introduced by terms $-k_1(\log \mathcal{P})^2 + k_2 \log \mathcal{P}$ it is easily shown that $\log C$ for a compound having $n+1$ CH_2 groups more (or less) than the optimum, falls below that for one having n CH_2 groups more (or less) by $(2n+1)k_1/4$ when each CH_2 group increases $\log \mathcal{P}$ by the maximum value of 0·5. This exceeds 0·5 for $k_1 = 0{\cdot}40$, 0·29 or 0·22 when n exceeds 2, 3 or 4. Examining Table IIc of the extensive review of antifungal activity compiled by Hansch and Lien (1971) one finds 3, 5 and 8 examples out of 33 equations where these figures are exceeded. A further 18 equations are listed where no $(\log \mathcal{P})^2$ term is called for. In only 16% of the organism–compound series combinations reviewed will $\Delta \log C$ exceed $\Delta \log \mathcal{P}$ within a range of 5 CH_2 groups on either side of the optimum. We have not examined the original papers to see in how many of these cases the experimental data cover this range. When $\log C$ does change more than $\log \mathcal{P}$ we feel that some other mechanism than partition change must be involved.

We have assumed, in this argument, that increase of $\log_{10} \mathcal{P}$ by a methylene group will not exceed the value of 0·5 for paraffin/water on the grounds that no lipid phase is ever so completely non-polar as paraffin. There is, however, the rather unexplored possibility that some aqueous phase in tissues may be "more aqueous than water". Acetone, for example, is not miscible with glycerol nor with 60% glucose in water and partitions from these liquids into paraffin more favourably than it does from water. The aqueous phase impregnating cell walls and cementing cells together is not pure water. Some type of cross-linking generally maintains it as a separate phase even in contact with water although its volume is mainly water. The remainder, in animals, is muco-proteins and, in plants, pectin-like substances. The latter, being poly-hydroxylic, have a glycerol-type solvent property and the cross-linked network usually carries many fixed negative charges, partly neutralized by cations, mainly Ca^{2+}. Not only does this create ion-selectivity by Donnan effects (Section 2.XVIII.C) but it will tend to "salt out" non-electrolytes and so accentuate the glycerol-type behaviour. This effect will be greater the more hydrophobic the compound. There will therefore be a tendency, as $\mathscr{P}$ (octanol from water) increases, for $\mathscr{P}$ (intercellular gel from water) to decrease. This possibility deserves experimental study.

VII. INFLUENCE OF SOLUBILITY

The Hansch approach, as its originator emphasizes, is only suitable for the comparison of chemically similar toxicants which are *presented in the same physical state.* The warning has not always been heeded. Preferably, as is usually done with culture-solution tests on bacteria, fungi, or even on higher plants grown in water, the compounds should be presented in solution. In some investigations no definitive steps have been taken to ensure that the solubility of some compounds is not exceeded. It is not safe to assume that all compounds will be water soluble at the few p.p.m. level nor that to dissolve them in acetone and rapidly dilute is a valid means of ensuring that they are. Even when no evidence of a second dispersed phase in the "solution" can be found (if looked for), many compounds which partition very favourably into oils from water will exist in solution in micellar form, particularly if the substitution that has given them the strongly hydrophobic property is a highly unsymmetrical one—a single paraffin chain. Several fungicides and bactericides come into this class and indeed their amphipathic (Section 2.XIV) nature, by membrane adsorption and disturbance of membrane function, probably contributes to the mechanism of toxicity. From such micellar solutions, the partition into an oil phase is difficult to measure and $\log \mathscr{P}$ will no longer be a simple additive function of the constituent groups.

"Contact" insectides and foliar-applied herbicides cannot usually be examined by application of dilute aqueous solutions. If wettable powders are examined, one cannot avoid a contribution of melting point to the overall effect and this property is a highly specific variable. Not only does this specific variable, by limiting solubility, introduce another factor into penetration rate but the area of contact with the cuticle will usually be much less for a solid than a liquid. Usually, not always; fine snow will make better contact with a polyethylene sheet than will rain. The essence of the matter is that the contact of a crystalline solid will depend on solubility, tendency to supercool and habit and size of the crystals. Application in a solution in a nontoxic oil can remove these variables and so enable the internal biophysical factors of the Hansch approach to be more clearly analysed.

This retreat from practicality must, however, always be appreciated as such. Choice of a compound which appears optimal in biochemical or cell-level tests without regard to problems of formulation and application, followed by formulation research as a separate exercise, can lead to waste of expensive effort when the compound is abandoned after unsatisfactory field performance.

VIII. PARTITION INTO OTHER PHASES

A wholly different approach to the paraboloidal toxicity–partition relationship is made by Higuchi and Davis (1970). It depends on a general principle which we feel is unfortunately rather obscured by an over-detailed presentation. Their model differs from others discussed in that it introduces more phases and assumes that the biochemical action which produces the symptoms measured goes on in a phase of polarity intermediate between water and the more extremely paraffinic tissues such as depot fats. It also assumes equilibration among the phases after introduction of a discrete dose. Response follows the attainment of a critical concentration of the drug in this biophase as was assumed in the Ferguson theory of narcotic action (Section 2.XI).

The relationship found by Collander (Section 2.X.C) between the partition coefficients between water and different solvents would, at its simplest, lead one to expect the partition in favour of the biophase to be some constant power (<1) of that in favour of depot fats. Suppose the three phases are present in equal volumes and partition coefficients are $\mathcal{P}$ in favour of fat and $\sqrt{\mathcal{P}}$ in favour of the biophase. The fraction of the discrete dose resident at equilibrium in the biophase would be $\sqrt{\mathcal{P}}(1+\sqrt{\mathcal{P}}+\mathcal{P})$, which has a maximum value for $\mathcal{P}=1$ and is symmetrical against $\log \mathcal{P}$ on either side of zero. The maximum would exist at some value of $\mathcal{P}$ for partition in favour of the biophase equal to $\mathcal{P}^n$ where $n<1$ and for any ratio of volumes. This 3- (or multi-)phase model is necessary to produce an optimum $\mathcal{P}$ behaviour in an equilibrium model for which Hyde (1975), without introduction of an extra phase, predicts monotonic (but asymptotic) variation with $\mathcal{P}$. It achieves this because, at one extreme of the $\mathscr{P}$ range the drug is wholly in the water and, at the other extreme, wholly in the fats.

The maximal response at an optimal $\mathscr{P}$ would not however be explained by this model if the toxicant were presented as a steady concentration in the ambient fluid. $\sqrt{\mathscr{P}}$, in our simplified example, would uniquely determine the ratio of biophase to external water concentration. It would also fail to show the optimum $\mathscr{P}$ phenomenon even for a discrete dose if the toxicant were consumed by irreversible reaction in the biophase. The "storage" phases would in this case have delaying action only and competition with some other rate, as already considered for the permeation models, would be necessary. The phase of intermediate partition coefficient could be a septum rather than a biochemical reactant, but again the optimum would then appear in the rate of access, the septum alone not directly governing the *amount* which can *eventually* reach the site beyond it.

11. Penetration of Pesticides into Higher Plants

I. Introduction 545
II. External morphology 547
A. The cuticle 547
B. Epicuticular wax 549
C. The cutin layer 550
D. Stomata, trichomes and grooves 553
E. Roots 555
III. Internal morphology 556
A. The translocation systems 556
B. Apoplast, symplast and free space 557
IV. Entry through stomata 559
V. Permeation of leaf cuticle 561
A. Introduction 561
B. Cuticular waxes 561
C. Permeation of detached cuticle by water 563
D. Swelling of the cuticle 566
E. Permeation of the detached cuticle by other substances 569
VI. Penetration into aerial parts 579
A. Relevance of detached cuticle studies 579
B. Measurement of uptake 582
C. Reversibility of uptake 587
D. Washing out by rain 595
VII. Uptake into roots and xylem 596
A. Entry into roots 596
B. Translocation from roots to shoots 600
C. Mechanisms of passage through selective barriers in roots 606
D. Transport in the xylem 610
E. Summary of the properties required for translocation in the apoplast 611
VIII. Transport in the symplast 612
A. Introduction 612
B. Mechanism of phloem transport 616
C. Active and passive transport 619
D. Sources and sinks 622
E. Transport in counter flow 623
F. Selective entry into transport systems 631

IX. Environmental effects 633
A. Effect of light 633
B. Effect of temperature on foliar uptake 637
C. Effect of atmospheric humidity on foliar uptake 639
D. Effects of temperature and humidity on uptake by roots 642
X. Effect of additives on penetration 646
A. Introduction 646
B. Humectants 647
C. Other auxiliary solvents 649
D. Surfactants 652
E. Nutrient and salt additives 654

I. INTRODUCTION

Systemic pesticides applied to crops to protect against chewing or sucking insects or, more recently, against parasitic fungi, are translocated within the receiving host organism before entry into the target organism. Should we include the translocation processes because they occur in the environment of the target organism, or exclude them because they are physiological processes within some organism? We must make some compromise and have considered translocation in higher plants more fully than that in insects or fungi, have concentrated on the physics of this process and have given more attention to translocation of systemic pesticides, which are desirably without action on the physiology of the plant, than to that of herbicides which at some stage must drastically affect it.

The initial penetration of chemicals into any organism is properly included in the field we are discussing. It is measured positively by its physiological effect or by amount (usually necessitating radiolabelling) found within the organism, or negatively, by measuring the decrease in the external deposit. It is perhaps important to emphasize that penetration is only one, and frequently only a minor one, of the processes by which pesticide is lost from the surface of an organism. For example Perry *et al.* (1964) showed that up to 50% of aldrin applied topically to houseflies could be lost by evaporation while Phillips (1971, 1974) demonstrated that under simulated tropical conditions, up to 90% of dieldrin applied to cotton leaves could evaporate in 24 hours. It should also be emphasized that removal of the residual deposit by some solvent washing process cannot make a clear distinction between washing off and extraction from within, an ambiguity discussed in a later section.

Penetration involves several different factors whose contribution is difficult to analyse quantitatively. The outer membranes of living organisms are complex heterogeneous structures, with macro and micro morphological features which can influence penetration profoundly. One has only to think of the various hairs, stomata, corrugations etc. found on plant surfaces to appreciate this point. Furthermore the passive movement of materials through the membrane may be affected by the simultaneous flow of solvents and by active processes involving metabolism by the organism. Penetration from discrete droplets into thin laminae cannot proceed far without lateral translocation from the very limited tissue volume under the droplet so that penetration proper and translocation are not easily separated. One is concerned with resistances in series in which a component of the chain with small resistance may have little effect on the overall process if the rate is limited by a much larger resistance elsewhere in the series. When one adds the complicating effects of any adjuvants present with the pesticide, the effects of environmental factors on the physiology (particularly permeability) of the organism and the large variations between different species, it will be obvious why generally applicable quantitative conclusions cannot usually be drawn. The best approach in most cases is to study the way in which variations in chemical and physical properties and environmental conditions influence penetration so that the factors involved and their likely effects may be identified for consideration in any particular case.

While it would seem self-evident that the performance of a biocide is influenced by the rate of penetration into the organism, the detailed nature of this influence is by no means obvious and the dangers of drawing simple conclusions must be stressed. The effects of rate of penetration depend greatly on the method of exposure to the toxicant, on the characteristics of the poisoning process, that is on whether it is acute or cumulative and chronic, and also on the properties of any detoxifying processes. These questions are taken up again at the end of this chapter, after the nature of the penetration and translocation processes has been considered.

Chemicals can enter plants by the roots or the aerial parts. Application both to soil and leaves is exploited extensively in crop protection. For obvious reasons the structure of the root is better adapted to the exchange of dissolved substances than that of the leaf which is better adapted for the exchange of gases.

The active tissues of the leaf are protected by a much thicker cuticle than the receptive tissues of the root. Barriers to free uptake of dissolved substances by roots are found in deeper layers of the tissue. The stages of uptake through roots are less easily, or usefully, separated. In much of the relevant experimental work only the amounts removed from the solution, or the amounts reaching the aerial parts were measured and there have been no studies comparable with the detailed investigations on separated leaf cuticles. We shall therefore consider penetration into roots together with translocation in the plant. The most important translocation process for materials entering the root is the upward movement which occurs predominantly in the xylem. However, it is not possible to give a comprehensive picture of redistribution in the plant without discussing the different translocation processes together so that this section will also include consideration of the other conducting system, the phloem.

Any treatment of uptake by roots under field conditions must be related to the movement and availability of the chemical in soil which are discussed in Chapter 5. That chapter attempts to define the characteristics of the supply of chemical following application to soil: a major purpose of this section is to consider how the plant is likely to respond to that supply, in other words to define the characteristics of the sink.

II. EXTERNAL MORPHOLOGY

A. The Cuticle

The leaf cuticle has the function of protecting the delicate photosynthetically active mesophyll from mechanical damage, excessive washing or drying-out and invasion by parasitic microorganisms, while permitting exchange of carbon dioxide and oxygen. Crop plants must be capable of rapid growth and their weeds would not be competitive if they did not also grow rapidly. This demands extensive exposure to light but rapid material exchange with the interior so that, between the main structural and transport vessels, the leaf is thin, containing typically about 20 mg of wet tissue per cm^2. The protective cuticle may comprise up to 10% of this. It covers the whole area of leaves and stems, usually extending over any protruding hairs which may be single- or many-celled. It also, in thinner and usually simpler form, lines the stomatal cavities beneath the general leaf surface. The cuticle has particular importance for our subject as being the organ which overhead spraying, dusting etc., must first hit and through which the pesticide must pass if it is to have systemic action.

The cuticle varies considerably with age, from species to species and between different parts of the plant, and is also affected by the environmental conditions under which it was formed. The structure and properties of plant cuticles have been comprehensively reviewed by Martin and Juniper (1970). Unlike the epidermis of mammals, which is formed of compacted and flattened dead cells, the plant cuticle is a non-cellular structure. In general the leaf cuticle has an outer layer of superficial wax consisting mostly of *n*-alkanes, straight-chain saturated ketones and alcohols and carboxylic esters, but including in some species other aliphatic and terpenoid constituents. This overlies a layer of cutin (a condensation polymer of fatty acids, dibasic acids and hydroxy carboxylic acids) containing platelets of cuticular wax. The cutin layer merges into a cutinized layer which generally contains cellulose. A layer of pectin cements this to the upper walls of the epidermal cells. The several layers vary greatly between species both in distinctness and coherence.

Laboratory methods of detaching cuticle are described by Martin and Juniper and include ammonium oxalate–oxalic acid solution and EDTA, both of which dissolve pectin by Ca^{2+} removal; and attack by pectinase enzymes, and strong hydrochloric acid–zinc chloride or cuprammonium solution which dissolve cellulose. The strong alkalinity of the last-named reagent certainly hydrolyses some cutin also. The firmness of attachment of the whole cuticle to the epidermal cell walls is very variable between species as is also the extent to which the cutinization persists into the anticlinal walls of the epidermal cells. The detached cutin is therefore in some cases smooth on the inside, in most cases shows a persistent pattern of the epidermal cell walls and in many cases is difficult to separate completely from cell residues.

There is a great variation in thickness. Martin and Juniper quote extremes for leaves of 0·8 μm (*Lamium amplexicaule*) to 14 μm (*Nerium oleander*). Most crop leaves have cuticles, if detachable, nearer to the lower figure. The present writers can now add the yet higher figure of about 30 μm for *Metrosideros excelsa*, a salt-tolerant evergreen. Most fruits have thicker cuticles than leaves. That of the apple (*Malus sylvestris*) attains 25 μm in the cultivar Bramley's seedling (Baker *et al.*, 1962) and that of the tomato is 23 μm (Bukovac *et al.*, 1971). (In quoting figures for thickness from measurements by mass per unit area a density of 1·0 has been assumed. Schönherr and Bukovac (1973) assume 1·1, which is consistent with the writers' observations, but the surface is often corrugated, particularly on the inside.) Martin and Juniper state that *Fragaria* and *Phaseolus* spp. yield cutin only in a disintegrated form while some resist separation by most techniques. At the other extreme, tomato skin is easily removed in the kitchen after the fruit has been submerged for a few seconds in boiling water. Cell debris is still attached but easily separated after soaking in ammonium oxalate solution, and integrity of the cuticle is maintained after treatment by a wide variety of solvents. The cuticle of the grape, on the other hand, apparently as tough when detached mechanically, disintegrates when treated with ammonium oxalate.

The writer has detached cuticles of both surfaces together from the apical half leaves of *Nerium oleander* and of *Metrosideros excelsa* after 10 minutes in boiling ammonium oxalate solution, like pulling off a tight fitting glove. It is even possible, with care, to peel it off inside out. The oleander leaf cuticle, coherent but rather papery in appearance, retains its integrity and is even reproducible in permeability after treatment with alcohols or carbon tetrachloride, but is made much more permeable or even disintegrated by a few seconds soaking in chloroform. This observation is recorded not only as an example of the mosaic nature of some coherent cuticles but to reinforce the objection expressed elsewhere (Dewey *et al.*, 1962) to the popular use of chloroform as a wax solvent. It is no better as a solvent for accessible paraffinic waxes than saturated hydrocarbons but has some strong polar solvent action in addition.

Recent chemical and biochemical research has concentrated on the cutin and wax components as being characteristic of the cuticle while the cellulose–pectin layers are common to cell walls and intercellular cement. One must not, however, forget that all layers are present in the cuticle *in situ* and make at least additive contributions to resistance to permeation. For obvious reasons most studies of plant cuticles have been made on those which are most easily separated in coherent form. Such cuticles may be atypical, particularly with regard to permeability.

B. Epicuticular Wax

It is unfortunate that "wax" and "waxy" are common words used loosely and with different connotations and that botanists, usually insistent on a specialized word of Latin or Greek derivation for all plant organs and qualities thereof, have allowed this to happen. It has given rise to serious misconceptions. The smearable greasiness of an apple skin and the water reflective glaucous "bloom" of a pea or cabbage leaf are both described as "waxy". Huelin and Gallop (1951) distinguished by acetone precipitation between a crystallizing "hard" wax and an oily "soft" wax in apple skin. The wax removed by very brief contact of solvent with the outside of a pea leaf shows no liquid properties below 60°C: its water reflective action is due to its sharply crystalline nature: in everyday physics it could be considered "hard" at its functional temperature but similar material obtained from grape (*Vitis vinefera* fruit) skin is called by Possingham *et al.* (1967) "soft" to distinguish it from the polycyclic oleanolic acid of even higher melting point.

The first effect of extracuticular wax "bloom" found on some leaves and fruits on spray chemicals is to reflect water drops entirely leaving no residue (adequate wetting causes retention). For this property a dense population of sharp edged crystals is necessary (Hartley and Brunskill,

1958) and is well provided by the long ($C_{29}-C_{35}$) *n*-alkanes and ($C_{20}-C_{30}$) primary alcohols. For typical bloomed surfaces under high magnification see Martin and Juniper (1970). Such waxes can be recovered by evaporation of solvent after rapid dipping in very clean condition, if aliphatic hydrocarbons (not chloroform or ether) are used.

Similar waxes can also form a smooth shiny deposit on leaf cuticles but this ability may require the admixture of less highly crystalline compounds. Such shiny waxy surfaces may show high contact angles with pure water but do not reflect droplets. Not all shiny leaf surfaces, even those showing high contact angle, have much extracuticular wax: the very glossy leaf of salt tolerant *Coprosma repens*, at a mean of $1{\cdot}7\times10^{-5}$ g cm^{-2}, has much less than the dull leaf of *Metrosideros excelsa* growing in similar exposure (4×10^{-4} g cm^{-2}): the apple (*Malus* spp.) fruit has a thick coating. The outer surface of tomato (*Lycopersicon* sp.) fruit shows a high contact angle even after repeated extraction with chloroform and Holloway (1970) found the contact angle unchanged or *in*creased by chloroform extraction in 6 out of 22 leaf surfaces examined.

The physical form of the extracuticular wax layer is of major importance for water reflection and attempts to recast it in reflective form by ordinary methods of deposition have failed (Hall *et al.*, 1965). Physical form rather than chemical nature is the most important factor. A soot deposit on glass or a magnesium oxide smoke deposit are excellent reflectors. Holloway (1970) stresses the importance of physical form particularly when the roughness is so extreme as to give a composite surface under a drop of water—i.e. retention of air (Section 8.IV), but finds that the contact angle on a smooth surface is fairly well reproduced by recasting dissolved waxes.

The epicuticular wax is in most species removed by immersion for a few seconds in a hydrocarbon solvent, but there is often a smaller further yield of similar waxes obtainable by prolonged soaking. In this process hydrocarbon solvents give a much cleaner product than chloroform when applied to intact, turgid leaves (Dewey *et al.*, 1962). This wax is considered to come from platelets buried within the cutin or cutinized layers. Observation of transverse sections in the polarizing microscope (Meyer, 1938) showed optical double refraction to be present of sign indicating arrangement of paraffin chains perpendicular to the plane of the lamina. Soaking in hydrocarbon solvent left an isotropic residue.

C. The Cutin Layer

Cutin itself is not a definite substance. It is a cross-linked condensation polymer of fatty acids and their hydroxy and dibasic derivatives. It shows, like lignin, the amorphous, stereochemically mixed structure of inanimate polymers rather than the unique regularity of cellulose or proteins. It is extensively dissolved by treatment with ethanolic potassium hydroxide. The constituent acids, recovered by ether extraction after acidification, are typified by the two most generally abundant, 10,16-dihydroxyhexadecanoic and 9,10,18-trihydroxyoctadecanoic (phloionolic) but numerous related compounds have been identified. Following Matic (1956) who first identified the major constituents, these mixed compounds are usually referred to as the "cutin acids". They mostly have chain length of 12–20 C atoms, i.e. the fat range, rather than the wax (24–35) range. Monocarboxylic acids of this type could form, by esterification, only linear polymers which would be completely soluble in an appropriate solvent. The extensive cross-linking, causing the natural product to be quite insoluble, could be due to the presence of dicarboxylic acids, some of which have been detected while others may have escaped detection (e.g. oxalic acid, in small concentration, would not be picked up by the usual procedure). Other linkages however are also known to occur, principally ether, C–O–C, and peroxide, C–O–O–C. Crisp (1965), who found successive treatments with sodium iodide and hydriodic acid solutions necessary to resolve the polymer completely, concluded that ester, ether and peroxide links were present in the ratio 70:1:20 in the cutin from *Agave americana* leaves.

Some unsaturated acids are present. Lee and Priestley (1923) and Priestley (1943) considered that cutin was formed by "inanimate" oxidation on exposure of excreted unsaturated acids by essentially the same mechanism as the hardening of poly-unsaturated oils formerly used extensively as paint vehicles or varnish. Matic (1956), and particularly Huelin (1959), emphasizing the importance of ester linkage, tended to discredit the "varnish" mechanism but the later discovery of peroxide links has restored it to at least a partial role. Martin and Juniper (p. 147) mention that the hydroxy acids tend to polymerize on standing and the writers found that the initially clear solution obtained by dipping *Brassica* leaves in cyclohexane deposited a flocculant solid on standing. They also found that isolated, dewaxed cuticles take up less volatile solvent and become more brittle after prolonged storage in the dry state, indicating an increase of cross-linking of some kind. Oxidation seems a more likely mechanism of further polymerization than one involving removal of water into an ordinary atmosphere.

Cutin contains unesterified carboxyl groups and, as such, has base-exchange properties. These have been examined in detail for the tomato fruit cuticle by Schönherr and Bukovac (1973) who find the content of dissociable acid groups to be about 1 m eq. g^{-1} but much of this is removed by acid hydrolysis and is attributed to attached pectin-like material and phenolic substances. The equivalent weight of the residual cutin was estimated to be about 2600 corresponding to 1 in 10 of C_{16} hydroxyacids having a free carboxyl group. The ionic content makes the cutin significantly swellable in water. Schönherr and Bukovac estimate that between 0·1 and 0·2 g of water is held per g of dry cutin, but that the degree of ionization, surprisingly, had little effect.

The presence in cutin of hydrophilic ionic groups, though well separated, has caused many workers to assume, without attempt at measurement, that cutin will be swollen by water and to neglect its probably greater swelling in less polar "solvents". If its composition were correctly represented as near to an indefinite polymer of dihydroxy hexadecanoic acid with nine-tenths of the CO_2H groups esterified and half the OH groups extended to peroxide links, its empirical formula would be $C_{160}H_{302}O_{36}$ with 13 out of every 16 C atoms as CH_2. A liquid of roughly the same composition and OH group content would be hydroxyethyl laurate ($C_{14}H_{28}O_3$) which is a feeble surfactant with its solubility strongly biased to the non-polar side. The polymer would be expected, in so far as its cross-linking permits swelling, to be more readily swollen by small molecule oily solvents than by water, with a probable maximal response around ethylacetate or butyl alcohol.

The composition reported by Schönherr and Bukovac for tomato fruit cuticle after removal of further protein and pectin "contamination" by means of 6 N HCl for 36 hours at 110°C, corresponds approximately (brought to the same C atom no.) to $C_{160}H_{252}O_{33}$. The major discrepancy is that the predicted ratio of H to C and O is too high. Inclusion of residual pectin (empirical formula $C_6H_8O_6$) in the cuticle could not correct this without making the oxygen prediction too high. The most likely explanation would seem to be that, in the polymerized form, more unsaturation exists than is found in the hydrolysis products. Hydrolysis could produce a vicinal di-OH compound from two conjugated double bonds. Epoxy compounds have been reported. This suggestion would remove multiples of H_2O from the predicted formula, leading to deficiency of O which could then be made up by inclusion of pectin or by short chain dibasic acids taking part in the cross-linking and escaping detection.

Consistent with the observed composition is the observed density. Schönherr and Bukovac report 1·1 for tomato fruit cuticle which the writer can confirm for this and for apple fruit cuticle while he found 1·18 and 1·20 for the adaxial leaf cuticles (oxalate detached, water washed and chloroform extracted) of *Nerium oleander* and *Metrosideros excelsa*. The abaxial cuticle of the latter had 1·38 and peach fruit (cv Golden Honey) 1·34. One would expect, for the theoretical polymer, about 1·03.

"Pure" cutin may therefore not be so paraffinic as the determination of ether-extracted hydrolysis products indicates but, according to both gross composition and evidence on chemical structure, it is predominantly hydrophobic. Some comparisons may be useful. Cellulose triacetate has empirical formula $C_{12}H_{16}O_8$, 20% less H and three times more O relative to C than tomato fruit cutin, but forms an almost water-indifferent fibre. The more soluble, non-crystalline ordinary acetate, produced industrially in large quantity in film form (and often used as a "substitute" leaf surface in spray experiments) is $C_{22}H_{30}O_{15}$ and, although passing water vapour, is little swollen by it and dissolves in acetone. No grouping in C, H, O compounds uses O more efficiently for promoting water attraction than does OH. *Even if* all O in cutin were in this form, the empirical formula quoted above could be written $C_{160}H_{219}(OH)_{33}$, each OH having to "carry" 5 C atoms in hydrocarbon structure. Even amyl alcohol, however, is less than 2% soluble in water. The ether group is very hydrophilic and a poly(ethylene oxide) chain $-(CH_2-O-CH_2)_n-$ is used as the hydrophilic part of many useful surfactants: if cutin were formed this way, the ethylene oxide would be associated with its own weight of hydrocarbon, which it fails to carry into water solution in the case of non-ionic surfactants. It fails to carry one extra methylene group since polypropylene oxide forms a useful hydrophobic part of some surfactants.

D. Stomata, Trichomes and Grooves

The stomata are specialized openings in the leaf surface which permit, when open, continuity of the air-filled cavity below them and the outside air. They are flanked by two guard cells which open the vent between them when fully turgid and close it when flaccid. The substomatal cavities are generally lined with an extension of the cuticle but this is much thinner than the main covering of the lamina. It probably offers little resistance to passage of small molecules and is easily ruptured by intruding oil. The exterior surface of the mesophyll cells is essentially hydrophobic: mobile non-polar liquids will perfuse rapidly and spontaneously the tortuous air space between the cells when a cut edge is dipped into the liquid, but water can only be brought in under external pressure (Lewis, 1948). The mesophyll cells, despite the unwettability of their surface, transpire water vapour rapidly. The main control of overall transpiration of the leaf is the state of closure of the stomata. The stomata also control exchange of carbon dioxide and oxygen between the outside air and the mesophyll air space but this control is relatively less effective for overall exchange of these substances than for water loss because there is virtually no barrier to water loss elsewhere while a large fraction of the resistance to diffusion of oxygen and carbon dioxide lies in the aqueous phase of the cells. The air space of the mesophyll is of course crossed by bridging cells but, between veins, is continuous. The cuticle is necessarily in contact with the epidermal cells.

Some authors, notably Franke (1967), claim to have distinguished, in a wide range of plants, specialized structures, the ectodesmata, which extend the protoplasm of epidermal cells into the cuticle; although they do not penetrate to the outside of the cuticle it has been suggested that they contribute significantly to the transport of materials through it. Ectodesmata are particularly numerous near the larger leaf veins, guard cells, and leaf hairs. There is some controversy about the nature of ectodesmata, some authors believing that they are artefacts produced during the preparation of specimens for microscopic examination (see Crowdy, 1972).

Many leaves are hairy. The botanist uses the word "trichomes" for all single-celled or many-celled protrusions from the surface. They are normally covered by a cutin layer and of course have a wide range of size and elaboration. Some are adapted to sting, others carry scent glands, others are hooked for climbing, others terminate sharply as defensive prickles. Some, described by Martin and Juniper (1970, p. 195), are specially adapted to intake of water but, in most crop plants, their significance for permeation of pesticides is incidental. Elaborate hairs (the aquatic fern *Salvinia molesta* is a classic example) may be so structured as to repel water and many densely hairy leaves will collect small mist droplets in the hair "canopy" outside the main leaf surface (Section 8.XI).

When leaves have been wetted, either because their surface is water-wettable or because the water is from a spray containing adequate wetting agent, the base of hairs will retain a ring of water after the plane surface between hairs has drained. The cuticle just around hairs will therefore tend to have much more than average residual density of applied pesticide.

The veins of most leaves form outstanding ribs on the abaxial (usually lower) surface and corresponding grooves on the adaxial surface. The grooves permit more rapid spread of a wetting liquid than the plane areas and provide preferential drainage channels in a wetted surface. Pesticide residues therefore tend to lie more over veins than interveinal areas.

E. Roots

It is doubtful (see Martin and Juniper, 1970, p. 107) whether any cuticle covers the young extending roots which are the main region of uptake of water and nutrients. Certainly it is very much thinner if it exists at all and is of no consequence for uptake. The major difference, however, between root and shoot for our subject is that the root is a much ramified rapidly extending organ specifically adapted to make good contact with the nutrient-carrying but very dilute aqueous solution in the soil. The leaves are freely in contact with moving air and do not need so much extension to make effective external exchange. The leaf surface is frequently designed to avoid real contact with liquid water. It denies entry to microorganisms, except the successful parasites which can pierce the cuticle, but has no special need in the ordinary natural environment to be impermeable to toxic chemicals since these are not present in most natural water. Many plants do excrete toxic substances as a weapon of competition but only through the soil to the roots and not, except via vapour, on to the leaves of their competitors. Resistance of the leaf cuticle to penetration by chemicals is therefore an incidental property.

Entry into the root would seem almost unavoidable in view of its ramification and function but root uptake has to be highly selective among natural solutes. The selection is made in its inner organized tissues rather than in an external cuticle. For this reason we consider it together with transport.

The areas of the root absorbing water and nutrients most effectively are generally located a few mm to cm behind the growing tip and are extended in most species by a dense population of transient hairs from about 10 μm to 1 mm long. The older parts of the root in many but not all species have mainly a conducting and anchoring function and become armoured with a corky "suberized" layer, which is very thick in major roots of many trees. Chemically, suberin is very similar to cutin. There is probably less difference between these two classes than between specimens of either from different species. Suberin builds up to a very much thicker layer than cutin, reaching several cm in the bark of the cork oak, *Quercus suber*, the source of the commercial product.

The suberization of old roots does not make them impermeable to some systemic pesticides. Treatment of trees with systemic insecticide is more efficient when these are applied near the base of the trunk.

III. INTERNAL MORPHOLOGY

A. The Translocation Systems

This brief summary does even rougher justice to a more complex subject but we hope it will be appreciated that we are trying to restrict ourselves to the features which directly influence translocation of pesticides.

Higher green plants have an elaborate transport system which lacks the circulatory character of the blood stream of mammals and is, in general, much slower. It is moreover adapted to carry only water and substances dissolved therein, principally sucrose (the energy and carbon source of metabolism subsequent to photosynthesis), and simple chemical "raw materials" such as mineral ions, nitrate and amino acids. There are in the flowing juices no oxygen-carrying compounds or organelles and no mobile fat globules as there are in mammalian blood, with their incidental ability to carry fat-soluble compounds. Although rubber globules which can absorb non-polar substances are found in plant "latices", these are contained in separate vessels not part of the long-distance transport system and only in some species permitting massive but still local flow.

The xylem transport system consists of long vessels connected to form continuous tubes which conduct water and dissolved salts from the roots to the peripheral areas of the leaves where most of the water is evaporated. This movement of water is called transpiration and it is mainly powered by evaporation itself, liquid water maintaining filled the fine capillaries from the dispersed ends of which evaporation occurs. The xylem vessel contents in tree trunks are thus under tension and their puncture, mechanically or as a result of pathogen attack, inactivates the tube system concerned. Observable bulk flow in the main xylem vessels of some rain-forest species may attain mean linear speeds of 100 m h^{-1}, but, in most crop plants, speeds of 10 m h^{-1} are the usual maximum. Water exudes from cut vessels which are at a low enough level, indicating that there is also some positive pumping action in the root. Some watery solution is also exuded through special openings (hydathodes) in the margins of leaves of herbaceous species. The phenomenon is known as guttation and the drops, forming at edges, are easily dislodged. The mechanism has probably a useful excretory function. Some involatile solutes, particularly potassium carbonate, find their way to the outside of the cuticle in other areas, from which they are washed off by rain and return to the soil (Section 11.VI.D).

The other conducting system, the phloem, is much more selective, has a finer structure and responds to metabolic demands in a more complex way. The main function of the phloem is to transport chemical raw materials and fuel from the regions of active photosynthesis to those of active growth. The most abundant of these materials is sucrose, present at concentrations of from 10 to 15% (much greater than in most cells), together with much smaller concentrations of amino acids. A metabolically powered loading mechanism must be involved in forcing sucrose against a concentration gradient into the phloem vessels. These are thin elongated cells, usually called sieve tubes, not forming a continuous tube but connected through porous plugs called sieve-plates. The system responds to leaf-illumination, or, more generally, to the provision of photo-assimilates and also to the demands of the "sinks".

The phloem vessels retain sucrose well and are consequently subject to high osmotic pressure: this is largely responsible for the extreme difficulty of examining the system microscopically in its natural, functioning state. The sieve tubes are closely associated with various other specialized components such as the companion cells and the integrity of the system is easily damaged during investigation. It has proved impossible to isolate the sieve tubes satisfactorily or to reach agreement about the details of their fine structure. Whereas the essential nature of xylem transport (i.e. a capillary flow induced by the pressure gradient caused by evaporation at the leaf assisted by positive pressure from the root) is reasonably well established, the phloem transport system has proved one of the more intractable problems in plant physiology. Work on the structure and function of the phloem, and the various hypotheses proposed to explain the mechanisms of phloem translocation are reviewed extensively by Crafts and Crisp (1971), Peel (1972) and Canny (1971, 1973).

Whatever the differences of opinion about the detailed structure and function of the phloem, there appears to be general agreement on one aspect of great significance in connection with translocation of exogenous substances. This is the important observation that the protoplasm of adjacent cells is connected *via* the plasmodesmata and the cells are linked with the phloem. The protoplasm of the whole plant therefore forms a continuous system, called the symplast.

B. Apoplast, Symplast and Free Space

The phloem and xylem both follow the branches of the root and stem systems and out along the spreading leaf veins. They are therefore closely associated in space. The phloem contents are eventually continuous with the *intra*cellular protoplasm, forming the symplast. The xylem vessel contents are eventually continuous with the *inter*cellular pectinacious fluid which is called the apoplast. The two systems, distinct within many parallel vessels in the main skeleton of the plant, become interwoven, but both continuous, in the chemically active cells of the leaf, rather like the air- or water-filled space in a sponge and its interlacing fibrous matrix. The concept of apoplast and symplast dates from Münch (1930). It has proved helpful in considering movement of systemic pesticides and has been extended by many authors, for example Crafts and Crisp (1971). Between the two systems there is always the semi-permeable plasmalemma.

In the case of uptake of ions into roots, two phases in the overall process can be distinguished. The first phase is a passive one whereby the chemical diffuses into the parts of the tissue which communicate reversibly and relatively freely with the external source. This phase is usually relatively rapid and is followed by a slower, more extended phase, in which the amount accumulated increases at an approximately constant rate and which can be shown by investigating the effects of metabolic inhibitors to be an active process requiring metabolic energy. The extent of the first phase can be characterized by the amount of solute accumulated in the initial period of rapid uptake and the apparent volume that this represents, as calculated from the concentration of the original solution, and has been referred to by such terms as "outer space" (Epstein, 1955) "apparent free space" (Briggs and Robertson, 1957) "diffusion free space" (Johnson and Bonner, 1956) or "effective volume" (Crowdy *et al.*, 1958). Without the qualification "apparent" or "effective" these terms imply equality with a distinguishable anatomical part of the tissues, whereas they must include not only the volume of water available for diffusion, but also the effects of ion exchange, adsorption or partition on to the surfaces bounding this volume and various other corrections which may account for much greater quantities of solute than the volume of accessible water itself (Crowdy, 1972).

Authors appear to differ in their exact definitions of the apoplast and how this is related to the free space. Crafts and Foy (1962) consider that there is no connection between the apoplast, even of the root, and external fluids except through the symplast. This view would clearly separate the apoplast and the free space. The highly selective uptake of nutrients into the xylem system of the shoot from the environment of the intact root and the ability of the root to create pressure in the xylem fluid provide evidence of this separation by metabolically controlled membranes. The barrier to the free movement of water and solutes can be attributed to the endodermis, a single layer of closely packed cells at the innermost region of the cortex (Bukovac, 1975). However, Crisp (1971) defines the apoplast as being bounded on the outside by the *ep*idermis and on the inside by the plasmalemma while Crowdy (1973) specifically includes the free space in his definition of the apoplast. While the concept of symplast and apoplast has many advantages, therefore, it is sometimes difficult to decide on a precise distinction. Furthermore the concept has the positive disadvantage that it has led some authors to separate symplastic and apoplastic movement of pesticides too sharply, whereas many materials can probably enter both systems to some extent even when one very definitely predominates.

IV. ENTRY THROUGH STOMATA

Alien gases can gain entry through open stomata. It has been shown that leaf damage caused by ozone and nitric oxide contamination of the air is correlated with stomatal opening (Dugger *et al.*, 1962; Otto and Daines, 1968). The structure and surface of the guard cells, however, are very well adapted to denying the entry of liquid water. Not only are the cuticular ledges, which come together to close the stoma, usually very sharp-edged, but they are also usually non-wettable. An advancing drop of liquid which forms a finite contact angle with the micro-surface will step, as it were, from one ledge to the next rather than turn through a very large angle (Sections 8.IV, 8.XII). Mobile non-polar liquids can enter stomatal cavities and cause extensive damage. Using fluorescent tracers Dybing and Currier (1961) showed that some penetration can be achieved with aqueous sprays if they contain enough good wetting agent. The observations were taken further by Currier *et al.* (1964). Schönherr and Bukovac (1972) have made a more exhaustive study of stomatal penetration in the single species *Zebrina purpusii*, measuring the external pressure necessary to force entry of applied liquid. They found a surface tension of 30 mN m^{-1} to be critical for zero contact angle and spontaneous penetration without pressure.

Uptake of some herbicides has been shown to be greater through stomata-bearing than through astomatous surfaces (Sargent and Blackman, 1962, 1965). This was true even when the stomata were closed and was held to be evidence of penetration being more facile through the cuticle over the guard cells than elsewhere. This could be due to the cuticle there having to expand and contract with change in turgidity of the underlying guard cells. Ectodesmata (minute channels from the interior of the epidermal cells which appear to terminate below the cutin) are particularly numerous over the guard cells (Franke, 1967). The cuticle is likely to be more porous when stretched (guard cells turgid, stomata open). Local penetration of tracers around stomata is therefore not necessarily evidence of liquid intrusion into the stomatal cavities and increase of general penetration when the stomata open can also have an indirect mechanism.

Sands and Bachelard (1973a) found that penetration of ^{14}C-labelled picloram (appied as triethanolamine salt in water) into leaves of two species of *Eucalyptus* was closely correlated with contact angle of solutions when different concentrations of different surfactants were added. Although the contact angle was reduced to zero only after washing the leaves with chloroform, which did *not* increase general permeation, they considered that actual liquid intrusion was probably involved. In discussion (1973b) they note that some authors rather uncritically quote Adam (1948) as having shown that sharp-edged small holes will not be entered by liquid, whereas Adam in fact concluded only that corrugations of concertina type in the bore of a tube would offer extra resistance to the progress of a meniscus when the contact angle of the liquid with the substance of the tube is finite. Adam's treatment is amplified in Section 8.XII. The criticism of Sands and Bachelard would have more force if the increased permeability had been critically dependent on zero contact angle. Actually they observed great increase over a short range of finite angles. There is need for more study of stomatal penetration. The detailed structure of stomata shows considerable variation between species. Observations quoted in this brief review refer to many species, but few have been common to different researches.

It has been considered that stomatal penetration, even if proved to occur, cannot be exploited in practice because (1) it would require spraying to be confined to the period when stomata are open, which depends on a circadian cycle modified by light intensity, humidity and temperature, (2) it would require, in the case of aqueous sprays, uneconomic rates of surfactants, (3) the residue, particularly from low volume and aerial spraying, would dry before penetration could be achieved, and (4) many crop plants and weeds have stomata only on the lower, less accessible leaf surfaces.

Since, however, it is established that penetration from leaf deposits is correlated with stomatal opening whether there is actual liquid intrusion or not, the climatic and physiological factors listed will always give rise to variations in performance. For the same reason, stomatal density is one of the factors in response to herbicides or uptake of systemic insecticides or fungicides. Most leaves have a higher density of stomata on their under surfaces and are, in any case, more permeable to compounds applied on this surface; we should give more attention, in the design of spray booms to the desirability of hitting this surface (see p. 766). In spraying forest or tree crops from the ground, lower surfaces are preferentially hit.

V. PERMEATION OF LEAF CUTICLE

A. Introduction

We are mainly concerned in this book with the penetration of pesticides. For obvious reasons, more work has been done on the penetration of water through the cuticle. This has significance for pesticides for two reasons. If water in the liquid state can get through fine capillary channels it could carry dissolved pesticide with it. If it can swell the cutin polymer this could make the latter more permeable. We shall not attempt a comprehensive review of water penetration but only of observations and theories which bear on pesticide penetration.

B. Cuticular Waxes

The wax embedded in the cutin has been held by many workers (see reviews by van Overbeek, 1956 and Crowdy, 1972) to exert the function of sealing unavoidable imperfections and providing a complete barrier to water permeation when the cutin is closed tightly around it. Swelling of the cutin (by water) then provides space for leakage. The barrier function requires the wax to be impermeable. This is consistent with the wax platelets (Meyer, 1938), embedded in the amorphous cutin, having their molecules oriented perpendicular to the plane of the lamina. Close-packed paraffinic molecules offer high resistance to permeation in the direction of the long-molecule orientation (Section 9.III.D).

It is also frequently assumed that these embedded waxes form continuous columns and can provide pathways for the diffusion of non-polar molecules. This is inconsistent with their crystalline nature and orientation. Crystalline fatty materials are not solvents even for non-polar compounds. This is clearly shown by the fact that the solvent power of natural fats for oil-soluble indicators gives a measure, consistent with more elaborate methods, of the non-crystalline fractional volume in the fat (Zobel *et al.*, 1955; Haighton *et al.*, 1971). Crystallinity of triglyceride fats is more complex than that of the simpler plant waxes, which have much fewer possible isomers and which are usually fully saturated.

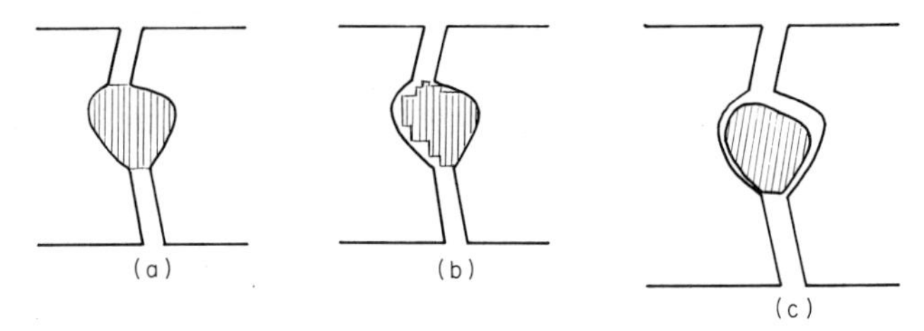

Fig. 1. Schematic diagram of pore through cuticle blocked by intrusion of crystalline wax. (a) Blockage complete since no diffusion through crystal. Transport possible only if (b) part of intrusion is dissolved, or (c) isotropic swelling of cuticle expands cavity away from non-swollen wax.

A saturated, crystalline wax column traversing a thin diaphragm with molecules normal to the diaphragm surface would not permit any diffusion through its substance. Transfer could occur only around the outer face of the block if it fitted its hole imperfectly or through the space it occupied if it were removed by solution (Fig. 1).

The concept that the pore-blocking waxes provide a "waxy pathway" for diffusion of non-polar compounds arises from the loose definition of wax about which we have already complained. The cutin oils could function in this way but not the hard waxes. Not only is the diffusion pathway concept inconsistent with the blocking property, but it also appears unnecessary since the cutin itself is an amorphous matrix through which diffusion is entirely to be expected.

The waxes are probably of special importance in the grape, the problems of commercial drying of which to raisins and sultanas has led to extensive study. The skin of the grape is, in common with that of most fruits, much thicker than is generally the case on leaves, but becomes incoherent when treated with pectin solvents. Its content of soluble waxes is very high ($1{\cdot}0$ to $1{\cdot}8 \times 10^{-4}$ g cm^{-2} in four varieties according to Dudman and Grncarevic, 1962). In their study of drying rate of sultana grapes at different air temperatures, Martin and Stott (1957) refer to the chloroform extractable material as "cuticle", a nomenclature which would not now be accepted but which, in the case of the grape, has some justification since the wax covering appears to be held in a predominantly pectinaceous and fibrous matrix.

The mature grape dries much more quickly after it has been dipped in alkaline emulsions, a practice dating from ancient Greece where olive oil and potash were used. This dip removes much less wax than a shorter dip in ethylacetate, which, however, according to Martin and Stott, accelerates drying less. Chambers and Possingham (1963) consider that the densely overlapping wax platelets delay evaporation by obstructing and making very tortuous the diffusion of water vapour: the dip makes the platelets wettable so that liquid water can move to the surface by capillary flow (see further comment, Section 9.IV.A). The constituents of wax on the grape have been identified by Radler and Horn (1965). Possingham *et al.* (1967) showed that the fruit carries from 8 to 10 times as much wax per unit area as the leaves and that 70% of this is the triterpenoid oleanolic acid, a very minor constituent of leaf wax. Holding the grapes for 10 seconds in the vapour of boiling light petroleum produced a 50% increase in water loss rate after removing only 26% of the "soft" wax and 1% of the "hard" (oleanolic acid) wax. A four-fold increase followed 5 minutes exposure which removed 85% and 12% respectively. This confirms the finding of Martin and Stott that "wetting out" the crystals is more effective than their partial removal.

Martin and Stott found that vine leaves, which lose water in the natural state 10–30 times as fast (per unit area) as the fruit, show an even greater increase on exposure to petroleum vapour—of between 10 and 40 times to a treatment which produces only a 2·5-fold increase in the fruit. Flooding of the pores is not involved in this case since it would require prior flooding of the air space of the spongy mesophyll which only occurs after extensive cell damage. The reviews by Martin and Juniper (1970) and Bukovac (1975) list other examples of acceleration of water loss consequent on removal of soluble waxes. Measurements have usually been made on isolated cuticle forming a septum in the air space above a reservoir of water, where flooding could occur only in capillary channels so narrow as to reduce significantly the vapour pressure of contained water. Of particular significance are observations on astomatous cuticle, e.g. that from the adaxial surface of ivy (*Hedera helix*) leaves through which Skoss (1955) found a 40-fold increase after extraction of waxes.

Flooding with alkaline emulsion would be a possible technique for accelerating entry if the leaf cuticle had the multi-tesselated wax platelet structure attributed to the grape. Acceleration by removal of embedded wax is less likely because this requires prolonged contact and the pesticide residue could become dry before this happend. Any such gross damaging treatment, even with herbicides, could defeat its own object by inhibiting translocation.

C. Permeation of Detached Cuticle by Water

Much work has been carried out to determine the rate of cuticular transpiration and its importance relative to vapour diffusion through stomata and in the stagnant air layer near the leaf surface. The sequence of various resistances to movement of water from soil to atmosphere was discussed by van den Honert (1948) and the leaf surface processes were particularly reviewed by Gates in 1968. While neither treatment is fully satisfying with respect to physical theory there is agreement that stomata add considerably to the total air resistance when very nearly closed and that with many crop plants, cereal and broadleaf, transpiration through the cuticle is sufficient to leave the stomata with only a partial influence on water loss.

Gates lists the cuticular resistance to water passage for several species, the data being derived from various sources, mainly by comparison of transpiration rates with stomata closed and open. The values are listed in $s\,cm^{-1}$ on a vapour basis—i.e. taking the concentration gradient as that between the vapour in equilibrium with the tissues on one side and that in the partly saturated air outside. The resistance on a liquid basis can be obtained by dividing by the ratio of water vapour concentration to actual liquid concentration (i.e. density) at equilibrium. At ordinary temperature this means multiplying by a factor of about 5×10^4. The permeance is the reciprocal of the resistance. A stagnant air gap of 1 cm thickness has a resistance of about $4\,s\,cm^{-1}$ on a vapour basis or about 2×10^5 on a liquid basis. The reciprocal of the latter, a permeance of $5 \times 10^{-6}\,cm\,s^{-1}$, measures the rate at which a column of water would sink if the only resistance to loss were evaporation from the surface of a finely porous base through a 1 cm air gap into dry air. $5 \times 10^{-6}\,cm\,s^{-1} = 0{\cdot}43\,cm\,24\,h^{-1}$. These relationships are fully discussed elsewhere (Section 9.I.B) but so many errors can be found in the literature arising from confusion of two different standards of measurement that it seems desirable to recapitulate.

Gates' list covers a range of cuticular resistances from 0·2–3·3 for several crop species ($s\,cm^{-1}$, vapour basis, cf. 4 for 1 cm air gap) to between 30 and 200 for conifer needles, 300 for the common oak (*Quercus robur*), with values up to 400 reported for bracken (*Pteridium aquilinum*).

Detached cuticle of the astomatous adaxial surface of ivy (*Hedera helix*) has been examined by Skoss (1955) and by Schieferstein and Loomis (1959). The former was particularly concerned with effect of light during development (and with removal of wax) and the latter with age of leaf and direction of flux. Skoss used an osmometer with one side of the membrane exposed to dry air and therefore measured directly the liquid-based permeance, the meniscus in the capillary stem being used to measure, with magnification, the linear rate through the membrane. The value for the cuticle from sun-grown mature leaves was 0·03 mm in 24 h or

$3{\cdot}4\times10^{-8}$ cm s^{-1} corresponding at 20° to a vapour-based value of $1{\cdot}7\times10^{-3}$ cm s^{-1} or a resistance of 590 s cm^{-1}, just beyond the range quoted by Gates. From the data of Schieferstein and Loomis for mature leaf under one year old we calculate a resistance of 340 s cm^{-1}. Three-year-old leaves had a cuticle 40% thicker but with a resistance probably not significantly (10%) greater. Permeance out→in was 30–50% greater than in the normal direction of transpiration. Skoss found that shade-grown leaves had a cuticle with less than one-third the resistance of sungrown.

It is of interest to compare these with values for other hydrophobic septa. The ivy leaf cuticle (sun-grown) is about 2·2 μm thick and the mean permeability of the substance (Section 9.I.A) of which it is made is therefore about $5{\cdot}5\times10^{-7}$ cm^2 s^{-1} (vapour basis). The last compares with 0·25 for stagnant air, 1×10^{-4} for stagnant liquid paraffins and $2{\cdot}5\times10^{-6}$ for polyethylene. At its best therefore ivy leaf cuticle would be a rather more water-proof wrapper than polyethylene film of the same thickness (the most usual thickness of small commercial bags is about three times as great). Close-packed paraffinic monolayers are much less permeable (5×10^{-8} to perhaps as low as 5×10^{-10} cm^2 s^{-1}) but, of course, very much thinner (Section 9.III.D).

Scheuplein and Blank (1971) found a permeance for human abdominal stratum corneum about equal to that quoted above for ivy leaf cuticle, but the statum corneum is about ten times as thick. It has about the same thickness and permeance as tomato fruit cuticle.

All the above figures refer to cuticle from which no attempt was made to remove the natural complement of oils and waxes. When soluble constituents are removed by soaking in chloroform the permeance in all cases increases, by a factor of from 40 to 50 in the case of ivy leaf cuticle and about 10 in the case of tomato fruit cuticle. Other examples of increase of permeance of plant cuticles are to be found in the reviews by Martin and Juniper (1970) and Bukovac (1976). Large increase of permeance follows de-fatting of the stratum corneum (Scheuplein and Blank, 1971). Soluble waxes play an even more important part in the water-proofing of insect cuticle (Section 12.II.A).

The evidence from permeation of dissolved substances (Bukovac, 1976) indicates that wax removal facilitates the permeation of polar more than of non-polar compounds, consistent with the latter being able to diffuse through the cutin itself. In the case of tomato fruit cuticle, however, extraction with chloroform still leaves the cuticle remarkably impervious. The ten-fold increase of water permeance is accompanied by a much smaller increase (Section 11.V.E) in the permeance of dissolved methylorange (sodium sulphonate of dimethylaminoazobenzene) and the permeance of the latter is still far lower than it would be were the water passing as liquid through a non-selective pore system. If removal of wax has increased water permeation by opening pores, these must be of size comparable with this fairly small organic anion, having room for very few parallel paraffin chains.

The physical organization and disposition of the wax is also of major importance. Baker and Bukovac (1971) attempted to recreate a low water permeance by deposition in paper of extracted leaf waxes from solution. The measured flow was, however, under a small pressure of liquid water, incomparably greater than that through intact cuticle, and must have been through relatively gross pores.

It is possible for mainly crystalline waxes to make paper almost waterproof but only by the hot sliding roller technique used in the preparation of commercial wrapping papers. The waxes now used for this purpose have a small proportion of non-crystalline components and usually of hydrocarbon polymers. The wax loading per unit area is some 100 times greater than that of the waxiest leaf surfaces examined Baker and Bukovac. Permeances lower than an equivalent thickness of polyethylene have been reported (Dijk and Kaess, 1947; Fasting-Jonk, 1959), but papers treated with straight waxes are made more permeable by creasing (Gruhn, 1971).

D. Swelling of the Cuticle

Although swelling has been considered to affect permeance by loosening the postulated wax plugs in pores, direct measurements of swelling, even by water, are lacking in the literature. The probable greater swelling in organic solvents does not seem to have been examined. One of us (G. S. H.) therefore carried out a few orienting experiments with the vapour of water, the lower alcohols and hexane using oxalate-detached cuticles from tomato fruit (cv Russian red), peach fruit (cv Golden Honey), apple fruit (cv Kidd's orange), leaves of *Nerium oleander*, adaxial surface and of *Metrosideros excelsa*, adaxial and abaxial. 0·3–0·5 g of folded dry cuticle was packed into baskets of stainless steel gauze which were suspended from a balance arm by a wire passing freely through a hole in the lid of a glass vessel. The vessel contained, in separate experiments: water with excess Na_2HPO_4; methanol, ethanol and butanol, each with excess potassium acetate; *n*-hexane or cyclohexane saturated with stearic acid and in a separate basket, silica gel. The object of the involatile additives was to reduce the vapour pressure and so avoid indefinite condensation on soluble matter in the cuticle. The cuticle specimens were exposed to some vapours before exhaustive extraction with boiling chloroform; other exposures were carried out after extraction. In some cases uptake was measured into the residue from the chloroform extract exposed as a smear on the inside of a miniature beaker. Results are given in Table 1 overleaf. A large fraction of each weight gain (and loss on ventilation) occurred in a few minutes but the results listed were after exposure of 24 hours or more.

In half the cases where it was examined, butanol was taken up in greater proportion than any other vapour offered. In all cases, hexane was least absorbed but there was always significant absorption. Cyclohexane appears to be less absorbed than *n*-hexane as would be expected (Section 9.II.F) for penetration between segments of macromolecules. Only in peach and apple fruit cuticles could the residue from evaporated chloroform extract account for a major fraction of the total uptake into unextracted cuticle. In some cases, notably apple fruit cuticle, the separate components take up more in sum than the composite cuticle. This suggests that the cross-linking limits the total extra volume which can be accepted. The opposite behaviour could be due to irreversible shrinkage following the removal of built-in waxes.

The higher absorption of butanol than of ethanol may have been due to its greater relative vapour pressure, K acetate being less soluble in butanol. Better controlled experiments are necessary but it is clear that these cuticles, even without their initial content of soluble waxes, are significantly swollen by small organic molecules, often more than by water. That alcohols are more absorbed into the mainly paraffinic cutin than are paraffins is probably due to the hydrogen bonds within the structure acting cooperatively as cross-links restraining expansion. Only when these bonds are loosened by the provision of mobile partners are the paraffinic regions free to expand. Studies, not yet made, of simultaneous uptake of water or alcohols and paraffins could throw light on this.

The composite cuticle has physical properties differing between the outside and the side formed by detachment from the cells. The differences are evident in wettability and swelling. The shiny outer surfaces of most separated cuticles accept water only with a finite contact angle. Removal of soluble wax does not always, as Holloway (1970) has shown, result in a decrease of contact angle. Schönherr and Bukovac (1973) found that treatment of tomato fruit cuticle in boiling 6 N HCl, which was shown to remove protein and further pectin, caused even the inner surface, previously freely wettable, to be met with a finite contact angle although this remained lower than on the outer surface. Tomato fruit cuticle, as soon

Table 1

Uptake from vapour into leaf and fruit cuticles before and after extraction by chloroform

Specimen	sp. gr. from buoyancy in water	% loss into chloro-form	Uptake in g per 100 g *initial* dry weight												
			Water		Methanol	Ethanol			*n*-Butanol	*n*-hexane			cyclohexane		
			b	a	a	b	r	a	a	b	r	a	b	r	a
leaves															
Met. ex. ad.	1·18	12·5	6·0	5·7	7·3	—	—	8·1	12·0	—	—	—	2·7	—	—
Met. ex. ab.	1·38	11·5	10·2	13·6	14·3	—	—	8·8	12·0	—	—	—	2·2		
Met. ex. both (1)	—	15·1*	—	—	—	11·6	1·9	8·3	—	6·3	1·1	3·5[a]	—	—	—
Met. ex. both (2)	—	14·8	—	—	8·5	—	—	—	—	4·6	1·0	3·9	—	—	—
Ner. ol. ad.	1·20	9·1	7·3	7·4	8·5	—	—	9·7	14·7	—	—	5·6	3·2	—	—
Ner. ol. both	—	23·0	—	—	—	8·7	1·7	9·2	—	5·0	1·2	3·9	—	—	—
fruits															
Apple	1·10	29·1	—	6·9	5·1	7·3	3·9	5·0	11·2	5·1	3·2	3·6	3·0	—	—
Peach	1·34	15·0	16·7	16·7	10·0	—	—	10·0	9·9	—	—	—	7·5	3·2	0·9
Tomato	1·10	1·6	—	16·8	6·7	—	—	8·6	7·8	—	—	—	—	—	2·1

Met. ex = Meterosideros excelsa; Ner. ol. = Nerium oleander; ad = adaxial surface; ab. = abaxial surface; b = direct before extraction; r = calculated from uptake into residue from chloroform evaporation; a = calculated from uptake into extracted cuticle.

[a] Further extractions of this specimen caused no more loss in weight, but uptake fell to 3·3, 3·0%. 10 days storage, dry, caused further fall to 2·7%.

as it is freed from cellular debris but while still wet, curls into a tight roll with the flesh side out. On drying it flattens. A small piece of flat cuticle placed on clean water, shiny outside down, usually remains floating, flat but in a slight depression in the water surface, for many minutes. If one edge is forced into the water, the specimen immediately rolls up tightly, outside in, and sinks. The flesh side expends laterally in water. Since the cuticle is swollen by organic solvents, presumably the hydrophobic outside more than the inside, one could anticipate the reverse curling when such solvents have access to dry cuticle and this indeed can be demonstrated, although, in the case of tomato fruit cuticle, the curling is less extreme than with water.

E. Permeation of Detached Cuticle by Other Substances

1. *Effect of Reservoir Volumes*

There are several accounts in the literature of measurement of permeation of detached cuticle by substances presented in aqueous solution or in agar gel blocks clamped to the surface. The latter method has obvious convenience if a sufficiently sensitive method (usually radioactive assay) is available, since leak-tight mounting of the cuticle specimen is unnecessary and replication is therefore greatly facilitated. The method is, however, virtually confined to dilute solutions. It does not appear to have been used for measuring permeation from solution in a formulating oil, although suitable gelling agents could be found.

Whether this method is used or the cuticle is mounted between gaskets and forms a septum between glass reservoirs adapted for sampling, it is essential to specify and, in calculations of permeance, allow for, the volumes on either side of the septum relative to the area of the latter. It is surprising how often these essential data are not given. A statement such as "only 2% of the antibiotic diffused through the membrane" is quite meaningless without this information. In at least one important paper the volumes are incorrectly allowed for because an equation relevant to a small volume exposed to a virtually infinite supply is used for calculations from experiments where the donor and receptor volumes are comparable. The correct equations for calculating permeance from concentration measurements over measured time intervals have already been given (Section 9.I.B).

$$\Delta C = (\Delta C)_0 \exp\left[-\left(\frac{1}{\mathbf{V}_{\text{don}}} + \frac{1}{\mathbf{V}_{\text{rec}}}\right) P\mathbf{A}t\right] \tag{1}$$

where ΔC is the concentration difference at time t and $(\Delta C)_0$ the initial concentration difference which will usually be $= C_0$, the initial donor concentration. $\mathbf{V}_{don}$ and $\mathbf{V}_{rec}$ are the donor and receiver volumes respectively, making exchange through an area $\mathbf{A}$ of the septum. If the reservoirs are cylindrical blocks of area $\mathbf{A}$ and thickness ℓ_{don} and ℓ_{rec} respectively, we may write

$$\left(\frac{1}{\mathbf{V}_{don}}+\frac{1}{\mathbf{V}_{rec}}\right)\mathbf{A}=\frac{1}{\ell_{don}}+\frac{1}{\ell_{rec}} \tag{2}$$

An equivalent alternative to (1), sometimes preferred, measures approach to the final equilibrium state when the concentration in both reservoirs must be the same and calculated to be C_∞. If C_{rec} is the measured concentration in the receiver and C_{don} that in the donor, the equation becomes

$$\left(\frac{1}{\mathbf{V}_{don}}+\frac{1}{\mathbf{V}_{rec}}\right)P\mathbf{A}t=\ln\frac{C_\infty}{C_\infty-C_{rec}}=\ln\frac{C_0-C_\infty}{C_{don}-C_\infty} \tag{3}$$

As long as ΔC has changed by only a small fraction of $(\Delta C)_0$, $(C_{rec} \ll C_\infty)$, eq. (3) can be put in the approximate form

$$C_{rec}=\left(\frac{1}{\mathbf{V}_{don}}+\frac{1}{\mathbf{V}_{rec}}\right)P\mathbf{A}tC_\infty=\frac{1}{\mathbf{V}_{rec}}P\mathbf{A}tC_0 \tag{4}$$

indicating that, in the very early stage of transfer, the donor volume is of no significance. It is also, of course, for measurements in the receiver, of no significance throughout if it is large enough compared to that of the receiver.

If an osmometer apparatus is used in which, in a fairly typical arrangement, $\mathbf{V}_{don}$ and $\mathbf{V}_{rec}$ are 10 ml and $\mathbf{A}$ is 1 cm^2, $(1/\mathbf{V}_{don}+1/\mathbf{V}_{rec})\mathbf{A}$ has the value $0{\cdot}2$ cm^{-1}. If the receiver concentration has increased only to 2% of C_0 in 24 h, we may apply eq. (4) to calculate the value of P as $2{\cdot}3\times10^{-6}$ cm s^{-1}, a value in the range observed for pesticides in water on either side of a leaf cuticle. Consider now that the large donor volume has been replaced by a spray drop spread to a mean depth of 40 μm, i.e. $\mathbf{A}/\mathbf{V}_{don}=250$ cm^{-1}. Even if $\mathbf{V}_{rec}$ remained large, ΔC from eq. (1) is seen to be at most $(\Delta C)_0\times\exp(-50)$, i.e. permeation is nearly complete, only an immeasurably small further change being possible. After only 1 h ΔC is $\exp(-2{\cdot}08)\times\Delta C_0=0{\cdot}125\,\Delta C_0$. Otherwise expressed, C_{res} has reached $\frac{7}{8}$ of its final value.

In practice, in a mesophyte leaf, the available receiver volume will also be small, being that of the apoplast, contributing at most about one tenth of the total mass per unit area and therefore an effective depth of about 20 μm. ($\mathbf{A}/\mathbf{V}_{don}+\mathbf{A}/\mathbf{V}_{rec}$) is therefore increased to 750 cm^{-1} and exp (−2·08) in the last calculation becomes exp (−6·2) = 0·002. Two per cent exchange in 24 h in the 10 ml/10 ml osmometer corresponds to exchange from a spray drop to a leaf interior having gone 99·8% of the way to completion in 1 h. "Completion" in the example chosen, is transfer of only one third of the drop contents to the inside. If the cuticle behaved in the osmometer according to our example, we should not conclude that permeation is negligible in the case of a spray drop, but rather that it would be so facile that the extent of permeation would be mainly limited by the ability of the leaf tissues to disperse the received pesticide laterally and/or accept it into the cell contents, thus raising the "ceiling" of one third.

A permeance of 1×10^{-6} cm s^{-1} would be high enough to make the cuticle resistance of little importance in practice, but at 1×10^{-7}, 10 h would be needed for the concentration difference between a 40 μm deep reservoir and one 20 μm deep to be reduced to 3% of its initial value (5 h for 17%). This indicates the range of permeance which is of importance. If below 10^{-8} m s^{-1} the uptake of the applied compound under practical conditions would be so slow as to be, in competition with mechanisms of loss, very inefficient.

In this brief account of the quantitative aspect of cuticle permeation we have so far assumed that the cuticle itself does not hold up an appreciable fraction of the compound applied. This assumption may sometimes be far from valid. The effect of significant cuticle capacity is to introduce a further factor of delay. In experimentation with large reservoirs the eventual steady rate of transfer is not affected but appreciable time may elapse before the steady rate is reached. The extrapolated lag-time (Section 3.V.F) is equal to the time which would be taken, at the final steady rate, to fill one sixth of the capacity of the septum. In experimentation with small reservoirs—e.g. thin agar gel discs—the cuticle capacity may alter significantly the final concentrations expected.

2. *Permeation by Ions*

Detached cuticles of apple fruits and of leaves of *Prunus lusitanica*, *Euonymus japonica* and apple were examined by Silva Fernandes (1965) for permeation of cupric acetate or sulphate and phenylmercuryacetate in aqueous solution. Astomatous adaxial (ventral) cuticles of the two evergreen leaves were compared with their stomatous abaxial (dorsal) cuticles. Apple fruit cuticles were examined at various stages of development and before and after extraction of waxes with chloroform. Permeation of astomatous leaf cuticles was not detectable by the methods used except when, in the case of some apple specimens, an area over a main vein was tested. The limit of detection for Cu was given as 3 μg, the apparatus had reservoirs of volume *c.* 20 ml on each side of a $0{\cdot}8\ cm^2$ septum and the donor concentration was 200 p.p.m. The longest period of contact was 96 h. The maximum value of P from eq. (4) comes to $6 \times 10^{-8}\ cm\ s^{-1}$ but could still be significant for uptake from spray.

The abaxial cuticles by contrast allowed measurable amounts through. The calculated permeances ranged from a little greater than the above limit to 60 times this value, the low values being obtained on specimens with closed stomata. The highest value is 50% greater than that taken in our hypothetical example (p. 570).

Apple fruit cuticle showed undetectable permeance except on fully mature fruits of one cultivar and on stored specimens of another until subjected to chloroform extraction when all specimens became permeable.

Even when no permeation was detected, significant adsorption of phenylmercuryacetate into the discs could usually be measured, often as much as was initially contained in a 1 cm depth of solution.

Yamada *et al.* (1964) examined the permeance and adsorption of several inorganic ions (radiolabelled) in tomato fruit cuticle. Adsorption of cations was much greater than of anions as would be expected on a matrix having many dissociable carboxylic groups, the estimated pK's of which, 2·8–3·0 in pear leaves (Bukovac and Norris, 1966), suggest pectic rather than fatty carboxyl groups. In order to distinguish adsorption on the two surfaces, each was exposed for 5 s only before blotting dry and counting. The outer surface of tomato fruit cuticle adsorbed $2{\cdot}7\ nmol\ Ca^{2+}\ cm^{-2}$ from a 0·1 mM solution and the inner surface $7{\cdot}9\ nmol\ cm^{-2}$. These would be contained in depths of 2·7 and $7{\cdot}9 \times 10^{-2}$ cm of offered solution. Since the cuticle is about 2×10^{-3} cm thick, there is at least a 50-fold enrichment on a volume basis even though adsorption would not, of course, be complete throughout the cuticle in this short period of contact. The total Ca^{2+} adsorbed in 5 s corresponds to $10\ m\,eq.\ \ell^{-1}$ which is only about one hundredth of the potential ionizable groups according to the later analysis of tomato fruit cuticle by Schönherr and Bukovac (1973). The chloride adsorption in 5 s was so small as to be probably insignificant.

The amount of Ca^{2+} passing through a 15 mm diam. disc from a solution of 0·1 mM in contact with the morphological outer surface into deionized water was 4 nmol in 20 h and 7 in 40 h from which we calculate a mean permeance of $3{\cdot}6 \times 10^{-7}$ cm s^{-1}. In the reverse direction, the permeance, falling off more steeply with increasing time of contact, was about half this value. Values for Cl^- (^{36}Cl in separate experiment) were about 0·5 and 0·8 on a molar basis, of the values for Ca^{2+} indicating that the Ca^{2+} must have been accompanied by unknown anions from the cuticle specimens. It should be noted that the amount of Ca^{2+} passing through 15 mm diam. disc in 40 h was $1{\cdot}4 \times 10^{-8}$ equiv., while the total ionizable content was 3×10^{-6} equiv. so that a rather small "impurity" of mobile ions could explain the discrepancy.

Most other information on permeance of cuticle to small ions comes from work on whole leaves and plants considered later. There is no doubt that some nutrient ions can penetrate effectively, though perhaps not through intact cuticle.

Reluctance to pass ionized substances is made evident by the comparison of permeance of detached tomato fruit cuticle to benzoic and phenoxyacetic acids carried out by Bukovac *et al.* (1971). Successive chlorination of phenoxyacetic acid produces compounds of increasing permeance. Successive chlorination of benzoic acids has the reverse effect. The measurements were carried out at pH 3. The phenoxyacetic acids are around 50% dissociated at this pH and the chlorination has little effect on dissociation. Increase of permeability is considered due to increased partition into lipoid tissue. The benzoic acids are stronger and increase considerably in strength on chlorination (Sargent *et al.*, 1969) so that increasing dissociation at pH 3 outweighs the increased lipoid solubility of the free acids. The ions evidently permeate more slowly than the free acids.

3. *Permeation by Small Unionized Molecules*

Yamada *et al.* (1965) included urea in their researches on tomato fruit cuticle and onion leaf cuticle, the latter bearing stomata. They found urea to permeate tomato fruit cuticle 10–20 times faster than simple inorganic ions, about 6 times as fast as maleic hydrazide and twice as fast as *N*-dimethylaminosuccinamic acid. Urea was examined over a wide range of low donor concentrations (0·01 to 10 mM) and the rate of permeation was found to be *less* than proportional to the donor concentration. At the highest donor concentration there was a marked increase of permeance with increasing time of contact. The approximately steady rate from a donor concentration of 1 mM yields a permeance of $1{\cdot}6 \times 10^{-6}$ cm s^{-1}.

Data for other small molecule compounds do not appear in the literature of detached cuticles but evidence from whole leaves (p. 655)

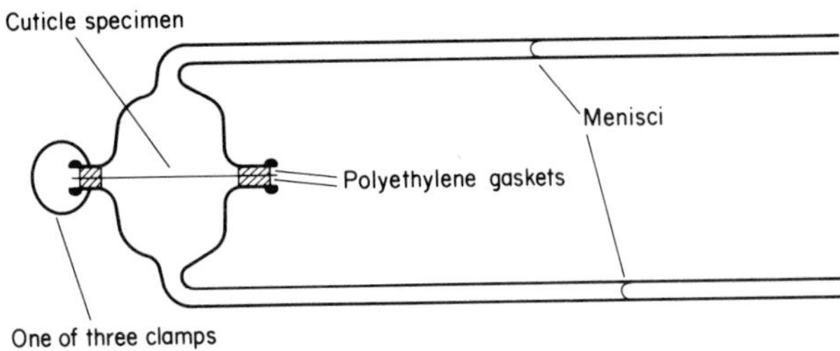

Fig. 2. One form of apparatus for measuring permeation of detached cuticle.

suggests that ammonia promotes rapid permeation. Hartley (unpublished work) has examined the mutual diffusion across cuticles between initally pure water on the morphological inside and pure alcohols or acetone on the outside. The reservoir volumes terminated in capillary tubes so that volume transfer could be measured directly (Fig. 2). The three lowest alcohols and acetone all penetrated more rapidly, on a volume basis, than water, the lower (organic) volume retreating while the upper (aqueous) increased against a small hydrostatic head. This head served to confirm absence of mechanical leakage when the apparatus was filled with the same liquid either side between runs. Simultaneous diffusion of water against the apparent flow could be calculated from total volume reduction (there being a decrease of volume on mixing all these liquids with water) and checked by density of the reservoir contents at the end of a run. The net flow organic → water increased in the sequence *n*-propanol < ethanol < methanol < acetone through both tomato fruit cuticle (detached by cold oxalate) and adaxial *Nerium oleander* leaf cuticle, (detached by boiling oxalate solution) but was in all cases more rapid through the latter, exceptionally so for acetone. The rate for acetone diffusing against water through the oleander leaf cuticle would correspond to a flow rate in the septum, if completely porous, of over 10^{-4} cm s^{-1}. Approximate initial rates are indicated in Table 2. These figures are not directly comparable with those obtained in dilute solutions. If we were to divide the acetone rate, the highest observed through tomato fruit cuticle, by the density of acetone, we should obtain a permeance of about $1{\cdot}5 \times 10^{-5}$ cm s^{-1}, nearly 10 times that of urea. A thermodynamic correction (Section 9.V) is, however, necessary to take account of the non-ideal nature of the solution. This would multiply the acetone figure by about 7, greatly increasing the deviation from urea. One must assume that acetone at high concentration, assisting to swell the cuticle, is facilitating its own permeation. Indicative of the same mechanism is the acceleration of the diffusion of water. When the same specimens of cuticle, mounted on the

Table 2
Interdiffusion of organic liquids and water. Initial rates in $(g\,cm^{-2}\,s^{-1}) \times 10^5$.

Liquid	Tomato fruit		Oleander leaf	
	organic	water	organic	water
n-propanol	0·80	0·93	2·07	1·20
ethanol	0·87	1·00	2·81	1·60
methanol	0·93	0·88	3·74	2·03
acetone	1·20	0·92	14·7	7·3

volumetric reservoir holding water against the morphological inside surface, were held so that the outer surface was in nearly direct contact with dried silica gel, the observed rate of water loss was $2 \times 10^{-7}\,g\,cm^{-2}\,s^{-1}$. Diffusion through 1 cm depth of stagnant air would reach $5 \times 10^{-6}\,g\,cm^{-2}\,s^{-1}$ in the absence of other resistance so we can assume the resistance to lie wholly in the cuticle. Similarly, the maximum loss through the cuticle into alcohols ($1 \times 10^{-5}\,g\,cm^{-2}\,s^{-1}$ for ethanol) must lie mainly in the cuticle since it can be estimated that a stagnant alcohol layer of about 5 mm thickness would be necessary to reduce the water transfer to this value. In the apparatus concerned the denser liquid lay above the lighter and the disc was inclined about 20° to the horizontal. Under these conditions the stagnant layer would be much less than 1 mm, but, in any case, any correction would only increase the 50-fold advantage which alcohol is observed to have over dry air in extracting water through the cuticle.

Clearly, alcohol in the cuticle must greatly accelerate the diffusion of water. Whether it accelerates it more than water itself would is not revealed by these experiments. As is discussed elsewhere (Section 9.II.D), a substance swelling a polymer matrix increases its own diffusion coefficient. In one case we have liquid water entering a fully swollen cuticle surface and having to diffuse to and emerge from a dry unswollen surface. In the other we have water entering a water-swollen surface and diffusing to and emerging from an alcohol-swollen surface. Further investigation of loss into a range of external humidities and/or diffusion of TOH in a fully saturated water system would be necessary to clear up this problem.

Measurements were made of the diffusion of phenol, chosen as the smallest stable molecule having a u.v. spectrum convenient for measurement. Less than 20% difference was found in the permeance of phenol diffusing to water through cuticle from water at a range of donor concentrations from 1 to 0·01%. The mean value was $2{\cdot}1 \times 10^{-6}\,cm\,s^{-1}$.

When methanol was used throughout as solvent, the permeance of phenol was about three times as great. In 50% methanol-water it was intermediate. When phenol was allowed to permeate from 1% solution in water in contact with the morphological outside of the cuticle into methanol on the other side the initial permeance was 12 times the water/water value. Since it was necessary in the set-up for the receiver solution to be above the septum, a stabilizing density gradient would in this case tend to establish, so that the true ratio for permeation of the septum itself could well be greater than 12. In the reverse case, phenol diffusing from methanol to water, the water/water value was divided by only 2·5. The discrepancy of the factors of increase and decrease is in the opposite direction to that expected from simple thermodynamic theory in a completely inert porous septum (Section 9.V) and indicates a high degree of interaction of the solvent gradient and the cuticle substance.

Much more investigation of other combinations would be necessary to elucidate this subject further. It is the permeation from a residue of formulating agents across the cuticle to the aqueous apoplast, rather than the more frequently studied permeation from a dilute aqueous solution to aqueous apoplast, which is the actual first stage in the entry of a systemic pesticide into the target organism.

It should be emphasized that the accelerating effect of lower alcohols and acetone was shown, by changing solvents on either side of the same specimens of cuticle, to be essentially reversible in both species. Leakage was not detected whenever the apparatus was returned to the state of water or alcohol in both compartments and ethanol/water behaved in the same way before and after runs with acetone/water. The very low permeance of ionic methyl orange (sodium sulphonate of dimethylaminoazobenzene) in 50% ethanol/water ($<3\times10^{-8}$ cm s^{-1} i.e. $<1/100$th of the water permeance) was not detectably increased after a whole series of tests. Increased water permeance following soaking the cuticle in petroleum ether or chloroform was, on the other hand, essentially irreversible (in the case of oleander cuticle the effect of chloroform is catastrophic, since the cuticle fragments in a fibrillar fashion).

4. *Permeance of Pesticides*

Several investigations have been reported of the permeance of herbicides through detached cuticle, a few of fungicides but we are not aware of any of insecticides. Schieferstein and Loomis (1959) measured the penetration of 2,4-D from a donor solution of 0·1 M in pH 6·2 buffer (personal communication) through the adaxial ivy leaf cuticle. For leaves up to 1 year old the permeance works out to about $1{\cdot}5\times10^{-7}$ cm s^{-1} calculated for the salt as such but would be about 1000 times this for the free acid if this alone were permeant. The rates for cuticle from 2- and 3-year old leaves were $\frac{1}{3}$ and $\frac{1}{30}$ respectively of the 1-year rate, although thickness increased by only 30%. The water permeance was slightly greater in the older leaves. On a liquid basis it was about 1×10^{-7} cm s^{-1}, i.e. less than the salt permeance. This could be due, as mentioned above, to the water permeance being measured from saturated to dry while solute permeance is measured in a cuticle saturated with water throughout its thickness. It could also be due to the ivy leaf cuticle having a coherent but flexible lipoid structure and the 2,4-D permeation being that of the free acid at a very high value through a matrix almost impervious to water.

The latter mechanism is indicated by the work of Bukovac *et al.* (1971) on a series of phenoxy acetic acids. At pH 3 they were found to permeate the tomato fruit cuticle increasingly rapidly in the order of increasing chlorination. Chlorination in the benzene ring will increase the partition into lipoid fluids without, in this case, any interfering effect on degree of ionic dissociation. Since it also increases molecular volume, increased partition into the cuticle matrix, which was confirmed in this work by direct measurement, is the only explanation of the observed trend. Estimation of the permeance requires some correction to the data as presented since more 2,4-D and 2,4,5-T appear to penetrate the chloroform-extracted cuticle than was actually offered. Correcting the offered concentration (personal communication) from $1{\cdot}6\times10^{-4}$ M to $6{\cdot}0\times10^{-4}$ M the values for 2,4-D become $0{\cdot}9\times10^{-6}$ cm s^{-1} for the enzymically detached cuticle and 11×10^{-6} after chloroform extraction. The salt is about half hydrolysed at pH 3 so these should be doubled if the free acid only is considered mobile. The values for mobile acid are considerably lower than that ($1{\cdot}5\times10^{-4}$) estimated above for ivy leaf cuticle. The tomato fruit cuticle is about 10 times as thick but in other cases there is little relationship between thickness and permeance. These authors do not record a water permeance, but the writer's own value, on cuticle of another cultivar detached by cold oxalate, was 2×10^{-7} cm s^{-1} on a liquid basis increased by a factor of 10 on chloroform treatment.

Similar investigation by Robertson *et al.* (1971) on diphenylacetic acid permeating tomato fruit cuticle yielded a permeance of 5×10^{-7} cm s^{-1}, about half that of 2,4-D despite the fact that, with two benzene rings, the molecule will be considerably more lipophilic. The amide, less lipophilic (Section 2.X.C), is significantly less permeant (about 2×10^{-7} cm s^{-1}) but the permeance of the dimethylamide has risen again to about $3{\cdot}5\times10^{-7}$. One must assume that molecular size is becoming important but is still less important than partition into lipids from water. In this series of experiments an appreciable time-lag is recorded before the initial rate becomes steady (at much longer times it would fall off again because of reduced gradient). The lag is of the order of 3–6 h. Three hours for a permeance of 5×10^{-7} cm s^{-1} and a septum thickness of 2×10^{-3} cm would, on the simple theory for a uniform septum (Section 3.V.F) indicate a partition of 15:1 in favour of the septum substance.

Kirkwood *et al.* (1968, 1972) studied many aspects of the entry of MCPA and MCPB into the field bean, *Vicia faba*, including tests on cuticle mechanically detached by Sellotape stripping: 2 mm thick agar blocks served as donor and acceptor. ^{14}C assay was made of each component of the sandwich, usually after a standard period of contact of 22 h. In the preliminary 1968 account data for donor, disc and acceptor were all recorded. A substantial fraction of the whole was found in the disc, always so much more, relative to probable thickness (*c.* 1/1000 × block thickness), as to indicate a very high partition in favour of the cuticle,‡ greater for MCPB than MCPA as would be expected. Only in the case of MCPA and high surfactant concentration does the ratio of ^{14}C content, lower : upper block, come near to the volume ratio (3:1) indicating nearly complete equilibration. Despite the higher partition in favour of the cuticle, MCPB had always penetrated less into the lower block. This could be mainly due to the lag period being a much greater fraction of the 22 h contact, but a standard period of water washing of the cuticle also removed relatively less MCPB.

Darlington and Cirulis (1963) had earlier used the agar block method to measure permeation of a series of *N*-alkyl-monochloroacetamides through detached adaxial apricot (*Prunus armeniaca*) leaf cuticle. ^{14}C assay was used. All layers were examined and negligible amount only found in the cuticle. What we call permeance was calculated from formula (3) but only the acceptor volume (determined by weight) was inserted. Since the donor volume was only *c.* $\frac{1}{3}$ of the acceptor volume, the permeance values must be reduced to $\frac{1}{4}$. So corrected, they range from 3×10^{-8} for the $NHCH_3$ compound to 8×10^{-7} for the NHC_6H_{13} compound. The general trend is to increase along with the measured partition coefficients in favour of chloroform, except that the unsubstituted NH_2 compound (1×10^{-7}) was *more* mobile than either $NHCH_3$ or $N(CH_3)_2$. Although this anomaly would not have been removed had a more inert non-polar solvent than chloroform been used, this would have been preferable since chloroform interacts strongly with amino groups.

Darlington and Cirulis included sucrose in their study and calculated a

‡ The ^{14}C was extracted from the cuticle with chloroform and the authors make the quite unnecessary assumption that it was therefore held in the wax.

permeance of 5×10^{-8} cm s^{-1} which we must divide by 4. They point out that a layer of stagnant water of the same thickness as the apricot leaf cuticle (2·6 μm) would have a permeance some 10^5 times as great but that rates 10^{-5} times that through the cuticle have been reported for the plasmalemma of algal cells. Darlington and Barry (1965) applied the same technique to investigate the effect of surfactants and prior treatment of the cuticle on permeance of *N*-isopropyl chloroacetamide. They found no effect of the surfactants but chloroform treatment of the cuticle increased the exchange to near equilibration in 4 h.

The agar block technique has advantages of convenience but the retention of a significant fraction of test substance in the cuticle as found by Kirkwood *et al.* must be looked for—as in any method involving small reservoir volumes. A quite different problem is that the agar gels, since their contents cannot be agitated, tend to reduce the observed overall transfer by slow diffusion within themselves. Darlington and Cirulis (1963) satisfied themselves that two blocks without the cuticle between came substantially to equilibrium within the time chosen for their main experiments. Gabbott and Larman (1968) attempted to allow for diffusion within the blocks by approximating the observed exchange without cuticle to an exponential decay and assuming that the time constant for the exchange through the cuticle would be the sum of that for the cuticle alone and for the blocks without the cuticle. This can only be a rough approximation (the authors claim no more) because the exchange between two blocks is not exponential. Initially it is proportional to $\sqrt{\text{time}}$, then changing through a complex function towards approximate exponential decay when most of the exchange has occurred. The time constant for two blocks is proportional to the square of the thickness, but Gabbott and Larman used 5 mm thick blocks instead of the 2 mm found to be satisfactory by Darlington and Cirulis. They were also investigating a cuticle provided with stomata in which the overall permeance for a triazine type herbicide was found to be *c.* 2×10^{-4} cm s^{-1}. The time for 90% equilibration between agar blocks in contact would be expected to be about 20 h for 5 mm thickness, 3·2 h for 2 mm. Failure of effective contact between agar block and cuticle could lead to a low estimate of permeance. Gabbott and Larman ensured that there was wet contact.

VI. PENETRATION INTO AERIAL PARTS

A. Relevance of Detached Cuticle Studies

Study of the detached cuticle can help in the analysis of the total uptake process and is of particular importance in the present context because it is only at this stage of the total process that physical methods of modification are likely to be effective. The foregoing review will show that permeance values of detached astomatous cuticle cover the range from values so high that the cuticle is probably no serious barrier to those so low that useful uptake must depend on specially permeable sites such as guard cells and hair bases.

The major doubt is whether the behaviour of the detached cuticle is representative of that in the attached state. As Gabbott and Larman (1968) put it, when attempting to separate the stages in penetration of the cotyledons of *Linum usitatissimum*, it is not always justified to consider that the whole is the sum of the parts—at least, not in a simple way. To take an extreme mechanical example, but one not without molecular analogues; if two perforated sheets of metal are glued together with holes in exact anti-correspondence, we could have an almost impervious sandwich although the separated layers would all be freely permeable. Not only does separation disturb an organization but frequently some components are inevitably lost. These tendencies would generally lead to the resistance of the whole being more than the sum of the resistances of those parts which can be isolated. It has, however, been several times suggested that the cuticle *in vivo* may be *more permeable* than when isolated. Plausible mechanisms for such increase have been invented and passed on from one review to another. They depend, with variations in detail; on the idea that an elastic mosaic under tension can have significant porosity when expanded by attachment to a turgid foundation but shrink to a non-porous state when relieved from this constraint. Swelling by water of the substance between wax platelets could increase permeability rather generally, but swelling by water of the foundation on which rest a series of plates normal to the gross surface could have a greater effect and this effect would be removed when the swellable foundation was removed. (See Fig. 6, p. 487.)

An evident defect of this picture as usually presented is that separation of the units in a super-molecular mosaic structure should result in a permeability pattern among different compounds characteristic of porosity. There would be a transition from a partition-dependent selectivity before removal of a constituent of the mosaic (e.g. wax platelets) to a coarser sieve-type behaviour more dependent on molecular size and shape subsequent to this removal. This may be true of the skin of the grape according to the study by Chambers and Possingham (1963) but this skin is certainly an abnormal one with a very high content of high-melting but soluble wax and coherence dominated by pectin. The writer's observations on the fragmentation of the otherwise tough cuticle of oleander leaves when soaked in chloroform provides another example of complete breakdown of selective permeance, but most investigations have revealed that permeation, though increased by treatment with wax solvents, is still very selective.

That the detached cuticle is misleadingly impermeable seems to derive mainly from the statement by Skoss (1955) that neither 2,4-D nor DNBP significantly permeate the detached astomatous adaxial cuticle of the ivy leaf but his evidence for this does not bear critical study and, in respect of 2,4-D was contradicted by that of Schieferstein and Loomis (1959). Also Kamimura and Goodman (1964a) claimed that permeation of detached apple leaf cuticle by several substances was negligible while they showed (1964b) that uptake into detached whole leaves was rapid. In the former work, however, they used apparatus described by Goodman and Addy (1963) in which the cuticle formed a septum between large reservoirs, the volumes of which they do not specify nor allow for, a correction later made by Solel and Edginton (1973).

The importance of turgidity in increasing permeability, associated with, though not necessarily supporting, the belief that the detached cuticle is less permeable, was emphasized by van Overbeek in his 1956 review. Evidence for the positive correlation between entry of applied herbicide into intact plants and the degree of turgidity of the treated leaves was mainly based on a brief summary of work by Weintraub *et al.* (1954), further details of which do not appear to have been published. Many other observations of this positive correlation have been attributed (Section 11.IX.C) to drying of the external deposit in air of low humidity. van Overbeek also quotes Hurst (1948a) as having shown that skins of tomato and grape are most permeable when in contact with liquid water. In the reference cited, however, Hurst was concerned with the permeation of insect cuticle by water (not by other substances as influenced by water) and refers to an earlier (1941) paper to justify a statement that fruit skins also pass water more easily from outside in than in the reverse direction. In the reference he himself cites, however, Hurst gives no details of the fruit skin work but only of experiments showing that water permeation (of insect cuticle) is increased if exogenous oils are present.

The evidence that swelling of the cuticle by water separates wax platelets is thus rather elusive. Crowdy (1959 and 1972) accepts the theory. Crafts (1961) postulates further that, at full humidity, the spaces so created provide continuous channels through which water-soluble substances can diffuse. Sargent (1966) accepts this view. The theory, of course, is plausible but the persistence of partition-dominated selectivity seems to demand that the mosaic structure is usually on a molecular scale.

As Barkas pointed out in 1948, anisotropic swelling can have complex and opposing effects. Their reality even in polymer films has already been exemplified (Section 9.III.A).

We may usefully add the results of some observations on tomato fruit cuticle *in vivo*. Twenty ripe fruits of cv moneymaker selected for uniformity of size (between 74 and 82 g) and without visible skin blemish were held in a desiccator for 24 h and weighed individually at intervals. The mean rate of loss was 0·26 mg min^{-1} and mean area (assumed spherical) was 86 cm^2 giving a mean permeance (liquid basis and assuming the loss to be of water only) of 5×10^{-8} cm s^{-1}. Groups of four were then treated as follows (a) returned to desiccator as controls, (b) immersed in 40–60°C petroleum ether for 5 min and returned to desiccator, (c), (d) and (e) placed individually in beakers of methanol, ethanol, and *n*-propanol, removed at intervals of a few minutes, blotted and reweighed. After about 2 h the water in the beakers was estimated by measurement of density. Behaviour in each category was remarkably regular. The petroleum ether soak increased the loss rate in air by a factor of 4. The specimens in methanol increased in weight: after about 1 h cracks appeared in the skin and fine air bubbles exuded from the pedicel junction: after about 2 h a considerable amount of perfectly colourless transparent jelly had oozed out from splits in the skin. The specimens in ethanol gained in weight slightly at first with a maximum of about 0·15 g in 20 min, then lost steadily. The specimens in propanol lost weight throughout, 1·6 to 1·9 g in 2 h, the skin becoming wrinkled: from one retained in propanol for 5 h, the skin was easily detached. The estimated rates of water loss were 5·5, 9·5 and $8{\cdot}3 \times 10^{-6}$ cm s^{-1}, compared with 8·8, 10·0 and $9{\cdot}3 \times 10^{-6}$ for the detached cuticles (from the same batch of fruit). The water loss into dry air through the detached cuticle was smaller, 2×10^{-7} cm s^{-1} but in this case four times as fast as from the intact fruit. Evaporation from the fruit will concentrate a pectin solution against the inside of the cuticle. With alcohol and pure water on either side of the cuticle, the alcohol-rich volume always decreased and the water-rich volume increased. Only with methanol was there net volume flow into the "intact" fruit—leading to its rupture.

B. Measurement of Uptake

Methods differ in the way the penetrating compound is applied to the leaf surface, in the way its entry is measured and in whether the leaf, while still assumed to be undergoing normal metabolism, is attached or not to the rest of the plant.

The method of application most relevant to normal practice is, of course, by conventional spraying or dusting but, for analysing the processes in the laboratory, this introduces too many variables at once. There is (1) statistical scatter of the applied dose, (2) the necessity for the spray to include some formulating additives and (3) evaporation of both carrier and active compound unless additional steps are taken to prevent this. Application by microburette in measured small volumes to selected leaf areas has been favoured in precise laboratory work and, for reasons of safety and economy, when using radiolabelled compounds. Almost always the locally applied drops have been in the range 5·0–0·5 μl volume and it should be remembered that 0·5 μl is already about eight times the volume of the largest drops (500 μm diameter) frequent in hydraulic-nozzle sprays.

Much laboratory work has employed larger donor volumes in order to eliminate effects of change of donor composition. This method has been particularly favoured by plant physiologists interested primarily in the influence of metabolism and its external controlling factors, particularly light, on uptake into leaf cells. The use of a very large volume of donor solution of constant composition necessitates, of course, that measurement be made on the receptor.

The method was adopted by Jyung and Wittwer (1964) to study the uptake of phosphate and rubidium ions, the leaves being wholly submerged under a perforated sinker plate in a relatively large volume of solution through which air was bubbled during the exposure period to preserve normal O_2 and CO_2 exchange (Fig. 3a). In slightly less extreme form, this convenient method was adapted by Baur *et al.* (1970) to uptake of dissolved herbicides. Detached leaves were arranged with their petiole-ends in water and the apical halves bent over to dip in the donor solution (Fig. 3b).

More extensively used have been short tubes clipped on to the leaf surface (Fig. 3c). Lanoline, petroleum jelly or silicone grease serves to form a seal at the ring of contact. A small volume of donor solution is then introduced into the tube and maintains a definite area of contact. The method was widely used by Sargent and Blackman (1962–1972) in their extensive researches into uptake by leaf discs, referred to later. These workers measured β emission directly from the discs after drying, the compounds examined being labelled by ^{14}C.

To reduce the donor volume to one from which significant loss of active compound can be measured, several workers, e.g. Darlington and Cirulis (1963), Gabbott and Larman (1968) and Kirkwood *et al.* (1972), have used discs cut from agar jelly cylinders. To avoid drying out the flat plugs

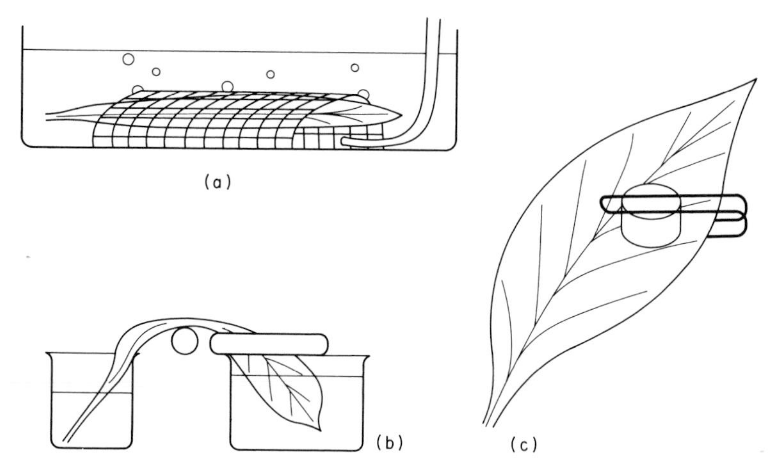

Fig. 3. Three methods of measuring entry into whole leaves: (a) leaf submerged and kept alive by bubbling air: (b) distal portion of leaf dipped in donor solution, petiole in water: (c) short cylinder to hold solution, or disc of jelly clamped to leaf surface.

the experiment must be carried out in a humid atmosphere which, of course restricts the situations in normal practice to which it is relevant.

The receptor units can be whole plants, detached leaves kept turgid by having the cut petioles in water or discs cut from the interveinal areas of leaves and kept turgid by lying on filter paper which dips into distilled water. The second is often preferred to the first purely for convenience of mounting and maintenance under controlled conditions in multiple replicated experiments, but has been used to limit basipetal translocation and facilitate collection and measurement of amount translocated. It may be questioned whether the translocation process in the severed petiole can be considered representative of that in the intact plant, but the work of McCready (1963) on polar transport through petiole sections inserted into agar blocks at both ends indicates that an active transport process can be maintained for at least 24 h. The leaf disc technique has been adopted to eliminate the effect of long-distance transport so that penetration into the substance of the leaf can be examined in isolation.

In view of the variety of native chemicals within leaf tissue, chemical estimation of the amount of some exogenous compound which has entered may be tedious and insufficiently sensitive. Application of chromatographic separation followed by electron capture or other sophisticated means of measurement has enabled the specific estimation of most pesticides in gross samples of plant tissue to be brought down to less than 1 part in 10 million, but for uptake analysis one needs to work with small samples of tissue. Radiotracer methods are therefore dictated but it must be borne in mind that total radioactivity measures the amount of the applied compound (the only radioactive one in the system) which has entered but does not, without further procedures, indicate that the activity comes only from the chemically unchanged compound nor whether some correction for loss (of evaporated $^{14}CO_2$) must be made. Where the distribution of the applied compound is to be investigated, the radiolabel is almost essential. Internal distribution is considered later.

After removal of a small donor volume, some washing procedure is necessary to ensure that a superficial residue is not wrongly recorded. Where the compound has been applied in aqueous solution or wet agar plug, a very rapid wash with water is adequate. Brian (1972) found 2×5 second dips in water adequate to remove excess freshly-applied paraquat and Jyung and Wittwer (1964) used momentary immersion six times in three successive portions of water to remove excess dilute mineral solutions. They admit that this procedure cannot preclude the removal of some "free-space salt". Yamada *et al.* (1964), examining leaves which showed a high degree of adsorption from aqueous solution, considered blotting of the surface with absorbent paper adequate. The problem arises of how to distinguish between washing *off* and washing *out*. Peterson and Edginton (1976) avoid this problem, when measuring uptake into potato discs, by use of as small a volume of donor solution, relative to that of receptor, as is consistent with good contact in an agitated vessel. Analysis of small aliquots of donor solution permits the time course of uptake to be studied. The method is limited to receptor units which are of suitable shape and rigidity to remain intact but not adherent and where overall contact is being investigated. It could be extended to petiole sections, sealed or not at their ends, such as were examined by Moffitt and Blackman (1972) by receptor measurement. Application to whole leaves would run into contact problems and, if applied to leaf discs, lateral entry could not easily be excluded.

Most authors ignore this problem, adopting some purely arbitrary standard technique in the hope that penetration is an essentially irreversible process and that there is a clear physical distinction between deeply penetrated substance which cannot be recalled and superficial substance which has not yet begun to enter. Many drastic methods are recorded in the literature, including 30 min agitation in water, 30 s in chloroform, successive minute-long washing in detergent solution, successive washings in methanol or in 50% ethanol and washing in xylene to remove oil-soluble esters. Most of these will probably remove significant amounts of compound which has already penetrated.

When a deposit has been allowed to dry, particularly if the formulation contains adhesive compounds, compounds leaving a crystalline residue or compounds of low water solubility dissolved in oil, the short water rinsing procedure may not be adequate. No simple answer can be given in these cases. They need adaptable experimental study, not routine procedures.

One should use a method no more drastic or prolonged than is necessary to remove deposit from an impervious model leaf. A convenient test is to apply the formulation plus a tracer (a soluble dye is an obvious choice) to model objects and pass them successively, according to the procedure being tested, through solvent in two beakers. If the procedure is adequate to wash off external deposit, tracer concentration will increase in the first beaker in a linear manner. Very little will appear in the second, the concentration in which will be proportional to the square of the number of operations. If the procedure is not adequate, the concentration in the second vessel will also increase nearly proportionally to the number of operations. If it is very inadequate, the concentrations in the two vessels will increase closely together. If this test is applied using a knife blade or a polyethylene spatula (several models should be tested) and a coloured solution as the source and five successive 1 second dips as the routine, the first behaviour is observed, but, if the source is coloured syrup and the wash vessels contain water, inadequacy of this routine is at once evident. If the model object and source are combined as a sheet of coloured gelatin, the two wash vessels increase in colour equally.

When the extent of penetration from a spray-type deposit is measured by washing off and estimation of the amount not penetrated, it is essential that other mechanisms of loss be examined. In outdoor experiments removal by run-off dew can be important. Biochemical decomposition is unlikely to occur outside the leaf tissue but photochemical decomposition is frequent. Evaporation, as evidenced in the introduction, is the most important source of loss which could, unless allowed for, be recorded as penetration. The mechanism of evaporation from a spot-wise deposit is considered elsewhere (Section 6.V). Allowance has been made by comparison of loss from impervious glass surfaces of similar shape and position to the leaves. Starr and Johnson (1968) compared the rates of evaporation of lindane from leaves and glass by estimation of the amount scrubbed out of a standard air flow. In this case there was no initial difference at a high local applicance but the rate from leaves fell off with time while there was a still significant residue presumably dissolved in cuticle components. Sharma and Vanden Born (1970) applied a correction in this way when measuring uptake of 2,4-D esters and state

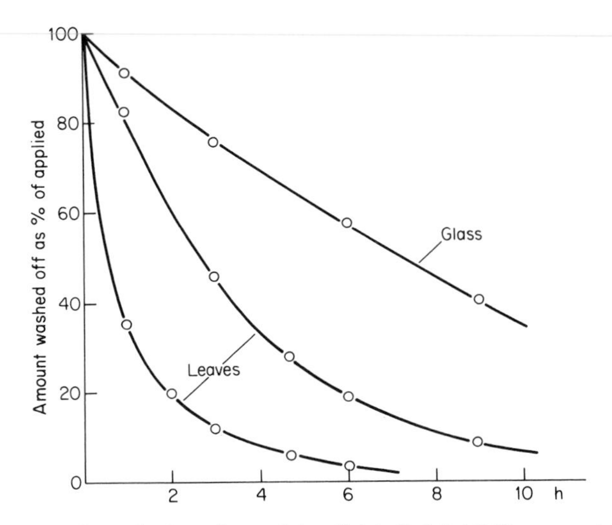

Fig. 4. Percentage of standard appliance (of radiolabelled 2,4,5-T) removed by standard wash from leaves of two ages of *Rubus procerus* and from glass plate similarly exposed. Redrawn from Richardson (1975).

that with salts the correction was negligible. Richardson (1975) gives curves for loss of 2,4,5-T ester from leaflets of *Rubus procerus* in comparison with glass and his data (Fig. 4) illustrate the need for this correction and its limitations. Residual deposits were estimated in wash water and, initially, the rate of loss from leaf deposits was greater than from glass, the difference being attributed to uptake into the tissue. From a deposit of 40 μg distributed over 4 cm^2, 80% was lost in 2 h from autumn leaves, in 6 h from the less permeable summer leaves, but was still not lost from glass in 10 h. The rate of loss however from glass (by evaporation only) was greater than from leaves in the later stages indicating that absorption or solution in the cuticle was reducing evaporation rate while still leaving the deposit available to wash water. Similar observations made by QueHee and Sutherland (1975) were recorded in a complex way (Section 6.V).

C. Reversibiltiy of Uptake

To help in the difficult problem of distinguishing between washing off and washing out, it seems desirable to consider the type of reversibility to be expected in simple model systems.

The simplest model is one in which resistance is confined to a very thin skin of low permeability and low capacity so that penetrant getting beyond the skin is then quickly distributed uniformly with negligible hold-up in the skin itself. If, as in some types of laboratory experiment, the applied concentration is maintained constant, tantamount to a very large value of V_{don} in eq. (1), Section 11.V.E, the concentration inside volume V_{rec} at time t_1 is

$$\mathbf{C} = \mathbf{C}_0\left[1 - \exp\left(-\frac{P\mathbf{A}t_1}{V_{rec}}\right)\right] \tag{5}$$

If washing were then started immediately and continued until time t_2, i.e. the external concentration held at 0 for $t_2 - t_1$, the concentration would fall to

$$\mathbf{C} = \mathbf{C}_0\left[1 - \exp\left(-\frac{P\mathbf{A}t_1}{V_{rec}}\right)\right]\exp\left(-\frac{P\mathbf{A}(t_2 - t_1)}{V_{rec}}\right) \tag{6}$$

The course of the expected curves for $\mathbf{C}/\mathbf{C}_0$ plotted against $P\mathbf{A}t/\mathbf{V}_{rec}$ are shown in Fig. 5. When this uptake ratio approaches its limiting value of 1, it is seen that the wash-out curve tends to an inverted replica of the uptake curve. In the early stage, where deviation of the uptake curve from the linear is only just becoming evident, the wash-out curve is much

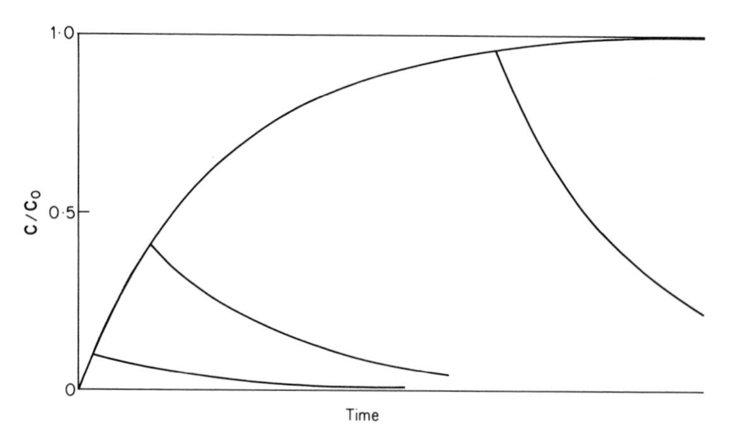

Fig. 5. Theoretical curves of amount penetrated or lost into a finite volume of freely diffusible (or stirred) solvent accessible only through a skin of negligible capacity but high resistance. Convex upward curve is exponential approach to limiting amount acceptable from donor solution of constant concentration. The three concave-upward curves show exponential loss after donor solution is replaced by pure solvent. The combinations illustrate full reversibility; measurements more accurate than is usually practicable and absence of complications such as arise from the capacity of the skin would be necessary to establish reversibility after a short loading time.

flatter. If $t_2 - t_1 = t_1/10$ and uptake is 90% complete, 21% of the amount entered will be lost. If uptake is 50%, loss is 7%. If uptake is 10%, loss is 1% of this. These curve forms would remain the same when the internal medium can reach a much higher concentration than the external due to some reversible partition effect, e.g. if the tissues contain oil globules and the penetrating substance is oil soluble. If the partition coefficient in favour of the whole tissue is constant and the rate of permeation is proportional to concentration difference in the continuous phase, eqs (1) and (2) remain valid but we insert $\mathscr{P}C$ for C and divide the ordinates of the curve by $\mathscr{P}$.

If the uptake curve proceeds until a "ceiling" is nearly reached we should expect the wash-out curve to copy it. If the wash-out curve then falls less steeply than the uptake curve rose initially and especially if it flattens off to a level above 0, some irreversible process must have retained or destroyed some of the compound penetrated. If, however, the uptake rate was constant, within experimental error, over the period of observation, the wash-out curve will not copy it: rather will it indicate no loss of penetrant over a similar period of observation. That it does so is no indication of irreversible binding. The quantity that has penetrated may be only a small fraction of that which could ultimately do so. "Irreversible uptake" is, however, a frequent misinterpretation (or an unnecessary interpretation) of the observations.

An interesting application of this type of relationship appears in the work of Sargent and Blackman on uptake of dissolved radiolabelled substances from relatively large reservoirs into discs cut from leaves of *Phaseolus vulgaris* through intact cuticle and subsequent extraction into solution of the "cold" substance in water. Figure 6 refers to 2,4-D (1962) and Fig. 7 to inorganic Cl^- (1970). The wash-out experiments were repeated after destruction of the integrity of the tissues by freezing and thawing. In neither case does the first wash-out curve differ significantly from a horizontal line, but, as the uptake curve does not differ significantly from a straight line, this is to be expected without invoking any special mechanism. It does not itself provide evidence of binding of the molecules to some component of the fixed tissue.

After the freeze-thaw, 2,4-D is only marginally more quickly released whereas Cl^- comes out catastrophically. The freeze-thaw process must destroy the integrity of some membrane or cuticle which was resistant to the passage of Cl^- ions. Such gross destruction, is unlikely to be selective and so the continued retention of 2,4-D is evidence of molecular binding. The work of Jenner *et al.* (1968) using a different method on *Avena* mesocotyls and of Moffitt and Blackman (1972) on internodes of *Pisum*

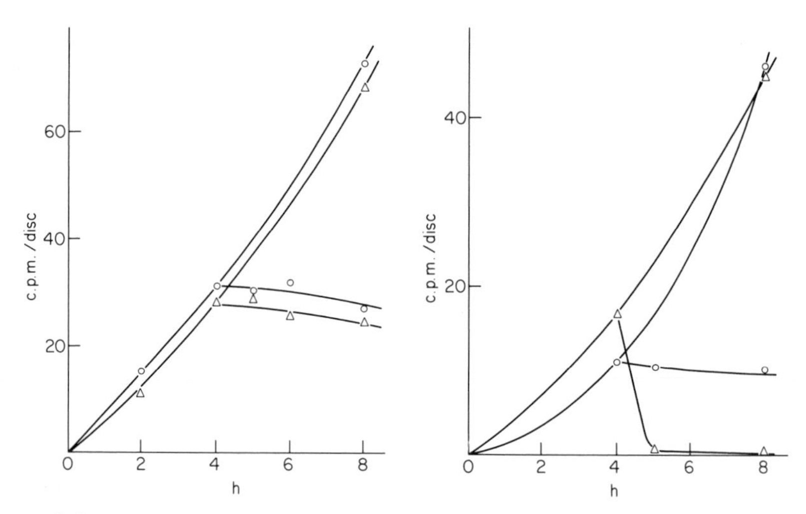

Fig. 6 (left), Fig. 7 (right). Experimental loading and loss curves for leaf discs of *Phaseolus vulgaris*, Fig. 6 for ^{14}C-labelled 2,4-D from Sargent and Blackman (1962), Fig. 7 for $^{36}Cl^-$ from Sargent and Blackman (1970). The circles show mean continued uptake and uptake for 4 h followed by loss into aqueous buffer or solution of "cold" Cl^- or 2,4-D (no significant difference). It is not possible to draw conclusions about reversibility, particularly as the loading curves appear concave upwards, contrary to expected behaviour in the simplest models. The triangles show behaviour when the specimens were frozen and thawed before the loss stage. This makes no significant difference to 2,4-D but causes rapid total loss of entered Cl^-. The authors plausibly conclude that the diffusion barrier is destroyed by freezing but 2,4-D must be retained by molecular binding.

sativum establish that in these tissues there was considerable accumulation, the amount of 2,4-D per g of wet tissue reaching in some cases 40 times the amount per ml of external solution.

Figure 5 represents the behaviour to be expected on simple theory when uptake is from a large volume or constant concentration and washing is into a stream at zero concentration—i.e. also an effectively infinite volume. The receptor capacity limits the process. Uptake may, however, be limited by a small external supply, e.g. small spray drop residues. If the internal volume is relatively rapidly available by internal transport processes and is, say, 99 times as large as the external volume, the course of internal concentration during washing would follow approximately (Section 9.I.C)

$$100\frac{C}{C_0} = \left[1 - \exp\left(-\frac{100P\mathbf{A}t_1}{\mathbf{V}_{\text{rec}}}\right)\right]\exp\left(-\frac{P\mathbf{A}(t_2 - t_1)}{\mathbf{V}_{\text{rec}}}\right) \tag{7}$$

According to this equation, if the uptake time allowed 89% of the applied dose to be taken up (i.e. 90% of what could eventually be taken up), then an equal washing period would remove only 2% or, washing for one-tenth of the exposure time, only 0·2%.

This appears to justify the assumption that the washing process need not be closely considered. The model, however, is a very extreme one. In fact a substance penetrating a leaf needs a long time to reach its final distribution in the plant. Even in the case of a mammal, although extensive distribution in the blood is achieved rapidly, diffusion into the greater volume (approximately four-fold) of the water of the soft tissues is a more lengthy process. The model is also seriously at fault in another respect. The compound is normally applied to a small fraction only of the total surface: washing, whether natural or artificial, covers the whole leaf area. The second factor in eq. (7) should therefore have a much larger value of **A** than the first. If the much greater volume of plant tissue not under the treated areas of cuticle is available through the transport processes of the plant, it is as directly exchangeable with the wash water over the extended area as was the restricted droplet area with the underlying tissues.

In the case of uptake of a stable substance restricted *only* by the resistance of a cuticle of zero capacity, time elapsed between uptake and wash-out is of no importrance. When distribution after passage through the cuticle takes considerable time, the efficiency of wash-out may be considerably reduced by increase of this time. We can illustrate this by another extreme model in which the only resistance is in depth, namely uniform diffusion into a thick sheet. Deliberate washing in uptake experiments is usually carried out in order to define the time allowed for entry. It can therefore follow immediately a short exposure time. When the times are so short (or thickness so great) that no significant amount of penetrant reaches the opposite wall of the diffusion layer (or the centre of it, if approached from both sides), the amount entering from a constant concentration C_0 in time t_1 is

$$\mathbf{M}_1 = 2\sqrt{\frac{D}{\pi}} \,.\, \mathbf{A}C_0\sqrt{t_1} \tag{8}$$

but, if t_1 is immediately followed by washing (i.e. surface concentration = 0) until time t_2 the final amount retained can be shown to be

$$\mathbf{M}_{1,2} = 2\sqrt{\frac{D}{\pi}} \,.\, \mathbf{A}C_0(\sqrt{t_2} - \sqrt{t_2 - t_1}) \tag{9}$$

for $t_2 - t_1 = t_1$ this represents a loss of 59%. For $t_2 - t_1 = t_1/10$ it represents a loss of 26% and even for $t_2 - t_1 = t_1/100$, 5% loss.

When the time of uptake is long enough for the thickness of the layer to be important, a more complex summation-of-series solution is necessary (Crank and Henry, 1949). If a period exists between t_1 and t_2 during which no uptake or release occurs, as could be approximately true were the whole of an applied dose accepted, the solution would be more complex still, except at the opposite extreme, when the "blank" period is long enough for the concentration in the layer to have become uniform. In this case, if the diffusion coefficient is constant, the wash-out process is an exact mirror image of the uptake process. If a dose is taken up in time, t, and then left to become uniform and represents a fraction, f, of the *ultimate capacity*, then the same fraction, f, but of the *amount taken up*, will be released during the same time, t, into pure solvent. The smaller the fraction of the total capacity which the applied dose supplies, the smaller the fraction of it which will be washed out *in the same time* as that taken for uptake.

To illustrate these behaviours we have plotted in Fig. 8 the course of the total loading in a diffusible sheet during uptake from a constant surface concentration and during release into washing solvent of zero concentration for the three situations indicated.

Reversibility of uptake and release in the simplest system can be upset if the diffusion coefficient varies with concentration. We saw (Section 9.II.G) that it is usual for diffusion in a polymeric matrix to increase very steeply with concentration. This increase leads to the absorption process being more rapid than desorption. The difference, however, is less than the variation of diffusion coefficient. Crank (1956, p. 153) computed an example where D is an exponential function of concentration and, at the highest concentration is 200 times that at zero concentration. The amount diffused into an "empty" block was found to be only about 10 times the amount diffused out from a saturated one in the same time. The times for equal amounts would be in the ratio 1:100. Typical experimental data quoted by Fujita (1968, p. 79) show less difference.

If resistance were confined to cuticle and diffusion within it showed the same variation with concentration, the rate of exponential uptake from the maximum concentration would be about 38 times the rate of exponential loss from a very dilute interior concentration into pure solvent. The times for given fractional gain or loss would in this case be in the same ratio. If uptake were maintained till half the ultimate loading had entered, only 2% of this would be lost in washing for an equal time.

There is no recorded evidence of such variation with concentration of a

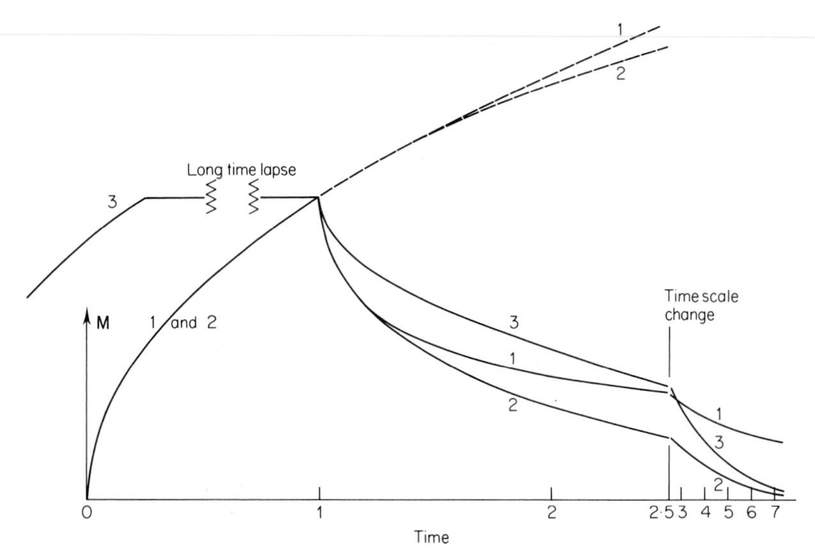

Fig. 8. Theoretical curves for uptake into block with constant diffusion coefficient from external solution of constant concentration followed by loss after donor solution is replaced (time = 1) by solvent. Curves 1 and 2, uptake followed immediately by washing out: for 1, the block is of semi-infinite depth, for 2 it is of such depth that half the final load is taken up at time 1. Difference in the continued uptake (broken lines) is not appreciable till later, and initial loss curves are identical, but then losses under condition 1 fall off and show a very long tail, since penetrant is still diffusing into further depth while the proximal zone is being depleted. Curve 3 as 2, but the block is isolated for a long period before solvent is presented. Penetration to greater depth during isolation makes initial loss less rapid than for 1 and 2 but the limitation of depth causes loss eventually to exceed 1.

solute in water permeating leaf cuticle but formulation additives could produce similar dependence since their assistance to permeation would apply only in the neighbourhood of spray residues and even there it would probably no longer operate after rain had removed the residual additives. That penetration may be more facile than washing-out could therefore be due to the former occurring under a high concentration of a.i. and additive while the latter must take place between very low concentrations of a.i. in the absence of additives. In the case of plants having water-reflective leaves, an extreme form of this mechanism would operate since rain drops make no effective contact in the areas not contaminated by spray.

Increase of permeability by swelling introduces another factor affecting availability and washing-out. A limited amount of swelling agent spread thinly over one side of a cuticle will be wholly absorbed before an appreciable fraction has reached the other side. The less the initial area density the lower the mean concentration reached and the lower the diffusion coefficient governing further advance. The fraction of the applied dose which has penetrated at any finite time is therefore less the lower the initial area density. Concurrent loss by evaporation will be less affected because the air permeability is not altered and back diffusion in the outer layer after free penetrant has disappeared will occur at a higher mean concentration than continued forward diffusion. The same in less extreme form, unless water greatly accelerates diffusion, will apply if washing occurs after a standard time. Localized, spot-wise application could therefore have advantage for penetration and resistance to washing of a systemic pesticide over the same applicance uniformly spread.

We can quote no direct experimental evidence of this effect but it can confidently be predicted for simple penetrants and polymer films from the established facts of their behaviour (Section 9.II.G).

We noted in comment on eq. (7) that the value of **A** in the second (washing) factor would generally be greater than in the first (penetration) factor. The effect just discussed is directly opposed.

The steady-state analogy applied in Section 3.V.C to penetration from the surface into soil predicts availance proportional to applicance, but this is valid only for diffusion coefficient independent of concentration. The effect we are discussing now arises from a probable great increase of diffusion coefficient with concentration which may well be assisted by formulation additives. Entry is facilitated by high local concentration and, if the pesticide is systemic, its loss from a large area by evaporation or washing operates at a very low concentration.

If the compound is not systemic and strongly partitions into the cutin its thin spread could still retard evaporation of the fraction which penetrates into the cutin. Availability to non-chewing insects could therefore be reduced by its being too thinly spread. The complex of opposing factors is in this case difficult to analyse but the present trend to small-drop application and the discovery (synthetic pyrethroids) of insecticides of very high activity make it advisable to give this subject closer experimental study. One ha of crop may receive less than 0·1 kg of pesticide on more than 10 kg of cutin.

This form of resistance to washing out arises from what we have referred to in Section 9.IV.A as "environmental" asymmetry, the inward and outward concentration differences lying in different parts of the concentration scale. We suggested also that environmental asymmetry of penetration of an exogenous solute could arise in a microporous cuticle from asymmetric flux of water, which could be particularly effective in pores diverging inwardly. There would be inward flux of water both under the spray drop and during rainwash which would assist entry and retard exit of the exogenous solute. Both fluxes would be of limited duration, the first because of evaporation of spray water, the latter because of turgidity limit. Persistence of diffusion through the cuticle matrix would make the former limitation less important. Variability in the latter could perhaps explain some of the contradictory results reported in the next section.

D. Washing out by Rain

Washing is not only an experimental tool but also a fact of nature for an agricultural crop. Periods of rain much longer than the time necessary for a spray drop to dry often follow application. $\mathbf{A}_2 t_2$ in our equations may therefore greatly exceed $\mathbf{A}_1 t_1$. Rain does of course reduce the effect of chemical treatment but, in general, its effect appears to be less than might be expected. Weaver *et al.* (1946) examined the effect of "artificial rain" (1 inch in 5 min) on soybean plants sprayed with 2,4-D amine (which amine was not specified) in water or 2,4-D acid in a non-emulsifiable oil formulation. Rain applied after $\frac{1}{4}$ h only failed to reduce the damage by the acid in oil, but reduced the ultimate effect of the water-based spray to a low level. A 48 h interval interval between spray and rain was necessary for the rain to have no significant effect. Elle (1951) carried out similar experiments with water-based amine 2,4-D on the more resistant sweet corn and found rain after 1 h to reduce the effect to an insignificant level while rain after 6 h left an effect not significantly different from no rain.

In these examples the "rain" was unrealistically heavy and of short duration. This could be more effective than real rain in removing some solid deposits, but, for removal of dissolved substances into wash water, which will have effectively zero concentration under any practical conditions, time wet will be more important than rainfall, as has been established for leaching of endogenous chemicals (see below). In field and laboratory studies by Linscott and Hagin (1968) on 2,4-DB, sprayed as aqueous salt solution on to various legumes and grasses, levels of 0·1, 0·5 and 2·0 inches were applied at 1 inch per hour so that they may be regarded as maintaining the leaves wet for periods of 0·1, 0·5 and 2·0 h. This treatment was applied immediately after spraying or 1, 3 and 7 days later. Chemical estimation in the harvested field plants was carried out and, in laboratory trials, ^{14}C was determined in water collected during 5 min washing. A large proportion of the herbicide content was lost by washing even after 7 days. Two hours rain removed at least 75% (mean about 95%) and 0·5 hours at least 20% (mean about 60). There was little dependence on interval between spray and rain, indicating that washing *out* was occurring.

Insensitivity to rainwashing may be due to early damage by herbicides decreasing the efficiency of the plant's transport system so that later uptake and redelivery to the surface may both be slowed down. This mechanism would not be expected to apply to systemic insecticides or fungicides which have no physiological effect on the plant. Washing out of systemic insecticides certainly does occur but systematic investigation does not seem to have been undertaken.

In the case of herbicides, it is generally the effect on the plant, not directly the fraction penetrated, which is measured. Where there is little reduction by rainwash, the biochemical damage may be initiated by early-entered and translocated toxicant. Rain may wash out a large fraction of the dose applied but most of this would not have contributed to further systemic damage.

Bovey and Davis (1967) examined the washing of young plants of several species after spraying with paraquat. Damage to peas was very little reduced when the interval was longer than 10 min but that to *Quercus virginiana* and *Ilex vomitoria* was greatly reduced even when 60 min had elapsed. Maleic hydrazide is generally believed to exert its full effect only if the deposit is not washed for several days.

Soluble minerals, especially potassium as carbonate, can be washed from leaves, as first shown by Arens (1934) who estimated that 60 kg ha^{-1} can be recycled in one season by a sugar beet crop. The process may be a necessary excretion and some species need washing to maintain health while others, adapted to an arid climate, are adversely affected. Carbohydrates can be lost and Dalbro (1956) estimated 800 kg ha^{-1} from an apple orchard but the part played by aphids was not determined. This quantity, unlike that of potassium, is much less than the resident pool. Wilting, even reversible, facilitates washing out of Ca and P (Tukey and Morgan, 1963) and the writer has found permanent wilting greatly to accelerate the leaching of sugars from ryegrass. The subject is reviewed by Tukey (1970). Many questions remain and the relationship to entry of exogenous compounds has not been explored.

VII. UPTAKE INTO ROOTS AND XYLEM

A. Entry into Roots

Much can be learnt about the nature of the uptake process by examining the time course of uptake. Many workers have shown that uptake of pesticide from nutrient solution, by both intact plants and excised roots or discs of tissue, decreases from an initial rapid rate which lasts for only a relatively brief period, to a much slower rate sustained for much longer periods (see, for example Reinhold, 1954; Crowdy and Rudd Jones, 1956; Crowdy *et al.*, 1956; Johnson and Bonner, 1956; Davis *et al.*, 1965; Prasad and Blackman, 1965; Moody *et al.*, 1970; Shone and Wood, 1972; Walker and Featherstone, 1973). Such kinetics could correspond with the model of an initial uptake into "free space" followed by slower accumulation by the less accessible parts of the plant which was discussed in Section 11.III.B. However, the concepts which have been evolved for ion uptake cannot be adopted unquestioningly for uptake of neutral alien molecules: at the very least different factors will influence the accumulation of solute by the surfaces bounding the "free space" and also the contribution of active metabolic accumulation processes must be examined.

Several investigations have provided indirect evidence about the mechanisms involved in the initial phase of uptake by the roots. Crowdy *et al.* (1956) studying the absorption of antibiotics from solution found that with whole plants and excised roots, the characteristic rapid initial phase of uptake was completed within 30 min for griseofulvin when the concentration in the roots approximated to that in the treated solution. Subsequently the further slower accumulation was faster for detached roots than for the roots attached to whole plants as might be expected since some of the griseofulvin taken up by the roots of whole plants would be translocated to the shoots. The overall accumulation by the intact plant increased linearly with time and hence with the amount of water transpired, itself a linear function of time in these experiments. Uptake also increased with increasing concentration of the antibiotic in the treating solution. More information about the initial phases of uptake came from investigations into the effects of metabolic inhibitors. Treatment of intact bean plants with either of the two respiratory inhibitors, sodium azide or 2,4-dinitrophenol, at concentrations which did not interfere with transpiration did not affect the steady accumulation during the second phase of uptake, but, rather surprisingly, markedly decreased the initial rapid intake. This inhibition did not, however, appear to prevent accumulation of griseofulvin in the roots altogether, because the rate of flow of the antibiotic to the shoots was less than in plants not treated with inhibitor: this was attributed to a greater loss from the transpiration stream to the roots of treated plants whose adsorptive capacity had not been already partially satisfied by a rapid initial phase of uptake. The exact mechanism of the inhibition of the initial rapid phase of uptake is not clear. Crowdy *et al.* (1956) point out that while the effect of inhibitors may indicate the operation of an active process, there was no evidence of accumulation of griseofulvin against a concentration gradient in their studies. They suggest that the high concentrations which accumulated in the roots were probably attributable to the preferential partition of the antibiotic into lipid rather than aqueous phases. The effect of the inhibitors may have been indirect, possibly altering the permeability of the cells to griseofulvin. Retention of material in the roots of bean plants appeared to be largely reversible. When bean plants were transferred to water after a period of 6 h in the antibiotic solution, the chemical which had accumulated in the roots was redistributed to the stems and thence to the leaves over a period of about 2 days, presumably by release to the transpiration stream. The degree of reversibility, however, appeared to depend greatly on the plant species: in a similar experiment with tomatoes relatively little redistribution occurred when the plants were transferred from the treating solution. In further studies Crowdy and Rudd Jones (1956) found that the uptake pattern for sulphonamides and the effects of inhibitors were similar in many respects to the results obtained with griseofulvin.

The release of chemical retained in the roots to the transpiration stream when bean plants were transferred to water from the treating solution reported by Crowdy *et al.* (1956) suggests that concentrations in the root reach some sort of equilibrium with the transpiration stream, depending on the concentration in the external solution. The measurements of the time course of uptake by Crowdy *et al.* gave no evidence for such an equilibrium because the concentrations in both detached roots and roots of intact plants increased throughout the 28 h period of the uptake experiment for solution concentrations from 3 to 25 $\mu g\ ml^{-1}$. Similarly, Shone and Wood (1972), who studied uptake of the herbicide simazine from solution by young barley plants, found that although the rate of accumulation by the roots rapidly decreased over the first 5 to 10 h of their experiments, the concentration therein was still increasing after 24 h. Moody *et al.* (1970) also found that uptake by excised roots of soybean plants (*Glycine max* (L.)) of the herbicides amiben, atrazine, chloropropham, EPTC and linuron mostly continued slowly throughout the 48 h of their experiments after an initial rapid uptake over the first hour, although in some cases, particularly at the highest temperature studied (30°C) amounts present in the roots passed through a maximum and then declined. However, in longer term experiments, Walker and Featherstone (1973) found that the concentrations of both linuron and atrazine in the roots of parsnip, carrot, lettuce and turnip plants growing in treated nutrient solution remained more or less constant between the first sampling time (3 days after the seedlings were transferred to the nutrient solution) and the final harvest 12 days after transfer. More detailed experiments with carrot and lettuce seedlings over 24 h showed that the concentration in the root reached an apparent equilibrium with the external solution within about 4 to 16 h. There were marked differences in the values for the different herbicides in the two plant species.

The general picture of the uptake by roots to emerge from these and similar studies, therefore, is a rapid initial accumulation, greater than can be accounted for by water uptake alone and presumably reflecting diffusion and partition on to the surfaces of the free space, followed by a slower further uptake reflecting movement to the less accessible tissues of the root and possibly passage across membranes. For the majority of neutral alien molecules the uptake process probably does not directly involve active metabolic accumulation (as discussed more fully below) and some sort of equilibrium, which may be largely reversible, is reached with the transpiration stream and the external solution, given sufficient time. The relative amounts of different compounds retained by roots depend greatly on the plant species and since it has been suggested (see below) that differential retention may contribute significantly to herbicide selectivity by controlling amounts translocated to the foliage, the properties determining relative retention of different compounds by different species are of considerable interest. Apparently there have been few attempts to investigate this question specifically. One exception is the study by Tames and Hance (1969) who followed up the suggestion by several workers that adsorption in tissues is involved in the uptake of organic molecules by plants. They determined the physical uptake, from aqueous solutions with a range of different initial concentrations, of various triazine and substituted urea herbicides by freshly killed roots of six different plant species. The relationship between amount of herbicide take up and the equilibrium external concentration was linear in all cases. For all the chemicals tested, uptake was greatest for bean roots. With the urea herbicides oats showed the smallest uptake capacity with pea, cucumber and radish intermediate, but with the triazines oats took up more than pea and cucumber. Uptake was apparently not related to the relative susceptibility of these plant species to the different herbicides. On average, amounts of the different compounds taken up were in the order atrazine = monolinuron < diuron < linuron < GS 14260 (an experimental triazine). Tames and Hance point out that this is broadly similar to the order found for adsorption by soil. As uptake of such unionized compounds by soil is determined mainly by the soil organic matter which is derived largely from plant remains, this result is not surprising, although uptake per unit weight for soil is greater than for the freshly killed roots. The order is also similar to that which might be expected for partition between organic solvents and water; the question therefore arises as to how far the uptake by the killed roots could be accounted for by adsorption on to surfaces in the roots and how far to the partition into lipid materials. Whatever the mechanism, these results throw some light on the factors influencing relative uptake of different compounds during the initial stages of uptake. Desorption experiments showed that the uptake by the killed roots was fully reversible so that presumably any material accumulating in the roots by this process would be available subsequently for translocation to the shoots when the concentration in the transpiration stream decreased, in agreement with the results obtained by Crowdy and co-workers (1956) with antibiotics.

While the bulk of the evidence for non-ionized pesticides suggests that metabolic processes are not involved in uptake by roots, it cannot be assumed that this applies for all compounds. In particular the effects of temperature on uptake suggest that some propionic, phenoxyacetic and benzoic acid herbicides which would be present partly in the anionic form are accumulated actively (Blackman and Sargent, 1959; Saunders *et al.*, 1966; Prasad and Blackman, 1965, but see remarks on pH, p. 615).

B. Translocation from Roots to Shoots

We turn now to considering translocation from the roots to the shoots. In the preceding discussion on uptake by roots, it was pointed out that in the first initial phase of uptake, the quantity of chemical absorbed was greater than could be accounted for by water uptake alone. Many studies have shown that following this the rate of uptake of pesticides is broadly correlated with the rate of transpiration (Minshall, 1954; Crowdy *et al.*, 1956; Crowdy and Rudd Jones, 1956; Sheets, 1961; Sutton and Bingham, 1969; van Oorschot, 1970; Walker, 1971, 1972; Shone and Wood, 1972; Walker and Featherstone, 1973). However, the solution which passes up the xylem is clearly very different from that presented at the root surface from which the plant draws the water transpired. This was demonstrated directly by Shone and Wood (1971) who measured the concentration of simazine in the xylem exudate of barley plants supplied with the radiolabelled herbicide in culture solution. To collect the xylem exudate, the roots and basal portions of the shoots were detached 0·5 cm above the seeds and the cut ends were inserted into glass capillaries. Comparison with the concentration in the external solution indicated that after an initial fairly rapid increase during the first 4 h of uptake, the concentration in the xylem increased only slowly over the remainder of the 24 h observation period, at which stage it was only about 0·8 of that in the external solution. In other experiments uptake by intact plants was measured and transport of simazine to the shoots related to the transpiration in terms of a "Transpiration stream concentration factor" (TSCF) defined as

$$\mathrm{TSCF} = \frac{\text{amount of simazine in shoots per ml water transpired}}{\text{amount of simazine per ml of culture solution}}$$

Under a variety of different experimental conditions, the TSCF was always less than unity, as studies such as those by Crowdy *et al.* (1956) have also indicated, for different compounds (see also p. 607). Lowering the temperature of the uptake solution, or adding the metabolic inhibitors 2-4-dinitrophenol or sodium azide considerably decreased the amounts of simazine taken up, but this was associated with an approximately equal decrease in the rate of transpiration so that the TSCF remained approximately constant at an average value of approximately 0·8. Shone and Wood (1972) concluded that the uptake and transport of simazine is largely a passive process closely associated with the movement of water. These authors also point out that the value of the TSCF probably depends at least partly on the extent to which the herbicide is retained in the root tissue. Other workers (e.g. Geissbühler *et al.*, 1963a; Walker and Featherstone, 1973) have suggested that by controlling the concentration of herbicide reaching the xylem in this way "root adsorption" may be an important factor influencing herbicide selectivity between different plants. The exact way in which such a mechanism might operate is not entirely clear. If "root adsorption" merely implies a reversible partition on to surfaces and into lipid materials with a finite capacity which the transloca-tion stream encounters on its way to the shoot (for which there is some experimental evidence as discussed above), then a steady state should be reached eventually when the non-aqueous phases have come to equilib-rium with the external solution. After this no further material should be removed by the root from the external solution which would therefore pass to the xylem unchanged, with a TSCF of unity. The experimental observation which would correspond to this mechanism would be that the amount retained by the root would increase to a limiting value while the xylem concentration simultaneously approached that of the external solution. The experiments of Shone and Wood were unfortunately not long enough to test this conclusively: both the amount retained by the root and the xylem concentration appeared to be still increasing slowly after

24 h when their observations were terminated. Under the practical conditions of a plant growing in soil, of course, periods of sustained uptake may never be long enough for such a steady state to be reached. The concentration of chemical in the soil solution is likely to fluctuate considerably, roots will grow to new regions of soil and their capacity to take up water and chemicals can be expected to change as they age. In such conditions large differences in the "adsorption capacity" of roots from different species could be expected to affect significantly amounts reaching the shoots and hence selectivity.

In considering sustained uptake over a prolonged period, the question arises as to whether the rate of uptake would not eventually decrease as a result of the shoot tissue becoming saturated. The probability of this happening must be assessed in relation to the nature of the translocation process in the shoot, which can be regarded as essentially flow of solution along a tube (the xylem) which terminates in a "membrane" permitting onward flow of the solvent but impermeable (or much less permeable) to the solute. In this situation the only way in which the solute could find its way back to influence accumulation in the shoot would be by back diffusion from the zone of accumulation against the transpiration stream. Calculations based on reasonable estimates for the rate of xylem flow and the diffusion coefficients for pesticides in solution, however, indicate that the effects of such back diffusion on the concentration at the base of the xylem would be infinitesimally small and can be entirely dicounted: it follows that accumulation at the ends of the xylem must go on indefinitely (resulting in deposition of solid if the solubility is eventually exceeded) unless the solute is lost by evaporation, by decomposition, by excretion or by return via the phloem system.

It is almost certainly a gross oversimplification to suggest that the TSCF is only influenced by reversible partition into non-aqueous phases in the root. We have already referred to the evidence indicating that the xylem fluids and the free space are separated by metabolically controlled membranes. Shephard (1973) studying the uptake of pyrimidine fungicides by different plant species concluded that there are at least two different and highly selective barriers impeding movement through a few centimetres of root to the base of the stem. The endodermis appears to be the likely location for one set of barriers (Crowdy, 1973; Bukovac, 1975).

The nature of these barriers and the mechanisms of passage through them are not well understood and while some inferences may be drawn from the results of physiological experiments, many uncertainties remain. Even the frequently observed proportionality between amounts of chemical taken up and volume of water transpired poses some difficult questions when coupled with a xylem concentration proportional to, but lower

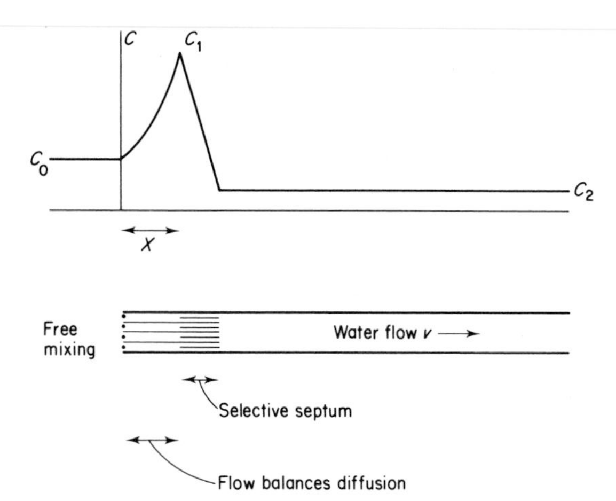

Fig. 9. Top, corresponding concentration–distance plot. Bottom, the model described in text.

than the external solution, as for example in the experiments of Shone and Wood (1972) or Walker and Featherstone (1973). This may be illustrated by the very crude simplified model of the root and xylem system shown in Fig. 9.

This model consists of a tube containing a selective membrane a short distance from the end at which pesticide solution is taken up from a reservoir which, at least in the case of nutrient solution culture, can be assumed stirred and large so that its concentration remains constant. Water can pass through the selective membrane more freely than solute, and if the solute has no direct physiological action on the root it is reasonable to assume that the rate of its passage across the membrane is porportional to the difference in concentration on the two sides, $C_1 - C_2$.

This model differs from that used by Nye (1967) (see Section 3.VIII.A) in that uptake per unit area we assume proportional to $C_1 - C_2$ while Nye assumes it proportional to what would be C_1 in Fig. 9. This is of no consequence for variation with external concentration in any system since both treatments make assumptions which lead to all concentrations being proportional to the external one. Our "permeance", P (see Section 9.I.A) and Nye's k (later α: K in our transcription) have both the dimensions of velocity, but must follow different functions of the other variables, v, D and $\mathcal{P}$. In the special case of no accumulation $K = v$ and $P = \infty$. A more important difference is that Nye (1967) considers the crescent state over a short period after a root has expanded into new soil while we consider an assumed steady state over a short distance. The approaches reveal different facets of a complex picture.

We assume that the "xylem" part of the tube, connected to the aerial parts is so long that back diffusion can be ignored (Section 3.VIII.A). The concentration for the greater length of this "xylem" from the septum thus remains at C_2 as a result of the flow of solution. The rate of transport of solute, **F** is given by

$$\mathbf{F} = \mathbf{A} C_2 v \tag{10}$$

where **A** is the cross-sectional area of the tube and v is the linear velocity of flow, here assumed constant.

On the other side of the septum the effects of diffusion cannot be ignored because the distance X is assumed to be relatively short. Hence, between $x=0$ and $x=X$, diffusion transfer must be added and the total rate is

$$\mathbf{F} = \mathbf{A} C v - \mathbf{A} D \frac{\mathrm{d}C}{\mathrm{d}x} \tag{11}$$

the solution to which, for $C_1 = C_X$, is given in eq. (63), p. 157.

Across the membrane itself

$$\mathbf{F} = \mathbf{A} P (C_1 - C_2) \tag{12}$$

Where P is the permeance of the septum. The rates in eqs (11) and (12) must be the same and solution to these equations, with $C =$ constant C_0 in the reservoir at $x=0$, is

$$\mathbf{F} = \mathbf{A} C_0 v \Big/ \left(1 + \frac{v}{P} \mathrm{e}^{-vX/D}\right) \tag{13}$$

For $X/D \ll 1$, eq. (13) approaches expected limits according to the relative values of v and P:

$$\text{for} \quad v \ll P, \mathbf{F} \to \mathbf{A} C_0 v$$

$$\text{for} \quad P \ll v, \mathbf{F} \to \mathbf{A} C_0 P$$

Where X/D is not $\ll 1$, that is when accumulation of solute on the outside of the selective membrane becomes important, the prediction of eq. (13) is less obvious. If v is the only variable, the denominator of (13) goes through a maximum value of $(1 + D/XP\mathrm{e})$ at $v = D/X$, between limits of 1 for both $v = 0$ and $v = \infty$.

$C_0/(1 + (v/P)\mathrm{e}^{-vX/D})$ is of course the concentration, C_2, obtaining in the xylem stream and eq. (13) thus predicts that the "efficiency" of uptake goes through a minimum with increase of v.

Taking D as $= 3 \times 10^{-6}\ \mathrm{cm^2\ s^{-1}}$ and bringing X down to a "cellular" distance, say 3×10^{-4} cm, we find the minimum efficiency at $v = 10^{-2}$ cm s^{-1} which is in the realistic range for the transpiration stream in plants. The minimum efficiency for $P = 10^{-3}$ cm s^{-1} is 0·21. For $v = 2 \times 10^{-3}$ cm s^{-1} it has the value 0·38 and for $v = 2 \times 10^{-2}$ cm s^{-1} the value 0·27. In this region, therefore, over a ten-fold range of transpiration rate a moderately low efficiency would show less than a two-fold variation.

Although Crowdy *et al.* (1956) and others have found that transpiration rates are more or less constant over short periods, in longer experiments where plants are growing in soil or nutrient culture transpiration rates must vary over a considerable range; the proportionality between transpiration and uptake therefore seems surprising at first sight.

The analysis above illustrates some of the factors involved in selective uptake, but is unlikely to be quantitatively accurate. Linear flow rates in xylem vessels are in the right range, but trans-membrane uptake will in general be spread over a much larger area than the root cross-section. In the region where the establishment of a diffusion gradient could be important, therefore, v will be greatly reduced and **A** correspondingly increased: vX/D is thus not likely to reach the values near unity assumed above unless X is much greater than seems reasonable for a root in water culture. Around a root in soil the "tube" on the input side of the barrier could include channels between the soil particles so that X would be much larger, but steady-state theory would not apply. Where comparisons have been made (e.g. Graham-Bryce and Etheridge, 1970; Walker, 1972) there is not much discrepancy between uptake of pesticides from soil and uptake from nutrient culture containing the pesticide at concentrations similar to those in the soil solution as calculated from adsorption behaviour; this implies that large values of X (i.e. long diffusion paths) are not generated.

Water uptake by roots is mainly confined to an area a few centimetres long close behind the growing tip. Nutrient ions are taken up by active processes, often against a concentration gradient, also mainly over a restricted zone. However, small alien unionized molecules such as those of systemic insecticides and fungicides, and herbicides which do not affect the root, appear to be taken up over the whole root length. The simplest improvement on the model of a single tube interrupted by a selective septum discussed above would therefore be a bundle of tubes originating in the water absorbing zone and forming a bulk flow system along the main length of developed root. This could be represented even more crudely as a single tube containing flowing water inside a permeable wall through which solute is diffusing. It can readily be seen that, if the flow rate is extremely slow, the full external concentration C_0 would be passed on to

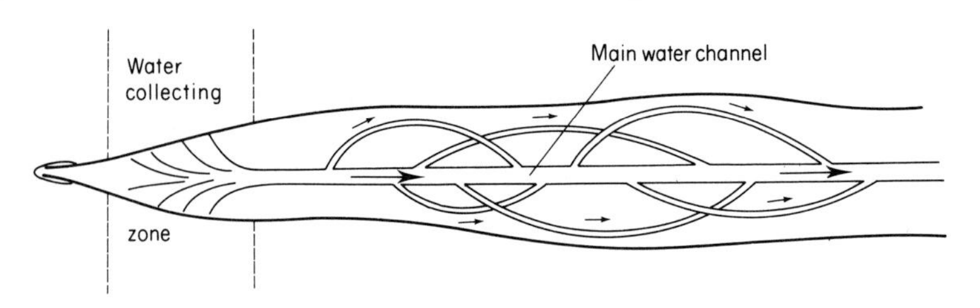

Fig. 10

the aerial parts of the plant. If the permeability were very small, the amount of solute passed on would be effectively independent of flow rate. At intermediate flow rates the concentration passed on to the leaf system would increase with decreasing flow rate, whether the limiting diffusion process was in the tube walls, giving concentration proportional to $(1-\exp(-K'/\Phi))$ (if small, proportional to $1/\Phi$) or in the tube contents when the concentration, if small, would be proportional to $(1/\Phi)^{\frac{3}{2}}$. To account for a rate of transmission which is proportional to flow rate and simultaneously of low efficiency (i.e. xylem stream concentration proportional to, but much less than external concentration over a range of flow rates) it is necessary to envisage a mechanism such as the anatomical division of streams shown in Fig. 10.

In this model, the minor streams which carry only a small fraction of the whole flow are fully exposed to entry by diffusion and thus reach the full external concentration C_0 whereas the main xylem stream in the centre of the root is protected. With this mechanism the main xylem concentration C_2 will remain proportional to C_0 but the ratio C_0/C_2 will be a constant less than unity whatever the overall transpiration rate, the exact value depending on the relative flow in the two types of channel. This model is, of course, hypothetical and as far as the authors are aware there is no anatomical evidence to support it, but some such mechanism must operate to account for the observed patterns of uptake.

C. Mechanisms of Passage through Selective Barriers in Roots

The actual processes by which organic solutes pass through the selective barriers have received some attention but are still far from clear. For nutrient ions, various pathways have been proposed for movement from the external solution to the xylem which involve essentially passage within the symplast or along the extracellular free space associated with the cell walls which would require entry into the symplast at some point such as the endodermis to account for active accumulation against a concentration gradient. Jackson and Weatherley (1962) considered that solute could also leak to the xylem by a pathway which was completely extracellular.

Useful information about the mechanisms by which solutes pass through the selective barriers in roots has come from studying the effects of pressure on permeation. Jackson and Weatherley (1962) examined the effects of applying external pressure to the culture solution on the composition of the xylem fluid forced out of the stumps of seedlings after cutting off the growing shoots. They found that K^+ flowed into the xylem at an increased rate, even when there was no K^+ in the external solution, for a short period after increasing the pressure. The polyhydric alcohol mannitol entered at a much reduced concentration compared with the external solution and at an absolute rate nearly independent of pressure and water flow, as did Ca^{2+} through tomato roots and Na^+ through castor bean roots after treatment with cyanide. This indicates passive diffusion, but in the case of Na^+ and Ca^{2+} through live roots, there was evidence for metabolic mediation.

House and Findlay (1966) found that fluid exudation from cut roots of corn (*Zea mais*) was temporarily decreased when the external solution was increased in osmotic concentration, but that there was then a recovery to a new, lower, steady value. They considered that the process controlling the osmotic flow was a metabolic loading of K^+ salts into the inner root tissue where Jackson and Weatherley's experiments showed them to be accumulated.

Perry and Greenway (1973) and Perry (1973) applied the method of external pressure on decapitated seedlings to uptake of polyalcohols, sugars and herbicides. At an external pressure of 2 bars on tomato seedlings, glycol came through in the collected "xylem" at 52% of the external concentration, mannitol, raffinose, diuron and linuron at 3, 1, 51 and 23% respectively. When the pressure was reduced for a period, the water flow showed immediate reduction to a lower level. During the reduced flow period the concentration of mannitol emerging increased (absolute rate remaining nearly constant) but the concentrations of glycol and diuron were not affected.

This anomaly is discussed below, but attention should first be drawn to another. Since living roots exert a pumping action, usually attributed to a metabolically controlled osmotic action, one would expect the flow rate to increase less than proportionately to the applied external pressure (total driving pressure $\mathbf{P}+\Pi$ when $\mathbf{P}$ is applied hydrostatic pressure and Π is the osmotic pressure of the internal liquid). Perry and Greenway's result shows the reverse effect, flow rates in mg (g fresh wt)$^{-1}$ min^{-1} being 33, 15 at $\mathbf{P}=4$ bars and 2·07, 1·77 at $\mathbf{P}=0{\cdot}91$ bars. This would still be anomalous even if the osmoticum was swept out of contact with the membranes by the improved flow (which calculation on the lines described in Section 9.IV.A indicates to be unlikely). The permeability may be reduced by compression, as is known to occur in artificial membranes but, until this effect is cleared up, one hesitates to accept that application of external pressure in a decapitated seedling is equivalent to increased transpiration in a whole seedling.

Constancy of concentration under different pressures would be expected if the inside and outside concentrations were the same, i.e. no rejection. If there is considerable rejection of a solute one would expect that the absolute rate of its influx would be the result of a passive diffusion under its own (high) concentration gradient, leading to constant influx rate, i.e. product of concentration × flow rate = constant, as was found for mannitol. Reduction of permeability by increasing pressure would exaggerate this effect.

This is a case when one should examine the behaviour of inanimate membranes under varying pressure. Fortunately, the commercial interest in reverse osmosis for purification of water has stimulated much research into this subject. Michaels *et al.* (1965) give a good description of the behaviour and probable mechanism of action of the cellulose acetate membranes developed for this purpose. If the outflowing water has a concentration of some solute which is only a fraction f of that on the supply side, $100(1-f)$ is the "% rejection". Cellulose acetate membranes have a low % rejection for NaCl at low pressures but this increases rapidly with increasing pressure and water flow, as would be expected if the NaCl passed through by a pressure-insensitive diffusion under its own concentration built up by the selective permeation of water. The driving pressure for water is $\mathbf{P}-\Pi$ and when this falls to zero, NaCl must eventually equilibrate.

Since hydrostatic pressure increases vapour pressure according to the equation

$$\frac{\mathrm{d}\ln\mathbf{p}}{\mathrm{d}\mathbf{P}}=\frac{\bar{V}}{\mathbf{RT}} \tag{14}$$

where $\bar{V}$ is the partial molar volume of the substance in the condensed phase over which the vapour is in equilibrium, it might be expected that applied pressure would accelerate the diffusion across the membrane not only of the water but of solutes dissolved in it. This mechanism alone would lead one to expect that most solutes, having $\bar{V}$ greater than that of water, would be subject to a greater diffusion pressure than the water itself. This is normally offset by the greater volume of the solute molecules decreasing, very greatly in a dense polymer matrix (Section 9.II.F), the diffusion coefficient. Where rejection occurs because of this lower diffusion coefficient a gradient of concentration is established which is more important than the hydrostatic pressure effect. A pressure difference of 20 atm acting on water of $\bar{V} = 18$ ml increases the vapour pressure by 1·6%. It would increase that of a solute of $\bar{V} = 200$ ml by about 17%. If the resistances were equal, the concentration of the larger molecules would actually increase on passage through the membrane.

Lonsdale *et al.* (1971) consider the pressure and concentration effects fully and they give one example—that of 2,4-dichlorophenol—where the direct pressure effect predominates and the solute is "negatively rejected", emerging at about 30% higher concentration in the eluent than the supply. The diffusion coefficient of this large molecule in the cellulose acetate must be much less than that of water but the solute is a solvent for cellulose acetate and is partitioned in a much higher concentration in the membrane. A membrane passing a major fraction of solute by this mechanism would yield increasingly enriched effluent the higher the applied pressure.

The % rejection of NaCl by a cellulose acetate membrane tends to a level value (in excess of 98%). The absolute rate of NaCl permeation therefore increases to a value much above its static diffusive value and proportional to the water flux. The water flux is mostly diffusive under a gradient of thermodynamic potential produced by the hydrostatic pressure. The simplest explanation of high but limited % rejection, and the one adopted by Michaels *et al.*, is that a small fraction of the area is porous, due to imperfections in the very thin filtering surface of the compound film. The pores permit independent diffusion of NaCl at low flow rates but contribute a faster leakage at high flow rates which is proportional to flow rate. This leakage is still, however, only a minor component of the total water flow.

In considering Perry and Greenway's results it should finally be mentioned that the finding of a ratio near 50% for substances so extremely different as glycol and diuron indicates two different mechanisms. Glycol is probably slipping through molecular holes incapable of passing mannitol while diuron is diffusing through the lipophilic matrix at a much higher concentration than is supplied in the external solution.

Hussain *et al.* (1974) have attempted to characterize the properties required for entry into the xylem and systemic movement in the cotton plant. Measured quantities of different phosphoramidothioate esters were applied to the petioles of carefully standardized cotton leaves maintained in controlled environment cabinets with the excised ends of the petioles dipping into vials of water. The amounts of chemical reaching the leaves were determined at intervals up to 12 h after treatment and the relationship examined with octanol/water partition coefficients ($\mathscr{P}$), solubilities in distilled water and steric substituent effects. Standard compounds such as dimethoate and paraoxon were included for reference purposes. In the limited case of *O,S*-dimethyl *N-n*-alkyl derivatives, there was a parabolic relationship between the logarithm of the rate of translocation ($\log K_r$) and $\log \mathscr{P}$. The correlation was significantly poorer when other *O,S*-dialkyl *N-n*-alkyl compounds were considered, in spite of their close structural similarity with the *O,S*-dimethyl compounds, while mobility of compounds was drastically reduced when branched *N*-alkyl substituents replaced normal ones. Even in this closely related series of compounds therefore, partition properties alone were quite insufficient to account for the observed differences in mobility. Specific dependence on molecular shape strongly suggests diffusion control in a macromolecular septum (cutin). This work receives more detailed comment in our discussion of the "Hansch approach" in Section 10.VI.

D. Transport in the Xylem

In the preceding sections we have examined the various barriers and restrictions which a solute must negotiate on its way from an external solution into the xylem. Although, by comparison, passage along the transpiration stream might seem relatively unimpeded, there are several ways in which solutes can be retarded during flow and it certainly cannot be assumed that any chemical reaching the xylem will be swept freely up to zones of accumulation where water is lost from the plant.

The walls of the xylem, like the surfaces bounding the free space, contain cellulose and smaller amounts of other materials such as hemicelluloses, lignin and tannins; they also carry pH-dependent negative charges associated with carboxyl and phenolic hydroxyl groups. Cationic pesticides or ones that can become protonated at physiological pH values are therefore likely to be adsorbed and retained by ionic forces. The retention of cationic molecules may be demonstrated by passing solutions of ionized solutes through excised sections of xylem (Crowdy, 1973). Retention does not appear to be confined to such charged species, however. Strang and Rogers (1971) showed by autoradiographic techniques that ^{14}C labelled diuron, supplied to the roots of cotton plants, accumulated in xylem vessels and on their walls but did not accumulate in the phloem. Shephard (1973) made direct measurements of the losses of experimental compounds of four different unspecified types during translocation through the stem. Solutions of the radiolabelled compounds were drawn through lengths of vine, apple or elm wood with a peristaltic pump and the effluent was sampled at intervals. The concentration in the effluent never reached that in the solution supplied, the reduction depending on the length of stem, the plant species and the nature of the chemical. When some pyrimidine compounds related to the systemic fungicides ethirimol and dimethirimol were passed through apple stems, the levels of radioactivity emerging at the outlet end were barely detectable. It is clear therefore that substantial changes in composition can occur as solutions move along the xylem.

E. Summary of the Properties Required for Translocation in the Apoplast

Having examined the different stages involved in apoplastic transport in some detail, we may now attempt to specify the properties which an alien molecule must have if it is to be translocated effectively. Following application to the underground parts of the plant, the chemical must be capable of entering the free space, passing through the barriers in the endodermis and then moving with the water flowing up the xylem in response to the demands of transpiration. As the system is essentially aqueous, the first requirement is that the chemical should have a reasonable water solubility if it is to move in significant amounts through the apoplast. It should be recalled however that, because pesticides are extremely potent, what is "significant" should not be judged by conventional chemical standards. A concentration of a few μg/ml may well be sufficient. The pesticide must be sufficiently hydrophilic not to partition strongly into the lipoidal materials which it encounters in its path. However, unless it can be transported by active carrier mechanisms such as those which operate for nutrient ions, it must have sufficient lipophilic character to pass through the barriers in the endodermis. The properties required are thus to some extent contradictory. In practice it appears that at least some mobility in the apoplast is found over a wide range of different polarities. Crisp (1971) reviewing the literature on translocation of insecticides, found that the great majority of those studied were reported to move in the apoplast and concluded that almost any compound could move from root to shoot by this pathway. Nevertheless some compromise between hydrophilic and lipophilic character is essential for optimum mobility, with a tendency for the hydrophilic property to be more important. For example, dimethoate which partitions very much in favour of water from non-polar solvents and has an octanol/water partition coefficient of 6·3 (Briggs, G. G., personal communication) is a very successful xylem translocated systemic insecticide. Lindane which is much more lipophilic with an octanol/water partition coefficient over 1000 shows very slight, but measurable systemic behaviour. There is, however, one group of compounds which manages to get the best of both worlds. Insecticides such as demeton, disulfoton and phorate are relatively lipophilic organophosphates and this property probably favours their initial entry into the plant. However they contain thio-ether and P=S groups which can be oxidized within the plant to give more polar substances which are, if anything, more effective toxicants and are much more readily translocated in the xylem. This conversion is considered more fully in Section 4.IV.C. Metabolism is also likely to affect the proportion of the parent compound translocated in most other cases, of course, but not usually in such an advantageous way.

In addition to partition into lipid materials, other restraining processes must be considered of which the best characterized is probably ion exchange. For high mobility therefore a pesticide should not be cationic. Overall, the apoplastic network may be regarded as a form of chromatographic system in which the flowing water carries dissolved materials with it, that can be retarded by interactions with the surfaces in the tortuous pathways. The endodermis provides an additional rather specialized barrier.

VIII. TRANSPORT IN THE SYMPLAST

A. Introduction

There has been much research on movement of dissolved substances in the vital tissues of the plant, particularly on transport by the specialized phloem system of leaf-generated carbohydrates to actively growing non-photosynthetic organs. With regard to exogenous substances such as pesticides there is frequent conflict of evidence and interpretation. These are not always easy for the reader to distinguish. Description tends to be coloured by the strongly polarized views of many contributors to this difficult subject and it is probable that in some experiments unsuspected determining factors are not measured or controlled.

Two texts by knowledgeable students of this subject have

described and commented on its present state of development. Peel (1974) considers transport in both xylem and phloem of nutrients of all kinds with some reference also to pesticides. Crafts and Crisp (1971) consider transport in the phloem only but give pesticides and herbicides a large share of attention. The review by Canny (1971) includes a warning against too-ready acceptance of any one view and recent collections of specialized articles (Zimmermann and Milburn, 1975; Wardlaw and Passioura, 1976) provide up-to-date reviews of many aspects of transport of normal constituents. Crafts (1964) describes the transport of herbicides and Hay (1976) contributes on this subject in a second edition of the same multi-author text.

The views of the two last-named authors on the much-investigated compound 2,4-D provide an interesting contrast. Crafts (p. 84) refers to this compound as "strongly accumulated within the symplast and moved solely in the phloem". Hay (p. 386) says "movement to the roots is negligible regardless of the flow of assimilates . . . once in the stems they (phenoxy acetic herbicides) move from the phloem to the apoplast and are carried back into the treated leaves or up the stem . . . ". The authors agree that 2,4-D is bound to tissue in the symplast and that it interferes with its own transport. Such interference is an almost necessary complication with herbicides (ideally it should not occur with systemic insecticides or fungicides) and the "hormone" herbicides are particularly complicating in this respect. It is, in retrospect, unfortunate that so much physiological work has been devoted to them. Neither author refers to the work of Blackman and his school, a surprising omission being that of the work on polar transport by McCready (1963 and his 1966 review). Apart from Hay's extensive reference to Craft's earlier chapter the two have, strictly, only one reference in common.

In a subject the extreme difficulty of which is well illustrated by this disagreement, two non-physiologists may be excused from attempting to adjudicate. We shall attempt only a further contribution to the discussion—on some quantitative aspects of transport in multi-flow systems which do not seem to us to have received adequate attention.

Much of the early work on pesticide transport was based on whole-plant autography after treatment with ^{14}C-labelled substances. The evident and elegant precision of side-by-side comparison of self-portrait and photograph makes it all too easy to assume that interpretation is unequivocal and straightforward. In fact it needs a more critical approach. A factor in the change of opinion mentioned above has been the increasing availability of extraction-analytical evidence from the use of radiolabels, in addition to autography.

Because of the long exposures required and the possiblitity of chemical interference with the sensitive film, radio-autographs must be taken from dried specimens. Some early autographs were misleading because of redistribution of water-soluble substances during drying. Crafts (1956; see also Yamaguchi and Crafts, 1958) therefore adopted the device of quickly freezing the plant material immediately after harvest, so that any movement of and in water was arrested, and then evaporating the ice *in vacuo*. This freeze-drying technique is now standard practice. It destroys fine structure but this could not in any case be registered by ^{14}C radiation in conventional whole-organism autography since lateral spread of electrons between specimen and sensitive film is several μm. Biddulph and Cory (1965) obtained excellent autographs of phloem bundles labelled with assimilates from leaf-applied $^{14}CO_2$ of very high activity by stripping and flattening the bark of bean (*Phaseolus*) seedlings. The resolution was about 100 μm. Improvement of an order of magnitude is possible with micro-autographic technique using stripping film or direct application of photographic emulsion to fixed sections. In the study of pesticides the relatively low activities obtainable in plant tissues have not encouraged the adoption of these more elaborate techniques.

Yamaguchi and Crafts (1958) note that a bean stem can absorb nearly all the radiation from a ^{14}C source, and, if it lies between the source and the photographic film, form a shadow. "Masking tape" also stops formation of an image but "ordinary cellophane" has hardly any effect. Hay (1976) also draws attention to absorption by thick specimens as a source of error, but critical experiments on this subject are lacking. ^{14}C radiation is almost completely stopped by a uniform layer of matter of mass-thickness 25 mg cm^{-2}. The dried interveinal area of most mesophyte leaves is around 2–4 mg cm^{-2} but leaf veins may be several times thicker. The area pattern of the autograph may thus be determined not solely by the area distribution of ^{14}C in the specimen but by its depth distribution as well. If ^{14}C could be uniformly distributed throughout a specimen of variable thickness, the image intensity would be least over the thinnest areas of the specimen, but its increase with thickness would become less steep with increasing thickness and reach a limit at about 30 mg cm^{-2}. If, on the other hand, the ^{14}C would be confined to a uniform smear on the remote surface of the specimen, the reverse image would result, the intensity falling off with increase of thickness, steeply at first but bending away to practically no response at 25 mg cm^{-2}. Most of the published autographs refer to specimens where mass thickness of 10 mg cm^{-2} will hardly anywhere be exceeded. A strongly contrasting pattern on the autograph can result only from localization of ^{14}C to the indicated areas, but poorly contrasting patterns or thick specimens need allowance for depth effects and additional experiments.

Canny and Askham (1967) consider that the frequently-observed faint general image of a leaf tends to be dismissed as insignificant without quantitative evidence to justify this assumption. Their finding of dark spots corresponding to aphid corpses on a leaf showing such faint background is clear evidence that the aphids had absorbed tracer. Whether the greater darkness of the spots was due to higher concentration in the live aphis than in the leaf or only to a higher area density on the surface where the aphis body, thicker in the live state than the leaf, was dried and flattened, is a question needing quantitative enquiry.

A further problem in interpretation, well appreciated, but easily forgotten when glancing through a number of clear autographic records, is that the visible record is of the ^{14}C (or other label) content of the applied compound, not just of the unchanged compound.

There seems no doubt that different compounds applied in the same way to similar plants can move initially in different directions. This must mean that exchange between the conducting systems is not free. The innermost layer of cuticle must be in contact with the apoplast between the epidermal cells, but the ectodesmata (Franke, 1967) although not penetrating the cuticle completely, could bring it into direct contact with the symplast. If the access to both were reversible, the cuticle could provide a means of exchange, albeit a slow one. Exchange within the active cells of the leaf between the terminal branches of phloem and xylem seems essential to function. It is widely believed that the specialized mechanism for loading sucrose into the phloem terminals against a concentration gradient may also accept, selectively, some exogenous compounds. The finding of Field and Peel (1971) that several exogenous compounds can be loaded into willow bark phloem against a gradient supports this view.

The phloem and xylem saps are both essentially aqueous. There are no mobile fat globules present so that strongly lipophilic compounds will not be rapidly transported. The phloem has a higher protein content and this may lead to specific adsorption effects. There is another general difference which could have important consequences. While the xylem fluid is generally appreciably acid, pH 5–6, phloem fluid is relatively strongly alkaline, pH 7–8 (Ziegler, 1975). The plasmalemma separating the systems is certainly much more permeable to lipophilic than hydrophilic molecules and where it contributes to the isolation of the two systems it is likely to be more selective in its permeability. If an organic acid of pK around 3 and therefore mainly dissociated in both fluids can permeate only in the non-ionized state, the total concentration (mainly ionic) in the two fluids (Appendix 3) could be between 10 and 1000 times greater in the case of the phloem "at equilibrium". This would not of course be a true physical equilibrium but an apparent one produced by whatever maintains the observed difference of pH. This factor alone could well account for the apparently advantageous effect on "mobility in the phloem" of carboxyl groups in the molecule (Crafts and Crisp, 1971) and for the corresponding partition into the phloem of water-favourable (Section 4.IV.C) degradation products of organophosphorus compounds.

Some authorities, particularly Crisp (1971), take the distinction of chemicals into xylem-mobile and phloem-mobile to extremes. The distinction is made mainly on the basis of behaviour in seedlings of soybean (*Glycine* max) which is selected because of the clarity of the distinctions which it allows. Crisp's dismissal of presently available systemic insecticides as "apoplastic" seems not to recognize that, of all systemic pesticide operations to date, the use of systemic organophosphorus compounds against *phloem*-feeding aphids is the most successful. Peterson and Edginton (1976) point this out and also that "apoplastic" urea and triazine herbicides must enter not only the symplast but the chloroplasts within it to exert their toxic effect. These authors are forced to distinguish between "eu-apoplastic" and "pseudo-apoplastic" compounds and to admit that 2,4-D is distinguished not by mobility, but by fixation, within the symplast. Their description might have been more easily understood had they not felt it necessary to preserve a distinction which their evidence shows to be unsatisfactory.

B. Mechanism of Phloem Transport

There is still debate among plant physiologists about the mechanism of transport in the phloem vessels. The extreme difficulty of microscopic examination of these vessels in their functioning state is the main reason for uncertainty. The vessels are intimately associated with other specialized cells and contain liquid at an osmotic pressure of several atmospheres. Controlled dissection is accompanied by uncontrolled disruption. The stylets of aphids are the most precise instruments available for tapping the liquid contents with minimal disturbance. Although only a few species on a few hosts can be "decorporated" *in situ* to provide a slow inanimate leak, examination of the honeydew after labelled compounds have been supplied to the host can provide useful information. We shall not attempt to review much original literature, but only to comment on that summarized in the texts referred to above.

Solutes could be moved in the phloem by actual flow of the liquid contents (unidirectional in any one vessel at any one time) or by a process of accelerated diffusion (Canny, 1971), e.g. multicyclic flow such as that made visible in some photosynthesizing cells by streams of chloroplasts. A main area of doubt is the relative importance of these processes. Crafts and Crisp (1971) came down very heavily in favour of mass flow in the main vessels. Unquestionably, unidirectional flow does occur under some conditions. It is necessary to replace liquid extracted by aphids. Exudation also occurs from cut stems and, although this is more often predominantly xylem sap or latex, phloem sap can be collected from some species, particularly cucurbits: this must be replaced by flow. That flow is occurring in an intact, functioning, series of phloem vessels does not however necessarily follow.

Evidence for diffusion is less direct. One argument in favour is that an aphis can be supplied simultaneously with one tracer applied above its location and another applied below. The possibility of flow from both directions to a common exit up the aphis stylet cannot, however, be dismissed. It depends whether normal unidirectional flow was faster than the supply rate to the aphis (Section 11.VIII.F). From more elaborate experiments, Biddulph and Cory (1960) deduced that, in mature phloem elements, flow transport predominated while, in juvenile elements, the mechanism was mainly diffusive. Although they later (1965) questioned their deduction, this view is worth emphasizing since many authors take strongly polarized views—flow *or* diffusion. Some apparently conflicting evidence could best be reconciled by recognizing that the very versatile phloem system can probably use either mechanism according to the interaction of changing stimuli and demands. What is quite certain is that the measured rates of transport of carbohydrate from synthesizing leaf to growing root tip demand that diffusion, if responsible, must be accelerated to many thousand times the rate of molecular diffusion. Another inescapable conclusion is that, in a simple tube, multicyclic motion (accelerated diffusion) must involve the expenditure of more energy for a given degree of transport than unidirectional flow. This follows because at least half of the solute at any one instant has a backward component of velocity, so that local forward velocity must be at least twice that necessary in unidirectional flow. The flow pattern being more complex, the work done against viscosity must be disproportionately increased.

Another possible mechanism is that solutes are transported not in the fluid itself but attached to fibrous material moving through the fluid. Thaine *et al.* (1967 and earlier references therein), using light microscope techniques, found evidence of strands traversing the length of the sieve cells and passing through the pores of the plates. Files of small particles passed in line through the pores. Thaine envisaged that solution and particles could be carried along *within* the strands (1969; see also Jarvis and Thaine, 1971) by a type of peristaltic pumping action. Fensom (1972) elaborated on the idea of multiple strands which are *in toto* stationary but force liquid through the pores by undulatory contraction. Canny took up the strand concept but envisaged the strands moving longitudinally in groups through the sieve plate pores, not necessarily in the same direction through different pores in the same tube. With Phillips (1963) he showed that reversible adsorption in counter-travelling strands could provide an accelerated diffusion mechanism.

In accelerated diffusion different parts of the fluid are considered to be moving at any instant in opposite directions or in a chaotic manner with no unidirectional component. The process is equivalent to eddy diffusion in the atmosphere (Section 3.IX.A) but the eddies are on a much smaller scale. The scale, however, will be much greater than that of molecular size and the accelerated process therefore diffuses all solutes together at the same rate. The Canny-Phillips strand mechanism could diffuse solutes at different rates because the fluid is no longer homogeneous and partition and adsorption effects could arise. The strands being of protein structure, the adsorption properties could be very specific and the mechanism could therefore account for different patterns of movement of different solutes.

Most physiologists discount the existence of permanent carrier strands and consider that the pores, in mature phloem elements, although they may have a complex lining, are open to liquid flow in their normal functioning state. Crafts and Crisp (1971, p. 284) admit that movement of or within strands may be necessary for exchange between the mesophyll cells and the terminal branchlets of the vascular system but that unidirectional flow is established as the only possible means of long distance phloem transport. The pores are easily plugged by slimy coagulate when excessive flow is induced and the sieve plates may therefore contribute to the function of protection against continued bleeding from damaged tissue (Eschrich, 1975). Other possible functions of these apparently unnecessary obstructions to a difficult transport process are prevention of entanglement of strands or conversion of chemical to kinetic energy as in the electroosmosis theory of Spanner (1970).

C. Active and Passive Transport

Only xylem flow, in so far as it is driven by evapotranspiration, is truly passive in the sense that energy conversion is external to the plant. All phloem transport is active in that chemical energy must be converted to dynamic energy within the plant but the mechanism and location of the conversion are still uncertain.

Quantitative theories of the Münch mechanism have been elaborated (Christy and Ferrier, 1973; Tyree and Dainty, 1975; Christy, 1976) in which osmotic flow through the sidewalls of the tubes is considered, but an important general property can be appreciated more clearly in the simplest possible model—two reservoirs sufficiently large for the concentrations in them not to change appreciably while a steady state is established in the tube connecting them, which we assume to have *im*pervious walls.

If there is uniform unidirectional flow at volume velocity Φ along the tube at a linear rate too great for diffusion to compete, the flux of solute is

$$\mathbf{F} = \Phi C_{\text{don}} \tag{15}$$

If there is no net flow but a multicyclic flow having the effect of diffusion

$$\mathbf{F} = D\mathbf{A}(C_{\text{don}} - C_{\text{rec}}) \tag{16}$$

The apparently characteristic difference, that diffusion transfers at a rate proportional to concentration difference while flow transfers at a rate determined by the donor concentration only, would disappear had we considered two tubes and allowed return flow in the second one to maintain the reservoir volumes constant.

In the Münch process the return flow is indirect, via the xylem, after unloading of sugar, so we will accept eq. (15). The flow, however, is produced by the osmotic pressure and, if we restrict our consideration for the moment to transport of the predominant natural solute, sucrose, the osmotic pressure difference is proportional to $C_{\text{don}} - C_{\text{rec}}$ so that the expected sucrose flux is

$$\mathbf{F} = K_1(C_{\text{don}} - C_{\text{rec}})C_{\text{don}} \tag{17}$$

Passioura (1976) tries to remove this complication from the more elaborate and more realistic treatment of Christy (1976) by consideration of the effect of increase of viscosity of the solution as the concentration and therefore osmotic pressure increase. This would compensate, and only roughly, in the 15–40% sucrose range which is unrealistically high. From a leaking osmometer immersed in pure solvent, the leakage rate of solute, without this compensation, would be proportional to the square of its concentration. Applied to the phloem initials it is more likely that the rate of loading, not the concentration, is controlled at the source and flow will then accommodate, along with concentration, until both are proportional to square root of the loading rate.

An accelerated diffusion process may also be powered, by mechanisms much less clear, by the sucrose concentration gradient, either down the tube or across its walls. These would lead to the equations for accelerated diffusive transport

$$\mathbf{F} = K_2(C_1 - C_2)^2 \tag{18}$$

or

$$= K_3\bar{C}(C_1 - C_2) \tag{19}$$

where $\bar{C}$ is a mean concentration.

For small concentration differences eqs (17) and (19) become indistinguishable and, if the differential equations are formed for a continuous process we should find the fluxes across transverse sections of the tubes to be proportional, for flow, to $C(\mathrm{d}C/\mathrm{d}x)$ and, for diffusion, to $\mathrm{d}C/\mathrm{d}x \,.\, C$ or to $(\mathrm{d}C/\mathrm{d}x)^2$, all very crude approximations. Distinction between mechanisms by analysis of concentration contours in the crescent state would be rather unrewarding.

It may help understanding of this very "non-Fickian" type of diffusion to point out that a process, essentially diffusive, in which flux, while probably not strictly proportional to $(\mathrm{d}C/\mathrm{d}x)^2$, increases more than proportionally to $(\mathrm{d}C/\mathrm{d}x)$, is commonplace in a crude form. If a dense solution is layered on top of a narrow column of solvent it proceeds downwards, diluting as it goes, by a multicyclic convection process which is more energetic the steeper the unstable density gradient.

The simpler, normal diffusion and flow equations, would be expected to apply to inert exogenous compounds at low concentration as long as assimilate movement remains constant. The relevant diffusion and flow parameters will depend on assimilation processes but will probably be more nearly proportional to the square root than to the first power of rate of assimilate export. Many herbicides behave passively during the early stages of systemic movement, but others, especially the "hormone" herbicides affect their own transport (and that of other substances) at an early stage due to interference with phloem mechanisms as discussed, with references, by Robertson and Kirkwood (1970).

Since the apoplast is considered to be continuous with the xylem and since the inside of the cuticle is continuous with the apoplast any compound applied to the outside of the cuticle and able to penetrate it would be expected to have free access to the xylem and therefore, in a transpiring leaf, to move initially towards the periphery, spreading outwards as it goes, from a droplet source near the base of the leaf. This is typical of what Crafts and Crisp call the apoplastic compound and the autographs obtained by Crafts (1967) show the behaviour very clearly in the case of ^{14}C monuron—wedge-shaped outward spread of leaf application, upward movement only from stem application and no movement out of the photo-synthesizing cotyledons of *Glycine max*.

One would expect this behaviour to be initially general, followed in the case of compounds able to enter the symplast, by transport to growing regions of root and shoot after some delay in penetrating the plasmalemma. 2,4-D, however, tends first to move basipetally after application to stem or cotyledon. Amitrole and maleic hydrazide move very rapidly throughout the plant, but with mature (exporting) leaves registering only very feebly on autographs. The behaviour of 2,4-D is confused by its fixation in tissues and interference with transport processes. The image produced by amitrole or maleic hydrazide is what one would expect (Section 11.VIII.E) at the steady state from rapid transport with somewhat hindered exchange. The evidence for initial preference for the phloem is not so clear but most workers in this field are convinced that there can be selective entry. For compounds having this property the barrier between apoplast and xylem must be more effective than that between apoplast and phloem making rather nonsense of the interchangeable use of the terms "apoplastic" and "xylem mobile".

The actual mechanism of loading of sucrose into the phloem initials against a concentration gradient is unknown. If this loading can be done directly and flow is then created at a corresponding rate the whole process is elegant and satisfactory but there are difficulties and inconsistencies. Field and Peel (1971) found that several quite unrelated foreign compounds were also loaded into the phloem from bark strips of willow. Monosaccharides do not appear in high concentration in the phloem but there is evidence that sucrose is inverted before loading as well as on leaving the phloem. The first observation suggests a very general mechanism: the second a highly specific biochemical one.

The general enrichment indicates an essentially physical process. If a peristaltic pump mechanism is feasible for creating flow in the sieve cells, is it not just as feasible for creating the pressure in the terminal units? The whole of the available solution could be pumped in to these units to high pressure, the water than leaking out through semi-permeable membranes, along with other small molecules having suitable partition properties. The pressure, mechanically generated, would, on this view, be the direct cause of flow and the loading against concentration of compounds to varying degrees would be a result rather than a cause of this pressure, associated with differential leakage.

D. Sources and Sinks

In the physics of diffusion much use is made of the concept of sources and sinks. A source is a region where diffusible substance is introduced into the system under examination in a definite amount, at a constant rate or at a rate maintaining a constant local concentration. A sink has the exact opposite properties or function. A "perfect sink" is a region where removal from the system is so efficient that the local concentration is reduced to zero. The physiologist more often uses the word sink to denote a region where the measured substance or label accumulates. If the accumulation results from chemical conversion of the mobile substance to an immobile one—e.g. sucrose carrying ^{14}C being converted to skeletal cellulose—it may not be useful to distinguish these meanings since the region of accumulation outside the physicist's "system" is intimately associated (geometrically) with the region from which mobile solute is removed. When, however, accumulation results from the filtering out of a mobile substance on the inside of a barrier which allows solvent to pass, it must be kept in mind that an entirely different process is producing much the same image on the autograph.

Hay (1976) distinguishes between "assimilate" and "transpiration" sinks, but we feel that these terms, though making the necessary distinction described above, are not well chosen. Transpiration is only one, although perhaps the most important, mechanism for accumulating mobile solute against a barrier. Osmotic "flow" or flow under pressure could produce the same effect. Phloem, if flowing unidirectionally, must dispose of its water, when relieved of sucrose by conversion to fixed tissue. There may be some leakage to the soil but physiologists seem mostly agreed that return is mainly to the xylem and we have seen that there may be barriers restricting return of some solutes.

The two kinds of accumulation register in the same way in gross autography of ^{14}C compounds. In fine structure they could perhaps be distinguished and in this tritium labelling would help, though needing much more elaboration to get records of other than superficial tissue. The pile-up against a barrier would have in general a more blurred outline on the entrant side. Fixation in the skeleton would show more anatomical detail. Additional procedures could be more definitive. The very useful practice of freezing before drying keeps the label where it was at the time of "harvest": comparison with similar specimens allowed to lose liquid water or soaked in water after partial drying could distinguish between mobile accumulations which would spread or disappear under the latter treatments and fixations in skeletal tissue which would remain well-defined. Mobile tracer could perhaps be "printed" from a dried specimen on to a diffusible jelly.

The two types of accumulation, indistinguishable on gross autographs without such additional procedures, are quite different in their dynamics of formation but they may interact. Skeletal fixation could be increased in the region of a barrier to solute flux simply because a higher concentration of mobile solute is available there. Accumulation within a previously homogeneous medium, against a solvent-permeable barrier, can only result from transport having a unidirectional component. It cannot be produced by diffusion but it can arise by partition into a separate phase. Fat-favourable solutes will therefore accumulate in stationary fatty globules and, in so doing, delay transport during the crescent stage. Accumulation of ionizable compounds could result from a maintained high pH (p. 616).

The autographs (reproduced in Crafts and Crisp (1971) frontispiece and p. 111) of *Zebrina pendula*, after application of ^{14}C amitrole and ^{14}C urea (the latter as progenitor of sugar) to a leaf, provide an interesting comparison. The young rootlets are in both cases strongly labelled. ^{14}C from urea will certainly be built into new skeletal tissue and we see that the root hairs are strongly labelled. The label from amitrole seems to stop short of the hairs as might be expected if an otherwise mobile compound is accumulated where exhausted phloem re-enters the xylem. Some label from amitrole may be built into tissue but other evidence suggests that the major metabolites are still mobile.

E. Transport in Counter Flow

Long-distance transport from transpiring leaves to growing roots, which must be effective over many metres in some forest lianes, requires that the main phloem vessels be remarkably impermeable. According to the majority view it also means that only unidirectional flow can be efficient enough but, whether flow is unidirectional or multicyclic, impermeability is essential for those products of leaf chemistry which are necessary for root construction. One could expect that small, foreign neutral molecules would permeate membranes lining the sieve cells as they do the membranes of other cells. Also, in the terminal branchlets of the system there must be exchange where loading and unloading complete the essential function of transport. Leakage in the growing tissue of young plants was demonstrated by Biddulph and Cory (1957). The type of distribution to be expected where two flows occur in opposite directions with slow diffusive exchange along their main length and more rapid exchange between their terminal branches is worth examination. Horwitz (1958) considered the effect of leakage, both reversible and irreversible, from a single flow into a static environment for both crescent and steady states. Canny and Phillips (1963) considered exchange between equal sets of carrier strands having equal but opposite movement in a limited static environment.

The simplest type of non-uniform diffusive exchange is where the main conduits are completely impermeable and exchange occurs only in terminal compartments. The model is shown in Fig. 11.

The tubes from $x = 0$ to X represent phloem and xylem vessels carrying solution in the indicated directions at volume velocities Φ_1, Φ_2 respectively. We assume that linear flow in the tubes is too fast for molecular diffusion to be significant in an axial direction. The rejected solute therefore accumulates in the terminal compartment of volume, $\mathbf{V}_L$, to a concentration, C_L, higher than the supply concentration, C_2, until it diffuses through the semipermeable barrier, represented by the broken line, into the "phloem" concentration, C_1, at compensating rate. Similar accumulation, to concentration, C_R, in volume, $\mathbf{V}_R$, occurs at the "root" end. These exchange volumes are considered to belong wholly to the donor tube and the concentration within them is assumed uniform: this is not wholly unrealistic, because they represent a complex of cells, plasmodesmata and microchannels forming a very thin layer perpendicular to the paper. The tubes being impermeable the concentration is uniform along each. The condition for the steady state of cyclic movement of solute in the system is simply

$$\Phi_1 C_1 = K_R(C_R - C_2) = \Phi_2 C_2 = K_L(C_L - C_1) \tag{20}$$

where K_R, K_L are permeation functions for the terminal membranes.

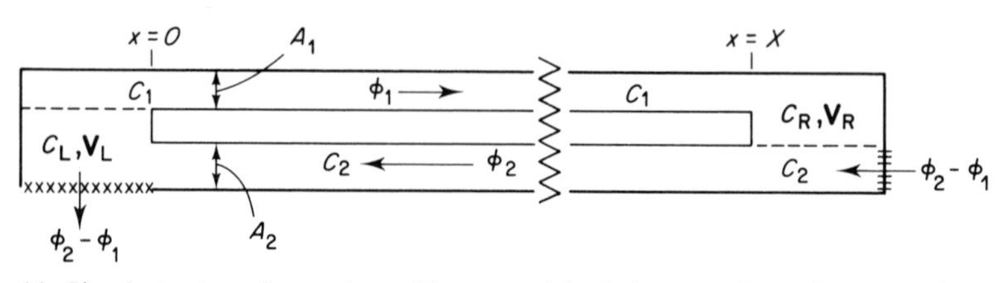

Fig. 11. Simplest return-flow xylem–phloem model. $\phi_2\phi_1$ are volume flow rates in xylem and phloem repectively. Broken lines indicate barriers through which water and solute can diffuse, the former more rapidly. +++++ indicates water supplied through root hairs. ××××indicates surface through which excess water evaporates.

The first point of comment on this multiple equation is that the "phloem" and "xylem" concentrations at the steady state are in the *inverse* ratio of the volume velocities. At the *steady* state (not unrealistic in considering autographs at a stage where the plant extremities have been reached) the more sluggish the phloem movement the greater the fraction of inert solute it contains. This is obvious when pointed out, for the system described. A strong phloem image on an autograph does not itself indicate rapid transport. The conclusion is not upset by consideration of cross-sectional areas because the solute content per unit *length*, whatever the shape, must be inversely proportional to the mean *linear* flow rate. The phloem, if isolated, would therefore normally register more strongly on the autograph than the parallel xylem under most conditions and the image would be stronger the slower the phloem movement. In gross autographs, of course, the two would register together except for self-absorption (of β ray) effects which would tend to favour registration of phloem.

Although we could not explain this way the occasional total by-passing of an exporting leaf by a compound transported in the main stem to regions of active growth, it is of interest that we should expect that the main flow in the xylem, on return by the phloem at low rate and therefore high concentration, would register more strongly in a non-exporting petiole than in one where the very active phloem returns the label at a low concentration.

It is evident also that decrease of permeability in a terminal compartment must increase accumulation in that compartment and simulate, on the autograph, a "fixation" sink. For purposes of illustration we have taken values for the controlling variables giving a reasonable degree of terminal accumulation. To calculate relative *amounts* in the various compartments, we have multiplied concentrations by volumes, assigning cross-sectional areas, $\mathbf{A}_1$ and $\mathbf{A}_2$, to the tubes. Our basic values, in arbitrary units, are, unless otherwise stated,

$$\mathbf{A}_1 = 1, \quad \mathbf{A}_2 = 5, \quad \Phi_1 = 0{\cdot}5, \quad \Phi_2 = 5,$$

$$X = 10, \quad \mathbf{V}_R = \mathbf{V}_L = 1, \quad K_L = K_R = 1$$

Figure 12 shows the distribution when Φ_1 is allowed to vary and Fig. 13 where K_L or K_R is allowed to vary at a constant value of Φ_1.

It is very unlikely that compounds which are unreactive with plant processes (as systemic insecticides should be), not excessively hydrophilic and able to permeate cell membranes, will be wholly unable to leak from phloem to xylem (or vice versa) in the stems. We must therefore introduce a diffusive exchange in the $x = 0$ to $x = X$ length governed by a third

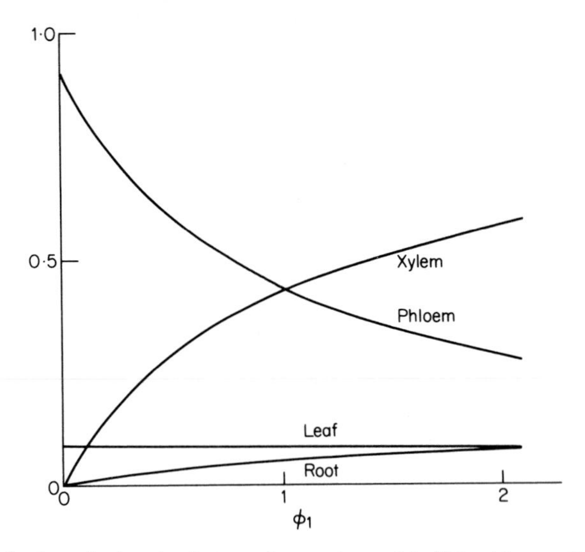

Fig. 12. Distribution of solute load at steady state in model of Fig. 11 according to eq. (20). Variation with ϕ_1 at constant value 5 of ϕ_2. Other values in text.

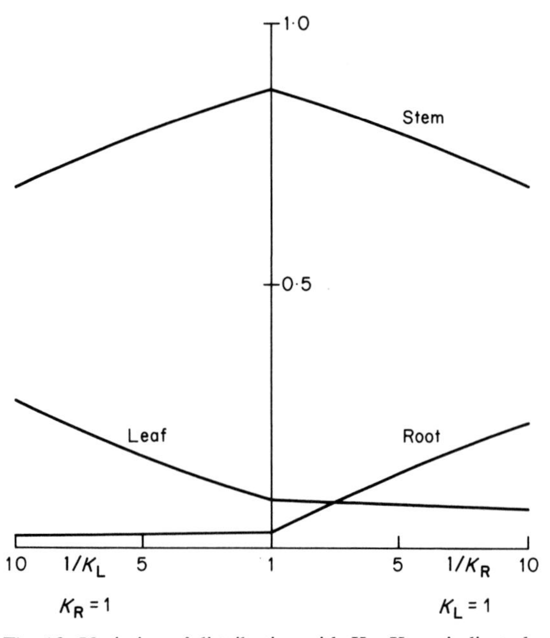

Fig. 13. Variation of distribution with K_L, K_R as indicated.

function K_s defined by

$$\text{rate of transfer } \textit{per unit length} = K_s(C_1 - C_2)$$

C_1, C_2 must now vary with x and, at the steady state where $\partial C/\partial t$ is everywhere $= 0$, the new condition imposed is

$$\Phi_1 \frac{dC_1}{dx} = K_s(C_2 - C_1) = \Phi_2 \frac{dC_2}{dx} \tag{21}$$

The general solution is

$$C_1 = \Phi_2\Psi + K_5 \qquad C_2 = \Phi_1\Psi + K_s \tag{22}$$

where

$$\Psi = \Psi_0 \exp(-K_4 x) \tag{23}$$

where

$$K_4 = K_s(\Phi_2 - \Phi_1)/\Phi_2\Phi_1 \tag{24}$$

and Ψ_0 and K_s are determined by boundary conditions.

The rate of transfer across any section in the direction of increasing x is $\Phi_1 C_1 - \Phi_2 C_2$ for the two tubes together and, from the equation pair (22) this is $(\Phi_1 - \Phi_2)K_s$. This could be finite at the steady state if the tubes connect two reservoirs of constant concentration or if one end receives a constant supply and the other is a perfect sink. This is the common situation in heat exchange problems and could, of course, be relevant to a herbicide irreversibly fixed at the receiving end or to a nutrient such as sucrose, but we are restricting our attention to the final distribution of systemic pesticides assumed to be inert.

One problem in the application of eqs (23) and (24) is, however, best illustrated when a steady net flux does occur. This is the limiting case when $\Phi_1 = \Phi_2$, for which the equations in their given form cannot be evaluated. It can easily be shown that the consistent special form has the concentrations linear against x and separated by the constant distance $2K_s$ on the concentration axis. This is the form at which Canny and Phillips arrived when considering equal carrier bundles moving at the same speed in opposite directions. Two straight lines, each terminating in an abrupt change at one end, offend the intuitive sense of what nature would arrange and it may help therefore to plot the theoretical curves from $C_0 = 1$ in a donor reservoir to $C_2 = 0$ in a receiver reservoir distance $X = 1$ away. This is done in Fig. 14, for $K_s = 4$ and the symmetrical flow combinations $\Phi_1 = 1$, $\Phi_2 = 2$; $\Phi_1 = \Phi_2 = \sqrt{2}$; $\Phi_1 = 2$, $\Phi_2 = 1$. The straight line now appears as the natural intermediate between concave upward

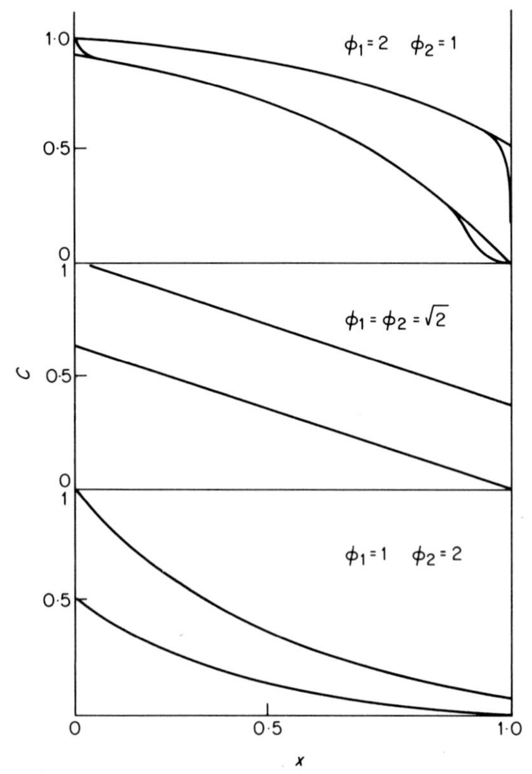

Fig. 14. Variation of concentration with distance for flow combinations indicated and other conditions as in text.

and convex upward exponential curves. The abrupt termination of one curve of each pair at one end is an artefact of our simplifying assumption that molecular diffusion (or thermal conduction in the heat-exchange case) has negligible in-line effect. While true, in gross, at normal xylem flow rates and distances, there must always be some rounding-off effect near entrances and exits as shown on upper curves only.

Another special feature of the limiting $\Phi_1 = \Phi_2$ case is illustrated by Fig. 14. If there is no net flow of solute at the steady state the concentration lines must be coincident and no exchange between them can occur. This would make any concentration change with time proportionally to its slope. The only steady state consistent with equal and opposite flows is therefore where the equal concentrations do not vary with x. Since no exchange between the tubes can occur we find that the state for $\Phi_1 = \Phi_2$

at any value of K_s, is also the limiting form of the $K_s = 0$ case already evaluated.

It should be noted that the concentrations within unbranched tubes vary proportionately and as an exponential function of distance. When flow balances axial diffusion in a single tube a similar function results (Section 3.VIII.A). In the present case the function is $\exp\left(K_s \frac{\Phi_2 - \Phi_1}{\Phi_2\Phi_1} x\right)$ while, for a single flow opposed by axial diffusion, we have $\exp(vx/D)$. Since $v\mathbf{A}$ in the latter would produce the same flux as $\Phi_2 - \Phi_1$ in the former the exchanging counter-flow produces the effect of diffusion with a coefficient $D = \Phi_1\Phi_2/K_s \,.\, \mathbf{A}$. If everything else remains the same and we increase both velocities proportionally we reduce the concentration gradient because D must increase as the square of the opposing velocity. If Φ_2 (xylem) only increases and is already much in excess of Φ_1, there is not much change of the coefficient of x in the exponential, but if the smaller Φ_1 increases this coefficient markedly decreases. Doubling Φ_2 in our example of $\Phi_1 = 0{\cdot}5, \Phi_2 = 5$ increases $(\Phi_2 - \Phi_1)/\Phi_2\Phi_1$ from 1·8 to 1·9 only. Doubling Φ_1 decreases this function from 1·8 to 0·8.

The general cyclic steady state distribution corresponding to Fig. 11 can be obtained from eqs (22) (with $K_5 = 0$), (23) and (24) with two three-fold equalities replacing the four-fold (20). These are

$$\text{at } x = 0 \qquad \Phi_2 C_2 = K_L(C_L - C_1) = \Phi_1 C_1 \tag{25}$$

$$\text{at } x = X \qquad \Phi_1 C_1 = K_R(C_R - C_2) = \Phi_2 C_2 \tag{26}$$

This leads to the following expressions for concentrations as factors of C_1 at $x = 0$

	Concentration	Volume
Phloem initial ($x = 0$)	1	
terminal ($x = X$)	$\exp(-K_4 X)$	
mean	$(1 - \exp(-K_4 X))/K_4 X$	$\mathbf{A}_1 X$
Root reservoir	$(\Phi_1/\Phi_2 + \Phi_1/K_R)\exp(-K_4 X)$	$\mathbf{V}_R$
Xylem initial ($x = X$)	$\Phi_1/\Phi_2 \exp(-K_4 X)$	
terminal ($x = 0$)	Φ_1/Φ_2	
mean	$\Phi_1/\Phi_2(1 - \exp(-K_4 X))/K_4 X$	$\mathbf{A}_2 X$
Leaf reservoir	$1 + \Phi_1/K_L$	$\mathbf{V}_L$

The volumes of the four main components of the system are shown together with the appropriate concentrations. The sum of the products is the total amount which enables the fraction in each to be calculated.

It is obvious that increase of K_s while other properties remain the same will increase K_4 and therefore reduce all concentrations towards the root (Φ_2 being assumed greater than Φ_1). $K_s \rightarrow \infty$ corresponds to completely free exchange and therefore transport exclusively in the xylem flow direction. It is obvious also that decrease of K_L or K_R must increase the accumulation in leaf or root respectively, as will, of course, increase of the volumes of the postulated exchange regions. Less obvious is the effect of the more probable change of all permeabilities together as we change from one compound to another. In Fig. 15 are plotted the values of the fraction of the recycling compound resident in the "root", "leaf" and "stem" (phloem plus xylem) in our model, for $\mathbf{A}_1 = 1$, $\mathbf{A}_2 = 5$, $\Phi_1 = 0{\cdot}5$, $\Phi_2 = 5$, $X = 10$, $K_L = K_R = 4K_s$, $\mathbf{V}_L = \mathbf{V}_R = 1$ for K_s ranging from 0·5 to 0. The "root" content increases monotonically, the "leaf" content goes through a minimum and the stem content through a maximum.

We have had to take arbitrary values to calculate and illustrate this trend but it will apply with alteration of relative magnitudes over a wide range of other values. The plant probably does as much selection as the

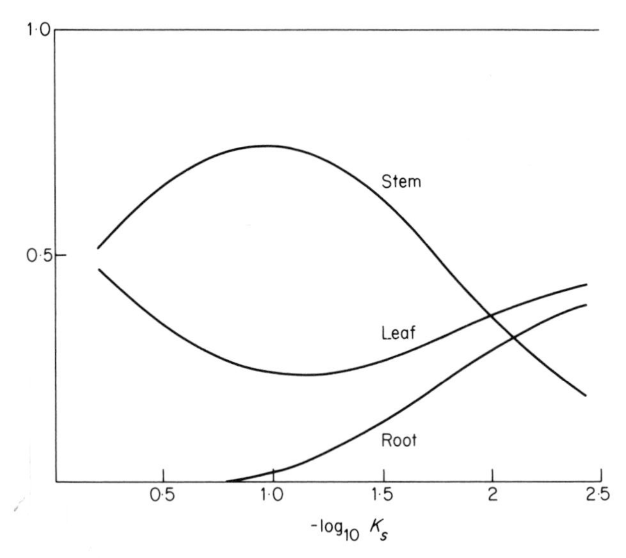

Fig. 15. Variation of distribution in the model of Fig. 11 when diffusive exchange between tubes is permitted with permeability K_S. K_L, K_R are assumed to be proportional to K_S, not constant as in Figs 12 and 13. Other values in text.

mathematician. The properties must vary with the growth form: a lettuce does not climb trees.

We felt it necessary to analyse this simplified two-flow system in some detail as the type of distribution to be expected at the steady state was not obvious. The steady state is unlikely to be realized with a metabolized chemical in a growing plant but the trend towards it may be expected. We see, however, that there is, as so often happens in this confusing subject, a conflict of tendencies. If a compound arrives via the xylem and is accumulated and returned in the phloem, return is eventually at a higher concentration the more slowly exchange occurs. In the early stages, however, the phloem concentration builds up more slowly. There must at some point be a cross-over.

We have assumed the more widely-accepted bulk flow behaviour in the phloem. The major difference had we assumed accelerated diffusion would be that accumulation at a barrier in the root would not occur and that, even at zero exchange between the two tubes, the stem content would decrease towards the root because only a gradient could produce the flux of solute.

F. Selective Entry into Transport Systems

An investigation of movement of a systemic insecticide recently reported (Galley and Foerster, 1976a; Foerster and Galley, 1976; Galley and Foerster, 1976b) is particularly important in the present context. Phorate [$(EtO)_2PSSCH_2SEt$] carrying ^{14}C in the methylene group was applied to roots (in water culture) and leaves of *Vicia faba* plants on which colonies of *Aphis fabae* had been established. Leaflets, aphids and honeydew were separately extracted and soluble contents partitioned between chloroform (containing the parent compound and most toxic metabolites) and water (containing most non-toxic metabolites).

The second paper presents clear evidence that the aphids plus honeydew contain several times more ^{14}C than the leaflet on which the colony was established and that this leaflet did not contain significantly less than a parallel non-infested leaflet, the compound having been applied to another leaf. The first observation is confirmed incidentally in the third paper which is mainly concerned with comparison of root and leaf application. When dosages were chosen to result in comparable aphis uptake, just short of that producing serious mortality, very much more (*c.* 500 times) entered the leaves from the roots.

That the aphids take up more than the leaf content without depleting the leaf after application of the compound to another leaf is clear evidence that they induce a long-distance flow from a richer source. It is at least consistent with the view that their resting content could have arrived by accelerated diffusion. A new "sink" at the end of the *same* transport process would cause local depletion. These points are rather overlooked by Canny and Askham in their parallel observation that aphids on a mature leaf produce a local image after application of $^{14}CO_2$ to another leaf, against a faint background from the infested leaf. Galley and Foerster found that less than 1% of the leaf-applied dose appeared in the extracts. They do not state how much entered the treated leaf, but it certainly seems that translocation (in the absence of aphids) from this source was very inefficient. Only a small fraction of what had entered the leaf more rapidly from the roots was available to aphids. Entry into the phloem either from outside the leaf cuticle or from the xylem inside the leaf was therefore very slow. In the time allowed (4 days longest, 18 h shortest) exchange can have gone only a very small fraction of the way to the steady state. The second paper compares the distribution of the chloroform-favourable and water-favourable labelled compounds. Exchange proceeded much further with the latter, as was confirmed again in the third paper. In the aphids plus honeydew was found only 0·4% of the amount of less polar compounds present in the remainder of the infested leaflets 18 h after uptake from the roots had commenced. Of the water-favourable compounds the proportion was 48%. Two days after uptake commenced the corresponding figures were 3% and 200%. The last figure does not necessarily mean that the phloem concentration is higher than the xylem concentration, although our analysis above indicates that this could be possible without metabolically activated diffusion. It is probably due to the aphids causing phloem flow from other leaves. After application to a leaflet, the water-favourable label was transported much more to other leaflets than was the chloroform-favourable label. As the authors point out, chemical conversion, producing compounds more favourable to water, occurs simultaneously with transport and confuses quantitative deductions, but the main conclusions are based on differences too large to be in doubt.

The result is entirely in accord with Crisp's (1971) conclusion from review of previous data, mostly autographic, that the presently available OP systemic insecticides are essentially xylem mobile. This suggests that it should be possible to design more efficient compounds and Crisp makes some rather elaborate suggestions in this direction which we consider in the chapter on chemical transformation. Work along these and other lines with the objective of modifying transport behaviour is certainly desirable, but knowledge in this very complex field, particularly when a series of chemical conversions is necessary in the right places and in the right sequence, is at present too incomplete for theorizing to be useful except as a stimulus to experiment.

It is possible that no plant-inert compound will be found to move mainly in the phloem and that rapid movement in xylem is a necessary part of the whole process. One would like to see investigation as detailed as that of Galley and Foerster carried out on the more water-favourable systemics such as dimefox which has also the advantage of greater stability and of better performance, *as a systemic insecticide*, than phorate. This compound has been phased out because of mammalian toxicity. Understanding of mobility of pesticides in plants has been hampered by concentration on labelled preparations of commercially successfully compounds. Investigation of model compounds selected to throw light on transport mechanisms might have taken us much further.

That compounds can pass down into and out of the root system is shown by the work of Mitchell *et al.* (1959, 1961) on methoxyphenoxy acetic acids. Physiological effects appear in untreated plants sharing only root space with plants treated on the foliage. The methoxy group appears to stop the lethal fixation of the phenoxyacetic compounds in the upper root tissue. The chlorobenzoic acids, which remain free to migrate from damaged tissues, behave similarly.

IX. ENVIRONMENTAL EFFECTS

The entry of externally applied substances into plants can be influenced considerably by the environmental conditions, particularly light, temperature and humidity. In practice the effects of these different factors are very difficult to distinguish, when observations are made on the intact plant. In the following sections, the various factors have been separated for convenience of discussion, but their important interrelationships should be borne in mind.

A. Effect of Light

Since the substance of green plants, even of non-green organs not exposed to light, is derived from products of photosynthesis in the leaves, it is to be expected that light will have a considerable influence on the translocation of exogenous substances which have gained entry. Several researches have demonstrated such influence on foliar-applied herbicides (e.g. Rohrbaugh and Rice, 1949; Barrier and Loomis, 1957; Clor *et al.*,

1963). 2,4-D was applied to one leaf or cotyledon which was excised after a standard period for absorption. Subsequent epinastic response and weight gain of the remaining plant over a further standard period gave some measure of translocation. Clor *et al.* used, additionally, radio tracer assay. If the leaf to be treated was kept in darkness for 24 h or more to become starved of photosynthates before herbicide was applied and then kept dark until excision, the effect on the remainder of the plant was much less than when the treated leaf was illuminated. Working with *Phaseolus* seedlings, Rohrbaugh and Rice showed that translocation from the starved leaf, while kept in darkness, was partly restored by prior external application of sugar solution or by cutting off the apex and dipping the cut edge in sugar solution. A similar, but less marked, effect was shown by Barrier and Loomis in *Glycine* seedlings when sugar was added to the 2,4-D solution.

Prior illumination can also have effect via the resistance of the cuticle. This was most clearly shown by Moffitt and Blackman (1973) examining uptake of various phenoxy acetic acids (^{14}C-labelled) into sections of stem of *Pisum sativum*. When the cut ends were sealed to restrict entry to the stem cuticle, uptake into etiolated stems was greatly depressed after they had been pre-exposed to light. It is a common observation in screening for new pesticides that plants grown in the glasshouse are usually more sensitive to applied chemicals than those grown outdoors but other factors—wind and fluctuating temperature—may be mainly responsible for the development of a thicker cuticle. Wind increases the crystal size and amount of wax per unit area of pea leaves (Martin and Juniper, p. 209). Skoss (1955) found leaves of *Hedera helix* to be more permeable when grown in the shade than in the sun.

Another effect of recent light history was revealed in unpublished experiments by Hartley which were aimed at elucidating the apparently "temperamental" behaviour of an experimental fungicide on cucumber leaves. Occasionally, a severe local scorch developed under and around each spray residue. When plants were grown and sprayed in high light intensity there was no scorch, not even on leaves fixed on an inclined rotor so that they faced fully into the sun all day. When leaves were kept shaded for 48 h, sprayed and then exposed to sun, local necrosis always resulted. When a temperature shock, estimated to be equivalent, was introduced via the ambient air without increase of light, necrosis did not result. Megalhaes *et al.* (1968) found that low light intensity treatment enhanced translocation of dicamba in *Cyperus rotundus* apparently due to "favoured herbicide penetration through the thinner cuticle".

In these different ways, involving different time-scales, the history of exposure to light can influence penetration. We have not included the most obvious way—by its partial control of stomatal opening—in which light could have such influence because the consensus of evidence (Bukovac, 1976) is that only in exceptional cases is direct liquid entry into the stomatal cavities important (see also discussion in Sections 11.IV and 8.XII). The effect of light (and other factors) on the stomata results from changes in the turgidity of the guard cells which probably directly effects cuticular permeation (Section 11.IV).

As Bukovac (1976) points out, the first effect of light on an applied chemical may be direct decomposition external to the leaf. In the absence of such photochemical reaction and independently of physiological effects of prior illumination, is there any effect of light *during* the process of penetration? It can be assumed that there will be no light effect on the permeation of transparent detached cuticle. The leaf disc technique of Sargent and Blackman (1962, 1965, 1972) gives information on total uptake into cellular tissue without complication of temperature or humidity changes or long-distance transport. In most cases the uptake of 2,4-D was greater in the light but by factors not often exceeding two-fold and with maximal effect at moderate intensity (*c.* 15 000 lux; Greene and Bukovac, 1971). This effect of light generally had no lag detectable over the minimum period of observation of about 30 min, but uptake by the adaxial surface of leaves of *Phaseolus vulgaris* showed a pronounced lag, of 2–4 h after illumination commenced, followed by a "surge" over a short period to a much higher level. The increased uptake in light may be attributable to increased binding of the compound to tissue components formed early in the photosynthetic process. Sargent and Blackman (1969) and Greene and Bukovac (1971) found that the increased uptake on exposure to light was eliminated by inhibitors of the Hill reaction.

The general (translocated) effect of several herbicides has been reported to be greater when a period of darkness follows application. The phenomenon appears contradictory to the more direct effects of light described above and to the stimulation of phloem transport by light. It is considered to be due to increased local damage to tissue when it is illuminated resulting in interference with the transport mechanisms or with diffusion to the conducting vessels. It was first reported for 2,4-D by Leonard and Crafts (1956) and Hay and Thimann (1956) and for DNBP by Meggitt *et al.* (1956). The effect was more pronounced with the dinitro compound, as later confirmed (Mellor and Salisbury, 1965) for 2,4-dinitrophenol. It tends to be confused by associated environmental factors. Baldwin (1963) noted a more extreme form of the effect with diquat, which has since been much studied together with the related paraquat, both of which exert their main toxic effect only in the presence of light. The greater effect with these herbicides than with those studied earlier offers more promise of practical application. The effect does not appear to have been reported for systemic insecticides or fungicides, nor would one expect this since they should not cause tissue damage.

Baldwin showed that applying diquat to one leaflet only of tomato seedlings grown under normal conditions and maintained in full light after treatment resulted in localized damage only. If the plants were held in the dark for 24 h after treatment no damage was apparent but, after exposure to light, general damage quickly resulted. Radio-autographs showed that, during the dark period, there was very little translocation but that the ^{14}C as well as the tissue damage spread rapidly after subsequent illumination. The observations and conclusions seem clear but Thrower *et al.* (1965) obtained results, by assessment of tissue damage only, on three leguminous species, which do not seem in agreement. Movement was extensive in the dark when the humidity was kept low and in the light when the humidity was kept high (both combinations, of course, are infrequent in the field). Damage was localized only if light was combined with low humidity and it was concluded that desiccation contributed to the inhibition of translocation. Baldwin (1963) considers movement of the bipyridyls to be confined to the xylem and that a reverse flow occurs from damaged tissue in high humidity (see p. 640).

Slade and Bell (1966) found that paraquat was more damaging when a period of darkness followed application although the damage did not become evident until after the plants were again exposed to light. They found also that a ^{14}C label moved more in plants subjected to darkness after treatment. Smith and Sagar (1966) concluded also that localized damage in the light was responsible for reduced translocation and that, for improved effect, darkness was necessary only in the treatment region. Davies and Seaman (1968) found no light-dark effect on the uptake of diquat into the submerged aquatic plant *Elodea canadensis.*

Brian has investigated the phenomena in much more detail. In 1966 he noted that the translocated action was increased by high humidity, particularly if there was moisture stress at the root, whether application was followed by light or dark and concluded that stomatal entry was not involved. He showed (1967a) that uptake into the treated leaves was increased by a dark period after treatment. The initial stage of uptake was observed to be very rapid and uptake was modified little by rain following shortly after treatment. Further studies (1967b) revealed the additional complication that the dark period had a greater effect when applied in the morning than at the natural time and that there was a diurnal rhythm in uptake persisting for 72 h in the dark. Although light is necessary for the toxic effect, Brian (1969) found that uptake in the light reached the higher dark value if CO_2 was removed from the ambient air. Uptake therefore appears to be impeded not by phytotoxic effects but by a high concentration of some product of photosynthesis.

The interactions of history of light exposure, prevailing light intensity (and the temperature and humidity changes it causes), light-induced tissue damage and variations in age and species of plant give a picture which is at present confused. Brian stresses that, for results to be interpreted, experimental conditions must be closely controlled and that time (of day and of year) at which the compound is applied is important, but in the same paper (1967b) he fails to state the age of his experimental plants. This is all the more regrettable since the only species he has in common with Thrower *et al.* (1965) is *Vicia faba* which had eight or more true leaves in the latter's work, rather unusually mature for laboratory experiments. A more consistent description, which may not, of course, be obtainable, would be necessary before the large differences of effectiveness sometimes resulting from different light regimes could be exploited in practice.

Sagar *et al.* (1968) conclude that transport of paraquat out of the treated leaf is mainly by a reversed xylem flow consequent on dessication, a suggestion first made by Baldwin (1963). Since extreme local desiccation can have the opposite effect a balance as delicate as that found in phloem transport of 2,4-D etc. is involved.

B. Effect of Temperature on Foliar Uptake

The effects of light, temperature and humidity are difficult to separate when observations are made on the intact plant. At low light intensity and at r.h. near 100% there should be no difficulty in examining the effect of temperature only, although the results would have only limited relevance to field conditions. The effect of temperature on the permeance of detached cuticle can certainly be determined unambiguously. Remarkably little definite information is, however, available in the literature. Field observations generally lead to the conclusion that effects via humidity (see next section) are more important than direct effects.

Norris and Bukovac (1968, 1969) examined the permeation of naphthalene acetic acid through isolated pear leaf cuticle with particular reference to temperature. The effect was fully reversible and the Q_{10} ratio was high, between 2·4 and 5·6 from 5 to 35°C with the major change from 15 to 25°C. These authors point out that metabolic processes cannot be involved and quote observations by van Overbeek (1956) and Sutcliffe (1962) that high temperature coefficients must be expected in organized lipid membranes and that a rather sharp transition will occur about the temperature at which parallel paraffin chains melt to a more chaotic state. High coefficients were found by Norris and Bukovac however, both before and after removing wax. Dewaxed cuticle is not usually optically anisotropic so that the chain-melting transition would not be involved. We have already shown (Section 9.II.E) that high temperature coefficients of permeability are found in isotropic polymer films. Even in liquid castor oil the diffusion coefficient of small molecules has a Q_{10} of about 2·5. In studies on the permeation of 2,4-D into *Phaseolus*, Sargent and Blackman (1962) maintained their leaf disc apparatus at five temperatures from 5 to 37°C and found Q_{10} values to be around 2·5 in the first three intervals dropping to 1·5 in the last. Gabbott and Larman (1968) quote Q_{10} values of about 1·8 to 3·6 for intact cotyledon (Flax) discs and isolated cuticle, the permeant being an azido di(alkylamino)-*s*-triazine.

Richardson (1975) applied ^{14}C 2,4,5-T butoxyethylester to confined areas of whole leaflets of *Rubus procerus* and estimated initial rate of uptake by the reduction in amount of ^{14}C which could be removed by washing at different intervals after treatment. Comparing leaves and glass plates (to correct for evaporation; Section 6.V) he obtained a very remarkable result for leaflets from 18 successive leaves down the stem. The rate fell rapidly from the latest expanded leaf to about the fifth (*c.* 8 days old) with a small Q_{10} of 1·3 between 18 and 30°C. From the fifth leaf to the eighteenth (increasing age) there was no systematic drift in initial rate at 18°C but a rapid rise to the eighth leaf at 30°C before the rate levelled off, corresponding to a Q_{10} of about 3·5. It would seem that the resistance must be caused by a different mechanism in the juvenile leaves which is *replaced* (not augmented) in the mature leaves. This work should stimulate more interest in isolated cuticles with more caution in generalizing from one species to another.

Herbicidal effects in the field do not show anywhere near so great an increase with temperature as the high Q_{10}s quoted above would at first suggest. This is undoubtedly largely due to the opposed effect (see below) of the lower humidity and greater internal water stress associated with high temperature in the field. Translocation rather than uptake into the treated leaves may often be the limiting factor in final effect. Morton (1966) studied uptake and translocation of ^{14}C 2,4,5-T (applied as acid plus surfactant in 50% ethanol) in *Prosopis juliflora* and found no difference in absorption between 21 and 30°C but significantly higher absorption at 38°C. This is a high-temperature plant. At 21°C there is no growth: translocation was basipetal. At 38°C there was rapid extension and translocation was acropetal.

C. Effect of Atmospheric Humidity on Foliar Uptake

In the penetration into tissues from external deposits the evaporation of water or other carrier can assume major importance. The first effect of evaporation is evident when only the a.i. and water are present in the spray solution. If the a.i. is a crystalline solid which does not supercool to significant extent when the water evaporates, the history of the penetration process will be as follows. The concentration in the spread drop will increase as water evaporates and/or penetrates. Penetration rate will increase in response to the increase of concentration gradient but will level off when saturation is reached and the a.i. begins to crystallize. At some stage, as drying continues, the penetration rate will fall because the area of liquid contact with the cuticle surface must decrease. Eventually only a group of crystals is left, making very restricted molecular contact. Further penetration at a reduced rate must proceed by vapour phase diffusion across the narrow separations, through water supplied by the leaf (p. 641) or by a partial replacement of the connecting water film by oily matter from the cuticle. The decrease in contact area and penetration rate in the drying stage will be strongly influenced by the crystal habit, small platelets making better contact than long needles. Rate of drying and presence of other compounds will influence the crystal habit. Many substances, particularly when in very small drops, may remain liquid—"supercooled" for a long time.

The evidence that a fully turgid state of the plant is favourable to entry of leaf-applied substances is far from clear although the statement is made confidently in several reviews. High humidity has been shown in many experiments to favour entry, but the effect of high humidity in keeping the spray deposit liquid is certainly operative and an influence on the cuticle rather than on the deposit is difficult to establish. One of the earliest investigations seems to establish the opposite effect. Hopp and Linder (1946) bent the growing tips of specimens of *Cuphea platycentra* so that they dipped in dilute aqueous ammonium sulphamate or sodium arsenate solutions. A consistent length of stem measured back from the solution level was killed in treated plants which had been left in the laboratory air at *c.* 30% r.h. for 14 days but remained alive in those which were enclosed together with, but not in contact with, an open surface of water. After painting the upper leaves of this plant and of *Coleus blumei* with both toxicants more damage was done at the lower humidity and addition of glycerol increased the damage. These results are consistent with low humidity favouring *translocation* from a *wet* spray application. This was exploited in a method for killing deep underground stems of *Convolvulus arvensis* by putting the ends of aerial shoots in acid arsenical solution when the plant is in a state of water stress (Robbins *et al.*, 1952).

Jordan (1977) has described a full investigation of the influence of temperature and humidity on the activity of glyphosate applied to leaves of *Cynodon dactylon.* Temperatures were 32 and 20°C and r.h. 100 and 40%. The herbicide was applied by overhead spray in $200\,\ell.\mathrm{ha}^{-1}$ at several concentrations. Some plants were held under controlled conditions for 6 days: others had foliage clipped after 24 h and were maintained under greenhouse conditions for 8 weeks to assess re-growth. Parallel ^{14}C assessments using labelled glyphosate were carrried out. The fraction washable off the surface after 48 h was at most only 9% and greatest at the higher temperature and higher humidity, but much more was retained in the treated leaf at low humidity and low temperature. Consistently, regrowth of clipped plants was much greater at low humidity and low temperature. The dominant effect of temperature and humidity was clearly, in this case, on translocation which was greatly favoured by high humidity.

The earlier results of Badiei *et al.* (1966) on the effect of 2,4,5-T on *Quercus marilandica* also indicated less effect on the plant and less translocation of a ^{14}C label under conditions of water stress, but the treatments were not such as to allow appreciable reverse xylem flow (p. 636). Very high volume spraying or sudden increase of humidity is probably necessary to induce the mechanism called on by Robbins *et al.* (1952).

Smith *et al.* (1959) examined the decrease of maleic hydrazide washable from leaves of tomato plants after spraying to run-off and maintenance under different temperatures and humidity. Salts of the acid a.i. with volatile amines left a residue of crystalline free acid and the K and Na salts a residue of salt crystals. Absorption was most rapid from the amorphous residue left by the diethanolamine salt and loss was said to be exponential and evaporation not to contribute appreciably. The half times for entry were 24 h at 100% r.h., 65 h at 75% r.h. and 100 h at 50% r.h.

Pallas (1960) examined the penetration of ^{14}C 2,4-D and benzoic acid (*not* chlorinated), presented as triethanolamine salts and found that both uptake and translocation were greatest at high humidity in apparent conflict with the results of Hopp and Linder (1946).

Luckwill and Lloyd-Jones (1962) followed the uptake and decomposition of 1-naphthyl acetic under varied conditions including different humidity. Uptake at 100% r.h. was about twice as fast as at 37% r.h. The slow uptake at low r.h. was increased by increase of humidity or a light spraying with water.

Prasad *et al.* (1967) examined the uptake of ^{14}C-labelled dalapon from leaf application in bean (*phaseolus*) and barley, measurement being made of the ^{14}C content of the shoot (after excision of the treated area) and of the root of plants that had been grown and treated under different humidity conditions. More was taken up during 6 and 18 h at 88% than at 28% r.h., with the uptake at 60% about half way between. Distribution within the plant was rather uniform indicating that the main effect was on the first stage of uptake. Plants grown at high humidity took up much more than those grown at low humidity when the treatment conditions were the same. Rewetting a deposit promoted further uptake and enclosure in a polyethylene bag increased uptake still more.

Hughes (1968) examined the effect of external humidity on the response of a woody species (*Tamarix pentandra*) to 2,4-D and 2,4,5-T both as butoxyethanol esters and K salts. High humidity did not in this case favour uptake. In the presence of a polypropylenediol mixture, it lowered it.

One must conclude that increase of permeability with increase of humidity by a plant physiological mechanism, as distinct from a physical effect on the external deposit, is not established as a universal trend. Nor is increase of translocation with increase of humidity universal. The sequential processes—permeation, local distribution in tissue, long-distance translocation—are not easily separated, and there is some evidence of dependence on history of humidity as well as on the immediate value. Maximal effect is probable when the plant is water deficient and the treated leaves are wet. There can be little doubt that the physical effect on the external deposit is important.

One would, at first, expect that the effect of external humidity on availability from a crystalline deposit would fall to a steady near-zero value when the humidity is so low that the deposit is completely dry. The limiting humidity would depend on solubility. For most organic salts the limit would be a rather high one. Saturated Na MCPA (25%) will be in equilibrium with about 96% r.h. and even ammonium sulphate equilibrates at about 80% r.h. but at least two findings in the literature quoted show a continuing effect of humidity down to less than 40%.

The internal leaf tissue is, however, in equilibrium with humidity above 98%. If an impervious platelet pressed on to a leaf is composed of substance deliquescent at less than this humidity, there could be liquid contact between a central area and cuticle surface. The extent of this wet contact will depend on the geometry of the cuticle plus deposit and on the presence of compounds exuding from the leaf, impurities in the formulation or deliberate water soluble additives.

An open network of needles on a rough leaf surface could fail to reach the critical humidity. Vapour transfer could, however, make a continued contribution except with compounds of very low volatility. The distance from crystal surfaces to cuticle will usually be less than one thousandth of the thickness of the stagnant air layer restricting outward evaporation. If the a.i. does not crystallize out—e.g. from deposits of triethanolamine salts of "hormone" acids—the rate of uptake might increase so far that there would be no reduction in uptake at low humidity. The evidence seems against this and reduction of water in the cutin itself may set a limit to the rate.

D. Effects of Temperature and Humidity on Uptake by Roots

The influence of temperature on the uptake of pesticides by plant roots is of interest from two standpoints: first as an important environmental factor which affects the practical performance of soil-applied herbicides and systemic pesticides, and secondly because knowledge of temperature effects can provide useful evidence about mechanisms of uptake. As has already been emphasized the effects of environmental factors are closely interrelated and in practice the influence of temperature on root uptake would inevitably be associated with corresponding effects of light, temperature and humidity on the aerial parts of the plant. Experimentally, however, much information has come from studying temperature effects on root uptake in isolation.

In considering these effects, it is helpful to recall some general features of root uptake, discussed in Section 11.VII. In summary, uptake of unionized pesticides commonly comprises a rapid initial phase of up to about 2 h followed by a longer second phase of slower, steady uptake; absorption does not generally appear to be an active metabolic process. Some auxin herbicides, DNOC (Bruinsma, 1967), and other herbicides under certain conditions (Moody *et al.*, 1970) show a different pattern of uptake: after an initial rapid absorption the rate of uptake declines to zero and eventually becomes negative as herbicide is lost from the root. There is also evidence that auxin herbicides may be accumulated actively.

Effects of temperature on uptake can be related to this general picture. Many investigations have shown that uptake by roots increases with increasing temperature. Experiments with excised roots demonstrate that accumulation by the root tissues themselves is greater at higher temperatures, the precise effect varying with the compound concerned. Thus Moody *et al.* (1970) found two patterns of uptake with excised soybean roots in culture solution: with chlorpropham and EPTC, the rate of uptake fell slowly after the initial fast phase lasting about 1 h, whereas the rate of atrazine absorption fell steadily leading to a loss of herbicide from the roots after 8 to 24 h. With all these compounds patterns of uptake remained essentially the same as temperature was increased from 5 to 30°C, but amounts absorbed were greater. Linuron, however, showed behaviour similar to that of chlorpropham and EPTC at 5, 10 and 20°C, but a pattern of uptake like atrazine at 30°C. Similarly amiben uptake continued steadily at lower temperatures but results at high temperatures were confused. In all cases the concentrations in the roots after 1 h uptake exceeded those in the external solution, and accumulation ratios increased with temperature. The highest Q_{10} value observed was 1·83, indicating that physical mechanisms were involved. Comparisons of short-term uptake by barley roots at 20 and 1°C by Shone and Wood (1973) also suggest that the initial phase of simazine, diuron and ethirimol accumulation is by physical processes, although absorption of 2,4-D may be influenced by metabolism. Bruinsma (1967) concluded from temperature studies that DNOC uptake by excised rye roots was not active, although subsequent excretion, which was markedly temperature dependent, was possibly driven by metabolic energy from inner parts of the tissue not affected by DNOC which is known to disrupt metabolic processes.

Many additional complications are introduced when uptake by whole plants is considered. Most obvious is the effect of transpiration. It is well established that rates of entry of water and its transfer through plant roots increase with increasing temperature (see, for example Kramer, 1940; Kuiper, 1964, 1972; Brouwer, 1965; Slatyer, 1960, 1967; Anderson and Reilly, 1968; Ginsburg and Ginzburg, 1971). The influence of temperature varies with the temperature range considered. The effects on water entry could be caused either by changes in permeability resulting from alterations in membrane structure or viscosity of cytoplasm or changes in the driving force for uptake, perhaps resulting from changes in the osmotic effects of active salt accumulation. It is difficult to distinguish between these two factors but Ginsburg and Ginzburg (1971) devised a system for measuring THO flow through maize root cortex at temperatures from 5 to 40°C under conditions of controlled driving force. There was little effect on THO flux up to 10–12°C but influx increased sharply between this temperature and 30–32°C after which the flux decreased slightly with further increase in temperature. A generally similar pattern of temperature dependence was observed by Anderson and Reilly (1968). Since the specific activity gradient for THO across the root cortex was constant in the experiments, Ginsburg and Ginzberg concluded that the effects of temperature were attributable to changes in permeability. They also suggested that the step-wise increase of flux with temperature indicated more than a change in physical properties such as cytoplasmic viscosity, unless a phase change was involved. Mean activation energies over the range 20–30°C were of the order of 20 kcal mol^{-1} possibly indicating that THO transfer was not by free diffusion through an aqueous medium. The results therefore supported other evidence that water moves through a cellular pathway, the rate of movement being limited by cytoplasmic membranes.

With whole plants, temperature will also affect transpiration rates by influencing loss of water from the foliage; these effects will be reflected in turn in the uptake and translocation of pesticides, although accumulation will not necessarily be directly proportional to amounts of water transpired. The exact effects of temperature on amounts absorbed and distribution within the plant will depend on the balance between the effects on the various component processes. Vostral *et al.* (1970), working with soybean seedlings in solution culture, found that both uptake of atrazine and water loss from the plants increased with temperature between 15 and 35°C, but the atrazine concentration in the roots was similar at all temperatures, suggesting that the major effect was on translocation. Similarly Penner (1971) found that raising the temperature from 20 to 30°C increased uptake and phytotoxicity of radiolabelled linuron and atrazine (although not phytotoxicity of trifluralin) to corn and soybeans in solution culture, but there was less accumulation of radioactivity in soybean roots from the atrazine treatment and in corn and soybean roots from the linuron and trifluralin treatments at 30°C compared with 20°C. Koren and Ashton (1973) reported that uptake of pyrazon by sugar beet seedlings followed the general pattern of two-stage absorption at 18·3 and 35°C, with amounts in the roots remaining effectively constant during the slower second stage. Both the level in the roots and the rate of translocation to the shoots were approximately doubled at the higher temperature; since the root concentrations reached a steady level at each temperature, the authors concluded that translocation rates, rather than absorption by the roots limited amounts of herbicide reaching the shoots.

Comparison of results obtained with whole plants by Moody *et al.* (1970) with the observations on excised roots discussed above provides further evidence about the differential effects of temperature on accumulation by root tissues and translocation. In these experiments, amounts of atrazine, chlorpropham and linuron in the roots of intact plants increased little between 5 and 20°C but rose markedly when the temperature was increased to 30°C. Uptake by roots of intact plants was generally somewhat greater than that by excised roots at 5 and 10°, although excised roots accumulated more at 30°C. Since transpiration is itself affected by factors other than temperature which are often not specified, results obtained by different workers are difficult to compare and interpret. The effects of relative humidity are demonstrated by measurements of simazine uptake by Sheets (1961). At a temperature of 37°C, the amounts which accumulated in leaves of oat and cotton seedlings were greater at 33% r.h. than at 66% r.h. but amounts in the roots of both species increased at the higher relative humidity, presumably reflecting a decreased transpiration. Wax and Behrens (1965) attempted to exclude transpiration as a variable when investigating the effects of temperature on ^{14}C-atrazine absorption and translocation, by maintaining the relative humidity constant in nutrient culture experiments with quackgrass. The amount of radioactivity in the leaves more than doubled when the temperature was increased from 17 to 24·5°C suggesting that factors other than transpiration can have important effects. The authors point out, however, that air movements were faster in the chambers at the higher temperature so that effects of transpiration may not have been eliminated entirely. Other evidence that increased transpiration is not the only cause of the greater translocation at higher temperatures comes from the work of Minshall (1969) who examined exudation from stumps of tomato plants grown in pots and cut 5 cm above the soil level. The soil was saturated with water after solutions of triazine herbicides and various nitrogeneous fertilizers had been added. Increasing the soil temperature over the range 10 to 30°C increased the rate of exudation. The concentration of atrazine in the exudate remained more or less unchanged in the absence of added fertilizer but applying potassium nitrate or urea increased the herbicide concentration, the effect becoming greater as the temperature was raised. Minshall suggested that the results could be explained by an increase in the osmotic pressure of the xylem fluids by solutes accumulated actively, which then caused more water to be drawn in together with associated solutes, including atrazine.

Interpretation of temperature effects is further complicated because chemical and biochemical activity, both that responsible for toxic effects and metabolism leading to degradation, are strongly temperature-dependent. For those compounds which are actively accumulated, effects on metabolism will include, of course, the uptake process itself which is therefore likely to be more sensitive to temperature than if physical processes are involved: this has traditionally been used as one diagnostic for active processes, although it should not by itself be regarded as definitive (Section 9.II.E). Thus Prasad and Blackman (1965) found that whereas the Q_{10} for the initial rapid phase of dalapon uptake by *Lemma minor* was 1·42 over the temperature range 17·5 to 27·5°C, that for the second phase was 2·38 suggesting that metabolism was involved. The second phase was also inhibited by metabolic inhibitors and the relationship of uptake to concentration was curvilinear. Other investigations have given equivocal results. The Q_{10} for uptake of picloram by oats was found by Isensee *et al.* (1971) to be 1·3, indicating passive uptake, but was 1·7 for soybeans which the authors considered could reflect a combination of active and passive processes. With some exceptions metabolic inhibitors decreased uptake into roots but effects on translocation were confirmed with some stimulation at low inhibitor concentrations.

Overall, the considerable body of experimental information on temperature effects allows few generalizations beyond the statement that uptake is likely to increase significantly with temperature in the range of practical interest and that in many situations translocation may be affected more than accumulation by the root tissues so that accumulation in the shoots may increase proportionally more than amounts in the root.

X. EFFECT OF ADDITIVES ON PENETRATION

A. Introduction

There is much evidence that uptake of many systemic pesticides and herbicides can be increased when they are applied together with certain additives. This is perhaps the clearest proof that permeation of the cuticle is of practical importance because it is only in the first stages of penetration that most additives are likely to have their effect. Once they are extensively diluted in tissue fluids, assuming that they can penetrate, they are unlikely, unless themselves physiologically active, to alter the behaviour of the active ingredient. More evidence is available for herbicides than for other pesticides because the herbicide must be systemic to be effective while many insecticides and fungicides exert protectant action only on the outer surface of the leaf. Also, with herbicides, one has the direct evidence of increased phytotoxic effect which is easily established and is usually, and probably correctly, assumed to be due to increased penetration.

Much of the literature of formulation additives is unsatisfactory. Protection of genuine advances by patent is difficult and much work in industrial laboratories remains unpublished. Commercial advantage is greatest with commodity active ingredients where sales competition is also keenest. Kanellopoulos (1974) has attempted a summary of the more reputable literature of additives to herbicides but has been unable to avoid reference to several code number proprietary products. Only additives of disclosed composition having effect on penetration will be considered in this chapter.

The additives can be classed as humectants, other auxiliary solvents, surfactants and a miscellaneous group comprising urea and several salts, not all nutrient. Sticking agents are frequently included in formulations but although they can reduce loss by wind and rain they are likely to have no useful effect on uptake except that which they may possess by being auxiliary solvents. We begin with humectants which naturally follow on the last section.

Richardson (1977) has reviewed the literature on factors affecting uptake of 2,4-D.

B. Humectants

Humectant is the name commonly used for a substance which can take up water from an under-saturated atmosphere. It can therefore cause a deposit to remain partly liquid in an atmosphere in which it would otherwise become completely solid. This may be of advantage where a liquid medium is necessary to provide a diffusion path between crystals and leaf surface. The humectant cannot, however, increase the thermodynamic potential of water. It cannot provide water from the atmosphere to an imbibing organ if the available water vapour pressure does not reach the limiting demands of that organ. If, for example, a soil is too dry to permit the swelling of certain seeds we cannot promote germination by surrounding the seeds with calcium chloride which will liquefy around them. The water may be even more difficult to extract from the calcium chloride solution than from the original soil. The humectant can only maintain a liquid phase in which diffusion can occur but the thermodynamic gradient must exist, as it will from solid a.i. to tissue fluids, if the liquid is to assist penetration. Other auxiliary solvents could provide the liquid and are used for this purpose but the humectant has the advantage of expanding its bulk by dilution with water from the atmosphere and in the case of saline pesticides it is usually a better solvent than many less polar ones.

Humectants frequently used have been glycerol, molasses and the lower molecular weight polymers of ethylene oxide. Molasses is readily and cheaply available in the sugar-producing countries where it has been chosen for that reason. It has several times been suggested that a sucrose-containing additive would have the further advantage that, since the active transport system of the plant is designed to carry sucrose, external application could accelerate transport. There is no clear evidence of this. The beneficial effect of active light on uptake of pesticides may well be due to stimulation of transport by endogenous sucrose, but sucrose does not readily pass through the leaf cuticle. Remarkably little sucrose is extracted from the abundant internal supply by washing of healthy leaves with water. Internal sugar is lost to rain appreciably only from over-ripe soft fruit or from dried leaves (e.g. from hay, not fresh grass) in which the protecting structures have been irreversibly damaged. Sugar is, of course, frequently deposited on the outside of leaves by excretion of aphids. The suitability of this honeydew as a culture fluid for fungi contra-indicates the use of sugar as an additive.

Glycerol has been most frequently used as a humectant in water-based sprays, but the clearest example of the increased contact effect was provided by Marth *et al.* (1945) in the case of dusts. These were made by incorporating free acid 2,4-D into talc by evaporation of alcoholic solution. Glycerol, carbowax® 1500 (polyethylene glycol *c.* $HO(CH_2CH_2O)_{34}$-(CH_2CH_2OH) and calcium chloride were also incorporated in some batches. The first two gave a marked enhancement in toxic effect on *Ipomea lacunosa.* Even calcium chloride, which might hold the a.i. as the even less soluble calcium salt, showed some enhancement. Where a spray drop leaves a crystalline residue by evaporation one can expect the crystals to accommodate to the geometry of the leaf surface more than will the preformed particles in a dust and it is not surprising that addition of gycerol makes less difference to the effect of an aqueous spray. Hopp and Linder (1946) found some increase when glycerol was added to a sodium arsenate spray. Humectants have generally less influence on the activity of organic toxicants, which have more possibility of direct molecular exchange with the cutin. Ethylene and propylene glycols were shown by Crafts *et al.* (1958) and Hull and Morton (1971) to give enhancement at low humidity. The polymeric forms have shown enhancement according to Ennis and Boyd (1946), Hughes (1968) and Pereira *et al.* (1971). The latter authors also examined sucrose and Fogg (1948) molasses. Holly (1956) found that glycerol in aqueous MCPA sprays gave some enhancement of uptake (amount not washed off) but that the highest uptake from deposits on undamaged leaves (amine salt plus glycerol) was still so slow that most of the a.i. was wasted.

C. Other Auxiliary Solvents

Whether a humectant will increase penetration more than can be accounted for by maintenance of diffusible fluid in contact with the sprayed area of the leaf depends on a balance of more complex factors. The general principles of auxiliary solvent effect on permeation have already been described (Section 9.V). The residual solvent could be a better one for the a.i. and therefore assist penetration if diffusive access to the leaf surface, from crystals of very low solubility, is a limiting factor. It could also, if it penetrates, facilitate diffusion of the a.i. through the cutin. Penetration could best be assisted by a good general solvent of small molecular size miscible with water but more compatible than water with the lipoidal cutin. Such would be the lower alcohols and acetone but these evaporate preferentially from water and whether they could usefully assist penetration before they evaporate is a matter for experiment. Norris (1972) has reported that isopropanol assists penetration as effectively as chloroform which is usually assumed to do so by removing waxes. These observations should be followed further. Isopropanol is very damaging to the leaf tissue locally and might therefore be restricted to application of herbicide for destruction of vegetation on industrial sites. It is also possible that the local tissue damage would inhibit translocation. Since the residual solvent composition at any time would depend on external humidity the behaviour might be complicated but also the complication might be exploitable.

Dimethyl sulphoxide (CH_3SOCH_3, DMSO) has received much attention stimulated by its effectiveness in human medicine to accelerate penetration through the skin of some pain-relieving drugs and of its action *per se* in some rheumatic complaints. After a brief period of magic mechanisms and extravagant claims, what remains valid and supported by documented research is what would be expected from an auxiliary solvent of small molecular size but less volatile than the lower alcohols. The molecule has a high dipole moment, the SO group is a good H-bond acceptor but there is no donor component. The molecule is therefore more attracted to water, and some other, molecules than to itself. Shellhorn and Hull (1971) found a very marked increase in the effectiveness of 2,4,5-T and picloram, but the spray they used contained 500 times as much DMSO as a.i. Olinger and Kerr (1969) found no assistance to the uptake of systemic insecticides either from nutrient solution or from cotyledon application although, as a solvent for topical application of carbaryl to beetle larvae, DMSO was better than acetone. Negative results were obtained by Pitre *et al.* (1972) in tests of DMSO as an adjuvant to insecticidal sprays. In studies with ^{14}C-labelled dicamba, paraquat and atrazine, Jones and Foy (1972) found little effect on the first but increased uptake and consequent acropetal movement of the others.

The polymerized ethylene and propylene glycols which have been mentioned as humectants could have auxiliary solvent effect of a less strongly polar type. They have OH groups at the ends of the chain molecules and the numerous ether groups, while maintaining water solubility indefinitely in the ethylene series and up to about five monomers in the chain in the propylene series, leave a residue at low humidity which has substantial solvent power for many organic substances of very low solubility in water. Glycerol, by contrast, is substantially aqueous in its solvent behaviour: indeed, in some respects "more aqueous" than water, acetone, e.g. being incompletely miscible. The polymers most used as additives have been in the 20–30 monomer (900–2000 mol. wt) range. Lower molecular weights, e.g. di- and tri-"ethylene glycol" ($HOC_2H_4OC_2H_4OH$ and $HOC_2H_4OC_2H_4OC_2H_4OH$) could well have received more attention.

The ethylene oxide surfactants so widely used in formulations can all function as residual solvents of this type. They are indeed very good general organic solvents, the ether-alcohol type solvent action and the paraffinic solvent action being able to function additively by the "solubilization" mechanism discussed in Section 2.IX.C. While, in dilute solution, they make their wetting-type contribution to the effectiveness of sprays, discussed below, they can also play a role as residual solvents which should be kept distinct in thought and planning.

If the lower alcohols are used as adjuvants to aqueous sprays they will under all ordinary circumstances have evaporated before the water and cannot be called residual solvents. According to the data considered in Section 2.V, the tendency to evaporate preferentially from aqueous solution increases with the length of the hydrocarbon chain, but the word *solution* must be emphasized. Thus 1% in water is reduced to 0·1% more rapidly, by free evaporation, as we ascend the series from methanol to hexanol, but if we attempt to apply this principle to, say, decanol it would be valid only below its very low solubility. One per cent would be an emulsion and the water would, in dry air, evaporate first. Preferential volatility even applies to the "cellosolve" series, $R–O–CH_2–CH_2–OH$, where R is an alkyl group. They are more soluble than the corresponding alcohols but disappear from aqueous solution during evaporation into air of ordinary humidity before the water itself has evaporated completely. The dihydroxy ethers such as "diethylene glycol"—$HOC_2H_4OC_2H_4OH$— and higher members of the series are not preferentially volatile and are left as residual solvents. The commonly used ethylene oxide surfactants are of abnormally high water solubility because of their aggregation in solution and are of very low volatility. They leave a solvent residue but it is possible, e.g. with the octyl member of the cellosolve series, octyloxyethanol, to have very good wetting action initially and leave no wetting residue after evaporation.

Important auxiliary solvents of a quite different class are the so-called non-phytotoxic or "phytobland" oils. They are virtually water-insoluble and must therefore be applied as emulsions if water is used as the main carrier. They were introduced in an effort to improve the effectiveness of sparingly soluble crystalline herbicides, principally atrazine, by foliar application. Their function is unquestionably to facilitate diffusion from the low-soluble crystals to the leaf surface. They are used rather than aromatic (benzenoid) oils, which are much better solvents, to avoid the local tissue-damaging effect of the latter which inhibits acceptance of the herbicides into the long-distance transport system.

Jansen (1966) showed that there was considerable variation between species in the enhancement resulting from adding oils to herbicide sprays and Burr and Warren (1971) using "isoparaffins" found a 16-fold enhancement of the effect of atrazine on *Ipomea hederifolia* but none on *Cyperus rotundus* or *Agropyron repens.* Isoparaffins are saturated aliphatic hydrocarbons with the C atoms not all in a straight chain: they are usually liquid even where the straight chain compound of the same empirical formula would be crystalline. Enhancement of dinoseb and linuron activity was greater when the isoparaffin was used as a carrier without water compared with the same rate applied as an emulsion. These authors confirm the finding of Banden (1969) that the oil can enhance the effect of atrazine even when applied up to 6 days earlier. This is interpreted as a plasticizing effect on the cuticle but it is not clearly established whether free oil is still present on the surface. Prendeville and Warren (1975) investigated the effect of oils of different viscosity and boiling range. ^{14}C-labelled dinoseb, terbacil, chlorpropham and 2,4-D were examined on *Phaseolus*, Cucumber, Morning Glory and Sorghum. They confirmed that the major effect of treating adaxial and abaxial surfaces indicated that stomatal penetration was not important. Spread of the applied oils on the leaf surface was mainly along the vein grooves and was arrested by vein ridges on the abaxial surface. Distribution of ^{14}C following uptake from oil drops allowed to spread was compared with that from drops confined by lanoline rings and showed that the spread assisted penetration and therefore allowed more compound to be translocated but did not alter the relative proportions in other organs. In these experiments the treated leaf was fixed in horizontal posture. Superficial spread along the petiole of bean contributed to local uptake. The authors refer to the suggestion of van Overbeek and Blondeau (1953) that external drainage of oil down the upright blade of grasses could assist the herbicide to gain access to sensitive tissue, but the experiments reported do not throw further light on this important possibility (Sections 8.V.C, 8.VIII). It is rather general experience that when a herbicide can be used in either oil or water formulation, the former is relatively more toxic to grasses.

D. Surfactants

Surfactants are added to many pesticide formulations to facilitate handling and/or application—to make a powder wettable so that it can be quickly dispersed in water for spraying, to make an oil solution emulsifiable for the same purpose, to reduce coagulation of the dispersions in the spray tank so that settlement which would lead to uneven applicance is avoided, or to eliminate reflection of water drops from extremely unwettable leaves. All these effects are directly related to surface activity. Some enhancement of penetration also results from increased spread of the drop on the target area. There is also some further enhancement, or occasionally impairment, of penetration which is probably not directly related to surface activity—if the surfactant is, at least when in contact with atmospheric water, a liquid, it can have auxiliary solvent action of the type discussed in the last section.

Several reviews of surfactant effects on pesticide performance have appeared (van Valkenberg, 1967; Foy and Smith, 1969; Hull, 1970, pp. 85–94; Kanellopoulos, 1974) and no attempt will be made here to give a complete list of original references. Little will be said at this point, because the effects dependent on surface activity itself are dealt with more appropriately, elsewhere—spreading is considered more fully from the physico-chemical aspect in Section 8.V, solubilization in Section 2.IX.C, formulation uses in the final chapter and auxiliary solvent effects in the preceding section. We shall therefore here draw attention only to some aspects of behaviour of commercial surfactants which tend to be overlooked.

Spreading of a series of pure liquids on a given inert surface is limited by the advancing contact angle which reaches zero at a critical surface tension (Section 8.III). The extent of spread under a given set of practical conditions depends also on evaporation of the liquid and, if it is water-soluble, uptake of atmospheric moisture. When the surface tension of the pure liquid is high (e.g. water) but reduced by the presence of a surfactant (usually at less than 0·1%), the relationship is never so simple, even on a smooth, inert surface (and leaf surfaces are, strictly, neither smooth nor inert). There are at least two complicating factors in the case of a dilute aqueous surfactant solution. (1) The surfactant is adsorbed on the hydrophobic material over which it assists the water to spread. Eventually, the spread would be limited by exhaustion of the solution, if other factors did not intervene. (2) Water, by diffusion in the gas or liquid phase, is more mobile than the surfactant but cannot advance over a hydrophobic surface until the surfactant has gone before. There need be no surprise, therefore, and no need for a specific biological mechanism, when it is found that increasing the concentration of surfactant, beyond that necessary to give minimum surface tension, continues to increase the spread of drops, particularly if they are subject to evaporation.

Several surfactants not only form highly viscous solutions when concentrated but their flow may be qualitatively anomalous. A gelatinous structure which the small shearing forces operating in a spread drop cannot overcome is common and such structure may actually be generated by very small shearing stress.

In theory it would be possible for a surface deposit to spread so thinly that penetration below the non-living cuticle would be actually reduced. It is easy to construct a physical model to show this but we know of no evidence for the effect being realized for entry of pesticides.

An ionic surfactant must always be associated with another (oppositely charged) species of ion. Interaction effects may sometimes involve not the surfactant ion itself but other accompanying ions. If 2,4-D, which has a sparingly-soluble sodium salt, is supplied as an amine salt concentrate and mixed with an anionic surfactant supplied as the sodium salt, we may expect a decrease of activity which would have nothing to do with the surfactant anion. A few anionic surfactants are supplied as the calcium salt which would depress solubility of the biologically active herbicide anion even more. The cationic bipyridilium herbicides could form a sparingly soluble salt with the surface active anion, a possibility examined by Smith and Foy (1966a) without revealing any clear inhibiting effect.

The surfactant may itself penetrate the cuticle. In doing so it may assist the penetration of the pesticide but may also have its own physiological effect. This effect is likely to be harmful to the plant and probably accounts for the existence of the optimum concentration of surfactants demonstrated by Evans and Eckert (1965) for sprays of paraquat. There seems to have been only one tracer investigation of the penetration and translocation of a surfactant itself, that by Smith and Foy (1966b) using Tween-20, a fatty acid ester of sorbitol condensed with ethylene oxide.

Two preparations were examined, one labelled with ^{14}C in the fatty acid, the other in the ethylene oxide. There was very little movement in either case from the treated part of the leaf and some evidence that the fatty acid was separated, retained more strongly in the treated spot but, in small amount, very generally distributed.

The structure of a surfactant molecule, dictated by its function of adsorbability at interfaces between water and less polar phases (air, oil or non-polar solids), would lead one to expect low mobility in complex tissues. The highly adsorbed surfactant would not easily be translocated but would be likely to cause local membrane damage. Brian (1972) found that ethylene oxide surfactants reduced the translocation of locally applied paraquat. His attached-reservoir technique eliminated effects of wetting and contact area. The chemically estimated movement was found to parallel biological activity. Some amine oxide surfactants did not show this inhibition of translocation.

E. Nutrients and Salt Additives

There have been several reports of increased herbicidal effect when ammonium salts, phosphates or urea have been added to pesticide sprays. The field evidence is conflicting. Some effects are probably attributable to change of nutritional status; these will be dependent on prior conditions in the crop. They are also outside the scope of this book. Recently several cases have been reported where the effect of foliar applied herbicides has been clearly enhanced in a manner which strongly suggests that the additives increase permeation of the cuticle in some way. It is not possible at present to generalize. In different combinations of plant species and chemical, different mechanisms, sometimes in conflict, may operate.

It is generally agreed that anionic herbicides such as 2,4-D, dinoseb and picloram permeate the cuticle more easily in the free acid, undissociated form than as salts, but the low water-solubility of the acids limits the practical advantage which can be taken of this. If the acid is neutralized by a volatile amine, any excess of which will evaporate preferentially from an exposed spray drop, a higher concentration of free acid develops than if a sodium or potassium salt is applied. If the acid can be removed from the drying deposit by permeation of the cuticle, more amine is volatilized and the permeation of free acid continues. For this reason ammonium salts are more effective than sodium or potassium salts (Crafts and Reiber, 1945). If the amine is more volatile *from* water solution, as is dimethylamine and, especially, trimethylamine, greater advantage may be expected, but will be limited again by solubility of the acid. Smith *et al.* (1959) showed that aqueous solutions of volatile amine salts of maleic hydrazide leave a residue of crystalline acid: sodium or potassium salts leave residues of the salt crystals while triethanolamine salts leave a syrupy non-crystalline residue. The last gives greatest penetration.

Turner and Loader (1972) showed that ammonium salts enhance permeation when added to ordinary alkali formulations of MCPA, mecoprop and picloram. The mechanism could be the same as that just discussed: with an excess of the added ammonium salt, the sodium or potassium in the ordinary formulation would not be important. The volatilization of base will cause acidification and, if an acid compound can also be removed (through the cuticle) the process will continue. However, the enhancement of activity by adding ammonium salts is found to be greater than can be expected from decrease of pH. This is particularly evident in the work by Wilson and Nishimoto (1975b) on penetration of ^{14}C picloram into detached guava leaves, following up (1975a) observation of enhanced activity in the field and a review of previous literature. Standard picloram formulation was examined with the addition, separately, of mono- and di-hydrogen phosphates of ammonium and potassium. Although this revealed a pH effect in the expected direction, the effect of substituting NH_4^+ for K^+ was greater than would be predicted for this alone. The authors found that drops at pH greater than 4 became more acidic on exposure so that independent investigation of pH effects was not possible by this method. However, the advantage of NH_4^+ over K^+ was established by the method of Sargent and Blackman (1962) where no significant evaporation occurred. When pH was changed from 4 to 7 by HCl or NaOH addition there was no effect of pH but a seven- to ten-fold enhancement by adding 0·5% ammonium sulphate. This is apparently inconsistent with the effects of phosphate which may therefore involve a phosphate effect. The enhancement of uptake with ammonium sulphate was evident after the shortest exposure time of 1 h. Uptake was measured by combusting the washed leaves. No other sulphate enhanced uptake. Ammonium nitrate and chloride, as well as sulphate and phosphate, but not carbonate (completely volatile) or molybdate, enhanced the effect of picloram.

Turner and Loader (1972) found that several herbicides could be activated by phosphate esters as well as by ammonium salts. This followed an investigation by Turner (1972), particularly with *S,S,S*-tributyl phosphorotrithioate (DEF), which is used as a defoliant in cotton crops and was shown to cause localized swelling of epidermal cells leading to bursting. Further work (Turner and Loader, 1974) concentrated on activation by commercial mixed butyl acid phosphates (BAP). Experiments were done on *Phaseolus* seedlings and one- to two-year-old seedlings or rooted cuttings of several woody species. The effect was assessed by epinastic response where suitable and by measurement of further growth after treatment. The ester additives were not effective with ester formulations of the herbicides. The work was mainly directed to improving water soluble herbicides, such as amitrole and glyphosate, which do not form esters. The activity of bipyridilium herbicides was not enhanced by either ammonium salts or phosphate esters. Generally, BAP was more effective when neutralized with ammonia than with soda, but this did not apply to amitrole for which only tributyl phosphate or sodium BAP was effective.

The greatest effect reported by Turner and Loader (1975) was when BAP in acid form was added to glyphosate applied to guava. Ammonium sulphate also increased the effect of this herbicide on guava but not as much as was found by Wilson and Nishimoto in the case of picloram. Turner and Loader found that ammonium sulphate enhanced activity of glyphosate on *Phaseolus* greatly and this test included urea which had comparable effect to ammonium sulphate.

Ammonium thiocyanate is an additive which must be regarded as a special case. It rapidly causes local cell damage and could be expected therefore to facilitate the first stage of entry by the mechanism proposed by Turner (1972) for DEF. At low concentrations it has been found to enhance the effect of hormone herbicides (e.g. Elwell, 1968). When added to amitrole it appears to increase translocation but there is an optimum concentration (Donalley and Ries, 1964) beyond which the effect is possibly inhibited by excessive local damage. Forde (1966), however, found that the translocation patterns of amitrole and thiocyanate are not the same. There remains considerable doubt about the mechanism of the amitrole thiocyanate synergism and it is likely that biochemical cooperation (van der Zweep, 1965) which lies outside the scope of this text, may be involved.

We have given preference in this brief discussion to effects on uptake which seem to be well established and confirmed. The very extensive literature on this subject contains many cases of effects reported and subsequently contradicted—in at least one instance by the same author. It is tempting to adapt the cynical remark of the classical physicist in the early days of the quantum theory—"Light behaves as waves on Tuesday and particles on Thursday". A very interesting observation by Allen (1970) makes the parallel by no means facetious and should be followed up. Examining the uptake of magnesium into apple leaves he found that, even using the deliquescent chloride, uptake ceased after a time when there was still excess outside on the mainly dry cuticle. After re-dipping the leaf there was a further period of uptake almost as rapid as the first. After a longer period of rest following one application, uptake could be resumed. The behaviour can only be explained on the assumption that only some areas of the leaf are adapted at any one time to take up magnesium ions and that the pattern of receptive areas changes over a longer period.

12. Penetration of Pesticides into Insects and Fungi

I. Introduction 658
II. Penetration into insects 659
A. Composition of the insect cuticle in relation to penetration . . . 659
B. General features of the penetration process 662
C. Effects of solvent 669
D. Effects of penetrant properties 672
E. Uptake of insecticide vapour 676
III. Penetration into fungi 679
A. Barriers to penetration 679
B. Routes to absorbing surfaces in fungi 681
C. Uptake of fungicides by fungi 684
D. The detailed pattern of fungicide accumulation 686

I. INTRODUCTION

The preceding three chapters are concerned with the general principles of permeation and with entry into plants. We turn now to consider the other two major targets for pesticides, insects (with which we include related invertebrate pests such as acarines) and pathogenic fungi. In keeping with the terms of reference we have defined, we concentrate on initial penetration rather than subsequent fate and action within the receiving organism. However, strict demarcation is artificial and undesirable and in re-stating this emphasis we should also recall that detoxification and redistribution within the organism have an important influence on initial penetration through their effects on concentration gradients. Furthermore, as we discussed in Sections 4.I, 6.V and 11.I, substances applied to the outside of an organism can be lost by decomposition and by evaporation as well as by passage to the interior. The observations by Devonshire and Needham (1974) that up to 65% of the dose of organophosphorus insecticides applied topically to aphids can evaporate within 4 h of treatment provide a further powerful reminder of the importance of these effects.

The definition of our field of interest in this book also implies that we devote rather less attention to insects and fungi than to plants, for reasons already presented in introducing Chapter 11. A further reason for less extensive treatment in this chapter is that many of the general physico-chemical principles which govern uptake by plants also apply to other organisms and have therefore been considered. There are, however, some important and obvious differences, not only in the composition of the outer barriers to penetration, but also in physiology and structural organization, which should be borne in mind when extrapolating from one group to another. These will become apparent in the more detailed discussion which follows, but at this stage we may note that the jointed nature of insect structures, the presence of specialized features such as sense organs and the generally greater mobility of insects tend to make their uptake of exogenous chemicals more complicated than that of plants, while uptake by fungi is likely to be simpler because of their much less complex structural organization.

II. PENETRATION INTO INSECTS

A. Composition of Insect Cuticle in Relation to Penetration

The vital physiological role of the insect integument (particularly for the water economy of the insect), interest in the phenomenon of moulting and the desire to optimize the performance of insecticides have resulted in extensive work on the cuticle which is summarized in several useful reviews (for example, Ebeling, 1964; Richards 1951, 1953; Wigglesworth 1957, 1965; Locke 1964; Neville 1967). In particular Noble–Nesbitt (1970) has considered structural aspects of penetration; the following description draws largely on his review.

Like plant cuticles, insect cuticles are complex heterogeneous structures which vary considerably within and between species. Typically the cuticle of insects is composed of three main layers: an inner endocuticle covered by an outer exocuticle which is in turn covered by the superficial epicuticle. In some insects the mesocuticle replaces the exocuticle or is present as an additional layer between the endo and exocuticles.

The *endocuticle*, which is composed mainly of protein and chitin (a polysaccharide consisting of *N*-acetylglucosamine units) is relatively soft and elastic and is the main component of the cuticle in larvae and soft-bodied insects. The chitin and protein are arranged as layers of microfibres with a constant orientation in the plane of each layer but with a regular change in the direction of orientation from one layer to the next. It is thought that the basic microfibre pattern is made up of chitin micelles to which the amino acids of the cuticular proteins are attached.

The endocuticle is not a constant structure, but can be added to or resorbed with changes in factors such as the nutritional status of the insect. During deposition it may be particularly difficult for substances to penetrate the cuticle, while more rapid penetration may occur during periods of starvation. These changes in the composition and thickness of the cuticle complicate the process of penetration and have the added implication that the cuticle cannot be regarded as inert and incapable of metabolizing penetrating chemicals.

The *exocuticle* is more important in hard bodied insects such as beetles, but may be very thin in many larvae and in the hard parts of soft-bodied insects. Unlike the endocuticle it has a homogeneous three-dimensional structure composed of chitin and quinone-tanned protein. The exocuticle tends to be less hydrated and less permeable than the endocuticle.

The structure of the *mesocuticle* is in some ways intermediate between those of the exocuticle and endocuticle. It has a weakly developed laminar structure but is less differentiated than the endocuticle and is stronger and probably less permeable.

The superficial *epicuticle* which can vary in thickness in different species from about 0·03 to 4 μm consists of an inner homogeneous layer and an outer cuticulin layer coated with "wax" and sometimes with a thin covering of cementing material on which there may be a further wax layer. The inner homogeneous region may contain chitin and can be considered as an extension of the exocuticle. The cuticulin layer which is itself multilaminate is about 15 nm thick and consists of tanned lipoprotein. The "wax" which covers the cuticulin varies in thickness and composition between species but generally contains alkanes in the range C_{25}–C_{31} and esters of saturated and unsaturated *n*-alcohols and acids in the range C_{26}–C_{30}; these are generally of a much more oily nature than the corresponding mixtures on plants and are therefore conveniently referred to as greases (see Section 11.II.B).

The cuticle is traversed by a network of fine-branched pores emanating from the epidermal cells and passing outwards to end at various levels in the epicuticle. These pores which vary between 10 and 100 nm diameter are probably extensions of the cytoplasm of the epidermal cells and are believed to provide a mechanism for continuously replenishing the cuticular wax. To the extent that they ramify through the cuticle, it may be said that the cuticle is living throughout and therefore capable of modifying materials passing through it.

As a barrier to the penetration of pesticides, the insect cuticle therefore has many similarities to the plant cuticle; both are multilaminate structures composed of both lipophilic and hydrophilic materials and both are traversed by specialized canals. Both vary considerably, particularly in the nature of the outer wax layer, from species to species and from place to place on any one organism.

The lipoidal epicuticle provides the main barrier to the passage of water and aqueous solutions, and this is probably its most important role in the physiology of the insect. Transpiration increases suddenly at a temperature which is thought to correspond to transition from the solid to the liquid crystal state where the paraffin chains become free to rotate and oscillate laterally. Removal of the wax layer by non-polar solvent greatly increases the transpiration rate. Generally the intact insect cuticle with its wax layer is less permeable to water than leaf cuticles, but without the wax layer, the insect cuticle is much more permeable.

The epicuticular pores may provide a route for the penetration of water and polar substances through the cuticle. Permeability generally increases when insects die and it has been suggested that this could be explained by the continuous production and outward passage of lipid micelles through the pores of living insects, which ceases when death occurs. The lipid micelles in the pores may provide a pathway for the penetration of lipophilic materials.

While this outline of cuticle structure has wide application as a general description, the detailed properties and hence characteristics as a barrier to pesticide penetration vary considerably not only from insect to insect, but even from location to location on the same insect. There is ample evidence that the rates of penetration differ according to the site of application on the insect body. For example, Ahmed and Gardiner (1968) found that penetration into locusts was much faster following application to the tegmen base than to the hind femur or tegmen middle, while Lewis (1965) showed that amounts of DDT absorbed through different sites on the blowfly differed considerably. Many other studies of relative toxicity following application to different parts of the cuticle also show that toxicity varies with site of application, although factors other than penetration such as proximity to the site of action may be involved (e.g. Klinger, 1936; Wiesmann, 1946, 1949; Lindquist *et al.*, 1951; Fisher, 1952; Gerolt, 1970). It is often suggested that penetration occurs more rapidly through thinner, non-sclerotized regions of the cuticle but as Wigglesworth (1942), Lewis (1965) and others have suggested, pore canals and other ducts may be much more important.

Failure to correlate the thickness of the integument at the locus of application with the speed of action of contact insecticides is one of the pieces of evidence which Gerolt (1969, 1970, 1972) uses to support his theory that such toxicants do not reach the site of action by penetration straight through the integument and circulation via the haemolymph, but follow a lateral route through the integument of the body wall and tracheae. Gerolt (1970) provides evidence that penetration through isolated integuments is extremely slow (although considerable quantities can accumulate in the integumental tissues themselves), and he implies that the rate of lateral movement may be very much faster. In later studies with a range of organochlorine, organophosphate and carbamate insecticides (Gerolt, 1975a), both the onset of symptoms in houseflies and penetration to the CNS were shown to be delayed by measures which reduced metabolic activity, namely depriving the insects of oxygen, or decreasing the temperature, indicating that the accumulation of insecticide involved active processes which require the expenditure of metabolic energy in living tissue. Gerolt speculates that the epidermis is the tissue likely to be primarily involved in this transport regardless of whether subsequent transport is accomplished by haemolymph circulation, as is commonly assumed, or by the integumental route which he favours. Further evidence against the haemolymph route came from investigations into the effects of ligaturing the "waist" of houseflies on the action of representative organochlorine, organophosphate, carbamate and pyrethroid insecticides (Gerolt, 1975b). This rather drastic treatment had little effect on speed of action, suggesting that the haemolymph is not important in the transport of these insecticides to their sites of action.

The interpretation of such results is uncertain (see, for example, Burt, P. E., in discussion of paper by Gerolt, 1970) and this remains an area of active controversy. However, at the very least the observations emphasize that there is still much to be learnt about the basic routes of penetration into insects, a conclusion to be kept in mind in any attempt to fit published work on penetration into a consistent framework, such as that which follows.

B. General Features of the Penetration Process

The general principles providing the background against which penetration into insects may be considered were derived in Chapter 9. We may recapitulate some of the essential points here as they apply to the subject we are now discussing.

The system with which we are concerned may be represented at its simplest as follows:

Solution of insecticide in solvent O	greasy outer layer ("wax")	cuticle	inner aqueous liquid

We designate the partition coefficient for the insecticide between the outer solvent and the grease as $\mathscr{P}_{gr}^{solv}$ in favour of the solvent. It is a key point that the toxicant is normally applied in a solvent as the solvent properties not only affect the ease of penetration, but may also determine the mechanism of penetration which occurs. This may seem a statement of the obvious, but the importance of solvent does not always seem to have been appreciated in comparisons of different investigations. The ways in which solvent can influence penetration, together with the operation of other important factors will be readily apparent from the following summary of the possible mechanisms of penetration, which are considered more fully in Section 9.V.

There are essentially three such mechanisms, which are more or less independent:

(1) If the pesticide is applied in a solvent which leaves the wax and cuticle unaffected and remains on the surface to provide a reservoir for uptake into the insect, then the rate of loss of penetrant from the surface reservoir will be approximately inversely proportional to $\mathscr{P}_{gr}^{solv}$. This is the thermodynamic potential effect discussed in Section 9.V. It is worth stressing that the relationship as stated applies to *loss from the surface*: some chemical is likely to be retained in the cuticle, and a proportion of both this and chemical penetrating further may be extracted by the rinsing procedures commonly employed to assess removal into the interior.

If the non-penetrating solvent evaporates rapidly to leave a residual deposit of pesticide, this will penetrate the grease and cuticle at a rate determined by its own properties alone, according to the general principles discussed in Section 9.V.

(2) If the wax layer is dissolved by the solvent used to carry the pesticide, then it is largely by-passed even if the solvent evaporates rapidly because it is then unlikely to reform in a coherent layer beneath the residue. The rate of entry is then determined by the degree to which the properties of the solute are appropriate for passage into and through the underlying cuticular layers.

(3) Solvents which can themselves enter the cuticular layers may modify permeability by causing swelling. Water-swollen tissues in particular may be more permeable as illustrated by the work of Scheuplein and

Blank (1971) who showed that hexanol, heptanol and octanol permeate detached stratum corneum more rapidly from saturated aqueous solution into pure water than when the pure alcohol is applied. We discuss this work more fully on p. 516 where we also point out that a water-dissolving but not freely water-soluble solvent on the outside of the organism could lower the water content in the outer layers of the septum so reducing permeability. This effect could, however, go in the other direction: the solvent could increase permeability by itself swelling the cuticle.

The nature of the solvent can have a further important influence on penetration. The rate of absorption depends on the area over which it occurs; this in turn will be influenced by the extent to which the applied solution makes contact with and spreads over the surface of the insect. These factors are determined by the physical properties of the solution and the nature of the surface according to the principles discussed in Sections 8.III, 8.IV and 8.V. Effective contact with the cuticle surface is clearly essential before penetration of dissolved toxicants can start, and adequate wetting can be difficult to achieve, especially for polar solvents because of the hydrophobic nature of the epicuticle and because of surface roughness. With suitable solvents spreading over the surface can be rapid, however (Lewis, 1962). This spreading is much faster with living than with dead insects, which Lewis attributed to microscopic movements of the cuticle which ceased at death.

The kinetics of uptake can be expected to reflect the mechanism operating in the particular case being considered. Again, we have considered the various possible time courses in Chapter 9. Real organisms differ in many important respects from ideal systems, however, and it will be clear from what has been said in this section that penetration through the insect cuticle can involve several processes which may occur simultaneously. Not surprisingly therefore the course of uptake may be difficult to interpret in some cases.

The most popular way of presenting results for the penetration of toxicants into insects is as a "semi-log plot" of the logarithm of the amount remaining on the outside (usually determined by a "surface rinsing" but sometimes by a more thorough extraction of the integument) against time on a linear scale. In some cases such plots have given a reasonable straight line; examples include measurements of the penetration of diazinon into the housefly (Forgash *et al.*, 1962), DDT into the cockroach (Lindquist and Dahm, 1956), various organophosphate insecticides and other compounds into the cockroach (Olson and O'Brien, 1963) and the synergist piperonyl butoxide into the cockroach (Schmidt and Dahm, 1956). In many other cases the rate of penetration as expressed in this way decreased rapidly in the early stages of the experiments before reaching a slower linear rate. This led the authors to distinguish two phases of uptake, the first much faster than the second. This pattern can be seen, for example, in the studies on penetration of malathion into the housefly and cockroach by Krueger and O'Brien (1959), diazinon into the housefly by Krueger *et al.* (1960), malathion into the cockroach by Matsumura (1963) and DDT, famphur and dimethoate into a range of species by Buerger and O'Brien (1965). Processes suggested to account for the first phase include initial carriage of the toxicant into the outer layers by the solvent (Buerger and O'Brien, 1965) or rapid "absorption" by components of the cuticle (Matsumura, 1963). This absorption presumably represents a rapid equilibration by diffusion with sorbing materials which are relatively accessible in the cuticle. Matsumura considered that these were probably proteinaceous and showed that if this protein were removed from the cuticle by extracting with hot water (it could not be with solvents which dissolved the wax), the remaining penetration followed eq. (1) below, reasonably closely. Elliott *et al.* (1970) distinguished three phases in the absorption of synthetic pyrethroids by mustard beetles which they interpreted in terms of an initial rapid diffusion into the insect, which slowed as the concentration gradient declined. Equilibrium was not reached, however, because insecticide was continually destroyed within the insect by metabolism. Consequently a more or less steady state was reached in which insecticide entered the insect at the rate needed to replace that lost by detoxification; this corresponded to the third linear stage of the semi-log plot. An essentially similar model, although not analysed kinetically in the same way, was suggested by Lewis (1965) to explain penetration of DDT into *Phormia terraenovae*. In this case passage through the layers just under the cuticle was considered to be the rate-determining process and the disposal of penetrated insecticide attributed to further redistribution rather than detoxification.

In considering these various interpretations, it is however important to examine the fundamental assumptions on which analysis by the semi-log plot, or by other kinetic relationships, is based. Where these assumptions are stated clearly (and this is by no means universally the case, it being taken for granted apparently that the simple linear semi-log relationship is an adequate diagnostic for diffusive uptake), they frequently rest on adoption of the equation for diffusive transfer from a supply reservoir (volume $\mathbf{V}_{don}$ concentration C_{don}) across a septum of permeance P into a receiving reservoir (volume $\mathbf{V}_{rec}$, concentration C_{rec}) (eq. (7), Section

9.I.B). The basic relationship between the rate of transfer of material ($d\mathbf{M}/dt$) and the concentration difference across the septum (area $\mathbf{A}$) is:

$$d\mathbf{M}/dt = P\mathbf{A}(C_{don} - C_{rec}) \tag{1}$$

The principle behind the semi-log approach and embodied in this equation is that the concentration difference ($C_{don} - C_{rec}$) falls exponentially as transfer occurs.

The solution to eq. (1) most commonly used in studies of uptake by insects (and fungi, see Section 12.III.C) is the limiting case (eq. (9) Section 9.I.B) which applies when either the volume of the supply can be regarded as effectively infinite or the concentration C_{don} effectively constant. This is usually expressed in the form:

$$\mathbf{M}_t = C_{don}\mathbf{V}_{rec}\{1 - \exp(-P\mathbf{A}t/V_{rec})\} \tag{2}$$

where $\mathbf{M}_t$ is the amount penetrating into the insect (volume V_{rec}) from the supply (concentration C_{don}) in time t and $\mathbf{A}$ is the area of cuticle exposed.

The conditions under which eq. (2) can be validly applied have not always been recognized, as we have already pointed out in relation to uptake by plants (Section 11.VI.E). The condition of constant C_{don} can be satisfied relatively easily in experiments with detached cuticles, such as those of Treherne (1957), by maintaining the external surface in contact with a large volume of stirred test solution. At first sight it appears doubtful, however, that it could apply when a finite amount of toxicant is applied to the outside of an insect, usually as a small drop. The requirement could be met, however, if the toxicant equilibrates rapidly within the epicuticle, or in a portion of the cuticle and subsequent movement through the remainder of the cuticle and the hypodermal cells is much slower. In this case the concentration in the outer equilibrated portion would remain effectively constant and the small amount reaching the haemolymph could be governed approximately by eq. (2). It is this type of mechanism which provides the basis of the two-stage models but it requires that the amounts extractable from the outer layers remain effectively constant. There is little evidence for this with any but the most non-polar toxicants: for example, in Matsumura's (1963) experiments, nearly one-third of the radiolabelled malathion applied to the outside of the cockroaches could no longer be extracted with an acetone rinse 1 h after application, while Olson and O'Brien (1963) found half-times of less than 1 h for penetration of all but the organochlorine insecticides in the group which they studied (see Section 12.II.C).

Where significant quantities of material are taken up from the surface, so that C_{don} cannot be assumed constant, its rate of decline (as measured by a true surface wash if this could be achieved) would itself be exponential if the internal concentration were so much smaller than the external concentration that it could be assumed negligible as a result of metabolism or excretion or because the effective internal volume was very large. Evidence for such a condition was obtained by Burt *et al.* (1971) who found that while pyrethrin I applied to the cuticle of cockroaches was lost progressively from the surface at a rate decreasing with time, the amount found inside (i.e. not removed by the surface wash) remained approximately constant at about 13% of the applied dose, following an initial period during which this steady condition was established. Most of the chemical found inside the insect, however, was adsorbed on the insect solids and concentrations in the haemolymph were too small to be detected chemically. Such studies emphasize the value of measuring the distribution of chemicals within the insect in addition to the amounts lost from the surface for penetration to be thoroughly understood.

Buerger (1966, 1967) has criticized the use of relationships based on Fick's law such as eq. (2) on grounds other than the inapplicability of the assumptions we have discussed above. He maintains that such relationships apply strictly only to isotropic materials, whereas integuments are composites of different immiscible layers, very far from isotropic. He developed an alternative mathematical treatment designed to take into account this organized structure and considers penetration as a sequence of steps representing transfer of material from each layer to those adjoining it. The necessity for this alternative treatment, and its validity may be questioned, however, as we discussed in Section 3.VI. Rigorous experimental evidence about the permeability characteristics of cuticles is limited, but Treherne (1957) found that diffusion of ethanol through isolated cuticle of *Schistocerca gregaria* from an aqueous solution supply to a large volume of aqueous saline could be described by an overall permeance which appeared independent of time and concentration of diffusant (Table 1).

Table 1
Diffusion of ethanol through isolated locust cuticle (from Treherne, 1957)

External ethanol concentration (molar)	Period of diffusion (h)	Mean permeance ($cm\ s^{-1}$)
0·1	20	1·12±0·14
1·0	1	1·23±0·28
1·0	20	1·29±0·29

Results from such studies with isolated cuticles must, of course, be extrapolated to intact insects with caution, particularly if as suggested earlier, the permeability of the intact insect is maintained metabolically.

The starting point for this discussion of the mathematical relationships used to describe the time course of penetration was a consideration of the popular semi-log plot. It may be concluded from our discussion that the rate of uptake plotted in this way can be expected to fall with time in many cases due to decreasing concentration gradients even in the absence of a multiphase mechanism. It can be very difficult to decide from a graphical plot obtained with variable biological material whether a reduction in rate of penetration with time represents different stages or is merely one continuous curve such as that found by Lewis (1965) for the absorption of DDT from relatively large reservoirs in the form of lanolin spheres applied to the cuticle of blow flies (see Fig. 1). Such considerations prompt caution in assigning mechanisms from analysis of penetration curves.

The interpretation which can be put on any results depends greatly, of

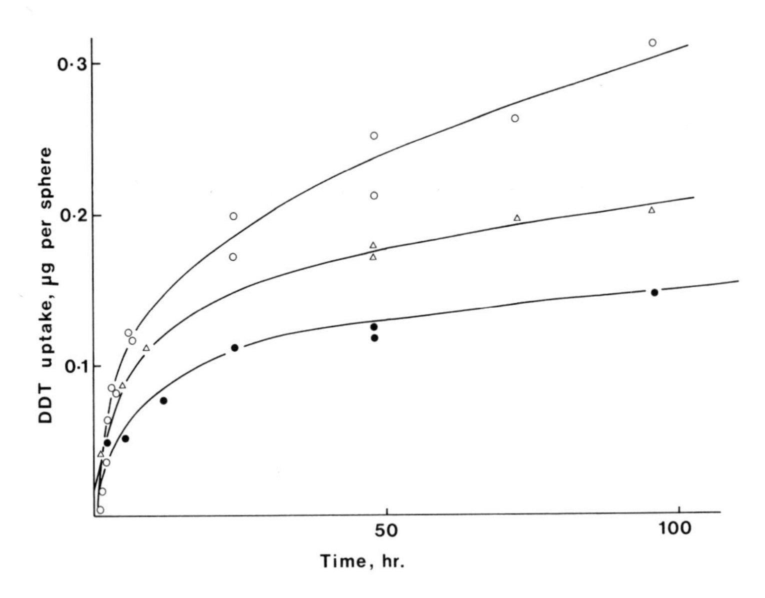

Fig. 1. Rates of absorption of DDT from spheres of lanolin solution applied to the head of *Phormia terraenovae* (from Lewis, 1965). —○—○— application to the antennae (living flies); —•—•— application to the antennae (dead flies); —△—△— application to the genae (living flies).

course, on the method used to obtain them and we emphasize again the limitations of the rinsing procedure, which is associated with unavoidable uncertainty as to what fractions are being removed. With insects it is often assumed that rapid rinsing with organic solvents removes only the epicuticular waxes at most, and Lewis (1965) gives evidence to support this for the case of diethyl ether, but it can never be completely certain that other underlying material is not extracted. Whatever the exact fractions removed, the rate of loss from the surface must be fastest immediately after application and must decrease with time thereafter as concentration gradients decrease. This measure of penetration therefore differs from that obtained by determining arrival inside the insect, the rate of which is likely to have a pulse form (i.e. passing through a maximum) when the septum has significant capacity. It is important therefore to bear in mind which definition of "penetration" is being adopted. For example, Sun (1968) analysed the effects of different insecticides on the housefly in terms of theoretical curves for penetration, activation and detoxication in which the rate of penetration rose rapidly from zero at the time of application to a peak from which it subsequently decreased; penetration was assessed, however, by the surface rinsing technique. Since the amount of activated insecticide within the insects was estimated from the superimposition of the rate curves for the various processes, this assumption could have led to considerable errors in the conclusions reached. Similar curves were plotted by Szeicz *et al.* (1973) although not surprisingly no points on the initial part of the penetration curves (before the supposed maximum) were obtained experimentally in this work or that by Sun.

In the light of this general background we shall now examine the influence of specific factors on penetration into insects in greater detail.

C. Effects of Solvent

From the foregoing discussion it will be clear that the nature of the carrier solvent can affect penetration of dissolved insecticides by determining the degree of spreading, which depends i.a. on viscosity and interfacial tension, and by influencing the route, mechanism and kinetics of entry which are related to polarity and affinity for the components of the cuticle. For both types of effect the volatility of the solvent will be important because it governs the period over which the solvent can exert its influence before evaporating.

The importance of spreading properties is well illustrated by investigations undertaken by Barlow and Hadaway (1974) and Hadaway *et al.* (1976) who measured the penetration of various organochlorine, cyclodiene and organophosphorus insecticides mainly into tsetse flies following topical application in a range of different solvents. The fate of the insecticide was examined at intervals after application by determining the amounts in an "external" fraction (obtained by rinsing the treated insects for 3 min in hexane) and in an internal fraction (obtained by homogenizing the hexane-rinsed insects in acetone). Recovery checks indicated that these fractions were reasonably distinct and that repeating the initial hexane wash before homogenizing removed very little further insecticide.

Uptake was strongly influenced by solvent, as shown by Table 2, which compares quantities of DDT recovered in the internal fraction 2 h after application (selected for illustration from the much more extensive results presented by Hadaway *et al.* 1976) and mortalities at 24 h with viscosities and boiling points of the solvents used.

Both penetration into the insect and toxic effect were greatest with solvents having the lowest viscosities, suggesting that spreading had a major influence on performance. Evidence to support this suggestion came from treating the flies with solutions containing fluorescent tracer or oil colour which revealed the degree of spread. The most effective solvents (di-isobutyl ketone (DBK), Shellsol N and xylene) spread rapidly over the whole dorsal surface of the thorax, reaching the edge of the tergites. A measure of the speed of this effect is given by the parallel observations that these solvents, being relatively volatile, evaporated from the surface of the thorax in less than a minute; the influence of the solvents on penetration continued, however, for several hours. Solvents having intermediate effect spread more slowly over the thorax and took longer to evaporate while Risella 917 and methyl oleate spread even less rapidly and remained as liquids. The retarding action of a persistent reservoir of solvent was further demonstrated by examining penetration from mixtures of the rapidly spreading, but volatile Shellsol N and the effectively non-volatile, viscous Dutrex 217. Acetone evaporated so rapidly under the conditions of these tests that little spreading occurred: the insecticide was left as a concentrated localized deposit, probably accounting for relatively poor penetration and toxic effect. The relatively low activity following application in acetone is worthy of note in view of the widespread use of this solvent in laboratory topical application studies. If such a volatile solvent were employed in outdoor spray application, it would rapidly volatilize from the drops to be replaced by water as a result of evaporative cooling. Other experiments demonstrated

Table 2

Penetration of DDT into tsetse flies from different solvents following application at 100 *ng per fly in* 0·2% *w/v solution* (*based on data given by Hadaway et al.*, *1976*)

Solvent	Chemical type	b.p. (°C)	Viscosity (25°C mN s m^{-2})	Amount in internal fraction after 2 h (ng per fly)	Mean percentage kill after 24 h
acetone	ketone	56	0·32	23	15
di-isobutyl ketone	ketone	168	0·95	85	72
Shellsol N	aromatic hydrocarbon mixture	167–262	1·14	60	79
xylene	aromatic hydrocarbon mixture	138–142	0·63	58	67
n-hexadecane	paraffin	287	3·0	36	58
isophorone	ketone	215	2·4	32	29
heavy aromatic naphtha	aromatic hydrocarbon mixture	184–280	2·1	31	44
diesel oil	aliphatic hydrocarbon mixture	170–360	3·6	31	29
Risella 917	saturated aliphatic hydrocarbon mixture	300–>365	22	10	4
methyl oleate	long chain ester	360	6·4	10	4
Dutrex 217	condensed aromatic hydrocarbon mixture	300–>365	53	—	—

that penetration from solutions in DBK increased as the volume applied increased (and concentration decreased) for a constant amount of insecticide.

The dependence on viscosity, the observed evaporation and the evidence from the tracer chemicals thus all point to spreading and persistence of the liquid reservoir being the dominant factors determining solvent effects in these experiments, as the authors conclude. There is no indication that chemical type or polarity had an important influence. All the solvents used were, however, probably sufficiently lipophilic to penetrate the superficial wax layers of the insects and in line with the principles discussed in the preceding section, the greatest advantage was given by solvents spreading furthest (thus increasing the area of uptake) and evaporating sufficiently fast not to leave a liquid reservoir having considerable affinity for the non-polar toxicant and thus holding it back. We are thus here concerned principally with mechanism (2) on p. 663.

The results obtained by Olson and O'Brien (1963) with a range of generally lower molecular weight solvents having a wider range of polarity probably reflect greater qualitative differences in penetration pattern. Uptake of various radiolabelled solutes by cockroaches (*Periplaneta americana*) was in general fastest after they had been applied in petroleum ether followed by acetone>ethanol>water. Penetration was assessed by rinsing the treated pronotum first with water and then with acetone to determine the percentage of the applied dose remaining on the surface. The results were consistent with the hypothesis that organic solvents which rapidly evaporate introduce the toxicant directly into the hydrophobic epicuticle from which diffusion occurs directly into the more polar exocuticle. Water, being much more polar, would not permeate the epicuticle and penetration of the dissolved toxicant would be delayed until the water had evaporated. This evaporation might be delayed by the rapid acquisition of a film of grease from the cuticle on the surface of the water. Until the water has evaporated, penetration of solute takes place effectively from an aqueous reservoir at a rate determined by its partition coefficient between water and the material of the outermost layer of cuticle according to mechanism (1) p. 663.

The specific consequences of such solvent effects on uptake depend, of course, on the nature of the solute, which we now consider.

D. Effects of Penetrant Properties

The rates of entry into insects of different penetrants will depend on their physico-chemical properties, which determine their affinities for the materials comprising the cuticle, and on the way in which they are supplied.

As we have seen, the solvent in which they are presented is particularly important as this can by-pass some of the contrasting barriers to penetration if it has the appropriate characteristics. The penetrant properties most favourable for entry therefore depend on the manner of application. We must stress again the importance of distinguishing clearly between loss from the surface and accumulation on the inner side of the cuticle, and the value of the balance sheet approach in which amounts retained by the cuticle itself are assessed.

There appear to have been few studies with isolated cuticles which allow penetration to be measured under strictly controlled conditions. However, in the studies to which we have already referred, Treherne (1957) investigated the permeability of isolated locust cuticles (both whole and "dewaxed" by washing with chloroform) to a range of closely related substituted ureas in aqueous solution. Permeance values for the wax layer were calculated by treating the composite cuticle as resistances in series, assuming the chloroform removed only the superficial cuticular wax and left the remainder of the cuticle unaffected. In view of the rather special solvent properties of chloroform, other solvents would perhaps have been safer for this purpose. There was little difference in permeance for the different compounds through the chloroform-treated cuticle, values ranging from $1{\cdot}06$ to $1{\cdot}91 \times 10^{-6}\ \mathrm{cm\,s^{-1}}$, with some tendency for higher values with the more polar compounds. Values for the whole cuticle varied much more widely however with a range from $4{\cdot}7 \times 10^{-7}$ to $1{\cdot}9 \times 10^{-8}\ \mathrm{cm\,s^{-1}}$. The calculated values for permeance of the wax layer showed a corresponding variation; the resistance to diffusion appeared to lie almost entirely in the wax layer for the less mobile compounds and even with the more mobile ureas this layer accounted for a large proportion of the total resistance (see Fig. 2).

The permeability of the epicuticular wax layer decreased with aqueous solubility (see Fig. 2). Although, as emphasized in Chapter 2, solubility is not a reliable guide to partition, it can probably be assumed to reflect the hydrophilic/lipophilic character within this series. Penetration through the whole isolated cuticle deviated increasingly below that of the wax layer as diffusion through the wax became faster (Fig. 2). This was attributed to the increasing importance of the resistance to diffusion in the cuticular layers below the wax. There was also a relationship with molecular size, smaller molecules tending to diffuse more rapidly as might be expected if the barrier had no specifically selective properties. Treherne found that the product $\mathscr{P}\sqrt{\mathrm{MW}}$ was approximately constant, although such constancy has strictly no theoretical basis. (See Appendix 1). By comparison with the value of $7{\cdot}05 \times 10^{-5}\ \mathrm{cm^2\,s^{-1}}$ for $D\sqrt{\mathrm{MW}}$ in water, Treherne calculated

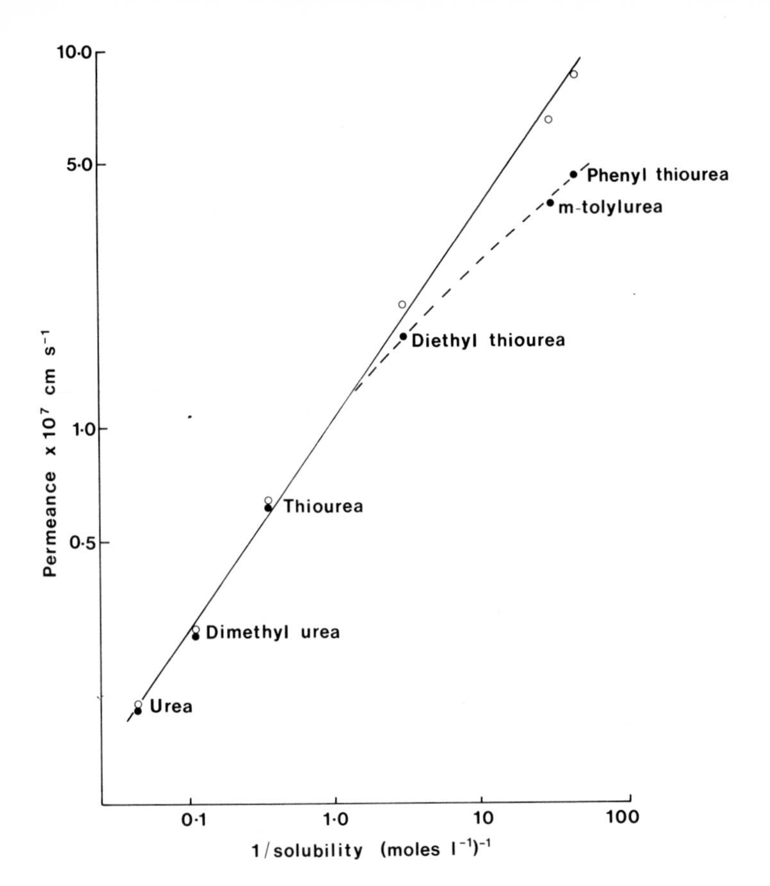

Fig. 2. Relationship between solubility and permeance of epicuticular layer (open circles) and whole cuticle of locusts (closed circles). From Treherne (1957).

that the permeability of the cuticular layers below the epicuticle was 0·24% of that of an equivalent unstirred water layer.

An opposite relationship between permeance and polarity was observed by Olson and O'Brien (1963) when they applied various solutes *in organic solvents* to intact cockroach cuticles. The rates of penetration of H_3PO_4, dimethoate, paraoxon, dieldrin and DDT decreased with increasing olive oil/water partition coefficient. Similar trends were recorded by Buerger and O'Brien (1965) and more recently by Szeicz *et al.* (1973) when they applied DDT, carbaryl, malathion and endrin to Tobacco Budworm larvae. Hadaway *et al.* (1976) found that various organophosphorus insecticides penetrated very rapidly into tsetse flies following application in DBK; in general accumulation within the insects was faster than with a series of less polar organochlorine and cyclodiene compounds although interpretation is complicated by effects of evaporation and decomposition.

These various results can be explained in terms of the mechanisms we have already discussed. As pointed out by Olson and O'Brien (1963), when the toxicant is dissolved in a suitable organic solvent, the barrier function of the coherent epicuticular wax is destroyed, so that penetration is determined by the rate at which the chemical moves from the wax through the underlying more polar layers: clearly this should be favoured by polar character. In contrast when the toxicant is applied in aqueous solution, diffusion through the lipoidal epicuticle is likely to be the rate determining step so that non-polar compounds will be absorbed from the supply most rapidly.

We should not overlook the fact that presence of a carrier solvent is not an inevitable feature of the reception of insecticides by insects in practice. In ULV applications the insect may receive undiluted technical insecticide while vapour transfer provides a direct route of uptake having special characteristics, some of which we consider in the following section. There seems to have been little systematic study of the properties determining relative penetration of different compounds in these situations but the available information appears to fit in with the broad picture outlined above. When insect eggs were exposed to the vapour of different insecticides with roughly comparable oil/water partition coefficients (Zschintzsh *et al.*, 1965; Bracha and O'Brien, 1966) uptake was directly proportional to the partition coefficient which would be expected as insecticide applied in this way must first move unaided through the lipophilic epicuticular layer. However, in line with the results quoted above, non-polar compounds penetrated further into the eggs only slowly and the *proportion* of the toxicant reaching the inside of the egg was much greater for polar materials. These results emphasize yet again the dangers of drawing conclusions on the basis of simple measurements of uptake.

One can conclude that for rapid penetration through the entire cuticle, conflicting properties are required so that, overall, chemicals offering a compromise between polar and non-polar properties are likely to be most successful unless applied in a solvent which by-passes part of the barrier. One way in which this difficulty could be overcome would be by conversion of a lipophilic chemical to a more hydrophilic but still insecticidal derivative in the cuticle. There is in fact at least one class of compound which could achieve this: the sulphur containing organophosphorus insecticides which are "activated" by oxidation to more polar derivatives as we discuss in Section 4.IV.B. If such oxidation occurred in the cuticle, it might substantially facilitate penetration. Hydrolysis of phenoxyacetic acid esters in the plant cuticle provides a parallel example for herbicides.

E. Uptake of Insecticide Vapour

Dichlorvos, not usually classed as a fumigant, is widely used for control of flies indoors by vapour. Vapour transfer over short distances and within folds of leaves is important with many insecticides in open agriculture but there has been little systematic investigation. The first publication on chlordimeform (Dittrich, 1966) clearly showed it to be effective against mites via the vapour and that the LC_{50} for 48 h exposure was only one-tenth of that for 24 h. Several later papers reported interesting and unusual effects in the field but made no mention of possible effects of vapour. In further definitive tests, Streibert and Dittrich (1977) showed that eggs of a noctuid moth succumbed to a shorter exposure (4–240 min) to saturated vapour than did mature larvae (60–600 min). Their tests included an attempt to eliminate vapour transfer by pulling air through the leaf from the treated surface. Graham-Bryce *et al.* (1972) found some evidence of significant vapour effects from phorate and disulfoton granules applied to field beans in controlled environment rooms.

The potential importance of vapour transfer, in relation to both efficacy and selectivity, can be demonstrated by calculations for idealized model systems. If an insect represented by a small sphere of radius a in a uniform vapour concentration C can retain all molecules arriving at its surface, the quantity of toxicant (**M**) acquired in time t will be given by:

$$\mathbf{M} = C(8\sqrt{\pi}a^2\sqrt{Dt} + 4\pi aDt) \tag{3}$$

where D is the vapour diffusion coefficient (see eq. (51), p. 144).

For times of practical interest, the first term becomes negligible so that the amount absorbed can be taken as $4\pi aDtC$. The saturated vapour concentration of phorate at temperatures normally applying in the field is approximately $1{\cdot}2 \times 10^{-10}$ g ml^{-1} and we can assume D to be about $0{\cdot}1$ cm^2 s^{-1}. For an insect such as an aphid having a "radius" of about 0·5 mm this gives a potential uptake over one hour of roughly 3×10^{-8} g, of the same order as the observed LD_{50}.

A surface coated with pesticide will saturate the immediately adjacent air and an approximately steady gradient will be established across the unstirred layer of air in contact with the surface. The factors determining the thickness of the unstirred layer are examined in Chapter 6; for our present purposes we may assume that under many conditions it will exceed the height of small arthropods such as aphids and mites which will therefore be exposed to a vapour concentration approaching the saturated value (SVC) for the pesticide concerned. It is of interest to compare the potential vapour transfer in this situation with the dose which would be received by direct interception of spray.

If the pesticide is irreversibly taken up by the object, then, for a small sphere of radius a, an availance I (where $I = \int_0^\infty C\,\mathrm{d}t$, see Section 3.V.C) applied in ambient air is equivalent to a dose **M** applied directly to the organism surface where

$$\mathbf{M}_a = 4\pi DaI \tag{4}$$

For the sphere exposed on the treated surface $I = Ct$, where C is the concentration in the unstirred layer and t the time of exposure. If, however, the sphere remains on the surface over the whole period during which the toxicant is evaporating:

$$I_{\text{surf}} = \mathcal{A}\frac{\delta}{D} \tag{5}$$

where $\mathcal{A}$ is the mass of toxicant applied per unit area and δ is the thickness of the unstirred layer. Combining eqs (4) and (5) gives

$$\mathbf{M}_a = 4\pi a\mathcal{A}D \tag{6}$$

whereas if the toxicant had been applied directly at the same rate to the target area (πa^2) of the sphere itself, the corresponding relationship would be:

$$\mathbf{M}_a = \pi a^2\mathcal{A} \tag{7}$$

The dose received by the vapour route is therefore $4\delta/a$ times the dose which would be received directly: for an insect of radius 0·5 mm in an unstirred layer of thickness 5 mm, this would have a value of 40. For a larger organism the figure would be smaller, both because a would be larger and because both C and I would be smaller as the mean distance from the surface would be greater.

In practice the organism will not of course be a "perfect sink" accepting from the vapour all the insecticide which could possibly reach it, but the less efficient the capture of vapour the greater the loss by evaporation of the directly applied insecticide. Unless there is a mechanism of direct liquid penetration, the *ratio* of direct to indirect offer to obtain the same accepted dose is unaltered.

This has important implications in relation to selectivity between pests and their natural enemies: there are often significant differences in size between predator and prey, as with aphids and their ladybird predators. The dependence on radius makes the difference between the availabilities of a pesticide to two species of different size greater by the vapour route than by direct interception. This is illustrated in Table 3 which gives theoretical calculations of the relative amounts of insecticide reaching spherical "insects" of different sizes by different routes (Graham-Bryce, 1975) using the relationship given above. For illustration it is assumed that the unstirred layer is such that the larger insect is exposed to a vapour concentration equivalent to half the saturated vapour concentration C_0 experienced by the smaller insect.

Table 3
Amounts of insecticide reaching spherical organisms of different sizes by different routes

	Route	
	vapour from surface deposit	direct interception
Amount reaching organism	$4\pi aDCt$	$\pi a^2 \mathscr{A}$
Amount received per unit volume	$3DCta^{-2}$	$\frac{3}{4}\mathscr{A}a^{-1}$
Amount per unit volume for small organism ($a = 1$ mm)	$3DC_0t$	$\frac{3}{4}\mathscr{A}$
Amount per unit volume for large organism ($a = 5$ mm)	$3DC_0t/50$	$\frac{3}{20}\mathscr{A}$
Ratio, amount received per unit volume small organism : large organism	50 : 1	5 : 1

Much larger insects rise above the unstirred layer so that other factors are introduced. It is worth noting that in the case where the insect remains on the surface over the whole period during which the toxicant is evaporating, the availance is independent of the concentration in the unstirred layer—in other words independent of volatility. The duration of this exposure is of course affected by volatility as is the dose reaching the insect over times shorter than that required for complete evaporation, and it must be remembered that all toxicants are tolerated indefinitely at sufficiently small concentration. The importance of the rate at which the chemical becomes available depends on the behaviour of the insect and on the mode of action of the toxicant.

Vapour effects are thus very dependent on the size of the insect, its location and behaviour on the leaf and on the nature of the toxicant. Volatility alone will not indicate the importance of vapour action, and it is unsafe to assume that because a compound is crystalline and moderately persistent, vapour transfer is unimportant.

III. PENETRATION INTO FUNGI

A. Barriers to Penetration

As with other classes of organism, penetration of applied chemicals into fungi is greatly influenced by the properties of the outer envelope which protects them from the external environment. There are, however, important differences in the nature of this outer envelope. Unlike insects and higher plants, fungi do not have a clearly differentiated cuticle, and fungi, together with algae, are believed to be the only group of plants which do not contain cutin or suberin in their outer layers. Nevertheless the nature of fungal structure and organization is such that the outer protective envelope probably plays an even more crucial role than with other organisms. A distinctive general feature of the vegetative phase of most fungi is a filamentous mycelium made up of a mass of branching tubular hyphae which continuously expands by apical growth and lateral branching. (For a general description see Hickman, 1965.) Various specialized structures may become differentiated and septa are regularly formed within the hyphae, but in most species these septa are incomplete and the protoplasm is thus continuous throughout the length of the hypha. Reproduction occurs after periods of vegetative growth with the formation of spores which can be uni or multicellular, and which are released from the parent to be dispersed by wind, water or by other organisms.

In the relatively simple organization of fungi, therefore, the cell wall provides the essential barrier between the external environment and the cell membrane or plasmalemma enclosing the protoplasm in which the vital processes occur. The cell membrane provides the principal molecular barrier to penetration, although the cell wall can have an important influence on uptake through adsorption or ion exchange effects. The cell wall construction largely determines the gross morphology of the fungus and this central importance of cell wall morphogenesis in overall morphology, together with interest in the relationships between wall composition and taxonomic classification has been the main stimulus for much of the work on cell wall chemistry and structure which can provide a basis for considering penetration of toxic agents. The nature of the cell wall varies with species, of course, and also with functional differentiation. For example, hyphae of many plant pathogenic organisms develop specialized organs called appressoria (Emmett and Parbery, 1975) for attachment to the host. These are localized swellings which adhere to the plant surface and provide a base for infection hyphae to penetrate the cuticle. The intercellular hyphae of many obligate parasites such as the powdery mildews also develop lateral outgrowths specially designed for absorbing nutrients, called haustoria. Before the brief general outline of cell wall composition which follows, it should be stressed that cell wall properties vary with these different functions and with different species and these variations correspondingly affect cell wall permeability and selectivity. More detailed treatments of cell wall composition and structure are given by Aronson (1965) and Bartnicki–Garcia (1968).

Structurally the fungal cell wall consists of a network of microfibrils, loosely interwoven and embedded in an amorphous matrix. Chemically it is mainly composed of polysaccharides, which typically comprise 80–90% of the total. The remainder consists of protein and lipid with significant amounts of pigments, polyphosphates, inorganic ions and possibly nucleic acids. In a few cases, composition differs substantially from this general pattern; for example the walls of some yeasts may contain up to 40% protein. Most true yeasts are also peculiar in that the microfibrils are composed of non-cellulosic glucans whereas chitin or cellulose fulfil this function in most other fungi. The presence of either chitinous or cellulosic walls has been established as a criterion for taxonomic classification although a third cell wall category has subsequently been recognized with the discovery of the simultaneous occurrence of chitin and cellulose in *Rhizidiomyces* species. The matrix material in which the microfibrils are embedded is thought to consist of proteins and various polysaccharides such as glucans, mannans, galactans and heteropolysaccharides. The cell walls also contain uronic acids which may impart the negative charge characteristic of the surface of fungal spores (Douglas *et al.*, 1959; Lukens, 1971). These negative charges can have an important influence on the selective permeability of fungal spores by retaining positively charged fungicides. The lipid constituents may also influence surface properties; it has been suggested that they contribute to the hydrophobic nature of spores. They may also have a structural role in the cell wall.

Electron microscope studies with several species have indicated that the cell wall may be multilaminate. Thus, for example, examination by Hunsley and Burnett (1970) of hyphae from three representative species treated singly or successively with different enzymes revealed various degrees of complexity. Wall thickness ranged from 155 to 220 nm. In all cases, the microfibrils, composed of either cellulose or chitin intermixed with protein formed the innermost layer, while amorphous glucans comprised the outermost layers. No other features were observed in walls of *Phytophthora parasitica.* However, additional layers of amorphous glucan and protein were present between the outermost and innermost layers in *Neurospora crassa* and *Schizophyllum commune*; in *Neurospora crassa* one of the protein layers contained a network of strands of reticulum.

B. Routes to Absorbing Surfaces in Fungi

Fungicides may reach the target organism in some cases by direct interception of sprays or by physical contact between the pathogen and residual particles, although in the last case there may be little absorption by the fungus. Where such direct transfer does not occur, the pathogen must accumulate the toxic dose by uptake from the vapour phase or from condensed phases in which the toxicant is dispersed. With systemic compounds these include the plant sap and internal tissues, although this method of administration does not preclude the chemical being available on the plant surface either before or after uptake and translocation within the plant. With residual protectant fungicides toxic quantities must clearly be taken up before infection starts.

The vapour effects of general soil-sterilant or stored product fumigants such as chloropicrin or methyl bromide are widely familar, although the final stages of the transfer of such substances to the receiving organisms may often take place in water films. We consider more fully elsewhere the vapour transport of fumigants (Sections 5.III.B, 5.III.C) and of less volatile compounds (Sections 5.IV, 6.VII.D). It may reasonably be assumed that several other obviously volatile fungicides such as naphthalene and pentachlorophenol are absorbed as vapours, and it has also been clearly established (Lindström, 1958, 1959) that redistribution by vapour transfer can play an important part in the improvement in performance which occurs when seed treated with organomercury fungicides is stored. For many other compounds whose vapour pressures are low compared with these substances or with many familiar volatile laboratory chemicals, vapour effects have been largely neglected. This neglect is not justified for there are several indications that vapour transfer, at least over short distances, and absorption of vapour by fungi can contribute significantly to the activity of residual and even systemic fungicides; this would certainly be expected on theoretical grounds as we discussed for insects in Section 12.II.E. Brook (1957) found that dry deposits of thiram (which is reported to have "negligible" vapour pressure at room temperatures (see Appendix 4)) could inhibit spore germination without physical contact and in the absence of free water, while Lukens (1971) reported that dichlone (which "sublimes at temperatures above 32°C") can be transferred as vapour to spores from residual deposits of dusts and sprays on leaves.

The short-range vapour action of fungicides against powdery mildews was investigated in detail by Bent (1967). He pointed out that whereas the spores of most pathogens require water for germination, and thus it is generally assumed that fungicides ultimately reach them by diffusion through water, powdery mildew spores are airborne and germinate in the absence of free water. Since sprayed fungicides do not leave a uniform cover on plant surfaces, vapour transfer is essential with non-systemic chemicals to protect the areas between the clusters of particles deposited by spraying against mildews and to ensure control in the gaps resulting from weathering or expansion of the leaf surface. Bent found that in addition to sulphur and lime sulphur, whose vapour action was already well established, fungicides based on drazoxolon, oxythoquinox, binapacryl, dinocap(dinitro-octylphenyl crotonates), and a phthalimidophosphonothionate prevented the growth of powdery mildews at distances extending several mm beyond the edge of localized deposits on leaves. Most of these materials were also active when separated from the leaf by glass cover slips, confirming that the effects were due to vapour movement. The tests were conducted on open glasshouse benches and blowing air from a fan shifted the zones of inhibition to the leeward side, but did not reduce their areas. Vapour effects were also demonstrated on mildew conidia incubated on glass slides treated with a spot of fungicide, and on infected plants placed in beakers coated on the bottom with fungicide. The volatility of these materials would normally be regarded as extremely small: Bent quotes figures of 4×10^{-6} mmHg at 30°C for both sulphur and drazoxolon, and 2×10^{-7} mmHg at 20°C for oxythioquinox.

Hislop (1967) found that several fungicides exerted a similar action against chocolate spot (*Botrytis fabae*) on broad bean leaves when infection was induced in the absence of free water. Phenyl mercury chloride, maneb, mancozeb, dichlofluanid and oxythioquinox were all shown to protect areas of leaf beyond the visible limits of fungicide deposits, and to diffuse from dry deposits on glass slides to inhibit germination of spores in water droplets placed at a distance from the fungicide source. In discussing these results Bent (1967) concludes that redistribution in the vapour phase may be involved in the control of "wet" diseases, such as *Botrytis* which require water for germination, under certain conditions, as well as with "dry" diseases such as mildews. With "wet" diseases, vapour movement would allow continuous movement of the fungicide from a deposit on a dry part of the leaf to infection droplets not directly in contact with the deposit. The effect depends on the nature of the chemical, however; although drazoxolon is active against a range of pathogens in addition to mildews when applied as a spray, it showed no vapour action from dry residual deposits against the "wet" diseases chocolate spot on beans, downy mildew of vine (*Plasmopara viticola*), scab of apple (*Venturia inequalis*) and brown rust of wheat (*Puccinia recondita*). It appears that powdery mildew spores can absorb drazoxolon molecules readily from the vapour, whereas the spores of the other pathogens take the chemical up effectively only when it is presented in aqueous solution; droplets containing spores of "wet" diseases do not appear to be sufficiently efficient at capturing the vapour to acquire a toxic dose. It may be concluded that important factors influencing the relative effectiveness of protectant fungicides against mildews and "wet" diseases are the relationship between their solubility and volatility from spray deposits, and the extent to which they are absorbed from air and water by the different pathogens. Uptake from solution is discussed in more detail later; there appears to have been little detailed study of the mechanism of uptake from the vapour phase. Presumably, however, the toxicant molecules are absorbed by the amorphous polymeric materials which form the outer layers of the fungal wall from where they move inwards, through the cytoplasmic membrane to the sites of action. Effective capture by the absorbing fungus and subsequent mobility within the organism would depend on the polarity of the penetrant in relation to the specific composition of the receiving species.

In later studies Bent (1970) found evidence for vapour action of the systemic fungicide dimethirimol against cucumber powdery mildew, using techniques similar to those employed in his investigations with protectant fungicides. Deposits on glass cover slips prevented growth over distances of 7–15 mm after incubation for 30 h at 19°C and 100% r.h. in the dark. Since root applications of these chemicals affect germ-tube emergence and appressorium formation from spores on the leaf surface, which occur before penetration into the leaf by the pathogen, the fungicides must pass from the leaf to the fungus either by direct transfer at points of contact, or by liberation of vapour from the leaf. The demonstration of vapour action would be consistent with this last mechanism.

While vapour transfer should not be underestimated, there are still very many situations in which the pathogen must acquire the toxicant from condensed phases. For lipophilic organic fungicides, the plant cuticle may be an important medium of transfer. Compounds of appropriate polarity may be absorbed readily by cuticular components from foliar treatments and are then available to the fungus as it penetrates this barrier. Lukens and Horsfall (1967) suggested that relationships between physico-chemical properties of folpet, captan and related compounds and their activities against mildew on snapbeans gave evidence for this route of uptake. A positive non-linear relationship between activity and partition towards oleyl alcohol from water was observed. It has already been pointed out that water is not required for mildews to germinate and penetrate the plant tissue. Lukens and Horsfall therefore concluded that this relationship indicated that the fungicide partitioned into the cuticular components from where it was taken up by the fungus. These fungicides are generally considered too involatile to have a vapour action, although the results by Bent (1967) and Hislop (1967) suggest that it is difficult to exclude this without careful tests.

In many disease control situations, water is an essential agent in the transfer of toxicant to pathogen, and uptake from aqueous solution the major route of accumulation. This certainly applied with some of the older inorganic fungicides such as bordeaux mixture. The redistribution of fungal spores and fungicide by rainwater essential for effective control in some cases, is discussed more fully in Chapter 13. There is evidence (reviewed by Lukens, 1971) that the solubility of copper may be increased by complexing agents secreted by the plant and especially by fungi. Although many of the newer organic fungicides have very small solubility in water, there is little doubt that their efficacy may also often depend on uptake from aqueous solution.

C. Uptake of Fungicides by Fungi

The uptake of fungicides from solution, particularly by spores, has been studied extensively, although there have been relatively few attempts to establish quantitative relationships to describe the uptake process. In many cases uptake has been inferred from the decrease in solution concentration after incubation of a spore suspension. The limitations of this type of approach have already been discussed, and were stressed by Somers (1968) who drew attention to the substantial errors which can occur with labile organic fungicides due to decomposition. Nevertheless it has been clearly established that fungal spores can accumulate fungicides, especially heavy metals, to concentrations as high as 10 000 times those in the external solution (see, for example Miller *et al.*, 1954; Marsh, 1945). Soluble salts of unsaturated fatty acids (soft soaps) were shown by Schmidt (1924) and Goodwin *et al.* (1929) to be effective against spores of several fungi, particularly *Botrytis* spp. Their effectiveness is also associated with a high degree of concentration. Free fatty acid accumulates in the spores leading to their disruption.

This pronounced accumulation of fungicides has been regarded as providing the most convincing explanation for the selective toxicity of many non-systemic fungicides. In reviewing the subject in the 1960s Somers (1962, 1968) pointed out that most of the fungicides then available were much less potent on a weight for weight basis than toxicants from other classes, and that many had rather general modes of action such as non-specific enzyme inhibition. This is illustrated by Table 4 which extends previous comparisons of different toxic agents such as that by McCallan (1957). Some recently introduced systemic fungicides are included; these are considerably more active than most older materials but even so are substantially less potent than the most effective compounds from other classes.

If the suggestions that the activity of many fungicides is due to selective accumulation are valid, it would be expected that in general fungi sensitive to a given toxicant would more readily accumulate the chemical than resistant organisms (Somers, 1968). Several investigations have given results consistent with this expectation. Voros (1963), investigating reasons why streptomycin was much more active against *Phytophthora infestans* and *pythium* than against a range of other fungi including *Botrytis*, *Verticillium*, *Aspergillus*, *Fusarium* and some yeast species in agar plate tests, found that uptake by pythium was much greater than by a representative group of the insensitive organisms. As the two sensitive species have predominantly cellulose walls, whereas most of the resistant species have chitinous walls, the results provide evidence for the possible influence of cell wall composition on permeability and hence on selectivity. Wescott and Sisler (1964) were able to relate the large differences in sensitivity to cycloheximide of *Saccharomyces pastorianus* (highly sensitive) and *Saccharomyces fragilis* (very insensitive) to differences in the ability to concentrate the antibiotic from solution. They attributed these differences in accumulation to the absence of internal binding sites in *S. fragilis*, however, rather than to impermeability of the cells. In other studies bearing on this question El-Nakeeb and Lampen (1965) showed

Table 4
Activity of fungicides compared with potent toxicants from other classes

Compound	Species	Approximate ED_{50} ($mg\ kg^{-1}$)	Reference
Tetradotoxin	Rat	0·01	Ogura (1971)
Tetrachlorodibenzo-*p*-dioxin	Guinea pig	0·0006	Schwetz *et al.* (1973)
Botulinum toxin	Human	0·00002	Simpson (1971)
DDT and analogues	Housefly	7–14	Elliott *et al.* (1974)
parathion	Housefly	2	Elliott *et al.* (1974)
paraoxon	Asparagus Beetle	0·03	Krueger and Casida (1957)
synergised decamethrin	Housefly	0·002	Elliott *et al.* (1974)
2,4-D	Tomato	10	McCallan (1957)
penicillin	Staphylococci	2	McCallan (1957)
sulphur	Bread mould	11 500	McCallan (1957)
mercury	Brown rot Fungus	2 830	Miller *et al.* (1954)
silver	Bread mould	165	McCallan (1957)
2-heptadecyl-2-imidazoline	Bread mould	5 800	McCallan (1957)
carbendazim	Bread mould	1–2	Clemons and Sisler (1971)
cycloheximide	Yeast	0·38	Wescott and Sisler (1964)
oxathiins	Various fungi	1–7	Marthe (1968)

that sensitivity to griseofulvin was correlated with its uptake by yeasts and by mycelium of other fungal species while Marthe (1968) found that uptake of oxathiin fungicides by the sensitive species *Rhizoctonia solani* and *Ustilago maydis* was considerably greater than by the resistant fungi *Fusarium oxysporum* and *Saccharomyces cerevisiae.*

D. The Detailed Pattern of Fungicide Accumulation

1. *Influence of Fungicide Properties on Penetration*

Since it appears reasonably well established that sensitive fungi readily accumulate a wide range of fungicides and that selective toxicity may be related to accumulation in many cases, the nature of the accumulation process is of considerable interest.

Selectivity between plants and fungi is clearly of paramount importance for fungicides which must control organisms living in intimate association with the plant. In comparing the accumulation of externally applied compounds by plants and fungi, the most immediately obvious anatomical distinctions are the more complex cellular and structural organization in the plant, and the absence of a cuticle or bark in fungi which protects the epidermal cells of the plant and which would be expected to retard penetration. Somers (1962) in his review of the selective toxicity of fungicides draws attention to the paucity of information on the comparative efficiencies of plants and fungi in accumulating fungicides. In one of the few studies yielding such information, Richmond and Somers (1962a) found that over the same time period (30 min) spores accumulated 7100 μg captan g^{-1} spores compared with only 1·1 $\mu g\, g^{-1}$ for apple leaves (dry weight), although when applied to the roots of bean and tomato grown in solution culture, which lack a cuticle, captan is appreciably phytotoxic (Lukens and Sisler, 1958). As will be clear from Chapter 11 however, the plant cuticle is by no means impermeable. The relative permeability of both plant cuticles and fungal cells to different externally applied organic compounds will depend on the polarities of the penetrants, although with different relationships in each case. Somers (1962) points out that selectivity between fungitoxicity and phytotoxicity may thus be achieved by having the correct lipophilic–hydrophilic balance in the fungicide molecule so that it can penetrate the fungus but not the plant cuticle. Results illustrating this were obtained by Wellman and McCallan (1946) who showed that with the 2-alkyl-imidazolines fungitoxity was greatest with straight chain substituents containing 17 carbons in the 2-position; whereas maximum phytotoxicity was obtained with the 11–13 carbon atom substituents. Similarly Ross and Ludwig (1957) found that fungistatic activity among *N*-*n*-alkyl ethylenethioureas was greatest for the octyl member while the amyl homologue was most damaging to plants. A further example is provided by the work of van der Kerk (1961) with organotin compounds: fungicidal activity of tributyl tin compounds substantially exceeds that of triethyl compounds but the triethyl compounds are more phytotoxic. The general importance of having the correct polarity for optimum fungicidal action is indicated by various other studies, for example Rich and Horsfall (1952) reported that the fungitoxicity of a limited range of alkyl-nitropyrazoles was related to their partition between olive oil and water and Lukens and Horsfall (1967) found that the fungicidal activity of a series of saturated cyclic imides increased linearly with partition between oleyl alcohol and water. Similarly, results obtained by Somers (1958) indicate that uptake of an homologous series of *n*-alkyl thioethers of 2,5-dimercapto-1,3,4-thiodiazole is related to both their toxicity and lipid solubility. Such relationships are not universal however: Richmond and Somers (1962a) detected little effect of polarity, as indicated by partition between oleyl alcohol and water, on the relative fungitoxicity of various *n*-trichloromethylthio analogues of captan. In a currently more fashionable approach to a related problem, Brown and Woodcock (1975) investigated the importance of polarity as part of a very comprehensive investigation of structure-activity relationships of fungitoxic *N*-(1-substituted-2,2,2-trichlormethyl) formamides and bis-1-(2,2,2-trichloroethyl) formamides applied against powdery mildew on barley. Using the Hansch approach which relates the molar concentration of toxicant causing a standard response to its partitioning and electronic properties, they showed that activity of aryl substituted compounds in leaf spray tests was influenced by both partitioning and electronic effects. The relationship with polarity was parabolic, regarded as typical of systems in which partitioning between hydrophilic and lipophilic phases occurs during transport from the point of application to the site of action. The significance of the apparent parabolic relationship is critically examined in Chapter 10. Polarity was measured as mobility on reverse-phase thin-layer chromatography and the optimum value differed slightly between the aryloxy and arylamino series of compounds. Brown and Woodcock interpret their results as indicating that these compounds partition between various hydrophilic and lipophilic phases in both plant and fungal cells during their redistribution from the initial deposit on the leaf to the target site. Somewhat surprisingly, activity following application as a root drench correlated with partitioning properties alone. Since there is no indication that the mode of action differs between root drench and leaf spray tests, the authors concluded that this result may have merely reflected the greater number of partitionings following root drench treatment. Mean polarity values differed for aromatic and aliphatic series of substituents; this again presumably reflects differences in partitioning behaviour en route to the site of action. Whatever the detailed explanation for the correlations observed, these extensive results clearly provide further evidence for the importance of effects of polarity on penetration in the activity of fungicides.

The fungicidal activity of surfactants may be related to this discussion. This activity is well established (see references in Clifford and Hislop, 1975) although the only surfactant in common use appears to be dodine (dodecylguanidinium acetate) for controlling apple scab. Certain other fungicides such as tridemorph (2,6-dimethyl-4-tridecylmorpholine) undoubtedly have surface active properties. Recently there has been increasing interest in the use of surfactants for controlling apple mildew (Frick and Burchill, 1972; Clifford and Hislop, 1975; Hislop and Clifford, 1976). In the present context the studies by Clifford and Hislop are particularly interesting. In evaluating different surfactants (mainly condensates of ethylene oxide with 4-octyl or 4-nonyl phenol and non-ionic surfactants based on primary alcohols, but also including some cationic and anionic surfactants) as dormant season sprays, they found that efficacy within a series of octylphenol ethoxylates increased with decreasing hydrophilic–lipophilic balance (HLB) number. Many factors contribute to efficacy with such treatments but these results provide support for the view that ease of movement into the fungal cell, as indicated by HLB number, has an important influence.

The consistent indications from the studies by Wellman and McCallan (1946), Ross and Ludwig (1957) and van der Kerk (1961) that maximum fungitoxicity requires less polar properties than maximum phytotoxicity appear surprising at first sight, as the constituents of the plant cuticle are more lipoidal than those of the outer cell wall in fungi. This presumably indicates that the polarity requirement for passage through the outer cell wall was not the dominant factor determining activity for the fungicides concerned. Nevertheless the composition of the outer wall can have some modifying influence on the ease of penetration as the studies by Voros (1963), already quoted, show. The chitinous composition of phycomycete cell walls would be likely to require different physical properties for optimum penetration compared with the cellulosic walls of other principal groups such as ascomycetes and basidiomycetes and this could at least partly account for the poor performance against phycomycetes of many fungicides effective against the other classes.

Furthermore fungal cell walls can have a substantial capacity for retaining positively charged fungicides. This property varies greatly with fungal species and with the nature of the cationic toxicant. Owens and Miller (1957) found that the walls of *Aspergillus niger* had much greater affinity for a series of metallic cations, and for the organic fungicides glyodin and dichlone, than those of *Neurospora sitophila*. For example the walls of *Aspergillus* spores accounted for 54% of the cadmium taken up from solution under conditions where none of that absorbed by *Neurospora* could be detected in the walls; the amount associated with mitochondria was also greater in *Aspergillus*, but there was more cadmium in the microsomal fraction of the *Neurospora* spores so that overall uptake by *Aspergillus* was about 1·8 times that of *Neurospora*. The influence of cation type is illustrated by comparing these results with uptake of silver: 29% of the silver absorbed by *Aspergillus* was present in the walls as against 5% for *Neurospora* while the total uptake by *Neurospora* was roughly 1·2 times that of *Aspergillus*. Studies on the uptake of copper by Somers (1963) also reveal marked differences between species. After incubating spores with copper sulphate in sodium acetate buffer at pH 5·6 for 30 min, 31% of the copper taken up by *Alternaria tenuis* was present in the walls, compared with 16% for *Neurospora crassa* and 32% for *Penicillium italicum*. Differences were even more pronounced for the surface active cation of dodine (*n*-dodecylguanidinium acetate) (Somers and Pring, 1966). The maximum uptake under the particular experimental conditions investigated was 450 μmol g^{-1} for walls of *A. tenuis* compared with 580 for the whole spores, whereas the walls of *N. crassa* absorbed 170 μmol g^{-1} compared with 1040 for whole spores.

Significant retention by cell walls has also been observed with more recently introduced systemic fungicides. Thus Gottlieb and Kumar (1970) found that over 8% of the thiabendazole taken up from solution by spores of *Aspergillus oryzae* was associated with a fraction containing cell wall and probably nuclei, obtained by grinding the spores with sand.

2. *Mechanisms of Fungicide Uptake*

In considering mechanisms of uptake, some distinction can be made between cationic and neutral compounds, although there are many factors common to the two groups. In describing their composition (Section 12.III.A) we pointed out that fungal cell walls contain ionizable groups such as phosphates and particularly carboxyl groups. These impart a negative charge to the surface at normal pH values which can be clearly demonstrated by electrophoresis (Douglas *et al.*, 1959; Hannan, 1961). It is not surprising therefore that accumulation of cationic fungicides shows many of the general features of ion exchange. The ion exchange characteristics of the wall are well illustrated by the investigations of Sussman and Lowry (1955) who used methylene blue as a tool to study the physiology of the cell surface of *Neurospora tetrasperma*. Methylene blue was strongly adsorbed with an isotherm of the Langmuir (p. 73) type, consistent with ion exchange and limited exchange capacity. The dye was not easily displaced from the spores by monovalent ions, but could be eluted by di- or tri-valent ions. In general the ease of displacement followed the Hofmeister (Lyotropic) series. Uptake decreased with increasing acidity, as expected for pH-dependent charges (Sections 2.XVIII.A,C), but was not completely eliminated, even at pH 1·0. There was no uptake of anionic dyes at any pH. A feature of the uptake of methylene blue paralleled in many studies of fungicide accumulation was the very rapid speed of equilibration: in most cases 95% of the total appeared to have been absorbed within the first few minutes, although the curves presented by Sussman and Lowry show some evidence for a much slower further accumulation. There was no indication that metabolic processes were implicated in the uptake: accumulation was largely unaffected by decreasing the temperature from 28 to 4°C or by killing the cells. These various lines of investigation therefore reveal an essentially surface ion-exchange, involving pH-dependent charges probably located largely in the cell wall.

The uptake of cationic fungicides generally corresponds with the pattern observed for methylene blue, with variations in detail. Dodine is taken up very rapidly from solution by conidia of *Alternaria tenuis* and *Neurospora crassa* (equilibrium completed within 2 min in stirred suspensions) with a Langmuir type adsorption isotherm (Somers and Pring, 1966). Uptake is similar for dead and viable spores and can be decreased by competing metal ions especially the uranyl ion which is known to complex phosphate and carboxyl groups. Decreasing the pH of the suspension also reduces uptake. Electrophoretic studies (Somers and Fisher, 1967) showed that as the concentration of dodine in suspensions of *N. crassa* conidia increases, the negative surface charge is progressively reduced to zero and then replaced by a positive charge, apparently attributable to the retention of outwardly oriented dodine cations, held by van der Waals forces to surface lipids or to the hydrocarbon chains of already attached, oppositely oriented, dodine molecules. At lower concentrations the negative charge on the surface of cell walls, or stabilized protoplasts was also neutralized. Although all this evidence points to a simple ion exchange process, at least for the initial stages of uptake, there are some indications that other mechanisms, possibly including covalent bonding, may also contribute to retention. For example although metal ions compete with dodine for ion-exchange sites, as discussed above, the dodine retained by cell walls is apparently not readily released by washing or exchanged with freshly added dodine (Somers and Pring, 1966) although it is difficult from the figures presented to determine the distribution between spores and solution which would be expected for true reversibility of exchange and hence the degree of hysteresis.

The uptake of copper by suspensions of *N. crassa* conidia also shows certain characteristics of an ion-exchange process (Somers, 1963). For example, uptake is associated with a release of cations, particularly hydrogen ions, from the cell; other cations can also compete with copper for uptake when included in the supernatant solution. However, accumulation of copper differs in some important respects from that of methylene blue and dodine discussed above. Somers (1963) found that equilibration of copper between spores and solution was much slower than these organic cations taking at least 2 h under the conditions of his experiments. This is also considerably slower than has generally been found for other heavy metal fungicides, although the rate of uptake varies widely with different cationic species. Thus Miller *et al.* (1954) found that maximum uptake of mercury required at least 16 min under conditions where silver and cerium were absorbed in only 30 s or less.

The uptake isotherm for copper determined by Somers (1963) also differed from those of methylene blue and dodine; the initial gradient was much less and there was no evidence for a plateau up to a concentration of 200 $\mu g\ ml^{-1}$ in the external solution. Spores killed by boiling took up nearly three times as much copper as live cells; retention by dead cells remained the same after washing them free of soluble lysates, indicating that the binding sites are associated with particulate material. A further difference from methylene blue and dodine was that uptake of copper was markedly dependent on temperature. Maximum uptake ranged from 1100 $\mu g\ g^{-1}$ dry spores at 4°C to 3000 $\mu g\ g^{-1}$ at 35°C. Although the uptake curves did not correspond with a first order reaction, Somers calculated apparent activation energies for uptake from the uptake rates during the first 5 min of incubation using the Arrhenius equation. The resulting value of 10·5 kcal is higher than would be expected for a simple ion-exchange process although not outside the limits found for molecular diffusion in some materials. Respiratory poisons such as potassium cyanide, sodium arsenite and sodium azide reduced the uptake of copper to a moderate extent, giving some evidence for the involvement of metabolic processes, while uptake was increased by 25% under anaerobic conditions.

Uptake of copper thus appears to involve more than simple surface ion exchange. On the basis of these various studies the accumulation may be envisaged as an initial ion exchange with surface negative charges, followed by a slow permeation to binding sites situated within the protoplast including those responsible for the non-specific enzyme inhibition, thought to be responsible for the toxic action of copper. The accessibility of the absorbed copper to subsequent extraction varies with the fungal species. Washing treatments with 0·1 N HCl or non-radioactive copper removed or exchanged 97–98% of radioactive copper accumulated by *A. tenuis* compared with only 65–68% for *N. Crassa* while amounts removed by washing with the complexing agent EDTA were 96 and 33% for the two species (Somers, 1963). This reflects the differences in amounts retained by the cell walls of these species, discussed earlier. It appears that copper associated with the cell surface may prevent germination of the spores, but not kill them.

It is not necessary to invoke active metabolic accumulation mechanisms to explain the concentration of copper and other heavy metal toxicants from solution by spores: passive diffusion to negatively charged sites present at high density within the organism would readily account for most of the features of copper accumulation, although a small proportion of the uptake does appear sensitive to metabolic inhibitors.

Although the uptake of certain other cationic fungicides such as dodine appears in some ways less complicated than that of copper, for example it is independent of temperature, much faster and is unaffected by metabolic inhibitors, there is no evidence that the toxic reaction is located at the spore surface. In the electrophoretic study of Somers and Fisher (1967), for example, the spores were completely killed before there was any perceptible reduction in mobility. Although as might be expected for a surface active agent, dodine has been shown to disrupt the permeability of conidia and release nucleic acid components, presumably by damaging the cell membrane, this only occurs at concentrations well in excess of the ED_{50} value (Somers, 1968; Somers and Pring, 1966). The simple surface ion exchange process implied by the results of studies on uptake must therefore be followed by a subsequent permeation into the cell to effect the toxic lesion, suggested by Somers and Pring (1966) to involve destruction of membranes associated with intra-cytoplasmic organelles. Again such a permeation could be achieved by simple diffusion without recourse to metabolic processes. However, changes in permeability to potassium and to sodium dodeceyl sulphate on treatment of spores with dodine led Somers and Pring to suggest that the initial reaction of the dodine cation with the phospholipid cytoplasmic membrane facilitated the entry of more dodine into the cytoplasm where it could exert its action on the intracellular membrane structure.

Compared with cationic fungicides, uncharged fungicides in general have relatively little affinity for the cell wall. For example, Marthe (1968) found only 2–5% of oxathiin fungicides absorbed by various fungi to be associated with the wall fractions. Few if any of the non-ionic fungicides appear to cause destruction of the cell wall (as distinct from interfering with its synthesis). Some may act by affecting the permeability of cytoplasmic membrane (Lukens, 1971) but most probably must penetrate into the cytoplasm to exert their effects.

Uptake of these fungicides from solution by suspensions of spores is completed very rapidly. The period required for equilibration clearly must depend on the chemical, the receiving species and the experimental conditions, but the following results may be quoted as representative: less than 2 min for 2-heptadecyl-2-imidazoline (Miller *et al.*, 1954); less than 10 min for thiabendazole by *Penicillium atrovenetum* and *Aspergillus oryzae* (Gottlieb and Kumar, 1970); 45 min for cycloheximide by *Saccharomyces pastorianus* (Westcott and Sisler, 1964); 30 min for carbendazim by *Ustilago maydis* and *N. crassa* (Clemons and Sisler, 1970); 30–90 min for oxathiin fungicides by *U. maydis* and *Rhizoctonia solani* (Marthe, 1968) and 70–200 min for captan by *N. crassa* (Richmond and Somers, 1962a).

The degree of concentration over the external solution tends to be rather less than for cationic fungicides. Careful studies on the uptake of organic fungicides have emphasized how misleading simple measurements of loss from solution can be in this connection. Richmond and Somers (1962a) found that captan was taken up from a $2{\cdot}5\times10^{-5}$ M solution by spores of *N. crassa* to the extent of over 2000 $\mu g\ g^{-1}$ spores (wet weight), equivalent to a concentration of over 300 times that in the external solution. When the spores were pretreated with non-toxic levels of thiol reagents such as iodoacetic acid and mercuric-*p*-chlorbenzoate, captan uptake was decreased to negligible levels without affecting captan toxicity (Richmond and Somers, 1962b). The authors suggested that most of the captan absorbed was decomposed by thiol groups distributed within the cells before reaching its site of action. Many of the more recently introduced fungicides, notably the systemic compounds, have much more specific modes of action than the older materials, particularly the heavy metals which acted as general metabolic inhibitors. They would therefore not depend in the same way on accumulation for their selective action. This is borne out by investigations such as those by Clemons and Sisler (1970) with carbendazim, Wescott and Sisler (1964) with cycloheximide, Marthe (1968) with oxathiins and Gottlieb and Kumar (1970) with thiabendazole which show that amounts taken up are much smaller than for most protectant fungicides.

With regard to the molecular properties required for facile penetration, we have already discussed one characteristic which appears important in many cases, namely polarity. Dependence on polarity has generally been associated with unaided diffusion through structured tissue containing hydrophilic and lipophilic phases (Chapter 10). Such dependence need not be confined to purely passive diffusion processes, however, but could also reflect the requirements for loading on to active carrier mechanisms. The experimental evidence indicates that the uptake of some non-ionic fungicides is essentially a physical process. For example, Marthe (1968) found the uptake of oxathiins during 30 min exposure periods to be directly proportional to concentration and he detected relatively little difference between heat-killed and living cells. However, metabolic processes are clearly involved in the uptake of other compounds. Richmond and Somers (1962b) found that the effects of oxygen supply, metabolic inhibitors, captan concentration and temperature ($Q_{10} = 1{\cdot}9$) on the accumulation of captan by *N. crassa* spores were all consistent with an active uptake process, and the involvement of reactions with thiol groups in the uptake process has already been discussed. The observation (Richmond and Somers, 1962c) that captan uptake (over 30 min, less than the period required for equilibration) by different spore species was unrelated to their surface areas provided further evidence against a simple diffusion process, although this result could also be due to other factors such as differences in permeability and binding sites, associated with differences in composition.

It is also perhaps significant that Richmond and Somers (1962a) found little effect of polarity on uptake of captan analogues, although as pointed out earlier, correlations between polarity and uptake are not conclusive evidence for physical penetration processes.

As the analysis of permeation processes in Sections 10.III and 12.II.B demonstrated, much further insight into the nature and significance of uptake can be obtained by considering the time-course of accumulation. In discussing the penetration of toxicants into insects (p. 666) we pointed out that the simplest relationship for diffusion into a receiving body of volume $\mathbf{V}_{rec}$ and surface area $\mathbf{A}$ from a constant external concentration of C_{don} at the interface is

$$\mathbf{M}_t = C_{don}\mathbf{V}_{rec}\{1 - \exp(-P\mathbf{A}t/\mathbf{V}_{rec})\} \tag{8}$$

Where $\mathbf{M}_t$ is the amount of material accumulated in time t. Where there are "binding sites" within the receiving organism or its composition is such that the concentration inside at equilibrium is given by $C_{rec} = \mathcal{P}C_{don}$, then the right-hand side of eq. (8) must be multiplied by $\mathcal{P}$. The quantity $\mathcal{P}C_{don}\mathbf{V}_{rec}$ is of course $\mathbf{M}_\infty$, the amount present inside the organism at equilibrium. In Fig. 3 the relationship between $\mathbf{M}_t/\mathbf{M}_\infty$ and $P\mathbf{A}t/\mathbf{V}_{rec}$ given by eq. (8), setting $P\mathbf{A}/\mathbf{V}_{rec} = 1$, is compared with the experimentally determined time course of uptake for various fungicides. To standardize comparisons, the time scales were adjusted so that all curves were coincident at the half-way ($\mathbf{M}_t/\mathbf{M}_\infty = 0{\cdot}5$) stage. The curve for uptake of copper by *N. crassa* (Somers, 1963) corresponds moderately well with the model curve for simple diffusion, particularly during the early stages

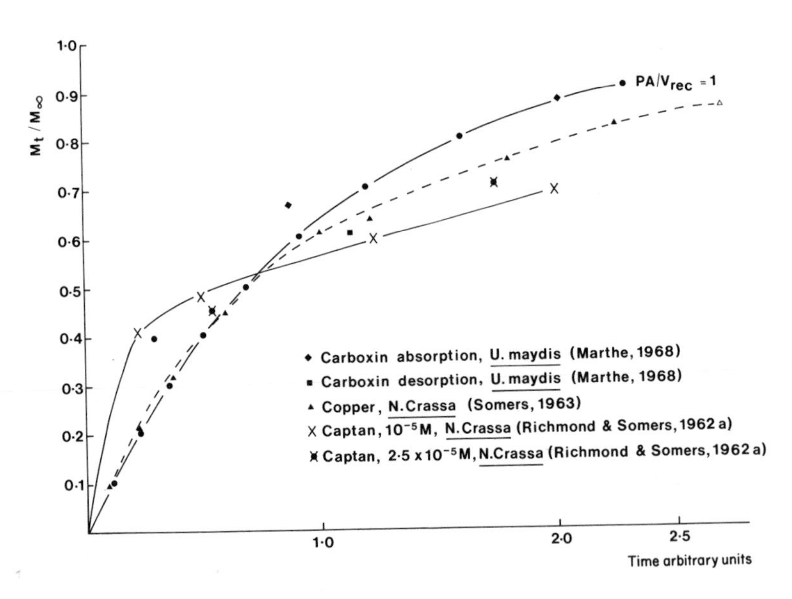

Fig. 3. Time course of fungicide uptake by fungi: experimental results compared with model relationship (eq. (8)).

of uptake. This is consistent with a relatively small contribution from metabolic processes. Similarly points from the curves for uptake of oxathiins (Marthe, 1968) fit the model reasonably well, although there were too few observations within the range of interest to allow detailed comparison. It will be recalled that Marthe's results give no indication of active processes. In keeping with all the evidence suggesting the involvement of complex metabolic reactions in the uptake of captan, the time course of its uptake by *N. crassa* (Richmond and Somers, 1962a) deviates markedly from the curve for simple diffusion. From their various studies on captan uptake, Richmond and Somers (1966) concluded that the energy dependent component of uptake is rate-determining and that this explains why the same external concentration of captan is required to achieve toxicity against *N. crassa* whether the cellular detoxication sites (i.e. the thiol groups) are blocked or free. Figure 3 gives some rather general support to this conclusion although much more complex diffusion patterns are possible than that depicted by the model curve, even in the absence of metabolic processes.

For many fungicides, however, a simple molecular diffusion through various barriers to "binding sites" within the cytoplasm is probably an adequate representation of their accumulation into the relatively simple fungal cell structure. The nature of these binding sites will vary considerably with compound and with fungal species. Where they involve simple ion exchange, partition or physical retention at readily available surfaces, uptake should be reversible. For a fungicide whose uptake followed the simple relationship given in eq. (8), release to a washing solution, after equilibration, should follow a time-course represented by the "mirror image" of eq. (8) (see Sections 3.V.J and 6.VI.C; sources and sinks). Most washing experiments have not been done sufficiently rigorously for the reversibility or the time course of release to be established. However, Marthe (1968) found that release of oxathiins was much slower than uptake; apparent "permeabilities" for absorption calculated by applying the relationship given in eq. (8) are several times those for release. Such results could imply covalent bonding at the site of action, but to explain them fully would require much more detailed investigation and a more sophisticated analysis than that based on the very limited relationship given in eq. (8).

13. Effects of Growth and Movement of Organisms on Interception of Pesticides

I. Introduction 698
II. Control of soil-inhabiting pests and pathogens 699
A. Introduction 699
B. Action of seed treatments 701
C. Interaction between biological factors and toxicant supply from seed treatments 704
III. Control of insects and pathogens attacking aerial parts of plants . . 710
A. Introduction 710
B. Pick-up of pesticide by insects from residual deposits 713
C. Effects of humidity on pick-up from deposits 720
D. Sites of uptake on insects walking on residual deposits 722
E. Uptake of pesticides from residual deposits by crawling insects . 725
F. Selectivity between different insect species 727
G. Interception of pesticide sprays by flying insects 730
IV. Movement of fungal spores in relation to contact with fungicides . 735
V. Attractants, repellents and related behaviour-controlling chemicals 740
A. Introduction 740
B. Directional response to behaviour-controlling chemicals . . . 742
C. Sensitivity of response to behaviour-controlling chemicals . . . 748
D. Practical use of behaviour-controlling chemicals in pest control . 754
VI. Growth of plants in relation to uptake of soil-applied pesticides . . 757

I. INTRODUCTION

The central theme of this book is how pesticides are transferred from the point of application to the receiving organism and how the nature of this transfer influences efficacy and selectivity. Preceding chapters have concentrated on the various physical processes involved in this transfer. There is a tendency to consider these physical processes to the exclusion of other mechanisms and to assume that they are by far the most important. This is often not so; movement of the organism itself can play an essential part in bringing the toxicant and the recipient together. This movement of organisms contributes just as much as the physical aspects of dosage transfer to the apparent differences between the activity of toxicants in primary laboratory screens, where they are applied more or less directly to the recipient, and their often seemingly capricious and unreproducible performance in the field.

In practice the biological and physical processes will be inseparably interrelated in determining the exposure of an organism to a toxicant not directly applied. As the organism approaches the toxicant source it will experience some kind of gradient of concentration. The dose available to the organism will depend on the nature of this gradient, determined by physical processes within the medium concerned, and on the pattern of locomotion. We illustrate these principles in greater detail in considering the movement of soil pests to treated seeds in subsequent sections. The two aspects of transfer may be usefully separated for ease of discussion, however, and our main purpose in this chapter is to review aspects of the behaviour of receiving organisms which have an important influence on their exposure to the toxicant.

Many existing pest control practices depend implicitly on "biological transfer"; some of the examples to be quoted may appear simple, but their significance and the principles they represent are not always recognized. In a substantial proportion of these examples, the significant behaviour of the organism is influenced by what might be termed generally attraction or repulsion. There is now considerable interest in the more specific and deliberate exploitation of such responses through the use of behaviour controlling chemicals; discussion of this approach is therefore included at the end of this chapter.

II. CONTROL OF SOIL-INHABITING PESTS AND PATHOGENS

A. Introduction

The control of soil-borne pests and pathogens by pesticides presents particular difficulties because the damaging organisms are often widely distributed in the soil. The problem is thus one of getting a lethal dose to many very small targets scattered within a large bulk of a material in which the amount applied is dissipated by chemical and microbial degradation and in which movement of the toxicant is inhibited by adsorption and by the restricted and tortuous nature of the pore system. Many of the most intractable outstanding problems in crop protection share this difficulty. These include the control of nematodes, slugs and a range of soil-inhabiting insect larvae such as the various dipterous pests of vegetables and diseases such as take-all of cereals.

Various aspects of this problem of getting the pesticide to the target are discussed elsewhere in this book. The simplest approach is to broadcast the toxicant on the soil surface, or incorporate it as uniformly as possible with the object of distributing it throughout the entire habitat of the damaging organism. However, this requires large rates of chemical and so can be prohibitively expensive, as with certain nematode problems, or can have undesirable effects on non-target organisms, particularly as the chemicals used must often be relatively persistent. The problems associated with the persistent organochlorine insecticides provide an only too familiar example.

An obvious avenue for potential improvement is to concentrate the chemical more in the critical region of the soil where the crop plant is at risk and thus take advantage of the movement of the pest towards its host. An appropriate candidate for such an approach is the wheat bulb fly larva (*Leptohylemyia coarctata* Fall.) which will be discussed as a representative illustration of the types of factor which may be involved. The adult fly lays its eggs during July or August in bare soil in uncropped fields, or fields where root crops or peas are growing; the ovipositing adult does not thus select a specific host. The eggs subsequently become incorporated within the soil by cultivation or natural processes and the larvae, which hatch from the eggs six months later in early spring, must locate a suitable host plant by moving within the soil if they are to survive. Many graminaceous plants including wheat, barley, rye and various wild grasses are attacked, although oats are completely immune (Gough, 1946). The larvae bore into the stems of the seedlings, entering wheat plants near to the swollen part of the shoot (known as the bulb) just above the germinating seed between 1 and 3 cm below the soil surface and killing the growing point. Protection of this region is therefore sufficient to prevent attack. The most convenient way of localizing an insecticide in this region is to apply it with the seed at planting. This can be done either by combine-drilling or treating the seed directly with the chemical. As might be expected, seed treatment, which confines the chemical specifically in the required zone is more efficient, but both methods achieve considerable economies in the amount of insecticide used compared with broadcast treatments as Table I, which also includes results for wireworms, shows.

Table 1
Influence of placement on effective application rate (g ha^{-1}) *of insecticides for controlling soil-borne insects* (*after Price Jones*, 1974; *Graham-Bryce, 1973a*)

Pest	Insecticide	Method of application		
		broadcast	combine-drilled	seed treatment
Wireworms	Lindane	840	280–420	56–84
Wheat bulb fly	HHDN	—	1120–2240	140

B. Action of Seed Treatments

To design the most effective treatments, it is necessary to know in detail the route which the larva takes between hatching and entering the plant. Observations by Gough (1946) showed that larvae can find host plants from eggs buried up to 45 cm deep in soil. Long (1958) and Way (1959) concluded from experiments in pots, seed boxes and field pots that the newly hatched larvae move upwards in soil at a fairly steep angle before searching for the host plant in the upper layers of soil. A seed treatment may therefore act by one or both of two mechanisms discussed by Way (1959) who also distinguished two phases of attack. First the larvae may be killed as they move towards the plant through the diffuse zone of soil containing insecticide around the treated seed. Larvae which survive passage through this zone may cause a primary attack on the plants. However, some chemicals may be absorbed by the plants and kill these larvae by a systemic action. Larvae not killed by this mechanism can migrate to a fresh plant after damaging the shoot and cause a second phase of attack. The relative importance of the different mechanisms depends on the chemical. Thus Way (1959) found that aldrin and dieldrin decreased both the primary attack, presumably by killing the larvae in the soil before they entered the plant, and the secondary attack, implying that the chemicals were taken up by the plant to kill larvae which survived the insecticide in the soil. In contrast, with lindane, control appeared to depend much more on the chemical killing the larvae in the soil or preventing them from feeding. Evidence obtained by Griffiths and Scott (1967) suggests that several organophosphorus insecticides also act mainly in the soil, preventing the larvae from entering the bulbs.

The distances over which these effects occur may be estimated theoretically by calculating the spread of chemical from the seed. Redistribution by molecular diffusion can be computed for an idealized spherical seed using the relationship

$$C = \frac{aC_0}{r} \operatorname{erfc}\left(\frac{r-a}{2\sqrt{Dt}}\right) \tag{1}$$

where C is the concentration at a distance r from the seed centre after time t, a is the radius of the seed and erfc denotes the error function complement. C_0 is the concentration at the seed surface, assumed constant; this would be a reasonable assumption for the situation where the amount of chemical on the seed greatly exceeded the capacity of the air, water and solid phases in the immediate vicinity to accommodate it so that the deposit on the seed acted as a reservoir maintaining a constant saturated concentration at the surface. Table 2 gives values of the concentration relative to that at the surface (C/C_0) after various times at distances of 0·67 and 1·0 cm from each seed. These distances correspond roughly with the mid point between cereal seeds drilled uniformly at 187 and 125 kg ha^{-1} in rows 17 cm apart. The value taken for D is 5×10^{-8} cm^2 s^{-1}, a reasonable average for soil applied insecticides and fungicides which are usually relatively strongly adsorbed (see Chapter 5).

Table 2
Redistribution of pesticides in soil by molecular diffusion from idealized spherical seeds

	$C/C_0 \times 10^2$			
r (cm)	14 days	30 days	90 days	150 days
0·67	1·8	4·5	8·6	10·1
1·0	0·2	1·3	4·8	6·6

The values in Table 2 emphasize again that redistribution by molecular diffusion is slow and that its effects are important only over short distances. Insecticide will also be transported away from the seed by downwash with percolating rainwater or by the bulk transfer associated with other water movements in the soil profile. The effects of such transport are difficult to estimate because the location of the seed in relation to the heterogeneous pore system in the soil and hence the pattern of water flow around it are difficult to specify and because of the fluctuation of water movements with weather. However, the general features of redistribution by bulk flow are reasonably predictable. Assuming that the predominant direction of the water movement is downwards, at least in the early stages of growth when the soil is usually moist, a chemical present initially as a point source will be transported downwards as an increasingly diffuse conical zone. The chromatographic nature of downwash through soil makes it possible to estimate approximately the rate at which the chemical is transported downwards from the seed. As discussed in Section 5.VII.C, simple chromatographic theory predicts that the main insecticide front will move downwards at a rate f times the rate of descent of the percolating water, where f is the fraction of the chemical which is not adsorbed and is freely mobile in the soil solution. If we again take as our representative conditions soil at 20% moisture content and a distribution coefficient for adsorption of about 20, which is reasonably typical of soil-applied insecticides (Hamaker and Thompson, 1972) then f is approximately 0·01. Under these conditions the vertical length of the cone of the treated soil below the seed would be about 1 cm following 20 cm of rain, which would result in a net 100 cm of water movement in the soil at 20% moisture content. The extent of lateral spreading, which determines the diameter of the base of the cone, depends mainly on hydrodynamic dispersion. It can be shown that the width within which half the content of an initial point source will be found after being washed downwards for a distance ℓ in a system with a dispersion coefficient D_{disp} is given by $y = 0{\cdot}94\sqrt{D_{disp}\ell/v}$ where v is the flow rate. Thus if we take as a representative value $D_{disp}/v = 0{\cdot}1$ cm at a flow rate of 1 cm day^{-1} (Leistra, 1973) y has a value of 0·31 cm for a depth of leaching of 1 cm.

This simplified picture of the shape and dimensions of the zone affected by chemical will be of course greatly modified by the heterogeneity of the soil around the seed and by the fluctuating water movements which occur in practice. Molecular diffusion will also spread the chemical above the seed to some extent, as even when the water movement is predominantly downwards, its rate is unlikely to be sufficient to completely overcome the omnidirectional nature of diffusion. In practice the zone is likely therefore to be roughly pear-shaped, the seed being situated somewhere in the neck of the pear. Whatever the shape of the zone, this discussion emphasizes again the very limited mobility of chemicals used as seed treatments. Direct confirmation of this limited mobility has been obtained in the case of the fungicide ethirimol by autoradiographic examination of cones of soil from the field containing seeds treated with ^{14}C-labelled chemical (Graham-Bryce and Coutts, 1971). These observations also reveal the importance of soil structural factors as well as adsorption in determining the extent of mobility and the shape of the treated zone.

C. Interaction between Biological Factors and Toxicant Supply from Seed Treatments

We can conclude with confidence, then, that when a pesticide is applied to the seed (or even to incorporated granules unless they are very small and numerous, see Section 14.VII.C) it will not become uniformly distributed by diffusion during the period of its desirable action. If chemical decay is significant the concentration will never become uniform. Such limited mobility is, of course, an essential requirement if the advantage of localizing the toxicant in the critical region for attack around the germinating seedling for an extended period is to be achieved. Any chemical which proves effective when applied as a seed treatment is likely to have this property. Our concern in this chapter is to consider how this non-uniform distribution influences efficacy against pests approaching the treated zone of soil. The problem is very complex, but some guidelines can be deduced by examining extreme cases (Graham-Bryce and Hartley, 1979).

We consider first the situation in which the pesticide is so soluble (or volatile) that the contents of each source (seed or granule) can be assumed molecularly dispersed at the time of application. This is the case of the "dissolved pesticide" discussed in Section 14.VII.D. At times sufficiently short for there to be no significant overlap of the concentrations diffusing from neighbouring sources, the concentration decreases at all times with increasing radial distance, r, from the source and the total dose (i.e. the amount dissolved, which remains constant) $\mathbf{M} = 4\pi \int_{\infty}^{a} Cr^2 \, dr$.

An organism wandering in the soil will experience an exposure $\int C \, dt$ over the period concerned. If it moves at a constant velocity, v, this can be expressed as $\int C/v \, dx$ where x is the distance measured along its path. If its movement is entirely random, it will sample all concentrations indiscriminately and the average exposure will be the average concentration, multiplied by the time. More extensive distribution of toxicant in the soil due to diffusion for a longer time will result in (1) less deviation of individual exposures from the average (2) a greater proportion of exposures being contributed by low concentrations.

Highly directional movement towards an attractive source, such as a "host" seed can be represented in extreme form as a straight line. The exposure becomes $\frac{1}{v}\int_{\infty}^{a} C \, dr$ if the zones can be assumed so localized that every organism starts outside the influence of any. Since the value of r at which any value of C is attained increases with time and $\int Cr^2 \, dr$ is constant (from the definition of the dissolved source) it follows that $\int C\,\mathrm{d}r$ must decrease with increasing time.

If the solubility, volatility and extent of adsorption of the pesticide are so small that the soil water at the surface of the source is held at the saturation concentration throughout the required period of action, the concentration at all values of r (except a) must increase with time and so therefore will the exposure of an organism making a direct line for the source. If this condition applies for only a limited period, until the pesticide at the source is all dissolved, the exposure will increase over this period and then decrease. The total amount dissolved, proportional to $\int Cr^2\,\mathrm{d}r$ must of course increase more rapidly since with increasing time, the more remote regions (with greater r) make an increasing proportional contribution.

In both dissolved and saturating cases, therefore, the ratio of exposure to amount of pesticide released (equivalent to that applied in the dissolved case) decreases with time so that action as rapid as the behaviour of the pest allows is desirable. If this behaviour sets a rather long lower limit to the necessary period of protection, it would be desirable (were there a free choice) to use a pesticide of solubility and at an applicance, matching this period. A compound of low solubility could well prove more effective than one which was very soluble. The characteristics of the various sources are illustrated in Fig. 1.

An additional factor which can influence the initial release of the toxicant and its behaviour in the immediate environment of the seed is the nature of the formulation and application procedure (see Section 14.IX.A). The optimum balance for the properties of formulation and toxicant to obtain most satisfactory practical performance will clearly depend on the detailed nature of the control problem and conditions in the field; while it is possible to indicate the factors to be considered, some degree of empirical investigation will almost certainly be needed to determine this balance. For example, Jeffs (1973) showed that adhesion of lindane powder could be greatly increased by pretreating seeds with suitable adhesives such as polybutenes or vegetable oils. However, effectiveness against wheat bulb fly was decreased by the polybutene treatment, presumably because the release of toxicant to the surrounding soil was inhibited. Furthermore, there is evidence that the effectiveness of some chemicals which act mainly in the soil may be limited by restricted movement, even without decreasing mobility further by formulation, especially when the seed is drilled deep and the treated zone of soil is below the level at which the larvae are most likely to attack. In such a situation chemicals which have systemic activity, such as dieldrin, appear

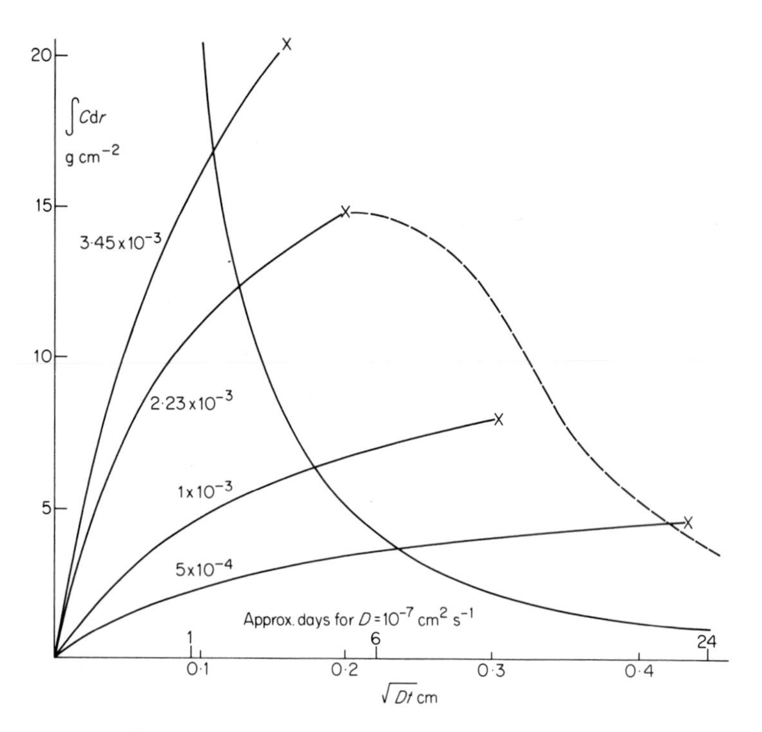

Fig. 1. Values of $\int_{a'}^{\infty} C\,\mathrm{d}r$ for direct radial approach to the surface of a spherical source over a short period but at different times (t) following the introduction of the source. The four ascending curves are for a source of radius 400 μm, the surface of which is uniformly covered with a pure substance of total solubility (including air space, water and adsorbed) as indicated on the curves in g cm^{-3}. The total amount of dissolving substance is in all cases 5×10^{-5} g. The points where the coating is just exhausted and the surface therefore ceases to be saturated, are marked X. The calculations are from an exact equation for the idealized conditions specified.

The descending solid curve is for a substance completely dissolved when introduced, calculated from the exact "point source" equation. More detailed computation shows that, for distances and times of practical interest, a source less than 1 mm diameter can be considered a point source. This equation gives $\int C\,\mathrm{d}r \propto t^{-1}$.

The maximum value attainable for the integral increases with increasing solubility but the time to reach this value decreases, the theoretical limit (origin of descending curve) being $\int C\,\mathrm{d}r$ infinite for zero time. It is evident that, once the saturating supply is exhausted, the value of $\int C\,\mathrm{d}r$ must fall, but the fall is initially less steep than in the point source case because the further diffusion is occurring from an already extended volume. A rough computation can be made by considering that, at each point X, the substance is uniformly distributed within, and only within, a volume of radius b where $\bar{C}b$ = the calculated $\int C\,\mathrm{d}r$ and $\frac{4}{3}\pi b^3\bar{C}$ = the constant amount (5×10^{-5} g). An exact equation is available for such a system (Crank 1956, p. 27) and has been applied to compute the broken line extension of one "saturating" curve after all the material in the source has been dissolved. This gives a rough indication of the broadly sigmoid, eventually asymptotic, approach to the point source curve.

to have advantages. Formulations which increase adhesion may also increase the risk of phytotoxicity.

The effects of formulation and application method are of considerable practical importance because there is much interest in improving present procedures, as we discuss more fully in Section 14.VIII. In commercial practice, a large bulk of grain must often be treated rapidly to minimize costs and process the necessary seed between harvest and sowing. Partly for this reason, methods developed for applying both liquid and powder treatments have had deficiencies (Lord *et al.*, 1971a, b; Graham-Bryce, 1973). With powders average loadings may be well below the target dose due to poor adhesion. With liquid formulations, average loadings tend to be nearer the target dose, but the distribution between individual seeds is very uneven. The consequences of such low loadings and uneven distribution depend on the way the particular chemical acts and on the nature of the biological transport effected by the movement of the pest and the physical processes of redistribution in the soil which have already been outlined. The effects of low loadings are indicated by the results of Griffiths *et al.* (1970) who investigated the effects of different doses of powder formulations of lindane, chlorfenvinphos and carbophenothion on seedling emergence and on damage by wheat bulb fly larvae. These insecticides were all ineffective at rates less than 10 μg per seed, but control of the larvae increased progressively with increasing rate of application. Such an effect would of course be expected for a chemical acting primarily in the soil if the increased amount on the seed resulted in a larger zone of treated soil or a greater average concentration within the treated zone. It should, however, be pointed out that if the reservoir of chemical on the seed is already sufficient to maintain the immediately surrounding shell of soil saturated with chemical, and if the rate of solution of the chemical is fast relative to the rate at which it diffuses away from this immediately adjacent region, then increasing the rate of application would affect neither the size of the treated zone nor the average concentration, until this reservoir was exhausted and all the chemical had dissolved. This is the "saturated" case discussed above. Evidence for such a plateau in the dose–response relationship was obtained in later work with dieldrin (Griffiths *et al.*, 1974). Above a threshold value of about 10 μg per seed, which in this case gave satisfactory control, there was little further improvement as rates were increased. This difference from the chemicals studied previously may be partly related to the different mode of action of dieldrin; it suggests that the performance of dieldrin should be rather less sensitive to variations in loading, which is consistent with the general observation that dieldrin has proved relatively reliable in practice.

Uncertainty about the dose–response relationship to be expected for a particular chemical contributes to the difficulty of predicting the detailed biological consequences of non-uniform distribution between individual seeds. Although it would appear obvious that best control would be obtained when all individual seeds carried the same amount of chemical, it is by no means clear at what point uneven distribution is likely to affect control significantly, particularly as cereal seeds are sown fairly close together. Even though mobility is limited, adjacent seeds would be expected to obtain some benefit from the chemical originally applied to their neighbours. The extent of this benefit depends on the biological transfer processes: for chemicals acting principally in the soil, on the exact route which the larva takes on its way to the host plant and for chemicals acting by systemic action on the extent to which the roots and other absorbing regions of each plant intercept the neighbouring treated zone. In the case of the systemic fungicide ethirimol, there are some indications (discussed by Graham-Bryce, 1973) of the contribution which this growth of the absorbing regions of the plant, which can be regarded as an important biological transfer process discussed more fully below, makes towards improving the availability of the chemical. Ethirimol is moderately strongly adsorbed and has very limited mobility in soil. However, there is indirect evidence that under field conditions adjacent seedlings may tap chemical originally applied to neighbouring seeds. Thus in trials comparing different loadings at a variety of seed rates, control of powdery mildew on barley was found to be determined more by the overall rate of chemical applied per acre than by the rate per seed. The results must be interpreted with caution, because in other experiments in which seeds treated with ^{14}C-labelled ethirimol were sown carefully by hand alternately with untreated seeds uniformly spaced at the average distance to be expected in the field, the radioactivity which appeared in the plants which had grown from untreated seed was only a small percentage of that taken up by plants from the treated seed. In this situation therefore, there was relatively little sharing. Under practical field conditions, however, where the seeds are sown much less uniformly and often cluster together, more sharing may occur, provided the seed rate is sufficiently high. In the case of wheat bulb fly control, there is little evidence that significant benefit is obtained from the chemical on adjacent seeds. In experiments with chlorfenvinphos in which batches of seed with different loadings were carefully mixed to produce different degrees of non-uniformity, control of wheat bulb fly decreased progressively as the distribution of chemical became less uniform (Lord *et al.*, 1971b). These experiments illustrated a further factor which can complicate the effects of non-uniformity, namely phytotoxicity. It was found that although control of the pest was improved by more uniform distribution between seeds, on some soils more plants were damaged with the more uniform treatments, so that best results were obtained when no insecticide was used. Clearly this suggests a need for less phytotoxic pesticides, although there could be some situations in which it would be acceptable to sacrifice a proportion of the plants by phytotoxicity if this eliminated a pest which would otherwise damage many.

In summary, therefore, seed treatments should ideally meet several different requirements, which may to some extent be mutually conflicting. The chemical used must be persistent, although not so persistent as to cause problems in the environment; it should not be phytotoxic and should have a limited but significant mobility in soil. The formulation must adhere well to the seed but not so strongly as to inhibit the release of chemical. Uniform distribution from seed to seed is desirable for good pest control, but may increase damage to plants if phytotoxic chemicals are used. It should certainly be possible to devise treatments which go a long way to meeting these requirements, but if this is not to be purely empirical, more detailed knowledge of the "biological transport" effected by the pest in relation to its susceptibility to different toxicants is required, as well as better understanding of the release properties of different formulations. This is undoubtedly true of most pest control problems: the characteristics of the target organism as a "sink" for the pesticide must be clearly defined before the most efficient treatments can be devised. In the case of soil-borne insect pests, remarkably little appears to be known about how the toxicant is taken up.

In spite of the difficulties and uncertainities which have been discussed above, seed treatment remains a convenient, economical and selective method of applying crop protection chemicals (Graham-Bryce 1973). Because it concentrates the chemical in the critical region to which the pest moves, it may be regarded as a form of toxic bait. There is evidence that selection of the host plants by wheat bulb fly is influenced by chemicals released by the plants into the soil (Stokes, 1956; Scott, 1974). The exudate from wheat has an arrestant effect whereas the exudate from oats, the only non-host among the graminaceous cereals, has an anti-arrestant effect. In the case of this pest, at least, therefore the "bait" relies to some extent for its effectiveness on the presence of a behaviour-controlling chemical. Manipulation of such behaviour-controlling chemicals offers interesting possibilities for exploiting biological movement more deliberately for pest control and is discussed below.

Control of wheat bulb fly has been considered at length to illustrate factors which must be taken into account in devising effective treatments for controlling soil-borne pests, and ways in which the movement of the pest may be exploited to bring target organism and toxicant together, thus combatting the problems associated with the pest being distributed widely throughout a large mass of intractable material. Certain other methods of application also offer a means of overcoming these difficulties by providing a concentrated localized source of chemical. Granular formulations are probably the most widely used. In some cases these are placed in the critical region of the soil to obtain best effect, but the optimum distribution will depend on the mobility of pesticide, movement and susceptibility of the pest and the characteristics of the crop to be protected. The use of granular formulations is considered in detail in Sections 14.II.F and 14.VII.

III. CONTROL OF INSECTS AND PATHOGENS ATTACKING AERIAL PARTS OF PLANTS

A. Introduction

In some ways the wide range of pests and pathogens which attack the aerial parts of plants might appear a somewhat easier target than those which are confined to the soil. However, even when a pesticide is sprayed on the foliage at high volume and to the point of run-off, the damaging organism is unlikely to be entirely eliminated by direct contact and the residual deposit of chemical is usually by no means uniform, certainly not on a microscale, as we discuss in Chapter 14. While the problems of bringing the toxicant and pest together may be less intractable than in the soil, they are nevertheless of considerable importance, and understanding of the biological component of the process is equally desirable.

For the purposes of crop protection, insect pests have traditionally been classified broadly according to their feeding habits and the nature of the damage which they cause to the plant, as these clearly greatly influence the type of chemical treatment which can be used for control. The main broad division is between chewing or biting insects and sucking insects. Chewing insects can be subdivided further according to the nature of the mouth parts and the type of attack on the plant; the subdivisions include leaf-eaters, leaf-miners, stem borers and groups which attack flowers, seeds or fruits. "Sucking" insects, which usually do not actually "suck" but exploit the internal pressure of the plants' tissues and vascular system to provide a self-powered supply of nutrients, have mouth parts adapted for penetrating plant tissue and tapping the juices. This group includes aphids, mealy bugs, capsids, whitefly and scale insects. Such a brief classification is of course far from complete and as with all classifications, there are pests which fall into more than one category or do not fit satisfactorily into any. Nevertheless it does provide a framework against which to consider the requirements and characteristics of different chemical treatments and may be related to a corresponding broad classification of insecticides according to the route by which they enter the insect's body. The largest category consists of "contact insecticides" which, as their name implies, are effective when contacted by the insect and are therefore generally considered to be taken up through the insect's integument into the haemolymph, although Gerolt (see Section 12.II.B) has suggested that certain compounds, traditionally regarded as contact insecticides, may also enter, via the tracheal system. Compounds such as the natural and earlier synthetic pyrethroids which are photochemically degraded rapidly and hence can only kill effectively outdoors by actually hitting the pest are termed *direct contact* insecticides; in some cases insects may acquire the lethal dose by flying through an airborne spray. Other compounds remain active for significant periods so that insects can be poisoned by contact action as they alight or move over a treated surface. Such chemicals protect the plant for a sustained period after application and are therefore sometimes termed *protective contact* insecticides. It will be clear that such compounds, of which DDT is the best known example, are effective against several of the groups of insect described above, particularly the chewing insects. As would be expected however, those which do not penetrate significantly into the plant are relatively ineffective against sucking insects.

A second category of insecticides comprises the stomach poisons, which act after being ingested by the insect during feeding. Many contact insecticides also fall into this group. Stomach poisons must be applied at an appropriate time to the parts of the plant which the pest consumes, must not repel the insect and must be capable of passing from the gut to the site of action within the insects body. Like the contact insecticides, such compounds tend to be most effective against chewing insects and are relatively ineffective against sucking insects which do not consume plant tissues. For the sucking insects, systemic insecticides, a group whose behaviour is discussed extensively in Chapters 9 and 11, clearly have great advantages. It is worth emphasizing again here that the degree to which insecticides are transported within the plant can vary greatly. Even where transport is limited to a translaminar movement from the side of the leaf receiving the chemical to the other, this can greatly help the toxicant to reach the target pest.

Significant quantities of some toxicants can be taken up via the vapour route. This has led to a further category of insecticide, those that exert their effects by fumigant action. As with the other categories, this group is rather unsatisfactory and ill-defined: insecticides have a very wide range of vapour pressures and while compounds like nicotine (vapour pressure $4{\cdot}25\times10^{-2}$ mmHg at 25°C) or dichlorvos ($1{\cdot}2\times10^{-2}$ mmHg at 20°C) undoubtedly exert their effects largely as vapours, it is very difficult to draw the line between fumigants and non-fumigants (see Section 5.III.A).

With all their shortcomings, these schemes for classifying insect pests and insecticides have a definite practical value in relating different types of treatment to different types of pest, and can serve as a background for our further discussion of biological transport. We shall be mainly concerned here with "chewing insects" and "contact insecticides" and with the contribution which insect behaviour makes to the pick-up of insecticide from treated surfaces.

There have been relatively few quantitative studies in this field; before considering these a few generalizations may be stated. We have already noted that in practice residual deposits are usually very far from a complete and uniform cover on the treated surface. More commonly they consist of a number of discrete spots. Neglecting for a moment any specialized aspects of behaviour or anatomy which may affect pick up, the amount of a contact insecticide picked up by an insect moving across the surface should be some function of the area of insecticide contacted and the time over which this contact occurs. While it might be expected therefore that the more mobile the insect, the more vulnerable it would be to a poorly distributed pesticide, there should also be a wide range of situations where there is a sufficiently high chance of contact between toxicant and pest for the longer period over which a slow moving insect remains in contact with a particular spot to balance the greater number of individual spots contacted for shorter periods by the faster moving species. It would be helpful to be able to define this range, because it could have important practical implications. For example, predators and parasites which contribute to biological control are generally considerably more mobile than the pests which they attack; the differences may be extreme in the case of natural enemies of effectively stationary sucking insects. Treatments should be designed to minimize damage to such beneficial insects. The argument presented above suggests that the beneficial species could conceivably be at *relatively* greater risk from a sparse population of spray residues. To define the range requires knowledge of the relative availability of toxicants from different types of deposit to the insects concerned and also knowledge of the movement of the pest in relation to different deposition patterns so that some kind of "mean free path" or probability of contact between pest and pesticide over a given period of time can be estimated.

B. Pick-up of Pesticide by Insects from Residual Deposits

The availability of insecticides from deposits has been investigated by several workers, notably Hadaway and Barlow. Their early studies included quantitative determinations of the weights of DDT picked up and retained by mosquitoes (*Aedes aegypti* (L.)) and testse flies (*Glossina palpalis* (R–D.)) exposed to various types of deposit (Barlow and Hadaway, 1952). The insects were active and walked over the treated surfaces during the exposure period. Under the conditions of the tests, pick up was rapid: at an area density of $2{\cdot}5\ \mathrm{mg\ m^{-2}}$ sufficient toxicant was picked up from some of the deposits within two minutes to kill all the flies in the test sample. The amount picked up increased progressively with exposure time up to 32 min, but at a decreasing rate. Tests with DDT particles carefully fractionated into the size ranges 0–10 μm, 10–20 μm, 20–40 μm and 40–60 μm showed that availability depended considerably on particle size. For a given area density and period of exposure the amounts picked up increased with particle size up to 20–40 μm but then decreased considerably again with the largest size range. However, there was an inverse relation between particle size and toxicity of the particles picked up so that overall effectiveness passed through a maximum at 10–20 μm. The relationships between particle size, pick-up and toxicity were broadly similar for tsetse flies. Rather larger quantities were picked up except for the smallest particles which were not so readily available, presumably because of tarsal differences between the two species. The 10–20 μm particles were, however, still most effective. One factor which contributed to the apparent toxicity of the particles picked up was the extent to which they were retained during the 24 h interval between exposure and assessment of toxicity. On the basis of measurements of the amounts of DDT which could be rinsed from the surface of the flies with benzene and observations of knock-down and kill, Barlow and Hadaway concluded that although the larger particles were more easily detached from the insects, this did not fully account for the differences in toxicity. They suggested that as the crystal size decreased, the DDT dissolved in the cuticle greases and penetrated to the site of action faster.

These chemical measurements of pick-up confirmed earlier bioassay results (Hadaway and Barlow, 1951) and also demonstrated the value of obtaining quantitative information about the distribution of insecticide in interpreting differences in effectiveness. The earlier work had shown that the influence of particle size on effectiveness varied considerably with the insecticide. While results obtained with ground crystals of the DDT analogues DDD and methoxychlor were similar to the findings for DDT itself, with γ-BHC and dieldrin which are more toxic to mosquitoes, there was much less dependence of effectiveness on particle size; the optimum size was much less clearly defined and even particles in the range 60–80 μm were reasonably effective. Hadaway and Barlow concluded that for these more toxic compounds either the pick-up of only a few large crystals would be toxic or their penetration through the insect cuticle might be so rapid that retention of the particles on the insect body was unnecessary. More recent studies (Hadaway *et al.*, 1970) with a range of carbamate and organophosphorus insecticides and a related strain of mosquitoes (*Anopheles stephensi*) gave results broadly agreeing with those found earlier. Toxicity of spray deposits on plywood and plaster panels was influenced by particle size to varying extents depending on the toxicity of the chemical and its coverage of the surface. There was a general tendency towards an optimum size range of 5–30 μm, although as with γ-BHC and dieldrin the larger particles of these relatively toxic organophosphates and carbamates were also effective. Hadaway *et al.* point out that it might be expected that the smallest particles would be most effective because they should adhere best to the insects surface and because their comparatively larger specific surface area should favour rapid penetration. They draw attention to two factors which should decrease the availability of the smaller particles: first the cohesion of the particles themselves and their adhesion for the surfaces increases markedly below 5 μm and second the smaller particles tend to form streaky deposits on plywood below 10 μm, filling the small crevices in the surface.

The important influence of the nature of the surface to which the insecticide is applied was observed in early studies with DDT (Barlow and Hadaway, 1952). Thus the 0–10 μm crystals were much more readily taken up from mud blocks than from plaster. It seems probable that differences in surface irregularities and in adhesion of the particles to the substrata were contributing factors. Such mechanical effects were further illustrated by comparing porous with non-porous materials. Barlow and Hadaway found that when aqueous suspensions were sprayed on to porous materials such as mud and plaster, the solution containing wetting agent penetrated into the pores relatively rapidly leaving the suspended particles readily available on the surface. On non-porous surfaces such as glass, the solution remained on the surface and held the particles firmly to the surface on drying, thus decreasing their availibility for pick-up by the insect. The importance of wetting was emphasized by results with two contrasting samples of compressed wallboard, one of which wetted easily, the other with difficulty. The amounts picked up and the toxicity of the deposit were much greater with the easily wetted surface.

Other aspects of the relationships between the physical state of deposits from various formulations of contact insecticides applied to different surfaces and their availability to insects were reviewed by Barlow and Hadaway (1953). Oil solutions of DDT were relatively ineffective on porous surfaces such as mud and plaster because the solution was rapidly absorbed. On non-absorptive surfaces such as glass, however, oil solutions were most effective, in keeping with the expectation that the lipoidal outer layers of the insect cuticle would have a high affinity for such solutions. Effectiveness decreased rapidly on evaporation of the solvent and deposition of a dry crystalline residue. In the case of DDT and possibly some other compounds, however, the position is complicated because evaporation can give a variable mixture of supersaturated or supercooled droplets (which remain relatively available to the insect) and true crystals, depending on the solvent and the conditions of evaporation.

Further variations in the possible patterns of behaviour are shown by certain fibrous absorptive surfaces such as wallboard. Initially the oil solution is absorbed into the fibreboard and its availability is thus low. Evaporation of solvent can lead to a supersaturated solution which can be induced to crystallize by mechanical means, including the movement of the insects themselves on the surface. With such substrata, the resulting crystals appear to project from the treated surface and therefore become more readily available to the insects. The effects of crystallization are therefore the reverse of those on glass surfaces where presumably the deposited crystals lie in the plane of the surface, in which orientation they are not so easily picked up.

The behaviour of emulsion formulations is in many ways similar to that of oil solutions although emulsifiers and stabilizers can affect adherence and availability and thus add further complications. Additives can also affect the availability of insecticides in aqueous suspensions. The influence of wetting agents has already been discussed. Inert diluents may also influence the availability of insecticides deposited from aqueous suspensions of wettable powders. Hadaway and Barlow (1951) found that the effectiveness of 20–40 μm DDT crystals on plaster blocks against *Aedes aegypti* decreased as the amount of the inert diluent, kaolin, increased. Such effects may be attributed both to a masking effect of the diluent covering some of the insecticide crystals and to adhesion of the inert particles as well as the insecticide to the insects. Another manifestation of this factor was observed in the field by Baldry (1963) who found that deposits of DDT lost effectiveness after becoming covered with dust from fibres; activity was regained after rain. It should be noted, however, that such effects are not always observed: in later studies with more toxic organophosphorus and carbamate insecticides. Hadaway *et al.* (1970) found that the ratio of active ingredient to inert carrier had little effect on effectiveness over a range of from 1:4 to 4:1. In interpreting such behaviour the importance of vapour effects should not be overlooked: for a volatile compound a small air gap is effectively contact.

The importance of the interactions between the toxicant, other components of the formulation and the surface, revealed by the studies of Hadaway and Barlow have been further emphasized by other investigations. Lewis and Hughes (1957) using carefully fractioned particles of 20 ± 3 μm mass median diameter and standardized conditions found that whereas the uptake of a lipophilic dye, BDH oil-soluble violet, by blowflies (*Phormia terraenovae*) was significantly greater from untreated filter paper than from filter paper coated with a hard leaf wax (carnauba wax, m.p. 84°C), uptake of a more hydrophilic dye, Casella diamine aldehyde blue, was much the same from both surfaces. This suggested that the hydrophilic particles were fairly indifferent to the lipoidal surface, but the lipophilic particles were more easily picked up from the fibrous surface because they had a much greater affinity for the wax. When the filter paper was coated with soft beeswax, uptake of both types of dye was much more variable but consistently less than from the hard wax. Lewis and Hughes postulate that the particles may have been pressed slightly into the surface by the rigid spines of the insects' tarsi, thus increasing the area of contact and hence the adhesive force between the particles and the substrate. Finally treatment of the filter paper surface with mineral oil greatly reduced uptake of the lipophilic dye compared with the pick-up of the dry particles from untreated surfaces.

Aspects of the factors revealed by Hadaway and Barlow and Lewis and Hughes to have an important influence on availability were studied further by Gratwick (1957). The uptake of DDT by blowflies from a dry deposit was found to be very similar to that of similar-sized particles of the BDH oil-soluble violet dye used by Lewis and Hughes (1957) suggesting that conclusions reached using the dye could be extended to uptake of DDT dust. The presence of oil reduced the uptake of DDT far more than the dye particles, uptake of the insecticide decreasing logarithmically with the amount of oil applied to the substrate. The viscosity of the oil had very little effect on uptake, which was also independent of the solubility of the particles in the oil for the limited range of materials studied. Gratwick therefore suggested that the differences between effects on DDT and the dye could be attributed to differences in the shapes of the particles: the needle-shaped DDT particles could have a greater area of contact lying horizontally on the surface than the isometric dye particles. The uptake of both the DDT and dye particles from filter paper and from the glass plates coated with hard carnauba wax was decreased by applying a wetting solution to the substrate immediately before the particles. The decrease was much greater with the waxy surface than with the absorbent filter paper, results which may be compared with those obtained by Barlow and Hadaway (1952) discussed above. When the wetting solution was allowed to dry before the particles were applied, there was no reduction in uptake on the absorbent surface and the reduction with the hard wax surface was much less than when the particles were applied before the wetting solution had dried. Gratwick concluded that the presence of oil or wetting agent on the surface increased the adhesion of the particles both by increasing the affinity for the substrate of the lipophilic particles studied and by increasing the area of contact. She suggested that the physical structure of the surface could be more important than the differences in the affinity of the particles for the material of the substrate in determining the amounts picked up by the flies. With insecticides sufficiently soluble in oil, increased uptake of material dissolved in oil might be expected to override effects on particle uptake in some situations. Direct uptake of solution will of course be less the more corrugated the surface and the thinner the oil is spread.

These and related studies did much to improve understanding of the availability of contact insecticides and make it possible to summarize the physico-chemical factors involved. The availability of particles in a deposit depends on the relative adhesion of the particles to the substrate compared with the adhesion to the insect's cuticle and on the extent of contact between the particles and the receiving surfaces on the insect. The morphological and other biological aspects influencing the extent of this contact will be discussed more fully shortly. The studies reviewed above show, as would be expected from general physico-chemical considerations, that the adhesion of the particles to the surface depends on the affinity of the insecticide for the material of the surface and on the physical structure of the surface which determines the orientation in which the particles can lie and their area of contact with the surface. The area of contact may be a dominant factor. The contact between relatively large irregularly shaped particles and hard flat surfaces is likely to be very small, thus limiting the forces of attraction (which decrease with the distance between the two materials), whatever the affinity of the two substances for each other. The area of contact will be much greater if the surface is deformable or if there is a liquid film between the surface and the particle (Section 14.V.E). Real contact area is not much dependent on particle size and its effect is therefore greater for small particles (see p. 468).

For any given area of contact or distance of approach, the affinity of the particle for the surface will be determined by the forces of attraction discussed in Section 2.III. These will include transient electrostatic forces, fairly generalized "long range" London forces and shorter range forces such as hydrogen bonding which may operate for appropriate materials. Some simple general conclusions about the affinity of the particles for the surface material are possible. For example, there is a tendency for like to be attracted to like so that particles of polar insecticides will adhere relatively strongly to polar surfaces such as glass, whereas lipophilic insecticides such as DDT have a greater affinity for lipoidal surfaces and are presumably slowly adsorbed by them. Both the affinity for the surface material and the physical structure of the surface can be modified by other components such as oils or wetting agents, although the influence of these is clearly less on porous surfaces into which they can be sorbed.

The adhesion of the particles to the receiving surfaces on the insect is determined by similar factors to those influencing adhesion to the substrate. However, the outer layers of the insect cuticle are usually deformable and oil films are often present, while penetration into the insect will also obviously affect the amounts retained.

In most of the studies we have been discussing, the search was for optimum pick-up under dry conditions from surfaces representative of those sprayed with residual contact insecticides for the control of disease vectors and for other public health purposes. In crop protection, we are more often concerned with pick-up from the aerial parts of plants. Adhesion to these surfaces has been less thoroughly studied. However, the same principles apply, although as with insect cuticles the deformable and lipophilic nature of plant surfaces may be important, and absorption into the plant, metabolism and dilution through growth may add further complications. For example, it used to be widely held by field-workers (although never clearly demonstrated) that there could be local drift damage after DNOC spraying caused by particles of the dried deposit being blown off by wind: the brittle deposit was discarded by the growing leaf (see p. 803).

In the application of insecticides to growing crops in temperate or humid climates, which is mainly directed against phytophagous insects in their phytophagous phases, strongly adherent deposits incorporating some protection against removal by rain are acceptable since mastication or digestion can release the active ingredient (p. 728) and it is a positive advantage to have the deposit ineffective against non-phytophagous predators. There are, however, cases where a phytophagous insect is only accessible during a non-phytophagous phase. The problem may then be more complex because the deposit should be easily picked up by insects but not washed off by rain.

This problem has arisen in attack on the grass grub (*Costelytra zealandica*) by photostable pyrethroids to which it is very sensitive. The damage done by this organism is mainly in the larval stage among grass roots and no satisfactory control agent has replaced the now banned application of DDT. The female adult beetle is periodically accessible above ground during a 3–4 week mating period where she is very vulnerable to a low applicance of emulsion provided this is recently applied (Lauren and Henzell, 1977). Biological effectiveness falls off more rapidly than residues of chemical recoverable from extracts. The insecticide thus becomes unavailable by contact. Lauren and Henzell consider absorption into plant "waxes" responsible (we should say, into cuticle and associated oils, Section 11.V.B) but this is not essential. Many corrugations on a grass leaf surface are on a smaller scale than the insects' tarsi and spreading along grooves could remove the oily active ingredient from reach as envisaged by Hartley (1966). In either case, there could be advantages in inhibiting spread. Viscosity, perhaps with a structural component, or absorption into a porous particulate solid like sepiolite are changes of formulation available, but their effects on pick-up and rainwash require careful experimental investigation.

It is perhaps worth pointing out that this problem—a deposit spreading in some way beyond accessibility to the moving target—becomes greater the more potent the insecticide and the smaller therefore the local density which must be applied. Ultra low volume (ULV) first came into prominence with malathion, an insecticide of rather low activity, popular because of its even lower toxicity to mammals. It may be necessary to dilute compounds of much higher insecticidal activity such as decamethrin or cypermethrin to prevent them being occluded in pores or corrugations. This is a conclusion, in reversed but relevant form, long accepted in timber impregnation which must be done with massive applicance, using extensive dilution, if necessary, to secure adequate penetration through porous absorptive material.

C. Effects of Humidity on Pick-up from Deposits

In practice, pick-up does not take place under the constant environmental conditions often maintained in laboratory tests. In particular, relative humidity can vary over a wide range and its effects have received much attention. Several workers have noted that variations in relative humidity can have an important influence on the availability of pesticides deposited on absorbent materials. These effects have been observed most typically on mud surfaces where for many insecticides, particularly organochlorine compounds such as DDT, dieldrin and BHC, the picture is relatively simple; effectiveness is much greater at high than at low relative humidity both in the field and in the laboratory. (See for example Burnett, 1956 and Barlow and Hadaway, 1958.) Laboratory tests have indicated that relative humidity can influence the effectiveness of deposits on materials other than soil (Gerolt, 1963; Kalkat *et al.*, 1961; Ebeling and Wagner, 1965; Barlow and Hadaway, 1968). However, the effect is not universal: for example Hadaway and Barlow (1963) found that while the availability of DDT, dieldrin and 3-isopropylphenyl *N*-methylcarbamate sorbed on dried soils increased considerably as atmospheric humidity increased, an increase from 43 to 80% r.h. during the contact period had no effect on the toxicity of wettable powder deposits on plywood of dieldrin, sevin and *O*-methyl *O*-(2,4,5-trichlorophenyl) ethylphosphoramidothiate. Related effects of moisture content on the activity of pesticides incorporated into field soils are discussed in Section 5.X.

In considering possible explanations for the effects of humidity on the toxicity of deposits of organochlorine insecticides, Barlow and Hadaway (1958b) distinguished two phases in the sorption process on soils. First the insecticide molecules move from the crystals lying on the surface to the internal surface of the superficial layers of soil where they are adsorbed. During this process the crystalline particles of insecticide on the surface can be observed to disappear. The rate of this initial sorption was inversely related to humidity, so that at high r.h. the crystals remained readily available as particles on the surface longer. This initial sorption process is clearly irreversible in that the crystalline particles cannot be reconstituted. The second phase involves diffusion of the insecticides further into the bulk of the soil. Unlike the first process, this diffusion was faster at higher humidities. This increased mobility at higher humidities was presumably due to competition of the water molecules for the relatively hydrophilic surfaces on the soil particles, which would displace (Sections 5.X, 2.XIII) the adsorbed lipophilic insecticide thus leading to an increased diffusion coefficient. Barlow and Hadaway concluded therefore that, once the initial phase of sorption was complete, the activity against insects contacting the outside of the mud was increased at high humidity not because the overall concentration there was greater, but because the insecticide already present was more mobile. While this increased mobility would result in more rapid diffusion into the bulk of the soil, it would also increase the potential for diffusion into insects resting on the treated surface.

Further information about the mechanisms of the humidity effects on other surfaces was obtained by Barlow and Hadaway (1968). They found that with aprocarb on cellulose, only aged deposits responded to changes in humidity, freshly sprayed deposits on cellulose and deposits on materials whose fibres are non-porous such as glass fibre filter paper being unaffected. They concluded that only the insecticide which penetrated into the fibres was affected by the presence of water adsorbed on the cellulose. Selective extraction of insecticides and changes in evaporation rates with time also indicated that the insecticide in aged deposits could be divided into two fractions. The first was superficial, easily extracted, evaporated more rapidly and its toxicity was unaffected by humidity. The other was less toxic to insects and more difficult to extract with solvents, evaporated more slowly and its toxicity varied with humidity. The alterations of toxicity with humidity were rapid, completely reversible and occurred at rates similar to the rates of adsorption and desorption of water by the fibres.

In contrast to the lack of effect of humidity on the toxicity of aprocarb on non-porous materials observed by Barlow and Hadaway, Gerolt (1963) found by direct radiochemical determination that the availability of dieldrin on glass surfaces was greater at high relative humidity for three species: *Musca domestica*, *Anopheles gambiae* and *Aedes aegypti*. The influence of humidity on uptake was greatest for *M. domestica* which walks most actively during exposure, intermediate for *Aedes aegypti* which walks less actively and least for *Anopheles gambiae* which seldom walks at all. Gerolt suggested that insects walking over the treated surface pick up the insecticide mechanically from a large area compared with that contacted by the stationary species. The rate of pick-up by the stationary insects would therefore depend more on the mobility of the insecticide molecules in the residual films so that a greater influence of relative humidity would be expected.

It is clear that the effects of relative humidity on the availability of surface deposits vary with the nature of the insecticide, the surface, the form and age of the deposit and with the characteristics of the insect. Availability can also be influenced by temperature (e.g. Hadaway and

Barlow, 1963). These environmental factors affect both the performance of residual deposits under field conditions and toxicity measured in the laboratory; it is therefore most important to take them into account when devising tests to compare the potency of insecticides and the susceptibility of different strains (Busvine, 1968).

While the studies reviewed above have done much to reveal how the nature of the deposit and its interactions with the receiving surface influence the availability of contact insecticides, the biological and particularly the behavioural aspects are far less clear. Some of the factors affecting adhesion to the receiving surface on the insect have already been considered; the effectiveness of the insecticide adhering to the insect must also obviously depend on its ability to penetrate through the cuticle. The properties required for penetration are discussed in Section 12.II. However, in addition to information about adhesion and penetration, full understanding of the pick-up of contact insecticides from residual deposits under practical conditions also requires knowledge of the likely extent of contact between the receiving surfaces of the insect and given patterns of deposit. For this, it is first necessary to identify the sites of uptake on the insect.

D. Sites of Uptake on Insects Walking on Residual Deposits

In the case of the blowfly (*Phormia terraenovae*) much information about these sites was obtained by Lewis and Hughes (1957), using carefully devised techniques for determining the amounts of dye picked up during settling and walking on treated surfaces. The quantities of particles picked up in the act of settling were much greater than the quantities picked up during subsequent steps and the rate of pick-up fell off rapidly during the first few seconds of exposure. For example, at an area density of $27{\cdot}5\ \mu g\ cm^{-2}$ approximately $2\ \mu g$ per fly was picked up on settling; after only 1 second corresponding to 6 footsteps (a footstep being defined as one movement of six legs) the flies picked up particles at a rate of only about $0{\cdot}5\ \mu g$ per step, only one quarter of the initial rate. Lewis and Hughes attributed these differences to an initial uptake of particles on the more favourable reception points on all the tarsal spines after which further uptake is necessarily slower. Detailed observations showed that most of the particles were picked up at the tips of the ventral tarsal setae, but some adhered away from the tips; this was explained by a momentary splaying out of the lower halves of the spines to lie flat along the surface during the walking movements. The particles from the deposit frequently adhered to other particles previously picked up to form aggregations. The frontal membranes of the contact chemoreceptors also became contaminated, but the intersegmental joint membranes are protected by large spines that become heavily contaminated. Lewis and Hughes concluded that "hairiness" and the pressure of segments against the surface are important variables affecting the relative rate of contamination of different tarsal segments. The particles initially picked up on the lower parts of the tarsi were transferred continually to the basal parts and to the head, wings and abdomen.

On transfer to a clean surface, particles were discarded from the tarsal setae by a reversal of the pick-up process. The setae recover the particles by cleaning from other parts of the body. An alternative but less important discarding process during cleaning involved the combing of fine particles into larger aggregates which could then be knocked off. Particles were lost relatively rapidly on initial transfer to the clean surface, but the rate of loss decreased progressively, becoming very slow when the quantity retained had reached about 10 μg per fly which took approximately 2 h under the conditions of the tests.

With these relatively high area densities, the flies had no difficulty in accumulating particles from the deposit, in fact their capacity for uptake was rapidly saturated. While this might suggest that efficient pick-up of insecticide could occur at much lower area densities, any significant interval between encountering particles would allow time for the discarding processes and the opposing factors would have to be carefully balanced to ensure that the "equilibrium adhesion" was sufficient for adequate toxic effect.

Measurements similar to those reported by Lewis and Hughes (1957) with the blowfly were made on a range of five contrasting species in a systematic study by Gratwick (1957b). The quantities of the lipophilic BDH oil-soluble dye picked up during the first few steps were remarkably similar for worker wasps (*Vespula vulgaris* (L.)), ground beetles (*Feronia madida* (F.)), soldier beetles (*Rhagonycha fulva* (Scop.)), cotton stainers (*Dysdercus fasciatus*, sign.) and plant bugs (*Notostira erratica* (L.)). These species were selected because of differences in size, hairiness of the tarsi, and numbers of tarsal segments. The uptake per step was somewhat greater for the larger insects, but the smaller insects picked up more particles per unit weight at each step. Thus ground beetles, with an average weight of 0·147 g, picked up 1·7 μg per step, giving 11·6 μg per step per body weight, while at the other extreme the mirids (*Notostira*) with an average weight of 0·006 g picked up 1·0 μg per step giving an average uptake of 167 μg per step per unit body weight. The ratios of the amounts picked up by the different pairs of legs were similar for all five species; in all cases the hind legs picked up quantities equal to or greater than the sum of those picked up by the fore and middle legs. Comparison of the relative quantities of dye picked up by the different tarsal segments of the five species suggested the collecting efficiency of a segment depends partly on the thrust applied to the substrate and hence on the way in which the tarsal segments are used in walking. The position of the particles on the tarsal segments indicated that the setae help to keep the particles away from the main surface of the cuticle which is thus not easily contaminated where the setae are dense. Gratwick also noted the importance of cleaning movements both in spreading particles picked up on the tarsi around the body and in enabling the insects subsequently to lose the insecticide which they have acquired, as observed by Lewis and Hughes in the case of the blowfly. Gratwick concluded that the effect of cleaning by insects densely clothed with setae could counteract any protection which the setae afforded.

Although the pick-up per step was found by Gratwick to be similar for this diverse range of insects, the differences in uptake per unit body weight could give rise to considerable differences in the relative potency of residual deposits to different species. Evidence has even been obtained indicating that differences in pick-up between strains of the same insect could be a cause of resistance. Bradbury *et al.* (1953) exposed resistant and susceptible strains of housefly (*M. domestica*) to deposits of ^{14}C-labelled BHC on filter paper in petri dishes for periods of from 15 min to 16 h. The radioactivity picked up was separated into an outer fraction, which could be rinsed off with carbon tetrachloride, and an inner fraction which had penetrated through the cuticle and was extracted by grinding up the tissues. The results showed that the resistant strain initially picked up considerably less insecticide than the susceptible insects: after 30 min exposure, the ratio for the two strains was approximately 1:4. The disparity progressively disappeared until after 4 h the amounts picked up by each strain were almost identical. Comparison of the total amounts picked up and of the distribution between the two fractions in the different strains suggested that the resistance could not be attributed solely to differences in penetration, and other evidence suggested that differential detoxification was unlikely. Bradbury *et al.* therefore concluded that the observed differences in pick-up could be an important resistance mechanism in this strain.

These various biological studies, together with the physico-chemical investigations discussed earlier, provide a reasonably extensive picture of the factors likely to affect the availability of residual deposits of insecticides to flying insects which have alighted on them, although there is much room for related studies on different formulations and species. However, a comprehensive understanding of the overall uptake process for such insects as outlined earlier in this section requires more knowledge of the other very important component of biological transfer: the pattern of alighting and movement on various different surfaces under natural conditions. This is of course difficult to study, but modern instruments capable of recording insect activity automatically may facilitate future progress.

E. Uptake of Pesticides from Residual Deposits by Crawling Insects

The problems should be less difficult for crawling insects such as lepidopterous larvae, which remain continuously for longer periods of time on the surfaces which they are attacking. As with flying insects, the effectiveness of residual deposits of insecticides against these larvae can be influenced considerably by particle size. Johnstone (1969a, 1971) investigated the relative effectiveness of 5 μm-wide size fractions of a wettable powder formulation of carbaryl over the range 0–40 μm against first instar *Pieris brassicae* larvae on cotton leaves. The spray was applied uniformly as 300 μm droplets on a square grid pattern using a laboratory technique developed for applying low volume spray patterns of variable form and density (Johnstone, 1970). Particles in the size range 15–20 μm were most effective giving LD_{50} values approximately half of those found with the largest and smallest fractions, which were about equally effective. Pick-up of similar size fractions of carbaryl and the fluorescent tracer salicyl aldazine was also examined using u.v. microscopic and photographic techniques. The u.v. photographs showed pick-up of the particles on the caterpillars' feet and claspers and visual counts confirmed the most effective particle size. It appeared that the larger particles were readily discarded as the caterpillars moved, while the smallest particles adhered too firmly to the substrate to be picked up.

The influence of particle size can interact with the effects of another most important variable governing performance of residual deposits: the population density of spots on the surface (spot density) for which in the case of caterpillars there is some useful information. Some further results obtained by Johnstone (1969b) in studying ULV techniques may be quoted. Measurements with a wettable powder formulation of carbaryl on cotton leaves gave the expected log-normal relationship between applicance and mortality of *Pieris brassicae* for spot densities from 64 to

4 cm^{-2}. Below this density, however, the relationship broke down and Johnstone concluded that mortality was determined more by the probability of the caterpillars coming into contact with the well-separated points of deposition rather than by the mass of toxicant on the surface. The exact relationship depended on the formulation used, particularly its particle size. One powder, with a mass median particle diameter of 5 μm, applied to cotton at 0·025 lb acre^{-1} gave greatest kill at low spot densities (0·25 and 1·0 cm^{-2}) and correspondingly high concentration; it was ineffective when applied at lower concentration to give the same applicance at 64 spots cm^{-2}. Increasing the rate of application to 0·1 lb acre^{-1} increased mortality at all spot densities but mortality was greatest at 4 spots cm^{-2} and decreased slightly as spot density was increased to 64 spots cm^{-2}. To explain these results, Johnstone suggested that the small particles adhered strongly to the leaf surface and were only readily available for pick-up by the larvae when deposited in sufficient concentration to overlie one another. A second formulation, with a mass median diameter of 18 μm gave similar mortalities at low spot densities, but was most effective at densities exceeding 4 cm^{-2}, suggesting that the larger particles adhered less firmly to the leaf surface.

Information about the factors influencing pick-up of residual liquid deposits by phytophagous larvae (which could follow very different rules from the crystalline deposits studied by Johnstone) was obtained by Polles and Vinson (1969) who investigated the application of different formulations of malathion to cotton for control of the tobacco budworm larva. At a constant applicance, the effectiveness of a ULV formulation applied by a spinning disc sprayer decreased with droplet size in the range 100–700 μm, although the amounts of malathion recovered by washing samples of several larvae from each treatment with carbon tetrachloride increased with droplet size. Polles and Vinson attributed these results to the decreasing probability of contact between larvae and deposited droplet with increasing droplet size; the increased diameter is likely to be more than offset by the decreased population density of drops on the surface. At the rates of application used, Polles and Vinson calculated that the spot density was theoretically 2·8 cm^{-2} for droplets of 200 μm and 0·1 cm^{-2} for 600 μm droplets. The larger amounts of insecticide extracted from the larvae exposed to deposits from larger droplets could be attributed to pick-up of relatively large quantities of insecticide by a few of the insects in the sample analysed.

With smaller droplets giving greater spot densities the probability of each insect receiving a toxic dose would be greater, but the total amount picked up could be less. Visual observation of the behaviour of the larvae on the treated cotton leaves suggested that the probability of contact was not determined merely by random movement on the surface. Untreated larvae appeared able to detect the presence of the insecticide deposits and attempted to avoid them. Clearly, the possibilities of such avoidance succeeding are likely to be much greater at low spot densities.

Johnstone (1969a) has pointed out that such reduced probability of pick-up at low spot densities provides some limitation on the extent to which the logistically highly advantageous technique of ULV application (which is discussed more thoroughly in Chapter 14) can be pursued and could result in reduced efficiency in some situations. The principal interacting variables to be considered in this technique are droplet size, volume application rate and concentration of active ingredient in the formulation: the first two will determine the spot size and density and thus whether the insect is likely to make contact with the insecticide. The concentration in the formulation will then determine whether the extent of this contact is sufficient to kill the recipient. These various factors are summarized by Johnstone (1973) who emphasizes that the optimum concentration is very dependent on the size of the insect as well as the intrinsic toxicity of the toxicant. He points out that the mass of the *Pieris brassicae* caterpillar increases in size by roughly 1000 times between first and fifth instar and that the dose required for kill can be expected to increase more or less correspondingly. From a knowledge of the pattern of locomotion over the sprayed surface, the efficiency of transfer between the deposit and the larvae and the toxicity of the insecticide it should be possible to calculate the spot density and concentration of toxicant required to kill larvae of any given size using the minimum quantity of pesticide.

F. Selectivity between Different Insect Species

The effectiveness of a residual deposit of a contact insecticide will also depend on its persistence, that is on the extent to which it is lost by evaporation or rainwashing, or by absorption into the treated surface. These factors have been considered earlier in this and other chapters. In the present context, however, some further discussion is necessary. Various attempts have been made to increase the persistence of residual deposits by the use of stickers, which bind the deposit to the receiving surface. These materials can undoubtedly extend the life of residual deposits; for example, Hartley and Howes (1961) described formulations involving addition of amine stearates to wettable powder formulations of DDT, designed to increase the rainfastness of the deposits on leaves by forming a matrix of stearic acid around the particles. Phillips and Gilham (1968) confirmed the effectiveness of the formulations for this purpose, but found a parallel decrease in contact insecticidal action. This illustrates a general dilemma: extended persistence and ready availability for contact action are likely to be mutually conflicting objectives. To offset this, the increased persistence need not be at the expense of stomach poison activity so that it may be achieved in conjunction with the great advantage of increased selectivity against chewing insects. An interesting approach to the design of formulations conferring this type of selectivity on insecticides normally having pronounced contact activity was suggested by Ripper *et al.* (1948). They postulated that by coating DDT particles with various hemicelluloses, it should be possible to obtain formulations active only against phytophagous chewing insects possessing the appropriate hemicellulases to degrade the coating. It is probable that a synthetic micro-envelope (Section 14.IX.E) could more simply achieve the same object if capable of disruption by the act of mastication. This concept was apparently not pursued in practice, presumably partly on grounds of cost, but the increased awareness of the benefits of selectivity, particularly in the context of integrated control, would seem to provide the necessary incentive for investigating the approach further. A more recent formulation technique which also seems particularly promising for decreasing contact activity and increasing persistence and which could have other advantages such as masking repellent chemicals is microencapsulation, whereby the toxicant dissolved in an appropriate solvent is enclosed in small spherical capsules of suitable materials such as cross-linked gelatin (for a fuller description, see Chapter 14). Contact activity can be eliminated with this formulation (Phillips and Gilham, 1968) or a sustained release can be achieved using capsules designed to leak slowly (Phillips, 1974). It should also be recalled that the physical properties of some residual insecticides may in themselves confer to some extent the type of selectivity we have been discussing. Many such insecticides are very lipophilic (in contrast with systemic compounds which generally tend to be more polar as discussed in Chapter 11). They are thus readily absorbed by the non-polar constituents in plant cuticles, decreasing their contact activity.

These various examples of selectivity are part of a general principle which applies much more widely. Selectivity between pests and beneficial insects, both pollinators and natural enemies of pests, has been an important objective of insecticide research. Such selectivity can be achieved either by using toxicants which are intrinsically more toxic to the pest (which was termed "physiological selectivity" by Ripper *et al.* (1951)

and which lies largely outside the scope of this book) or by arranging that a much larger proportion of the applied toxicant reaches the target organism than reaches non-target organisms, which Ripper and his co-workers defined as "ecological selectivity". It will be seen that the methods of decreasing contact toxicity discussed above are examples of ecological selectivity which in practice is achieved by adjusting the timing, the method of application or the formulation to match the behaviour of the pest rather than its natural enemies; in many cases this involves maximizing the biological transfer which can be accomplished by the target organism.

Consideration will show that these principles are implicit in many well-established pest control practices. The use of winter or early season treatments against orchard pests provides a familiar example of the use of timing. Localized treatments such as placed granules or seed treatments illustrate methods of application which not only give considerable selectivity, but by concentrating the toxicant in the critical region, also allow considerable reductions in the amount of chemical used as we discuss in Section 13.II.B.

Another general method of application offering very considerable selectivity, now taken for granted, is the use of systemic insecticides. The advantages of systemic insecticides for controlling sucking insects without harming predators such as coccinelids were discussed almost 30 years ago by Ripper *et al.* (1951). It is important that the pest killed by a systemic insecticide should not be eaten by the predator, or not poison it if consumed. Biochemical selectivity in addition to placement is therefore desirable, even though the dose transferred to the predator would normally be much less than that reaching a sucking insect passing plant sap through its body. Clearly, the most favourable selectivity is achieved when the systemic chemical is applied in such a way that beneficial insects are not treated directly or allowed to come into contact with the insecticide. This can be achieved by application to the soil or seed for uptake by the underground parts of the plant. Such treatments are unfortunately not always practicable. However, even when the insecticide is applied to the aerial parts of the plant at a time when both pest and beneficial insect are present together, suitable formulations can provide a significant measure of selectivity. This may be illustrated by the observations of Free *et al.* (1967) on damage to pollinating insects on field beans. Infestations of aphids on field beans in Britain often reach levels at which control measures must be applied at the time of flowering. Insecticidal sprays give good control but can cause severe harm to bees if applied during this period. Table 3 shows that such damage can be effectively eliminated using granular formulations. The very transient effect with the spray suggests that only those bees which are hit directly are damaged; subsequently the residual action appears to have little effect, although giving good aphid control.

Table 3

Dead worker bees (Apis mellifera, L.) collected in traps by hives adjacent to fields of flowering field beans treated with different insecticide formulations (data from Free et al., 1967). Numbers collected are totals for four colonies for each treatment

Treatment	Mean number collected per day		
	Within 24 h of treatment	From 1 to 6 days after treatment	From 7 to 14 days after treatment
Oxydemeton-methyl spray	1310	41	35
Phorate granules	28	16	7
Disulfoton granules	60	11	7

Insecticides applied as granules to field beans in this way as a top dressing reach the pest by a variety of routes (Graham-Bryce *et al.*, 1972). After application, a proportion of the granules lodge on the leaves or in leaf axils, while the remainder fall through to the soil. The chemical is released from the carrier by evaporation, by the leaching action of rain or by molecular diffusion in water films. After release the chemical may kill the aphids by direct vapour transfer, by the contact effect of surface films or solutions on the foliage or it may act systemically after uptake by the leaves, stems or roots. The relative importance of these different pathways depends on the initial distribution, on the physical and chemical properties of the chemical, the granular carrier and the soil and on the weather.

Other aspects of selectivity based on differential reception of toxicant vapour are discussed in Section 12.II.E.

G. Interception of Pesticide Sprays by Flying Insects

In the control of flying insects by pesticide sprays, it is clearly desirable to optimize direct interception of the droplets by the pest. This requires sound knowledge of the movements of the target species and thus comes squarely within the province of this chapter. It is first necessary, however, to discuss some general characteristics of spray droplets.

An important factor limiting the useful size of a droplet in a spray of a non-systemic insecticide is the dose necessary to kill the insect with certainty. If a single drop of 100 μm diameter is enough, then over 90% of the content of a 300 μm diameter drop would be wasted even if it contacted the insect, except for chemicals acting mainly by pick-up and penetration through the cuticle, where a large deposit could conceivably be contacted by several different individuals. Where contact occurs in flight, unnecessarily large drops might be redistributed and wastage reduced by wing movement as has been shown to occur in locusts (Wootten and Sawyer, 1954). For foliar deposits of stomach poisons, however, the whole overdose is likely to be consumed before the insect is affected. The larger residue is, on the other hand, more likely to be avoided, as Polles and Vinson (1969) found when observing the movement of the tobacco budworm among an array of residual droplets. On balance it is a reasonable generalization that drops carrying much in excess of a single lethal dose would be wasteful.

Simple statistical argument sets the desirable size even lower. If all insects in a population have equal chance of contacting all droplets directed into their habitat, the ratio of probabilities of 0, 1, 2 etc. contacts form the Poisson distribution (e^{-n} times successive terms of the expansion of e^n) where n is the average number of contacts. For $n = 1$, the probabilities are 0·368, 0·368, 0·184 etc. For $n = 8$ they are 0·003, 0·0027, 0·0107 etc., with the sum for 1–4 contacts being 0·100 and for 5–11 contacts (mean $8 \pm 37\%$) 0·788. For $n = 2{\cdot}3$, the probability of no contact is 0·100 and of 1 contact 0·231. If therefore we spray for the maximum number of single contacts, we achieve only 37% of these with 37% insects escaping and 26% overdosed. If we reduce the size and increase the number of droplets so that the lethal dose would be in the 8 ± 3 bracket only 10% of the insects would escape and 11% be overdosed. If we had increased the number of drops without reducing the size so that only 10% of the insects escaped contact, 66% would have been overdosed. The trend towards more efficient kill, i.e. reduced wastage by overdosing for a given low percentage escape, continues indefinitely with increase in the mean number of droplets necessary to ensure kill of an individual.

This argument assumes that there is equal probability of contact of the insect with all droplets available, whatever their size. This is justified provided the insect's "collision diameter" is much larger than that of the droplet, while the droplet is still large enough not to be steered aside by movement of the target with respect to the air. It also assumes equal probability of access of all droplets to all parts of the habitat, and that there is no avoiding action dependent on drop size. These various assumptions must be considered in relation to drop size, but in general the last two will shift the balance in favour of smaller droplets, while the first may set a limit to the desirable drop size.

Interception by flying insects and the relationship of drop size and dynamic behaviour has been studied most extensively in the case of locust swarms. The magnitude of the problem, the urgency of dealing with the swarm and the relative simplicity of the factors—a fully exposed population of similar individuals moving uniformly and accessible to attack from the air—have all contributed to this attention. A good illustration of the uniqueness of this situation is that up to 6% of the insecticide delivered on to a swarm can be recovered from the insect corpses (MacCuaig and Watts, 1963; Rainey, 1974) while most insecticide usage is several orders of magnitude less efficient. However, Rainey considers that certain other insects may be collected into suitable airborne targets by convergent winds. Wootten and Sawyer (1954) and MacCuaig and Yeates (1972) discuss the dynamic factors involved in such aerial applications. The smaller the droplets, the longer they remain in the gross volume of the swarm and therefore available for contact by the insects, but also the more easily they are steered aside by the air currents generated by flight. There results a rather broad plateau of optimum size about 100–300 μm. The upper limit of lethal size is unimportant in this case. Most appropriate insecticides are effective in the range 10–20 μg per g body weight, equivalent to 25–50 μg per adult locust. A 200 μm drop of pure insecticide would contain about 5 μg; in practice the content is usually less because many insecticides must be dissolved in solvent to remain liquid. A single drop carrying a lethal dose would therefore be larger than the size preferred on dynamic grounds.

The locust is, of course, a very large insect. The doses of methyl parathion needed to kill 50% of some other insects by topical application in the laboratory are given by Himel (1969a) as 200–400 ng per insect for bollworm, *Heliothus zea* and 30–60 ng for the boll weevil, *Anthonomus grandis.* Himel and Uk (1972) give figures of 43–57 ng for chlorpyrifos and 143–216 ng for endosulfan against the female adult housefly (*Musca domestica*), the range covering measurements for vapour exposure (lowest), topical and spray (highest) application. Even the housefly is large compared with many insects of agricultural importance such as whitefly and thrips. As a 100 μm droplet of pure insecticide could contain about 600 ng, the requirement that droplets should contain only a fraction of the lethal dose indicates that normal hydraulic sprays could be very wasteful against such pests.

The effectiveness of residual deposits of insecticides on crops can in principle be increased by reducing droplet size because very small drops penetrate the canopy better, for reasons discussed in Section 8.IX.B. Himel (1969b and other references cited in this publication) considers that the main mechanism of transfer for an airborne insecticide to a representative phytophagous forest pest (spruce budworm) is by direct hit rather than consumption of residues and that for this purpose, droplets larger than 50 μm diameter are completely wasted. Only smaller droplets, by zig-zag motion in slight air currents, can penetrate the canopy, except in open spaces where larger drops would fall uselessly to the ground.

The evidence for these conclusions was based on the use of insoluble fluorescent particles (Cd – Zn sulphide) suspended in the spray liquid. The concentration in most cases was such that a 21 μm droplet had a most-probable content of one particle, a 100 μm droplet therefore had about 110 particles etc. Examination of dead larvae and of other treated objects such as pine needles in u.v. light showed the number of particles in each residual spot, and hence the size of the original drop. Analysis of dead larvae (Himel and Moore, 1967) indicated that 83% had been hit by no droplets larger than 30 μm and 93% by none larger than 50 μm. Calculation from the data (Himel, 1969b) suggests an average dose per dead larva of 0·8 ng insecticide (Zectran$^{(R)}$) but topical application gave an LD_{50} of 150 ng per larva.

Himel attributes this very large discrepancy to the contribution of yet smaller droplets containing no fluorescent particles or to more efficient penetration from very small droplets into the insect. The latter explanation seems unlikely in view of the good agreement in other cases (Himel and Uk, 1972) between LD_{50} values obtained by topical, spray and vapour routes. The former can be discarded on the basis of statistical considerations. The observed mean number of fluorescent particles corresponding to 0·8 ng insecticide was 9. In a uniform dispersion the "unrecorded" volume necessary to bring 0·8 ng up to 150 ng would contain 160 particles. However this unrecorded volume arrived at the insect, the probability that it would contain no particles when the mean number was 160 is e^{-160} or 10^{-70} for one larva, and over 1000 were examined. There are three possibilities not discussed by Himel. First there could have been substantial uptake by consumption of residual deposits, when the ingested fluorescent particles in the gut would not have been visible. Secondly gross segregation of the particles could have occurred in the spray tank and thirdly the particles might have become clumped together during fall or absorption of spray droplets by the insect so that the apparently separate sources of fluorescence were composed of many initially separate particles (1 μm diameter). The gross disagreement between laboratory and field observations was not found with bollworm or boll weevil on cotton. Clumping of particles, which occurs frequently in suspensions even in the absence of the complicating factors introduced in this work, is likely to depend greatly on the formulation and insecticide used.

Despite the quantitative doubts raised above, there is much evidence that ULV spraying has technical advantages over higher volume spraying in some situations. It always has economic benefits. Its advantages are not confined to aerial application but also include some ground operations (Matthews and Mowlam, 1974) and apply not only where the insecticide contacts the pest directly but also to residual action (Polles and Vinson, 1969). The subject is well reviewed in the symposium volume published by the British Crop Protection Council (1974). Application and formulation aspects of the technique are discussed more extensively in Chapter 14, where the considerable attention given to the potential problem of drift is reviewed. Surprisingly, however, there appears to have been almost no enquiry into a problem which could be at least as serious: the disturbance of predator–prey relationships. Our reference to ULV methods has emphasized the direct contact of spray droplets and insects. Although the advantages of the technique are not restricted to this mechanism of transfer, the evidence for improved residual effect against phytophagous insects is far from conclusive. In general, predatory insects are more mobile than their prey. Adherent residues on foliage are generally relatively more available to phytophagous insects than to walking insects, whereas contact by direct spray is likely to reduce predator populations more than those of phytophagous insects. We have already noted in considering selectivity that these dangers were well appreciated early in the development of modern synthetic pesticides (e.g. Ripper *et al.*, 1948) but have been to some extent obscured by the more recent swing of popular attention to "the environment" and to man and "wildlife" (mostly avian) assumed to be its most important inhabitants. The problem of how CDA and ULV techniques affect predator–pest relationships is not simple; for example the narrow spectrum of reduced droplet size could be expected to increase relative activity more to small organisms than large. However, it certainly deserves more attention.

The relatively large potential effects against small organisms suggest that ULV would have advantages for protecting plants against germinating fungus spores. No major success appears to have been reported so far. Frick (1970) comparing droplets of 100, 175 and 400 μm diameter against apple mildew at the same (low) area density of deposit on the leaf (LAD) of water and dinocap, found that 175 μm droplets gave better control than either of the other sizes. The smaller droplets were thought to become tangled up in the hair mat (see p. 455) of the abaxial surface, where the fungus is most resistant to attack. In general, it could often be impracticable to use such a high LAD that a large proportion of spores were hit directly by very small droplets: 50 μm diameter droplets at an LAD of 10 $\ell . ha^{-1}$ in a uniform array would be spaced with their centres 275 μm apart. In this situation uptake by spores would depend on some physical transfer, either vapour phase diffusion or spreading of the small droplets.

There has been some investigation of the possibilities of ULV application of herbicides. The obvious difficulty of delineating the target area restricts the technique to targets such as rangeland and forestry: use in mixed arable cultivation would be too dangerous. The ability to control weeds with selective herbicides by drift in a forest plantation without a spread boom is an advantage which can be exploited in suitable weather (Brown and Thomson, 1974). As would be expected, there is no economy in applicance. Although no examples can be quoted of reduced penetration into the plant due to the chemical being spread too thinly over the adsorptive cuticle, this phenomenon has been demonstrated in insects (O'Brien and Danelley, 1965). The ratio of cuticle to internal tissue in most crop plants is at least as great as in most insects. It thus seems probable that comparable retention in the plant cuticle has not been observed because it has not been investigated.

IV. MOVEMENT OF FUNGAL SPORES IN RELATION TO CONTACT WITH FUNGICIDES

Since the effectiveness of many protectant fungicides against foliar diseases depends on their being present on the leaves before the invading fungal spores arrive, it is evident that the movement of spores in air or droplets of water plays an essential role in bringing toxicant and target organism together, just as the movements of insects ensure the transfer of residual contact insecticides. More specifically, the relatively localized movements of spores in relation to the distribution of fungicide residue on the foliage can have an important influence on disease control. This is well illustrated in the discussion by Courshee (1967) of potato blight control using copper fungicides.

Common sense would suggest that the effectiveness of residual fungicide deposits would increase with the uniformity and completeness of the coverage on the foliage, and that good spray cover should always be the aim in practical application of fungicides. Certainly this could be expected with sparingly soluble, relatively involatile materials such as the traditional copper fungicides. In practice, however, effectiveness often appears surprisingly independent of the uniformity of spray cover and even relatively poor distributions such as those obtained by applying the chemical with a watering can may give adequate control. Such effects have been attributed to redistribution of the fungicide from the original deposit by water which is essential for germination of the spores. Some important features of this postulated redistribution were revealed by Courshee's studies.

The effects of spray droplet size and extent of coverage on the performance of copper fungicides against potato blight were investigated in various laboratory and field experiments. Effectiveness was assessed by applying a suspension of *Phytophthera infestans* sporangia to the leaves as a fine mist after the fungicide deposits had dried. The degree of protection attained was found to be proportional to the spray cover. However, redistribution of the fungicide achieved by exposing the treated plants to either natural rainfall or artificial rain produced by a distilled-water rain machine gave no improvement in performance as indicated by the same bioassay. Further laboratory work to determine the smallest dosage needed for adequate control when coverage was complete provided the explanation for the discrepancy between the results of these experiments and practical experience. In the later laboratory studies the suspension of sporangia was applied to the treated leaves as single drops using a platinum loop, in order to standardize the spore loading in each experiment. When these spore-bearing drops were applied to suitably inclined leaves, they rolled to the lower edge, which was just the region where the copper had accumulated in retention experiments.

These observations emphasize that the significance of redistribution depends greatly on the nature of the infection and the mobility of the spores. Following redistribution by rainwashing, there may be effective quantities of fungicide along the lower edge of a leaf, even when most of the main area of the leaf is clean enough to become infected. In Courshee's original experiments, the spores had been applied as an airborne spray which did not run to the same point as the redistributed copper. Based on these results Courshee distinguishes two extreme kinds of infection, wet and dry, with the full range of intermediate types. With regard to potato blight, he points out that water as rain plays an essential role in initially delivering spores to the plants through its action, described by Hirst (1958), in collecting them from the air. This is consistent with the infection of leaf edges and the preponderance of spores on the upper surfaces of leaves in the early stages of blight attack. Later in the infestation "pepperpot" infection occurs in which the leaf is covered with several separate lesions resulting from the arrival of relatively dry spores from nearby fruiting lesions.

Glendinning *et al.* (1963) found that few spores survive the short period of desiccation which would occur during aerial transport from outside the crop. The survivors require liquid water for germination, while fresh spores can germinate on a dry leaf in high humidity. This explanation has not been widely accepted but is consistent with the conclusion of Courshee (1967) that complete coverage is necessary to arrest local infection. Complete cover is impracticable in the field and could not be maintained under weather conditions suitable for development of the disease. Evans (1968, p. 224) considers that practical measures in the field only delay (but with economic advantage) the onset of the disease and that timing of spray is the most important factor.

The importance of rain splash in the dispersal of crop pathogens was recognized many years ago by Faulwetter (1916a, b). In field studies on the dissemination of angular leaf spot disease of cotton and related laboratory experiments on the mechanisms of water droplet splash, he established conclusively that the disease organisms could be carried in splash droplets from a source to adjacent plants and he identified many of the factors which influence the pattern of dispersal. The potential of the process for dissemination may be judged from Faulwetter's observation that a single drop (3·4 mm diameter) falling 4·9 m on to a wet horizontal surface in still air gave rise to 800 splash droplets of different sizes. These travelled distances up to 144 cm from the source, the larger drops travelling further than the smaller ones. In contrast, in the moving air produced by an electric fan, the small drops, evaporating to still smaller residues, were carried furthest, up to 5·5 m from the source.

Splash dispersal has been comprehensively reviewed by Gregory (1973). When a drop falls gently on a dry, flat surface it spreads out transiently to a disc and then recoils without disruption. If the energy of impact is great enough, however, the disc assumes a dish form, the rising rim of which thickens by surface tension and becomes varicose (see p. 453 and Fig. 14 thereon). The thickened portions, having greater persistence of velocity, give to the whole the appearance of an ornamental "coronet", the jewels of which later fly off as splash droplets. When a drop falls on a deep liquid a transient crater is formed in the surface, the rising walls of which again form an unstable coronet, but another mechanism of droplet formation comes into play. The first stage in levelling out the crater forms, by over-correction, a central "rebound column" rising above the original surface and breaking up to one or more drops larger than those spreading out from the coronet. The crater formed by a large enough drop falling from great enough height will close at the top to form an air bubble before the rebound column breaks through.

The transient flattening of a drop falling without splash on a dry surface can "puff off" a small cloud of loose dust, including unwettable fungus spores if any are present. Wettable spores and bacteria can be dispersed in splash drops during rain and it is noteworthy that much less energy (smaller drops) is necessary to produce splash from a surface already wet, especially when the resident water film is thin. Faulwetter thought that the splash droplets in this situation were formed only from the film water but Gregory *et al.* (1959) demonstrated, by colouring the drop and film differently, that droplets were formed from both liquids, separately and mixed. Only if the incident drop is very small (less than about 200 μm diameter) is there no mixing since the drop is reflected totally without real contact (see p. 794).

During splash from a wet surface, the surface film is displaced outwards when the drop impinges and flattens. The liquid of the drop forms the inside of the expanding dish and it is presumed that, in the outward and upward movement, the liquid entrains some of the surface film together with any suspended spores it may contain. During the break up of the coronet to form the splash droplets the liquids would become intimately mixed so that droplets would contain spores whether they came predominantly from the incident drop or the surface film.

The dispersal of spores of *Fusarium solani* from thin horizontal films by water drops falling in still air in the laboratory was studied in detail by Gregory *et al.* (1959). The total number of droplets produced by a single splash and the total number of droplets carrying spores increased with increasing diameter and velocity of the incident drop and with decreasing film thickness. One splash produced between 100 and 5000 droplets with a total volume from one quarter to over twice the volume of the incident drop. The splash droplets had a wide range of diameters from 5 to 2400 μm. The smallest droplets were most numerous, 50% having diameters less than about 80 μm for most of the conditions tested. However, few of the smallest droplets contained spores; the median diameter of spore-carrying droplets was between 140 and 150 μm depending on the experimental conditions. The splash droplets travelled horizontal distances up to about 125 cm but 50% reached only 8 to 21 cm from the source. As Faulwetter had found, the smaller droplets were soon retarded by air resistance and fell closer to the source than the larger droplets, which were the main agents for carrying spores. The vertical range of the splash droplets was roughly half the horizontal range.

In the field, many additional factors will complicate the process of splash dispersal. The receiving surfaces may be rough, irregular and inclined at various angles. Distances travelled by the splash droplets, particularly the smaller ones, will be increased by wind, which will also increase evaporation. However, in reasonably intense rainfall, significant proportions of the droplets may be collected by falling drops, thus reducing the extent of wind-borne travel. On the other hand, spores deposited on exposed sites are available for further splash dispersal by other raindrops. The simplified laboratory picture probably resembles most closely the redistribution caused by drip from vegetation. Wind speeds are slower within the canopy and drops falling from vegetation are relatively large and uniform in size. Drip from vegetation is probably particularly important in forests where up to 90% of the rainfall may be intercepted, much of this eventually falling in drops of fairly uniform size and larger than can fall as rain without bursting. These larger (although slower) drops are more effective for splash dispersal.

The splash dispersal mechanism can of course affect any suitable unbound material on the receiving surface: this includes protectant fungicides. Not surprisingly therefore rain splash has been considered to contribute to their redistribution. Thus Yarwood (1950) attributed the control of mildew on the lower surfaces of hop leaves by fungicides applied to the upper surfaces to transfer of the fungicide by rain splash, which also caused deposition of spores on the lower surfaces. Rayner (1962) proposed a similar explanation to account for the control of coffee leaf rust which infects only the abaxial surfaces of leaves by fungicides deposited mainly on the adaxial surfaces.

The simultaneous splash dispersal of fungus spores and fungicides was investigated in the laboratory and greenhouse by Hislop (1969). In the laboratory experiments, drops of simulated rain containing conidia of *Botrytis fabae* were applied to leaves from broad bean plants (*Vicia faba*) using a specially designed apparatus, and the infections produced by splash dispersal on receptor leaves recorded. Redistribution of fungicides applied to the target leaf in a settling tower, a spinning disc apparatus or a micrometer syringe was assessed in terms of the infectivity of the spores in the splash droplets. On this basis, the amounts of copper oxychloride redistributed increased with the fungicide deposit density while the relative redistribution of different materials was related to the tenacity of the deposit as indicated by separate chemical analyses. Redistribution of copper oxychloride was influenced by the addition of surface active agents to the formulation applied to the leaves, probably through effects on tenacity. In other tests, the effects of washing with simulated rain before applying the spore suspension were investigated. Although the amount of copper removed from the initial deposits for redistribution increased with increasing length of the preliminary rainwash, the amount retained on the receptor leaves decreased because it was itself removed from receptor leaves by splash droplets.

Hislop's results may thus be seen as complementary to those of Courshee. They emphasize from a somewhat different standpoint the interplay between dispersal of disease organism and redistribution of fungicide in bringing toxicant and target together and underline that increasing the tenacity of fungicide formulations will not always be beneficial because the reduction in redistribution could diminish activity.

V. ATTRACTANTS, REPELLENTS AND RELATED BEHAVIOUR-CONTROLLING CHEMICALS

A. Introduction

In considering how pest movements contribute to the transfer of pesticides between point of application and recipient (Section 13.II.C), we referred to examples such as the use of seed treatments against soil insects, where advantage was taken of pest behaviour to improve the utilization of pesticides by localized application. In many cases such behaviour is strongly influenced by chemical messengers which in theory could be manipulated more deliberately to provide improved methods of controlling pests or managing beneficial insects.

The basic concepts are not new. Certain volatile chemicals or plant extracts have long been used on or about the person to protect against irritating or biting insects. It has also long been known that other compounds, acting initially on the organism through some chemoreceptive mechanism (which in the human species we call the sense of taste or smell) assist insects to locate suitable food sources, egg-laying sites or the opposite sex or have other effects such as regulating population densities. The importance of such behaviour-controlling chemicals is particularly well illustrated by social insects such as honey bees and ants where they play an essential role in the organization and functioning of the colony, for example in trail following, caste recognition and defensive activities. The use of toxic baits for controlling leaf-cutting ants of the genera *Atta* and *Acromyrmex* represents one of the most clear-cut and best developed examples of the deliberate manipulation of biological transfer to bring pest and toxicant together. Similar techniques have been employed against related pests such as the imported fire ant in the USA. Leaf-cutting ants, which can cause serious losses to cultivated crops, pasture and forestry in South and Central America, form nests which can extend to several metres across and are often situated in difficult terrain. For effective control it is necessary to kill a major proportion of the nest inhabitants, particularly the queen who is normally very well protected and inaccessible within the nest mound. Many insecticides are toxic to the ants, but general application of sprays or dusts is likely to be ineffective because insufficient toxicant reaches the critical sites, is potentially non-selective and is likely to distribute residues widely in the environment. In contrast, toxic baits, which are collected by the foraging ants and conveyed back to the nests, can be spread sparsely in the ants' habitat so reducing environmental contamination and improving selectivity while at the same time greatly improving efficacy. The food-seeking and collecting behaviour of the pest concentrates the toxicant to lethal levels within the nests. (For a review of the advantages of the method, see Cherrett and Lewis, 1973.)

The carrier or matrix material of the bait should preferably be positively acceptable to the ants, or at least non-repellent so that arrestant or attractive substances can be incorporated which will induce the ants to pick up the particles and carry them back to the nests. A variety of cheap and locally available materials have been employed as combined carriers and arrestants, of which dried citrus pulp is the most effective and widely used. Inert mineral materials such as vermiculite treated with orange juice make satisfactory alternative carriers (Phillips *et al.*, 1976). While both citrus pulp and these impregnated mineral matrices are very acceptable when encountered by the ants, there would be advantages (particularly for aerial baiting) in having baits which were attractive from a distance and the incorporation of attractants which would increase the effective diameter of the baits is being investigated. Further important requirements are that the toxicant used in the bait should have a delayed toxic action over a reasonable concentration range, that it should be non-repellent and that it should be readily transferable from one ant to another to ensure good redistribution within the nest. The use of protective formulations such as microencapsulation (see Section 14.IX.E) can help in these respects. Etheridge and Phillips (1976) and Phillips *et al.* (1976) describe recent work on the formulation and evaluation of baits for leaf-cutting ants, and give references to earlier studies.

Terminology in this field is far from satisfactory. The term pheromone is now almost universally accepted for "substances which are excreted to the outside by an individual and received by a second individual of the same species in which they release a specific reaction, for example a definite behaviour or developmental process" (Karlson and Lüscher, 1959). Dethier *et al.* (1960) put forward a comprehensive system of terminology for the different ways in which a moving insect may respond to chemical stimuli. However, compounds playing the two most familiar roles in behaviour patterns are usually called attractants and repellents. It is not possible to avoid these established terms, but they must be used with caution because they represent an oversimplified description of an erroneous interpretation of their function.

Repellents have obvious uses and their success in the very limited field to which we have already referred is well established. Attractants on the other hand have little use as such to the human operator, and the methods conceived for employing them in crop protection are indirect. Their ability to bring the target to the source can be exploited in population studies by locating the source in a suitable trap. However, the trap must also contain some form of lethal device if it is to kill the attracted target as a means of control. To be successful for this purpose, the artificial source must compete with more numerous natural sources, or the "attractant" compound must be used in some quite different way. The highly specific response to attractants, particularly sexual attractants, and their very great sensitivity, response being elicited by extremely small concentration–time products, has naturally aroused much research interest. So far, however, this has been biased too far towards the chemical identification of natural attractants, stimulated by major recent advances in techniques such as chromatography, nuclear magnetic resonance and mass spectrometry used in this branch of analytical chemistry.

B. Directional Response to Behaviour-controlling Chemicals

The aspect of attractants and repellents of most interest in the present context is the problem of how a stimulus which is itself completely non-directional can elicit a directional response, movement bringing the insect towards or away from a source of vapour. Of the main senses, sight is the most positively directional: light coming in a straight line from the source to the eye conveys a message of direction immediately. Hearing is less clearly directional because sound waves can move round opaque objects to a much greater extent than light, but an organism possessing two receptors can nevertheless locate a source of sound fairly accurately.

The bat, with relatively large, focusing ears receptive to very high frequencies is much better at this than the human. Direction finding by both sight and hearing can, of course, be confused by reflecting surfaces.

The sense of touch or oral taste can convey limited directional information because objects must contact a receptor for any stimulus to be received and there is a mechanism of choice based on acceptability. The responses of the olfactory receptors to stimulus are not so clear, being confused by molecular diffusion and local eddies. The sense of smell is thus, by itself, the least useful for indicating distant direction or exact location. It is extremely sensitive and, in the case of insect sex-attractants, extremely specific; its diffuseness has an advantage in revealing the presence of a hidden source but it is not itself directional. It must thus be integrated with some other information system to produce the complex response of attraction or repulsion.

With the sense of touch or oral taste, there can be no attraction from a distance because there can be no stimulus at a distance. The organism must search by random movement or by movement which, if systematic, uses some other sense for organizing the system. When it chances on the source of the required sensation it ceases, at least temporarily, to explore. The effect is usually said to be "arrestant" rather than "attractant". Generally a further response is elicited by the arrestant or some other associated compound, the most common being feeding or oviposition. An extreme form of arrestant behaviour is exhibited by free-swimming barnacle larvae seeking a suitable site for attachment. Crisp (1975) showed that selection of sites is influenced by an arrestant substance consisting of a layer, at most a few molecules thick, of an insoluble protein laid down by established residents. It is still recognized when the adsorbed layer is presented in the donor protein solution (Crisp and Meadows, 1962) and when it is rendered completely insoluble by formaldehyde tanning (Knight-Jones, 1953). In such an extreme case it is doubtful whether the sensory perception should be called touch or taste. Touch as a selective sense must involve appreciation of three dimensional structure and/or deformability on a scale comparable with the organism's receptors. Selective taste involves some kind of appreciation of molecular structure.

An arrestant component can clearly contribute to the mechanism by which surface-sources of volatile or soluble attractants are selected, but these are also potentially able to elicit response at a distance. It has been established in several cases that the directional response to the arrival of the non-directional stimulus is to move upwind: positive anemotaxis. Flügge (1934) seems to have been the first to appreciate this and

Kennedy (1939) showed that mosquitoes in flight select the upwind direction by visual observation of ground markings. Kellogg *et al.* (1962) showed that this applied to fruit flies seeking the source of rotting banana scent in a wind tunnel. In a uniform scent, the flight was steadily upwind. As in Kennedy's experiments, an inappropriate response could be induced when a pattern of projected light was moved on the otherwise featureless floor of a tunnel. When the scent from a localized source enters a tunnel with a stationary pattern on its floor, the generally upwind flight is in detail more tortuous. When the insect flies out of the scent plume it executes a sharp turning movement with a downward bias. In this way a source on the floor is soon located but one emitted from a downward pointing tube near the ceiling is not.

The experiments of Kellogg *et al.* (1962) were designed to test the significance of the irregular "whispy" filamentous structure of a smoke plume trailing downwind from a localized source (see Wright, 1964, Chapter 4). It was thought that the insect might discern the spacing or degree of contrast as it flew through filaments of more concentrated vapour and would keep in the direction of increasing contrast or decreasing spacing and therefore towards the source. The actual behaviour did not support this theory but indicated that the general direction guidance was anemotaxis. The insect responds to the presence of airborne scent by flying upwind. It responds to loss of scent by sharp turning. There was a reaction time of about 0·2 s in the latter response which could be longer than the time from one filament to the next in a smoke plume. Daykin *et al.* (1965) and Daykin (1967) showed that the "homing" of female mosquitoes on to the skin of the blood-meal supplier is a combination of upwind flight in response to increased CO_2 content of the air and sharp turning movements with a downward bias when the insect leaves the column of warm, moist air arising from the host.

Farkas and Shorey (1972) returned to the idea that insects could sense the direction of the source of a plume of vapour from the variation of concentration pattern within it. They generated plumes in a 7 cm s^{-1} wind which was then stopped before moths (*Pectinophora gossypiella*) were introduced. They observed more moths flying through a fixed hoop towards the source than away from it. Grubb (1973) questioned the statistical validity of this finding in comments replied to by Farkas and the question remains rather contentious. Both Farkas and Wright are somewhat in error in assuming that a smoke plume structure will be repeated invisibly in the vapour plume. The smoke particles have effectively zero molecular diffusivity while the diffusion coefficient of the evaporated attractant in air will be at least 0·03 cm^2 s^{-1}. Molecular diffusion which occurs in directions of decreasing concentration could therefore substantially decrease concentration differences within plumes of molecularly dispersed attractant compared with the patterns observed with smoke. In a plume which started downwind initially made up of separate layers, each 1 cm thick, composed alternately of clean air and air containing a uniform concentration of attractant vapour, the maximum difference of concentration between different layers after 20 s would be only about 20% of the mean. After 40 s variation of concentration between layers would not be detectable. This correction does, however, extend the possible ways in which insects might sense differences between parts of the plume near to and distant from the source. It may be noted that Kennedy (1966, p. 102) considered that the Wright and Kellogg "whispy" filament theory was not disproved by their own experiments.

In a more recent paper Kennedy (1976) brings together much of the previous work in developing further a comprehensive theory to explain distant attraction of flying insects to odour sources. One starting point for this was the early hypothesis of Wright (1958) which he subsequently withdrew (Wright, 1964) but which has been more recently revived by Shorey (Farkas and Shorey, 1974; Shorey, 1973). It was postulated that while the insect experienced an increasing or steady concentration it flew straight, but if it experienced a decrease in concentration it would begin a programmed sequence of zig-zag flights. Kennedy points out that the difficulty with this theory is that a flying insect would experience a decreased concentration if it moved out of the odour plume in any direction, so that the zig-zags could themselves occur in any direction relative to the plume and its source. The general consensus of those observing these movements is, however, that although the zig-zags may be somewhat irregular they do not occur in all directions, but preferentially across the plume's axis and along it towards the source. Consideration of various lines of evidence led Kennedy to conclude that decrease of the chemical stimulus, resulting either from overshooting an odour source, or moving beyond the margin of the odour plume results in a reversing orientation *across the wind* (anemomenotaxis). When flying upwind towards the source, one anemomenotactic reversal would usually be enough to bring the insect back into the plume where it would be stimulated by the odour and hence cease its searching or casting behaviour until again leaving the plume. Kennedy considered that reversing anemomenotaxis could also occur in response to the decreases of concentration experienced within an odour plume, thus accounting for the observed irregularity of zig-zagging.

Two essential components of the mechanism by which flying insects are attracted to distant odour sources thus appear to be the positive anemotactic orientation in response to the onset and maintenance of odour stimulation and the reversing cross wind orientation in response to decreasing odour concentration. Kennedy believes that these are combined with other mechanisms of airspeed control to give an integrated system capable of producing the zig-zag flight up a wind-borne odour plume and the searching behaviour after losing the plume.

Walking insects may follow terrestrial trails by responding chemotactically to the steep odour gradients at the margins of the line of deposited material (Hangartner, 1967; Dethier, 1957) but these are effective only over very short distances. For example, ants and termites will lose track of pheromone trails when more than about one inch away. As with flying insects, there is clear evidence that movement towards distant sources in response to airborne plumes involves anemotaxis. Kennedy and Moorhouse (1969) established that perception of a concentration gradient by widely spaced antennae was probably not the principal mechanism by fixing the antennae in a crossed posture or by inactivating one or both by covering with vaseline. These operations impaired, but did not prevent the anemotactic response of starved locust hoppers to grass odour. Kramer (1975) reported observations on the response of walking silkworm (*Bombyx mori*) males to wind containing female sex pheromones which suggest the occurrence of reversing anemomenotaxis as with flying insects. The sensing of wind direction is in principle simpler for a walking than for a flying insect. Kalmus (1942) considered that a direct tactile sense, for example at the base of antennae flexing in the wind or a sense of temperature difference due to unequal evaporation of water, could be the mechanism. An airborne insect must assess ground speed by visual markings, and to do this must remain in visual contact with the ground. In attempting to reach an attractant source it flies in the direction in which it makes slowest progress: if it cannot make progress in the slowest direction, i.e. if the wind is too strong, it does not fly. There are of course many other situations in which insects remain airborne by flying, but allow themselves to drift with the wind; indeed this is an important mechanism in insect dispersal.

There is little evidence that direct sensing of concentration gradients contributes generally to the location of odour sources, but it could possibly operate close to the source. A static gradient could be perceived if two sensors were each compact but well separated as are the antennal knobs of a butterfly. The spacing could be replaced by a separation in time as in the rapid sideways alternating flight of the German wasp. Both these mechanisms of orientation (tropotaxis and klinotaxis respectively) were observed by Martin (1965a,b) who investigated the orientation of walking honey bees to steep concentration gradients of volatile plant constituents at the junction of a Y-tube produced by allowing different vapours to pass down the two arms of the tube. Two sensors could in theory indicate the wind direction of a structured plume by a mechanism analogous to detection of the passage of a shadow (Hassenstein, 1959) since the interval between the passage of a given pattern past each sensor is longest when their disposition is in line with the wind (Kennedy, 1966).

Repellency also is not so simple as it first appears. Repellents cannot be regarded as negative attractants as they do not drive the insect a long distance downwind. Attraction results from an appropriate sequence of responses to stimuli, not all olfactory. Inappropriate responses, failure to respond or responses made in the wrong order can frustrate the approach and so have the effect of repellency. Daykin *et al.* (1965) showed that the most effective and commonly used mosquito repellents act by upsetting the normal responses to an ascending column of warm, moist air.

It should not be assumed, however, that all repellents act in the same way. If some insects in some situations can respond directly to a gradient of attractant, the same could be true of a repellent. The common mosquito repellents show little effect when the insect can choose only between vapour-laden or clean air, but they stop the insect locating its blood meal. Some compounds, however, e.g. oil of citronella, do deter settling near a high concentration. Hartley (unpublished results) carried out tests using symmetrical boxes, opposite walls of which contained gauze-covered "windows", one backed by a source of vapour and the other by active charcoal. There was no avoidance of dimethyl phthalate vapour, but mosquitoes alighting on liquid-treated gauze left immediately. However the insects definitely avoided citronella vapour. They usually steered away without alighting. The result could be explained by a low concentration causing more active flight, but response to a gradient cannot be ruled out. When active charcoal was omitted, the concentrated vapour knocked down the insects.

We are on the borders of a large and complex area of insect physiology that is beyond the scope of this book, but have felt it necessary to trespass so far and establish that the physical direction indicators are several and not the same for all insects and situations. The best tactical use of an "attractant" or "repellent" must depend on how the olfactory stimulus is combined with other stimuli and processed in the insect's nervous system into a directional response. The use of the single oversimplified terms attractant and repellent to describe different mechanisms works against successful practical exploitation because it leads to the unconscious tendency for preliminary testing in the laboratory to be too stereotyped and for each new compound to be evaluated in the field in the same manner as the last even when a different insect species is involved. This may be quite inappropriate to its real mechanism of action.

C. Sensitivity of Response to Behaviour-controlling Chemicals

Common to all mechanisms is some response to change of concentration of the stimulating vapour either with time or space. The olfactory sense, like other senses, responds to increase of stimulus on a logarithmic scale in humans, according to the Weber-Fechner law. As a quantitative instrument it records very crudely. A factor of 2 or 3 in concentration is usually only just noticeable, but the range is enormous, the smell of ethyl mercaptan according to Wright (1964, p. 49) being just detectable at about 10^{-8} of the concentration which is almost intolerable. There is always a fatigue effect. After several hours exposure to a strong smell even this becomes unnoticeable and certainly a much lower concentration would be undetected although a fresh nose would detect it at once. Recovery of sensitivity after over-exposure may take many hours—this is certainly true, in our experience, with chlorocresols. Insects may respond to smaller differences of log (concentration) as experiments by Wright (1966) on *Drosophila melanogaster* suggest. Ignoffo *et al.* (1963) found that male *Trichoplusia ni* moths lost response to strips treated with sex attractant after 2–5 min and needed a rest of several hours to recover.

Quantitative experimental data directly relating to threshold gradients in space or time are lacking. The full exposure to air of the receptors projecting from the antennae of insects eliminates delays which occur in mammals due to rhythmic breathing and absorption in nasal liquid. It is probable, however, that between two receptors or between the sensation in the same receptors at different times, the concentration of stimulating vapour must differ by a substantial fraction if it is to be perceived. We may assume a value for $\Delta \ln C$ of 0·2 (about 20%) which is certainly beyond the discrimination of the human nose. The change perceived will have to occur within about 2 cm at most except in insects with extremely long antennae. The reaction time recorded by Kellogg and Wright for mosquitoes to turn after leaving a stimulating concentration was about 0·2 s. It would seem that $\mathrm{d} \ln C/\mathrm{d}x$ must exceed about $0{\cdot}1\ \mathrm{cm}^{-1}$ or $\mathrm{d} \ln C/\mathrm{d}t$ must exceed about $1\ \mathrm{s}^{-1}$ to produce a response which would be useful for direction-finding.

Any source of vapour will produce a gradient of concentration until the source is exhausted or the environment saturated. Molecular diffusion and eddy diffusion both tend to expand the volume in which gradients can be found and at the same time reduce the steepness of these gradients. The general principles have already been considered (Chapter 3). Here we will deal only with aspects particularly relevant to olfactory stimulants. Molecular and eddy diffusion act on different scales. For large objects at great distances this does not matter but insects close to a source are operating to some extent within the eddy structure, where an eddy can bring a sharp concentration difference to a local region previously homogeneous. By contrast, one is accustomed to thinking of molecular diffusion acting only to reduce gradients. Material is carried by diffusion down a gradient, but in general this creates a smaller gradient in advance. The simplest ideal system for theorizing about diffusion is that where a limited plane source, containing an amount **M/A** per unit area of freely diffusible substance at time zero is dissipated by diffusion in one direction or two opposite directions. At time t and distance x the concentration will be (see p. 139, eq. (43))

$$C = \frac{\mathbf{M}}{\mathbf{A}} \cdot \frac{1}{2\sqrt{Dt}} \exp(-x^2/4Dt) \tag{6}$$

The gradient $(\partial C/\partial x)_t$ goes through a maximum with increasing x, the magnitude of which decreases as it moves further from the source. According to the Weber-Fechner law however, sensory awareness of a gradient should depend on relative not absolute, change. The gradient of log C is easily derived from (6) as

$$(\partial \ln C/\partial x)_t = -x/2Dt \tag{7}$$

This quantity decreases with increase of time as we should expect, but, at any instant, *in*creases indefinitely with increasing distance from the source. Before we attempt to bring this intuitively unacceptable prediction into line with common sense let us derive similar equations for simple diffusion in different geometrical situations.

When the plane source, instead of containing a limited amount of diffusible substance, is held at a constant concentration C_0, (or when a block of air at concentration $2C_0$ is suddenly brought up against a block of clean air) the concentration varies according to:

$$C = C_0 \operatorname{erfc}(x/\sqrt{4Dt}) \tag{8}$$

from which we find

$$(\partial \ln C/\partial x)_t = (1/\sqrt{4Dt}) \exp(-x^2/4Dt)/\operatorname{erfc}(x/\sqrt{4Dt}) \tag{9}$$

which again *in*creases with increase of x at any one t value, less steeply however, than in eq. (7) because it has the finite value $1/4\sqrt{Dt}$ at $x=0$. Equation (9) approaches eq. (7) as $x/\sqrt{4Dt} \rightarrow \infty$.

If the source is a small sphere of radius a the outside of which is maintained at constant concentration C_0, the value of $(\partial \ln C/\partial x)_t$ is the same as in eq. (9) with the addition of the further negative term $1/(a+x)$ where x is now the radial distance of the point considered, measured outward from the surface (i.e. $r=a+x$). In this case a minimum value of the gradient of ln C is predicted. The large (negative) value at the sphere surface remains constant. The distance at which the minimum is found increases with time.

In all these cases, and one may therefore accept it as generally valid, eq. (7) represents the limiting behaviour approached as $x/t \rightarrow \infty$. The absolute concentration however, approaches zero, so the Weber-Fechner law and the existence of a minimum threshold of detection are in conflict. In the language of communication theory the relative change of signal strength (with time or distance) is greatest when the signal strength is below the noise level and its strength cannot therefore be appreciated. There must be a practical optimum which can only be determined by experiment and which may be very different for different organisms and stimuli. Data on this subject are at present lacking.

Some very important practical qualifications must be made about the gradients so calculated but, before considering these, it is instructive to compare the pure diffusion gradients with gradients of intensity of light or sound with which we are much more familiar and unconsciously take as standards. For both light and sound there is no gradient of intensity with distance from a very extensive plane source, unless the medium can absorb energy, in which case the gradient of ln I (I = intensity) is constant. A better example for comparison is the point source with no absorption when the intensity is inversely proportional to the square of distance, giving

$$\partial \ln \mathrm{I}/\partial r = -2/r \tag{10}$$

When divergent diffusion from a point source has reached a steady state, concentration is inversely proportional to the first power of radial distance, so that

$$(\partial \ln C/\partial r)_t = -1/r \tag{11}$$

Light or sound would therefore provide a better directional guide were they called on to do so by a gradient mechanism instead of by the more direct effect of unidirectional propagation of energy.

A major difference arises because the gradient of light or sound

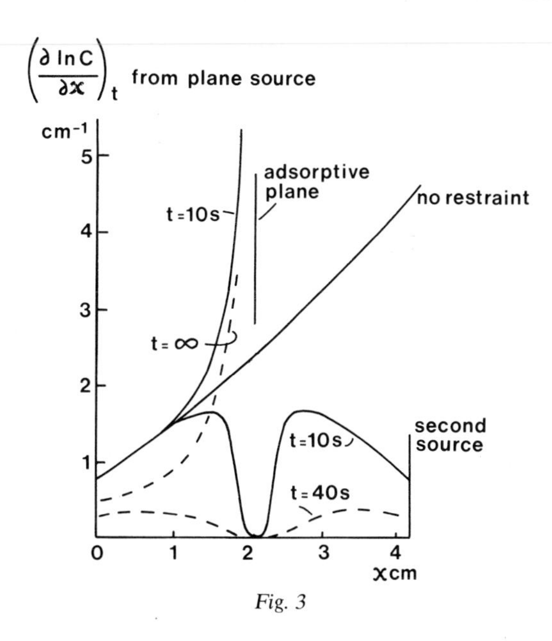

Fig. 3

more likely to be adsorptive. The upper deviating solid curve in Fig. 3 is for the condition that the plane at $x = 1{\cdot}5\sqrt{2}$ is a perfect sink. The effect is to increase greatly the value of $\partial \ln C/\partial x$ as the plane is approached. In this case the steady state is that of constant $\partial C/\partial x$ and the steep gradient near the sink therefore persists. The corresponding broken curve is calculated for the steady state ($t = \infty$).

The sink condition is by no means unrealistic. A depth of 2 cm above a plane surface, if saturated with a 16 C atom alcohol of vapour pressure 4×10^{-5} mmHg, would contain less than 1% of the content of a compressed monolayer on the surface area, and it can be calculated that, if the initially empty surface had to be supplied by diffusion across a 2 cm gap from a saturated source, more than 3 h would be necessary to build up the monolayer. The adsorptive capacities of plant cuticles and their associated oils and waxes are much greater than the equivalent of a monolayer on the apparent plane area. Gradients of insect pheromones within a crop will depend more on the extensive sinks than on the local sources. It is doubtless for this reason that "calling" females often place themselves in exposed positions on, e.g. the apices of upright vegetation and that for an appropriate range of wind speeds, traps baited with attractant are best located above the main part of the canopy. The canopy may have more than an interfering function. The ground dwelling male may be effectively protected from odour fatigue by the adsorptive vegetation. His introductory stimulus may not be simply the arrival of odour but of a steep local gradient of odour.

Eddies will confuse the diffusion process just as diffusion will blur the "wispiness" of eddies, as we have pointed out above. The interior of a plume of vapour spreading downwind from a local source will be much more uniform than a plume of non-diffusible smoke, but puffs bursting out from the uniform interior will be frequently creating new sharp gradients on the fringe of the plume. It is here that direct awareness of gradient in the open is most likely, consistent with Wright's observations on homing mosquitoes. Remarkably, however, it is the negative gradient (with time, i.e. $\partial \ln C/\partial t$ negative) on departure from the plume that causes response although the factor of fatigue would lead one to expect a weaker signal than on entering the plume. The flying insect, of course, creates its own temporal gradient by pulling back air from in front but this again would lead to a sharper transition in the positive direction. The insect's olfactory apparatus is presumably specially adapted to perceive, despite fatigue, the negative gradient which is more important for its behaviour. In interpreting experiments on the orientation to scent of walking honeybees, Kramer (1976) was forced to the conclusion that bees "are capable of memorizing odour concentrations with a high degree of accuracy".

D. Practical Use of Behaviour-controlling Chemicals in Pest Control

Three principal methods of using sex attractants have been attempted: (1) trapping for population surveys, (2) trapping for destruction and (3) reduction of mating by confusion of signals. When traps are used for population surveys it is not always necessary for them to be very efficient but they must not be subject to unequal environmental effects: design, posture and location must be standardized. In some cases, however, particularly when very early warning of an initial infestation is required as in the control of pea moth on peas for processing when even very slight lowering of quality can result in rejection of the crop, it is essential to detect the presence of the pest at low levels and traps must be very efficient. A wide variety of different designs have been used by those working on attractants, but there has been almost no systematic investigation of the factors determining trap efficiency, at least in relation to the pattern of attractant release and dispersal, and no framework of general principles has been established. Recent work by Lewis and Macaulay

(1976) using smoke plumes has indicated substantial differences in patterns of release from different designs which can be related to differences in performance, emphasizing the need for further detailed investigation. It is important, however, to recognize the limitations of particulate smoke as an indicator of the behaviour of true vapour, which we discuss on p. 744. Furthermore, to minimize misleading effects, the smoke should be released isothermally, should be sufficiently dilute (less than about 100 mg m^{-3}) and its sharpness must be accepted as an exaggeration.

If used, generally in combination with insecticide, for direct control, traps must either be more numerous or more attractive than the natural sources—food or females. This is a difficult requirement but greater attractiveness seems to have been achieved in control of the oriental fruit fly by porous blocks soaked in methyl eugenol plus the insecticide naled (Steiner *et al.*, 1965) and of the melon fly by similar use of cuelure (4-acetyloxyphenylbutan-2-one) (Cunningham and Steiner, 1972). Neither of these compounds is the natural attractant and, although the attracted flies avidly feed on them it is mainly males which are attracted. Relatively large porous blocks (*c.* 5 × 5 × 2 cm) are used for convenience of distribution and to reduce the evaporative loss of the rather volatile lures. Although large blocks may be much easier to find than many point sources, their efficiency when used in some cases at less than 8 blocks per ha must be mainly due to the chemical accident of the synthetic lure producing a much greater directional stimulus than the natural attractants.

In the third method of use, attractant is released into the air at a much greater rate than is the natural product by the resident population of females. The object is to "jam" the communication system and so prevent the males from finding the females, the added uniform concentration being so great that local gradients, spatial or temporal, are reduced below a non-response level. The method has had limited success in a few cases and has been totally ineffective in others. One might expect it to be most effective on low-level populations where successful search involves a long directed path and to be unsuccessful against very high populations when other mechanisms could bring the sexes together, yet one of the most effective disturbances was reported (Farkas *et al.* 1974) in a population of the looper moth, *Trichoplusia ni*, at sufficient density for 100–150 males to be caught each night in each trap baited with a live female. Remarkably the catch dropped to 1% of controls when additional sources of pheromone put out only 25 mg ha^{-1} per night from stations spaced 200 m apart. This increased to 5% when the spacing was doubled while the output per whole area remained the same. It rose to 16% when the spacing was kept at 200 m but the output per whole area reduced to a quarter. That so coarse a pattern of emission can successfully confuse, strongly suggests that gradients of low concentration, remote from calling females, must play an important part in the locating mechanism. High logarithmic gradients near each female would not be much affected but those near to foliage sinks would be increased by the artificial emission.

Confusion of signals by excess of stimulant, apparent repulsion by another substance which interferes with normal response and competition by another substance evoking a "wrong" response are processes which can supplement or interfere with one another. More needs to be known of the detail of directional response if costly field experiments are to be replaced by more efficient screening.

Recent work has tended to neglect the possible role of choice. While it is true that many commands given to insects via the olfactory sense are absolute "dos" or "don'ts" and in experiments insects have been commanded not to feed on an appropriate food or to feed on a completely unsuitable one, there are probably more cases where significant selection is made between alternatives. Eggs may be laid and larvae develop on several varieties of the same plant species but where a choice is available more will be laid on some varieties than on others. This gives rise to what Painter (1951), in the absence of a single word denoting the opposite of "preference", called "non-preference resistance" to distinguish it from resistance due to toxic properties, mechanical defence or tolerance of injury. In a study of this phenomenon in radish subject to attack by the root-fly, *Erioischia brassicae*, Ellis and Hardman (1975) find clear evidence of non-preference for certain cultivars. They say, however, "To be of practical value, non-preference resistance must be maintained... where the fly has no choice of host crop".

This seems to us to be failing to grasp the opportunity to revive in more effective form the old practice of trap-cropping (see Martin, 1973, p. 343). One should present the fly with a choice of say, four rows of a non-preferred cultivar and one of a preferred cultivar, only the latter being heavily treated with insecticide and destined for rejection. It is clearly more economic, particularly now that the pre-packaging trade demands a high standard of freedom from blemish, to have an easily-selected 80% of the crop quite clean rather than have perhaps less than 20% unmarketable but randomly distributed. The radish may not be the right crop for such a policy nor the root-fly the right pest, but the work quoted demonstrates the existence of choice. We suggest that the choice should be turned to advantage and more effort be put into the search for "non-preference".

VI. GROWTH OF PLANTS IN RELATION TO UPTAKE OF SOIL-APPLIED PESTICIDES

So far in this chapter we have considered the contribution which movements of pests and pathogens make to bringing toxicant and receiving organism together. Compared with these organisms, plants appear very static if one excludes seed dispersal which is outside the terms of reference of this book, and there is a tendency to overlook the role of plant growth in the dosage transfer of both herbicides and systemic insecticides and fungicides applied to the soil for uptake by the underground parts of plants. In general terms, amounts taken up are related to the area of root and stem exposed to treated soil and the average concentration of pesticide in these absorbing regions. In considering the behaviour of pesticides in soil in Chapter 5, we discussed the factors determining the value of this concentration following different initial distributions. We stressed that many soil-applied chemicals are relatively immobile: the extent to which roots and stems intercept treated zones of soil and hence the area of plant surface over which uptake occurs therefore often depends to a large extent on the growth of the plant. "Biological transfer" by the growing plant can thus have a considerable influence on efficacy following different methods of placement. This has an important bearing on the selectivity of soil-applied herbicides, particularly the so-called depth protection (see also Section 5.IX) which can be achieved with certain pre-emergence applications.

The depth from which seeds can emerge varies greatly with species and soil conditions. As a generalization and for obvious reasons, the larger the seed and the greater therefore its food reserves, the greater the depth through which it can push its shoot upwards without undue damage or exhaustion. Many crops grown to provide food in the form of seeds, including cereals and many grain legumes such as peas and various species of bean, have naturally large seeds, whose size has been increased further by breeding. Certain other crops in which the seed is not harvested directly for food also have large and vigorous seeds. Cotton is a familiar example. Most of these large-seeded species will produce a thriving stand when sown at depths of 5 cm or more from which many weeds, having much smaller seeds, will not survive. Crops propagated by tubers, such as potato and sweet potato, can also be planted at depths too great for many weed seeds. In contrast many green food crops (for example lettuce) have small seeds which must be sown near the surface.

Where there are appropriate depth differences these can be exploited to secure selectivity of soil-applied herbicides, or at least to enhance a selectivity based on biochemical differences which is insufficient in itself to give reliable results under the variable conditions experienced in the field. The localized application of a relatively immobile compound in the surface layers of soil can achieve a selective availability to the shallow-seeded plants. As we discuss below, however, successful practical applications of this principle have been relatively few. The list of successes is not longer for three reasons. In many species of plant, the underground shoot is capable of significant uptake, so that even deep-seeded plants can acquire a damaging dose as they grow through the treated layer. In most species the main root descends very rapidly after germination, allowing weeds germinating near the surface to grow away from the treated soil. Finally in many species the roots branch laterally almost immediately after germination. These various factors tend to blur the distinction between effects on deep- and shallow-seeded plants: to benefit from sowing a crop deep, a herbicide not taken up by the shoot from the soil and a root that grows down rapidly with little initial branching are required.

Apart from the cases of selectivity based on differences in the depth location of the very sensitive coleoptile node, which we discuss later and which can be regarded as a very effective exploitation of depth protection, the only important practical examples of the method appear to involve the use of triazines and substituted ureas on potatoes, various herbicides including diuron, trifluralin and formerly DNBP on cotton, triazines on field (broad, *Vicia* spp.) beans and to a lesser extent various herbicides on snap and soya beans (*Phaseolus* spp. and *Glycine max*). Excessive rain can sometimes lead to crop damage by causing downwash of the herbicide but the effects of rainfall are not always simple. Thus Davis (1956) reported that 1·2 cm of artificial rain applied after treating the soil surface with dinoseb caused more damage to cotton sown 2 cm deep than either lesser or greater amounts. There was less damage to seed sown 4 cm deep. Presumably the greater dilution by the larger quantities of rain more than compensated for the deeper penetration of the herbicide band. If a crop can usually escape damage by deep seeding, the method is safer in arid regions if furrow irrigation is used, as the resulting water movement keeps the herbicide near the surface. The use of amiben derivatives in such situations is discussed by Linscott *et al.* (1969). Another facet of depth protection is demonstrated by the investigations of Linscott and Hagin (1969) into double-band methods of applying herbicides before seeding alfalfa (*Medicago sativa*). It was found that in combination with inter-row bands of triazines, EPTC could be safely laid *below* the seed, giving more effective weed control.

Depth protection can sometimes be supplemented by the use of active charcoal which can afford some protection to crop seedlings by adsorbing the herbicide. To be effective the charcoal must be applied as a band in the soil above the level of the seeds and, of course, shoot uptake must not be significant. As early as 1945, Hartley (unpublished work) showed that brassica seedlings could be protected from DNOC by this method. Unfortunately, however, weed seeds in the row, where they are most troublesome, were also protected so weed control was not good enough to justify the cost. Rising labour costs and the development of more sophisticated machinery have led to an increased interest in the approach. Ripper and Scott (1957) advocated its use in sugar beet. Kratky and Warren (1971) achieved effective protection of single cucumber seeds by filling a charcoal/vermiculite mixture into the holes created by inserting the seeds. Burr *et al.* (1972) and Lee (1973) claim that the method is economic in grass seed crops using diuron as the herbicide.

The divaricating root system grows very rapidly and in an opportunist manner, exploiting soil zones which can supply water and nutrients. Some herbicides have a local damaging or arresting effect on root growth, and if their distribution in the soil is localized, their systemic uptake can be arrested. Lyndsay and Hartley (1963) demonstrated very clearly with pea seedlings growing in water culture in two adjacent vessels, one containing approximately one-third of the roots and the other two-thirds, that MCPA had little effect on the growth of the aerial parts when it was presented to the one-third portion of the roots only, but a lethal effect when one-third of the concentration was present throughout. With atratone there was no difference in toxic effect on top growth when a similar comparison was made. MCPA produced severe local damage to roots: atratone did not. Corresponding results with linseed were even more extreme. These workers later (Lyndsay and Hartley, 1966) confirmed that the effect could also be manifested in soil. Pans were packed with treated and untreated soil separated by a vertical divider until pea seed was sown. Seedlings originating in clean soil, only 2 cm away from treated soil, grew normally but, when the pan contents were knocked out, no fibrous root mat held the soil together in the treated half. In control pans, the root system had spread throughout. Uptake of MCPA by the plant from a non-uniformly treated soil is therefore more a function of root growth than lateral diffusion. When the soil is uniformly contaminated with MCPA the seedlings die because root uptake of water and nutrients is inhibited.

This effect could be important when herbicide is applied in a coarse granule formulation (see Section 14.VII) and there are indications that root-growth inhibition could be exploited to protect from other herbicide action. Thus Nishimoto and Warren (1971) used diphenamid to inhibit adventitious root growth in pot experiments. When confined to the shoot zone, diuron, ametryne and terbacil were less effective in the presence of diphenamid which made no difference to the effects of EPTC, trifluralin, chloropropham and DCPA. O'Donovan and Prendeville (1975, 1976) showed that trifluralin and nitralin reduced the effects of triazine and urea herbicides simultaneously applied to the shoot zone of the soil by inhibiting adventitious root formation. Arle (1967) had previously shown similar protection of cotton by suppression of lateral root branching. It seems clear that both true adventitious roots and branch roots can be suppressed. The need for reliable distribution of chemical under variable weather conditions is likely to restrict practical exploitation of these effects: the chief importance of the function of healthy rootlet growth in collecting herbicide is that it makes absence of effect of the herbicide on root growth a generally desirable property. Nevertheless a severe depressing effect on root growth without systemic effect is useful in the particular situation of keeping underground drains free from obstruction by roots. Duffy (1972) found caustic soda preferable to copper salts for this purpose.

Uptake by underground shoots can cause severe effects with some herbicides on many plant species. Its significance has become much more appreciated during the study of several wild-oat herbicides and has been the subject of numerous investigations. Many but not all grasses are vulnerable by this route. Vapour phase diffusion usually plays a more important role in the process than diffusion in the soil water, probably because the shoot surface is not easily wetted, it tends to emerge through loose soil with which it makes poor contact and, unlike the root, has no hairs to enhance contact. Parker (1966) demonstrated these effects by growing individual oat seedlings against inclined glass plates in which herbicide could be localized within separated zones of damp soil and vapour transport reduced by thin plastic films. Barrantine and Warren (1971) used a similar localized placement technique to demonstrate shoot uptake of trifluralin and nitralin. In some species adventitious roots may confuse the picture; Nishimoto and Warren (1971) were able to clarify their contribution in the experiments to which we have already referred by using diphenamid to eliminate their formation without injuring the shoot.

Several of the factors we have been considering can play a part in the selective control of weeds in rice. Baker (1960) examining the selective activity of herbicides against barnyard grass (*Echinochloa crus-galli*) in rice showed that the weed extends its first internode so that the coleoptile node reaches the surface layers of the soil while in rice it remains on a level with the caryopsis. This behaviour gives rice an advantage which increases with depth of sowing, not because the weed cannot germinate and emerge from depth, but because it sends upwards a very sensitive organ, direct access to which appears to be essential for the action of some herbicides. Weiss (1962) found that *E. crus-galli* and *Paspalum dilatatum* sown at various depths from 0·6 to 5·0 cm in soil treated with 1 kg ha^{-1} simazine on the surface showed damage "varying inversely" with depth; in this case however there is no direct effect on the coleoptile node. Knake *et al.* (1967) and Chem *et al.* (1968) found *Setaria viridis* and *E. crus-galli* to be most sensitive to herbicides applied near the surface, the latter finding that injury by molinate to rice decreased with depth of sowing.

Related effects have been observed with other crops. Burnside (1964) found *Sorghum vulgare* to be more sensitive to herbicide above the seed than below it and Banting (1967), consistent with the other observations we have discussed, found that diallate incorporated into the top 5 cm of soil killed *Avena* spp. seedlings even when they emerged from 15 cm depth but that wheat was safe below 6·5 cm. Barrantine and Warren (1971) and Rahman and Ashford (1970), examining selective toxicity to *Sorghum vulgare* and *Setaria viridis* in wheat showed that, as with rice, the coleoptile node of wheat remains near the caryopsis while that of weed grasses extends to near the surface.

Although the most effective shoot-active soil herbicides attack a local and vital organ and are most effective at an early stage of growth, there can be a contribution from shoot uptake to the absorption of herbicides acting mainly by root uptake. Walker (1973) made a careful study of the effect of depth of incorporation with several herbicides and plant species. Nearly all the herbicides, including triazines, were active when confined to the shoot zone only, but all except pronamide and chlorpropham were more active in the root zone. These two compounds showed no activity when confined to the root zone.

We can find no information on the effect of population density on phytotoxicity, but one would expect it to differ according to route of uptake. The root systems of most seedlings spread very quickly. The roots of neighbouring seedlings will have interpenetrated well before the aerial canopy is closed. At this stage, whether true interpenetration occurs or the root-spread of each seedling tends to be arrested by that of its neighbours, the downward moving dose of herbicide eventually available to each seedling will be proportional to the overall applicance divided by the population density. If the phytotoxic effect is not dependent on population density, this can only be a compensating effect of competition—the plant in a crowd succumbs to a lower dose because it is weakened by the competitive effects of its neighbours. This would not be unreasonable: the total yield of a clean crop does not depend greatly on seeding rate over a rather wide range and the crop may therefore respond as a whole organism on a mg kg^{-1} basis. The shoot, however, is not a divaricating collector. Access to it from the soil can only be by physical transfer (unless adventitious roots are formed) giving advantage to volatile compounds and with the supply coming from a very limited volume of soil. Uptake, therefore, at any normal field density of crop and weed seedlings (1000 seedlings m^{-2} corresponds to 3·4 cm average spacing) will not depend on population density. Competition could therefore act without compensation and the herbicidal effect would be expected to increase with population density.

In attempting to devise the most effective methods of using herbicides these various aspects of the growth and behaviour of crops and weeds must be related to the patterns of redistribution of soil-applied herbicides discussed in Chapter 5 and Section 14.VII.B. The examples we have discussed, however, demonstrate forcibly that although plants are much less mobile than many insects and fungal pathogens, the patterns of their growth and extension have an important influence on the uptake of herbicides, and thus biological processes make an important contribution to dosage transfer in this sphere also.

14. Application and Formulation

I. Introduction 764
II. Application methods 765
A. Dispersion and localization of spray 765
B. Formulation for spray 766
C. Dusts. 768
D. Fumigation 768
E. Smokes 769
F. Granules 770
G. Foam 773
H. Baits. 774
III. Limiting factors in formulations. 774
A. Introduction 774
B. Melting point and solubility limits 775
C. Chemical stability 777
D. Corrosion and compatibility 778
IV. Uniformity and control of applicance 781
A. Metering 781
B. Uniformity across the swath. 784
C. Uniformity along the swath 790
D. Uniformity of penetrated spray 791
V. Retention 792
A. Introduction 792
B. Reflection of liquid drops 793
C. Retention of solid particles 796
D. Electrostatic attraction 798
E. Adhesion of particles 800
F. Weathering of deposits 803
G. Selective adhesion 806
H. Adhesive formulations 809
J. Selective availability 811
VI. Control of drift 812
A. General 812
B. Avoidance of small drops 814
C. Spray thickeners 815
D. Deposit stickers 818
E. Reduction of vaporization 818
F. Reduction of reception 818
G. Drift of dust and micro-granules 819

VII. Application of pesticides to soil 820
A. Introduction 820
B. Release from initial source and its significance 821
C. Patterns of distribution 823
D. Diffusion from sources distributed in limited volume 824
E. Effect of chemical decay 831
F. Leaching for discrete sources 835
G. Accelerated solution of active ingredient 837
H. Granule–soil-water contact 840
J. Internal and external delay 841
K. Loss from surface application 846
VIII. Seed treatment 849
A. Introduction 849
B. Methods of applying pesticides to seeds 851
C. Evaluation of seed treatment processes 853
D. Effects of seed treatment formulation on biological activity . . 854
E. Pelleting and coating 856
IX. Controlled release 857
A. The advantages 857
B. Retardation of release—physical 860
C. Retardation of release—chemical 864
D. Delayed release 866
E. Encapsulation 868

I. INTRODUCTION

All sections of this book have considered some aspect of the relation between gross field dosage in agriculture, i.e. kg a.i./ha of land, and actual availability to the real target of the operation. Application devices, the skill with which they are used and the formulation of the products used in them, can have some influence on the efficiency with which a material distributed by an essentially wasteful method actually reaches the target organism and on the selectiveness of essentially indiscriminate operations.

Most research in the past into machine performance has been directed to improving convenience, reliability and cost and uniformity of distribution. Most research into formulation has been directed to ease of mixing, storage stability and packaging. These are important matters but lie outside the scope of this book. Our main concern is with the aspects of application and formulation having most influence on biological effect. Since this influence is more difficult to establish clearly than more purely physical behaviour, it has received scant attention. A leaking drum is an obvious loss to all concerned: a five-fold increase in the effect on pest relative to that on predator is difficult to prove and probably expensive to achieve but could have far-reaching consequences.

In this chapter therefore we emphasize the factors most affecting biological response and selectivity but will first deal, in outline only, with the mechanical and industrial problems in order to establish the practical limitations within which biological improvement must be sought.

II. APPLICATION METHODS

A. Dispersion and Localization of Spray

The use of pesticides in agriculture demands some means of dispersal for the purely practical reason that the product must come from factory to farm in a small container and be distributed over a large area. The application machine (we will call this the *applicator* as distinct from the *operator* who controls it) must therefore travel over the ground and measure and disperse. It must also aim, even if only in the very crude sense of discharging its contents on to the crop and/or ground rather than on to a wider environment. Dispersal and aiming are necessarily to some extent conflicting, particularly when uniformity is considered an aspect of aiming—very fine particles are better dispersed but their trajectory is more influenced by gusty wind.

For this reason, mass is usually deliberately added, the product being sprayed in larger droplets after dilution with water rather than released as a concentrated aerosol or dust. This increases momentum. The particle can be thrown rather than allowed to drift. Another reason for added mass and volume is visibility. This enables the operator to see more clearly how his machine is performing and what it has immediately covered.

The added mass of diluent water, however, itself creates problems: it means more weight to be carried over the ground, with considerable increase of cost if the applicator is airborne and with considerably more wheel damage to the crop and soil if it is ground-borne.

On these purely physical factors of safety, convenience and cost, we already have requirements in conflict and some compromise must be sought. Steps can be taken to reduce some conflicts. Devices built in to the applicator can monitor the functioning of conducting lines and orifices so rendering visibility of the discharge unimportant. By suitable design and use of special additives the proportion of very fine droplets in a spray can be reduced so that volume rate reduction does not force so sharp a conflict between good coverage and risk of drift. Drift is generally less serious with insecticides or fungicides than with herbicides and therefore lower volumes are often acceptable with the former.

Faced with these problems the designer is naturally reluctant to introduce others. The evidence must be good and the demand insistent. The improvement wanted may, however, not always be difficult to achieve. The goal of uniformity is often accepted uncritically. Uniformity in a gross sense is always desirable. It is undesirable for one half of a field to receive twice as much as the other, for the uphill swaths on a sloping field to receive twice as much as the downhill or for the whipping motion of a long boom to result in irregular variations by a much larger factor from one metre to the next (van der Weij, 1970). Whether we need 100 particles to the cm^2 or 1 particle per 10 cm^2 should, however, be a question answered by the biologist's research rather than the engineer's imagination or aesthetic sense.

For some purposes controlled non-uniformity is deliberately sought. Mechanical hoeing is still a cheap way of eliminating weeds except when they are very close to the crop lines. Expensive selective herbicide can then be "band-sprayed" on the crop lines. Paraquat is applied on weeds between fruit or ornamental shrubs by hand devices, visually controlled, and between orchard trees using flexible booms. A device to spray horizontally into a collector from which liquid is returned (McWhorter, 1964) confines herbicide to foliage tall enough to intercept. Discovery of the very effective translocation of glyphosate from foliage to root has renewed interest in this method (Evetts *et al.*, 1976; Boyles *et al.*, 1978) and stimulated development of machines (Wyse and Habstritt, 1977) which drag over the weeds a supported wick from which herbicide is wiped on to protruding leaves, mainly and advantageously on to the abaxial surfaces. The methods are limited to such combinations as tall grass weeds in beans or tall inedible weeds in grazed pasture. Avoidance of harmful wastage on the crop makes variable excess applicance on weeds acceptable.

B. Formulation for Spray

Most spraying (or other liquid distribution) is carried out with water as the mass-providing diluent but, with increasing trend to very low volumes, other diluents become economic and may be essential for satisfactory performance. Some pesticides, notably the acidic herbicides used in saline form, are sufficiently soluble in water or a water-miscible solvent to be formulated as a concentrated solution (soluble liquid, s.l.). Others require to be suspended in the water in particulate form, sometimes as a preground solid supplied as a wettable powder (w.p.), sometimes dissolved in oil formulated as an emulsifiable concentrate (e.c.) to form an emulsion on simply pouring into water.

There has recently been a revival of interest in concentrated aqueous suspensions, now called "flowables" or, more satisfactorily, suspension concentrates (s.c.). They are made possible by the wide range of suspending agents now available and are already flocculated so that further settlement does not occur and they become pourable on shaking with less risk of personal contamination than with dusty w.p.'s.

Diluted suspensions of any formulation of insoluble a.i. are not indefinitely stable. Slow sedimentation, upwards or, more usually, downwards, occurs and means of agitation are therefore provided in most applicators for large-scale operations. Sedimentation can be accelerated by reversible clumping (flocculation) of the particles. This may give rise to more permanent increase of particle size and, in the case of emulsions, to fusion of globules, but the flocculated suspension is much more easily resuspended if a tank load has had to be left at rest. A non-flocculated suspension settles much more slowly to a hard clay.

Machines which depend on direct hydrostatic pressure to disperse the liquid as a fine spray necessarily have fine orifices through which the liquid must pass. These are vulnerable to disturbance of function by even partial blockage with solid matter and must therefore be protected by filters placed in the flow-line behind them, with mesh size smaller than the orifice to be protected. Eventually the filter may block. This is frequently a source of delay and more attention could be given to self-cleaning or quickly replaceable filters.

The particles in a satisfactory formulation (i.e. the ready-to-dilute concentrate as supplied by the chemical manufacturer) should not of course themselves block the filters, but this requirement necessarily restricts the type of suspension which can be used in a fine-orifice machine. Blockage of filters is usually initiated by particulate matter in "open" water supplies (rivers etc.) and the most troublesome components are of vital origin—filamentous algae, diatoms, fallen seed-pappus, insect wings etc. (Naughton and Swinstead, 1955). When a filter is liberally coated with such debris it may begin to pick up the solid matter of the formulation itself.

In air-assisted as opposed to hydraulic nozzles the liquid is shattered by a high-velocity airstream and does not have to be forced through fine orifices. A rotating disc sprayer flings off the liquid, and any suspended solid, centrifugally. Much coarser suspensions could be handled through such machinery but the limitation imposed by the more widely used hydraulic nozzles has inhibited development of coarse suspensions.

C. Dusts

Large-scale dust applicators are essentially dispersive machines with very little aiming ability. Solid spherical particles of the size of the liquid drops common in hydraulic sprays (100–500 μm diameter) have little tendency to lodge on plain surfaces: mostly they cascade through the crop on to the ground except for a small proportion caught up in crevices formed by stipules, auricles etc. Dust particles are therefore preferred in the range 5–20 μm: they are carried on the wind and penetrate considerably into dense canopies, eventually falling under gravity in local still-spots and collecting mainly on upper surfaces.

Since the dust has to be fed into the dispersal device—usually just a rapid airstream—it is desirable that it should have acceptable flow properties. Only solid active ingredients are therefore easily formulated as dusts. They are usually ground together with some fine mineral expander, to prevent re-adhesion of shattered organic crystals, and then diluted with some locally available fine powder.

There have been suggestions to improve the adhesive powers of dusts by formulating so that they become sticky on exposure. Milk powder and non-granular "instant" coffee behave this way as they contain a high proportion of deliquescent saccharides. Such powders however cannot be liberated by an ordinary air-blast applicator pumping external air. This invites clogging up the whole machine with syrup. Either dried air must be used or the formulation must contain a powder which takes up the first-arriving water vapour while remaining dry—e.g. anhydrous sodium sulphate. Provided powder is then metered into the airstream at a suitably matched rate, there is no development of stickiness till the increased water supply in the outside air is reached. These are possibilities but are usually considered to require too sophisticated control, particularly as dust, formulated with a local diluent and distributed from a home-made shaker, is mainly used in regions of more primitive agriculture.

D. Fumigation

Fumigation is used to some extent in field agriculture to treat soil, prior to growth of expensive crops, to destroy some micro-fauna, especially nematodes. Mechanical dispersion is not required because the gas, delivered as such or as a very volatile liquid, does its own dispersing. It is injected at discrete stations or along lines, the lead-tube following a coulter or by special devices in subsoil plane. Limitation is mainly of chemicals available. They must have small molecules and low water solubility. Methyl bromide, ethylene dibromide, dichloropropane and related compounds, carbon disulphide and very few others are used. Fumigation is more often used in closed agriculture (glass houses) and stores. As discussed in Sections 5.IV, 6.VII.D and 12.II.E vapour transfer is often important for pesticides not generally considered to be, or not used as, fumigants.

E. Smokes

Droplets in the 20–50 μm range can be produced by air-assisted spray devices but at increasing cost of wasted energy as the size decreases. Particles in the 0·1–5 μm range can be produced efficiently only by a build-up from the vapour state rather than shatter of a condensed state. Such dispersions are called smokes and enable rather involatile pesticides to have some of the self-dispersing, penetrating properties of fumigants. The compound to be dispersed must be sufficiently stable to withstand a short period of high temperature during which it is evaporated. The vapour is then mixed with a great volume of air as it cools. The particles are produced, like "steam" from a kettle spout, by condensation and are, initially, usually below 1 μm diameter, but increasing by coagulation.

A simple and primitive smoke generator is provided by metering liquid into the exhaust pipe of an internal combustion engine (tractor or aircraft). The sprayed liquid has a fraction of a second's contact with hot gases which volatilize it before dispersal in the cold air. An advantage of this system is that the hot gases contain little oxygen and therefore are less chemically destructive than normal air at the same high temperature.

Self-contained smoke generators are based on pyrotechnic compositions. An intimate mixture of crystalline oxidizing agent and oxidizable compound (e.g. sodium chlorate-sugar) together with some initiator (e.g. elementary silicon) and retarder (e.g. ammonium chloride) is itself mixed, in a coarser grain structure, with the compound to be converted to smoke. The temperature of the reaction and the large volume of gas generated volatilize the compound and force it at high speed into the outer cold air, made turbulent by the gas jet, where it is cooled and condensed to smoke. The composition is adjusted to provide a gas of slightly reducing balance, i.e. a deficiency of oxygen so that there is some CO as well as CO_2, the other components being H_2O and probably N_2 (if a nitrate oxidizer is used). Excessive CO can lead to ignition of the issuing gases. The method, when optimum composition is worked out by trial and error and prior experience, can disperse a surprising number of compounds without serious chemical degradation.

Smokes are mainly used, for obvious reasons, in closed situations. The self-contained generators have a delay fuse ignition system so that a train of them can be set off in a closed glasshouse or store, giving time for the operator to vacate the building before the smoke is generated. The smoke is readily dispersed on slow air currents and penetrates well on these currents within the rather open canopy of leaves, but true diffusion of the smoke particles by Brownian movement is very slow, so that small stagnant crevices such as between leaf or flower bud scales are not penetrated. Deposition by diffusion on surfaces is correspondingly slower than that of a soluble or adsorbable gas, but there is a pronounced migration from warm air to cold surfaces which is not however of importance except in heated buildings. Collection by impaction from a moving airstream is inefficient. Upward facing surfaces collect more smoke deposit than vertical or under surfaces (Harris, 1964), sedimentation being more important than diffusion in the final process.

F. Granules

The application of pesticides in granular form has become an increasingly common practice over the last 20 years, stimulated mainly by the development of pre-emergence herbicides. Granules and smokes stand on opposite sides of dusts. Granules are too large and insufficiently adhesive to be of much value in direct application to plants except where an upwardly open funnel structure or flat rosette form ensures collection. They will trickle through most open vegetation to the ground (or water) level. This property is advantageous where action is required via soil or water and where hold-up on the vegetation would be at best wasteful and perhaps harmful. For application of a primarily soil-acting herbicide in an established crop or for control of aquatic insects in overgrown waterways granules are particularly advantageous.

Often however the main reason for their use is convenience and even fashion. They have a considerable convenience advantage in small packs for domestic garden use in that a "pepper pot" type of dispenser makes no mess and can be picked up and put away in a moment with no cleaning out. They have a convenience advantage for agricultural use in combined operations such as seeding, fertilizing and laying down of pre-emergence weed control in one pass over the land. The localized drift from a spray included in such a multiple application combines with dust from the soil to coat working parts with mud.

An advantage often claimed for granules is that there is no need to seek and transport water. This is no economy. The granules contain a more expensive diluent than water and it has had to be carried much further. There is, however, the genuine advantage that appropriate dilution has already been made and a ready-to-use product needs application at a clearly stated amount per hectare: there is no need for field arithmetic and less possibility of error. This ready-to-use convenience cannot in general be achieved by supplying an aqueous product, which would be no more massive, because in many cases chemical hydrolysis and/or physical sedimentation would render the stored product unserviceable.

The ready-to-use advantage is, however, only achieved at the expense of some versatility. A spray concentrate has to be diluted in the field but it can be applied, for different situations, at low or high volume, in a fine spray or coarse. The granules must be accepted as supplied and only the total amount per area is adjustable.

Granules can be made of different sizes and with different disintegration or release properties. They offer scope for delayed release which a spray cannot do. This important development is deferred to Section 14.X. In this section we describe the main types of "ordinary" granules which the discussion in Section 14.VII shows to have only limited control of rate of release.

There are essentially three different types of granule which may be classified according to manufacturing process and resulting properties. In the first, the a.i. is applied to the outside of an effectively impervious base, together with some adhesive. The a.i. in such a product is almost as instantly available in the soil as is that from a spray drop of comparable content absorbed into a soil crumb. Many different granule bases can be used, the most common being limestone chips—a screened-to-size by-product of limestone grinding for preparing road stone and input for quicklime production. The a.i. is applied by the same technique as is used for coating of pharmaceutical pills—powdered a.i. and sprayed adhesive are alternatively introduced into an oblique, rotating, baffled drum (concrete mixer) until the required amount of dry adhesive coating is attained.

In a second category the a.i. in a suitable solvent is sprayed on to particles of a non-disintegrating porous matrix. The solvent is evaporated after the liquid has been absorbed. Availability now depends on porosity and on whether there is any less porous surface coating.

The most widely used granule base is a particularly bulky clay known as attapulgite. Attaclay granules are obtained by breaking down the native deposit or by re-aggregating fines by the paste method described below. If the clay granules are merely dried, they will disintegrate when wet, but, if fired, they become permanently hard. The a.i. can only be introduced into the fired product by the solution absorption process. Most granules of commerce are made by the absorption of solution into pre-formed granules, the chemical formulator taking his ready made blank granules from a large-scale manufacturer who does not have to adjust his process to different mixes.

There is obviously some gradation between the surface coating with an "insoluble" a.i. and absorption of an a.i. solution into a porous base, depending on solubility and porosity. Some granule bases, although leaving most of the a.i. outside, do allow some absorption. Such is the conveniently low density product made by grinding walnut shell and used extensively for air-blast cleaning of engine parts. Dried corn-cob and some other agricultural by-products can be crushed and sieved to form granule bases, such materials usually being made and treated for a local market.

The third type of granule is the agglomerate, which can be made by incorporating the a.i. into a paste or powder mix. To form granules from a "setting" paste such as plaster of Paris or hydraulic cement the paste is forced through a sieve of appropriate mesh size which is vibrated in its own plane. This builds in planes of weakness, which tendency can be increased by blowing dust on to the exuding paste. After setting, the mass is broken up into fairly uniform granules on another screen. The performance of this type of granule depends on its porosity and ease of disintegration when wet.

A recent development in granule technology has been reduction of particle size to one comparable with the small end of the drop size spectrum from fan-jet nozzles. Granules with effective diameter in the 100–300 μm range are usually called "microgranules". They were developed mainly to reduce drift and to make soil-applied herbicides of low solubility more rapidly effective (p. 824) but are finding other uses also. Microgranules have moved part of the way towards dusts, but are prepared free of particles much smaller than the main content. They can reduce the drift hazard of dust application, standing in the same relation to the latter as controlled drop application (CDA) does to hydraulic sprays (p. 815).

The methods of production suitable for larger granules become more difficult and costly when the size is much reduced but spray-drying in some form is opened up as a practical alternative. This is made possible for microgranules because of the shorter drying time and lower terminal velocity. It is practicable to keep the drops of paste formulation suspended in the upward hot air current long enough for them to become non-adherent. Spheres or wrinkled or saucer-shaped granules are formed according to whether the drops form a skin before the interior is dry. Incorporation of a fibre-forming polymer in a paste slurry should enable micro-rodlets to be produced by the fibre-chopping technique used to produce artificial velvet.

Claims have been made that microgranules can replace spray for leaf application but without information on shape it is difficult to assess these. The importance of shape for retention is emphasized on pp. 797, 820 and the possibilities of electrostatic dispersal of awkward shaped particles are mentioned on p. 800).

G. Foam

Another method of laying down diluted toxicant in the field has been to incorporate foaming agents into a suspension in water and pass this out through a nozzle which injects air and so delivers a band of foam. Some very specious claims have been made for this method—that it reduces drift, delays evaporation, holds the toxicant for a longer period in contact with the vegetation and even that the foam is heavier than spray. The last claim is of course totally ridiculous: a bubble will always float away on the wind for much longer than would its liquid contents collected into a single spray drop. The band of foam can be laid down with no spray formation but, unless it contains much more water than a normal band of spray, it is so light as to be easily blown away in bubble clusters by the wind. The bubble structure collapses mainly from the outside: a bursting bubble produces spray droplets which can drift but the main liquid released cascades down the outside of the band on to the ground. If the vegetation is more completely wetted it is generally because a much greater quantity of water is applied, although the foaming agent may also assist wetting.

The performance of foam has been critically examined by McWhorter and Barrentine (1970) with several herbicides and no biological advantage over spray could be demonstrated. The only useful function which could be served is that a dense foam, opaque white as a result of internal reflections, is clearly visible. This is very useful in spot spraying of weeds in pasture. The introduced air provides the cheapest marker and one which can be made no more persistent than is necessary to ensure that plants are neither missed nor treated twice. Foam is used in some wide-boom crop sprayers to make an intermittent mark at the swath edge.

H. Baits

With respect to dispersion and localization, application of toxicants in bait form represents extreme contrast to the distribution of mist, smoke or dust in open agriculture. The most extreme case is perhaps the use of cyanide sealed into glass ampules and inserted in raw meat as a bait for marauding leopards. Intermediate is the rat poison tucked into holes accessible to these animals but not to domestic animals, additional selectivity being sometimes obtained by sealing the poisoned food in polyethylene sachets which the rats, with misplaced cunning, will gnaw open. Baits, however, need not be disposed so locally in such large units. One may rely on attractive substances to concentrate the target towards the weapon and distribute the latter rather broadly, as done so successfully with methyl eugenol against the tropical fruit fly and is widely done in small granules to be picked up by fire ants and taken to their nests. From the point of view of application and formulation baits present such specific problems that they cannot be usefully discussed in general terms unless, as with the fire-ant granules or bran-metaldehyde granules for use against slugs, they come under the title "granules".

III. LIMITING FACTORS IN FORMULATIONS

A. Introduction

Most ready-to-use products are in solid form—granules or dust, although self-dispensing aerosol packs, restricted by cost to domestic use, contain dissolved a.i.'s. Products to be diluted on the field may be soluble solids, wettable powders, concentrated solutions (s.l., see p. 766), emulsifiable concentrates or suspension concentrates.

The s.l.'s, s.c.'s and e.c.'s are normally mobile liquids, although some stiff concentrated emulsions and paste suspensions are marketed. The other concentrates are all solids.

The concentrate must be capable of safe packaging and there must not be undue deterioration in storage of either package or contents. The package must be easily and, of course, safely handled at the user end and the contents easily dispersed in water (or occasionally oil) under field conditions. The available water may have high Mg and Ca content and may be appreciably acidic or alkaline. The diluted spray must be stable for a reasonable period (although for purely economic reasons the operator will want to spray out his load as soon as possible) and must not create a deposit of any kind which will build up to troublesome amount after several loads.

These requirements place some inescapable restrictions on the formulator's choice of solvents, inert solids and additives and may even restrict him to one type of product—s.l., e.c. or w.p. The properties of the a.i. greatly influence the width of choice. The properties of most importance are m.p., solubility and chemical reactivity. Although chemical reaction is often the most important limitation it is also the most complex and we will deal with the factors in the order given.

A very general point should, however, first be made. The behaviour on storage of homogeneous solutions (which include emulsifiable concentrates) is always more easy to predict with certainty from short-term tests than the behaviour of a concentrated emulsion, suspension concentrate or any solid product. Physical changes can occur in solid or dispersed products even though the a.i. remains chemically unaltered, and these changes may have serious consequences, be difficult to predict and, when established to occur, difficult to eliminate. The experienced formulator knows that every new a.i. may bring some new and unexpected source of difficulty. If solubility and chemical stability permit, it will be easier to provide a reliable and quick formula for the e.c. than for the w.p. or s.c. The following experience provides a good example. A new a.i. was required for water spraying. It was very stable chemically, of high m.p., low solubility and easily purified. The w.p. was the answer dictated by the low solubility and rendered appropriate by the other properties. Its preparation was easy but it was quite unserviceable after 1 month's storage. The compact micro fragments of crystals had grown in storage to very long needles ("whiskers") which made the product almost impossible to get out of the container, let alone through a spray nozzle.

B. Melting Point and Solubility Limits

The difficulty just mentioned with the storage of a w.p. is unusual, certainly in so extreme a form. Much more common is sticking and caking of the product if the m.p. is low. A compound with m.p. within the range of normal storage temperatures is generally worse for a w.p. than one which is always liquid, since liquids can be absorbed into spongy mineral fillers like kieselguhr and sepiolite. The substance of m.p. about 25°C will be frequently changing from liquid to solid and this provides a mechanism for binding particles together. Accelerated storage tests should always include a wide range temperature cycle to anticipate such trouble.

Low m.p., however, means high solubility in some solvent (Section 2.VIII) and an e.c. or s.l. then may be satisfactory. If the a.i. is soluble enough in water at spray dilution a solvent may be found for the concentrate which is also water-soluble. This may indeed be water itself. Thus MCPA as the sodium salt is soluble in water to about 25%. A mixture of K, Na and NH_4 salts gives even higher concentration or less risk of crystallizing out at low temperatures. Other compounds may be soluble enough at spray dilution but not soluble enough to provide an economic concentrate. The NH_4 salt of dinoseb is a case in point. Organic salts are often much more soluble in alcohols, acetone etc. than in water, often even more soluble when the auxiliary solvent contains some water. Wet acetone would provide an answer in this case but its flash point is low and there are therefore fire-risk restrictions on transport. Some amine salts are sufficiently soluble in water to be formulated satisfactorily without organic solvent.

If the a.i. has too low a solubility in water to be sprayed in this form but is soluble enough in a water-immiscible solvent to provide an economic concentrate, the addition of suitable emulsifiers will enable a satisfactory e.c. to be made. DDT, 25% soluble in methyl naphthalenes, can be formulated this way.

Difficulty arises in intermediate cases. An a.i. of intermediate polarity may be insufficiently soluble in a water-immiscible solvent to form an economic e.c. and insufficiently soluble in water for an additional solvent to be unnecessary at spray dilution. It may be soluble enough in an intermediate solvent such as isophorone to provide an economic concentrate but extensive dilution with water dissolves the solvent and the a.i. comes out of solution. With judicious mixtures there may be some way out of the difficulty but the formulator is severely restricted. A formulation may be serviceable for high volume spraying, where both a.i. and solvent have dissolved, or for low volume where both are held as transient emulsion, but emulsions of partially water-soluble oils are rarely stable. At intermediate volume the formulation could be quite unserviceable. Norris (1973) describes a good example of this problem.

A crystalline organic compound can be ground to a fine powder but the presence of a powdered mineral extender is usually necessary. This prevents direct contact between crystal faces which can result in permanent adhesion during storage. During the grinding operation itself local heating can be intense and transient fusion of crystal faces limits grindability unless some foreign particles of much less fusible material reduce the frequency of contact. The necessary wetting agents, particularly if of the liquid polyethylene oxide type, can create caking problems. Generally speaking, the higher the m.p. and the greater the purity of the a.i. the less mineral extender—such as kaolin or precipitated silica—is necessary. An inorganic pesticide such as copper oxychloride needs no extender, but only wetting agents. Carefully purified DDT can be prepared as a w.p. with 80% a.i. With the less pure and lower m.p. organophosphate, dimethoate, it is unwise to use more than 50% a.i.

Wetting and dispersing agents are necessary to make the powder freely dispersible in water and to keep it in suspension for adequate time as well as to improve retention and spreading on the target. The mineral extender usually contributes more to the "milkiness" of such a suspension than the a.i. itself and may give a false impression of non-sedimentation since it is also usually more finely divided. Suspension must therefore be judged by analysis, not visually. If the a.i. is fairly soluble in water, visual judgement could make the opposite error. The w.p. is then really a soluble powder as far as the a.i. is concerned but the mineral extender, necessary to get the a.i. in a rapidly soluble form, remains suspended. In such cases, a soluble mineral extender such as anhydrous sodium sulphate might with advantage be used.

If enough mineral extender is used and it is of a porous absorptive type, such as kieselguhr, meerschaum or sepiolite, even a liquid a.i. can be put into the w.p. form. Water will usually displace the organic liquid from the pores when the powder is wetted and the final suspension is therefore often an emulsion in which the mineral extender acts as a solid stabilizer.

C. Chemical Stability

Many pesticides are very stable substances and present no particular problems in storage. MCPA and 2,4-D, although subject to bacterial attack in fertile soils (Audus, 1964), have been marketed for years as aqueous solution concentrates of alkali and amine salts without any chemical storage problems.

DDT, despite embarrassing persistence in the environment, is subject to catalysed dehydrochlorination by active centres in calcined clays and the reaction of the hydrogen chloride liberated with the iron of containers can initiate further decomposition. Other chlorinated hydrocarbon insecticides show similar weakness but the problem is satisfactorily solved by treatment of the "inerts" with deactivators, as has been mentioned (Section 4.III) in considering the effect of adsorption in soil on reactivity and as is discussed and referenced by Polon (1973).

Rather more general are hydrolysis reactions. Their importance is increasing as the impact of the very stable pesticides on the environment is forcing increased development of less stable compounds. Fortunately improved water-impermeable packaging and availability of anhydrous wetting agents and other formulation auxiliaries is generally keeping pace. Hydrolysis eliminates the use of water-based or water-containing concentrates in many cases—e.g. sodium trichloroacetate is soluble enough but inescapably decomposes slowly to chloroform and bicarbonate with development of pressure. Hydrolysis has also severely limited the marketing of concentrated emulsions and requires that anhydrous auxiliaries be used in the preparation of emulsifi*able* concentrates. This restriction is often irksome because many organic salts, including ionic wetting and emulsifying agents, are more soluble in organic liquids when these contain some water and the physical performance of a "wet" formulation is often better.

The decomposition of most hydrolysable organophosphate esters and carbamates is alkali-catalysed and such compounds must not be formulated as w.p.'s on basic "inerts" such as precipitated calcium carbonate or some silicates. Alkaline centres in clay are less easily deactivated than acidic ones and calcined gypsum or various forms of silica are preferred "inerts" when alkaline hydrolysis is important.

Chemical degradation in a homogeneous solution (s.l. or e.c.) follows a simpler pattern than that of an a.i. adsorbed on a solid. The latter is often accelerated by the development of liquid decomposition products which provide micro-regions in which a different type of reaction can occur. Moreover the effect of decomposition in storage may be more important in the w.p. and s.c. than in the e.c. In the e.c. it will usually represent just a simple loss of a.i. In a w.p. it may make the product unserviceable through change of physical properties. The probable extent of decomposition in homogeneous solution over a prolonged period at ordinary temperatures can be predicted reasonably accurately from measurements over a shorter period at two or three elevated temperatures. Such prediction of the complex physically-modifying changes in the w.p. and s.c. is much less reliable.

D. Corrosion and Compatibility

Corrosion of containers is a decreasing problem as more sophisticated packaging methods are adopted. It is a problem under the control of the chemical producer who must find the appropriate answer within the extensive expertise at his disposal. More serious is corrosion of metal parts of application machinery and the related problems of swelling and disintegration of plastic tubes and other components. This is more difficult because a variety of materials must be used in the construction of applicators and a variety of chemicals passed through the machine. It is not just a one formulation–one container relationship and is made more difficult by the fact that products from many chemical manufacturers pass through machines from many machine manufacturers.

The chemist has carried the greater burden of research and development to solve what are joint problems, probably because there are greater profits in chemicals and the chemical firms in the pesticide business have been able to spend more on research than the engineering firms. A good example of chemical limitation imposed by engineering is provided by the avoidance of ammonium salts in the formulation of herbicidal acids. The formulation desirably contains a slight excess of ammonia. This is drastically corrosive to zinc and zinc alloys. The brass gauze filters, widely used in basket form in tank filling ports, and zinc-based die-cast plugs are rapidly made unserviceable or unsafe. While the biologist has made (Section 11.X.E) a fairly good case for ammonium salts being better able to penetrate into leaves, the chemist has moved backwards, withdrawing ammonia formulations, rather than the engineer being encouraged to move forwards to replace zinc alloys.

The problem is confounded by the overriding necessity for the chemical manufacturer to sell his products. He tries to make his products serviceable in all machines. The machine is, initially, a more expensive item. The farmer with an expensive machine avoids the product which gives trouble in it, although this may in the long run be a false economy, rather than replace the machine. Only when the properties of the chemical are so outstanding as to make a clear demand, as is the case of paraquat and machinery for direct drilling of unploughed land (an example *not* concerned with corrosion), do we find machinery developed to exploit a chemical.

The herbicide glyphosate has rapid and deep systemic effect when applied to foliage, little biochemical selectivity and generally little action via the soil. This combination makes the compound very suitable for localized application and research into means to exploit this is being pursued. Devices to smear liquid on weeds standing above the crop were mentioned on p. 766 but their use is restricted to a few crop-weed situations. Development of better hand tools is likely to follow and could help towards a more sophisticated approach to machine design.

Progress is being made through committees set up by the British Crop Protection Council and corresponding bodies in other countries towards agreement between chemical and machinery manufacturers on a common set of "do's" and "don'ts". It is, however, still true that a formulation development instigated by biological research has at present little chance of commercial success if it creates difficulty with existing machines. The case must be a very strong one indeed and preferably demand a wholly new type of applicator, so that there will be no malfunction of an existing one to inhibit the development.

There are obvious economies, in time, labour and mechanical damage to crop, in applying several a.i.'s together in a single spray, or granule mixture, but compatibility problems can arise. We are not in this chapter concerned with biological compatibility—i.e. with whether the timing is right for attack on both pests or whether one herbicide may antagonize another—but with adverse physical or chemical reactions between the formulations or active compounds.

Most marketed formulations are used in many different machines. Many are also mixed, in the spray tank, with other formulations. The mixing of some a.i.'s may be detrimental to their desired effect or damaging to the crop. Where this is due to biochemical effects at cell level it is not our concern in this book, but the incompatibility may arise at some more external stage. A common cause of phytotoxicity in mixtures of insecticides and/or fungicides is increased uptake when an oil-containing (e.c.) formulation is mixed with a w.p. which is supplied as such, as are many carbamate insectides, because of potential phytotoxicity. The user who mixes them with an e.c. is doing what the manufacturer is trying to avoid and warning against this danger should be given on the label.

Incompatibility may also start in the spray tank by physical or chemical interaction. A clear example of the latter arises from the strong alkalinity of lime-sulphur fungicide. Many other a.i.'s, particularly some organophosphorus and carbamate insecticides, are rapidly hydrolysed in alkaline solution and lose their potency. This restriction is not very onerous since lime sulphur severely damages most foliage and is used therefore mainly in winter before bud-burst.

Physical interaction is more frequent. When a surfactant is introduced as the calcium salt an active organic anion may be precipitated (Section 2.XVII). The manganese salts of hormone acids are of low solubility and this sets a limit to the application of this trace metal supplement along with a herbicide spray. It is one of the manufacturer's responsibilities to warn users against such incompatibilities.

Surfactants may be anionic, non-ionic or cationic. Although some combinations of cationic and anionic, judiciously used, are very good emulsifiers, the uncontrolled mixture of these two types is generally disastrous, producing a waxy, filter-blocking precipitate with loss of stability of the emulsion or suspension. This is one reason for the manufacturer using, as exclusively as possible, the non-ionic surfactants. If an ionic addition is necessary, it is preferably anionic. The manufacturer might feel that there could be a sales policy reason for marketing products declared to be incompatible, but he usually accepts that tank-mixing outside his control is inevitable and prefers not to invite complaint.

Many self-emulsifying oils are compatible as are most wettable powders but it is preferable (as well as easier) to pour them separately into an agitated tank of water than to mix them first. In some cases the stability of the resulting emulsion or suspension may be reduced but still be acceptable. Concentrates of saline water soluble ingredients, such as salts of herbicidal hormones, particularly in low volume spraying are very likely to cause precipitation of powders or breaking of emulsions.

The probability of a customer making such tank mixes and the reluctance of the sales department to receive complaints about them tends to make the formulator work within a rather conventional framework.

Most granules would be compatible but, without suitable machinery, their intimate mixing is difficult.

IV. UNIFORMITY AND CONTROL OF APPLICANCE

A. Metering

The distributing machine—the applicator—must incorporate some provision for metering the liquid, dust or granules so that a controlled amount is delivered per unit field area. This quantity, frequently called the dose, we prefer to call the "applicance". Its relationship to the true dose actually received by the target organism is indirect and involves the factors, some uncontrollable, discussed at length in other sections of this book. It is, however, the quantity that the operator aims to control, an intention which the term "applicance" emphasizes. A very strong reason for aiming at a specified applicance is that of cost and accounting. The grower or contractor must budget for a given amount of material and does not want to rush in extra supplies to a machine which is delivering more than intended nor to have budgeted chemical left over and deduce that he may have underdosed. We shall consider in the rest of this section how the applicance is controlled in practice.

The principle adopted in most agricultural applicators is to operate the vehicle, ground- or airborne, at as uniform a speed as possible and to dispense diluted chemical at an independently controlled time-rate. Usually neither control is very exact and most operators make fine adjustments on the day. The amount of chemical used in treating a measured area is recorded and the rate of application then stepped up or down according to how the actual applicance compares with that intended. With conventional liquid sprays, the rate of output is controlled by a pressure gauge in the pipe supplying the nozzles and adjusted by a valve in a by-pass returning to the tank, the pump having excess capacity. The return to tank serves a second purpose in agitating the tank contents. It should discharge to the bottom of the tank so as to avoid generating froth. The rate of output of a simple, filtered, liquid through an orifice can be controlled by pressure down to very low rates. This enables the capital cost of the spraying machine to be reduced by combining the metering and distributing function in the nozzles. The nozzles are chosen and the chemical concentration and pressure set to give the intended chemical applicance in the desired spray pattern. The need for occasional adjustment arises from wear (corrosion and erosion) of nozzles and pump; it is best to standardize on a satisfactory spray pattern and adjust the applicance by changing the concentration.

Simple combination of metering and distribution is not possible for dusts or granules. Dusts are usually metered into the distributing air flow through a controllable gate in the bottom of a hopper, the contents of which are prevented from bridging by an agitator paddle. Granules were generally dispersed from numerous outlets along a boom, the feed to each being commonly metered on a rough volume basis by a dimpled roller driven from the land wheels. Some machines, however, spread the granules by air flow or centrifugal action in which case a fixed time-rate from a hopper can perform the metering function. With the advent of microgranules the distinction between dust- and granule-applicators is becoming less clear: the main difference is the behaviour of the particles after leaving the machine. Dust and granule applicators which depend on a controlled volume-rate for metering require a product of standard bulk-density which is not subject, as is so often the case with wettable powders, to packing down to smaller bulk during storage and transport. There are, of course, metering devices used in handling solids in factories which control a mass rate but they depend on some balancing mechanism which would not be practicable in a machine travelling over rough ground. Mass rate devices employing control via the torque necessary to drive a rotating distributor at constant speed could doubtless be designed to get over this difficulty but we are not aware of any in practical use.

Spinning-disc (or dish) distributors for liquids, now being developed bring sprayers into the same category as dusters and granule distributors with regard to separation of the metering and distribution functions. In practice the ability to produce drops with a closely defined size range which these machines provide has been exploited principally in two directions: first, the distribution of very small droplets ($<100\ \mu m$) against mobile insects, and second the distribution of herbicides in drops (200–300 μm) smaller than the average output of hydraulic nozzles but without increase of drift hazard since the discharge has no significant content of drops much smaller than the chosen size. Provided the discs are not overloaded their performance is independent of output rate so this can be controlled separately while the distribution is maintained at its optimum.

Ideally a machine should meter chemical at a rate proportional to the forward speed. If the swath width is constant, the applicance must then be constant also. Such a design may seem easy in principle, but there are serious practical problems which have retarded development for field spraying. If a land wheel is used to drive the pump, wheel-slip may create difficulties: geared to the driving wheel the output would be excessive on soft ground; geared to a trailing wheel the reverse would apply. In either case, if the pump supplies diluted spray to the nozzles the spray pattern will alter with the speed although perhaps not seriously if the speed range is small. Pump wear will have a more direct effect than in a conventional machine but nozzle wear will be less important.

These difficulties could be avoided by having separate metering and distributing devices because distribution would be independent of feed rate (within limits) and metering, by itself, has very little power demand and could be operated reliably by a trailing wheel. It is possible to have the mechanisms separated in hydraulic nozzles by supplying water alone under optimum pressure and feeding in concentrated chemical at a rate proportional to land speed. Unless, however, the concentrate is separately piped to a position close behind each nozzle there is a necessary delay between change of speed and change of output concentration. Where speed changes are rapid the compensation can easily make the applicance *less* uniform.

Distance-control of chemical output has been successfully exploited in railway track spraying where very different conditions apply. A much greater variation of speed, due to traffic control, makes track-distance metering more necessary and the slowness of speed change, due to high inertia-to-power ratio, makes the pipe line delay unimportant. Moreover, the greater mass and friction enable more power to be taken from track wheels without slip.

Concentrated chemical can be fed in at a speed determined by other means than a metering pump—it can, for instance, be supplied at constant pressure through a variable orifice the resistance of which is controlled by land speed. Amsden (1970) has reviewed the advantages and disadvantages of the various systems. The controlled drop applicators (CDA) (Taylor *et al.*, 1976) could more easily exploit land speed metering of the sprayed liquid. The only variation of output produced by an increase of rate of feed (short of overloading) is an increased number of drops of unaltered size following the same trajectory. The ground coverage should therefore be identical, in respect of number, size and distribution of drops and amount of chemical per unit area, if the feed rate is proportional to land speed. There being no change of concentration, response in output would be immediate. This potential has not yet been exploited.

Separation of the metering and distribution functions has a potential advantage in the case of sprays in that individual, fine-orifice nozzles are vulnerable to blockage by foreign particles in the liquid. Adequate filters, preferably one protecting every nozzle, are essential. The present tendency to reduce spray volume and therefore use finer nozzles makes filtration increasingly important and also replacement or cleaning of filters more frequent. A spinning disc distributor throws off any solid particles and the feed to it can be at low pressure using a much larger orifice than a corresponding high pressure nozzle.

In the case of granule distribution the metering and distributing functions are necessarily separate but there is a rather different advantage to be gained by separating them much further than seems absolutely necessary. Continuous reliable flow of granules cannot be controlled down to very low rates as is possible with a well-filtered liquid. A much more reliable control is possible if the flow rate is substantially faster than is necessary for a single outlet on the boom. The problems to be solved are those of uniform subdivision and lateral spread of the rapid feed, considered in the next section.

B. Uniformity Across the Swath

A low-volume spray cannot be discharged from nozzles spaced so close along the boom that deliberate lateral spread is unnecessary. A swirl-chamber nozzle produces a hollow-cone jet which, if held stationary over a horizontal surface, produces a ring of wetness. Moving across a diameter, the "sprayfall" plotted at right angles to the movement—i.e. across the swath—shows a double peak (Fig. 1). The fan-jet tends towards the same behaviour since the spreading triangular film of liquid tends to

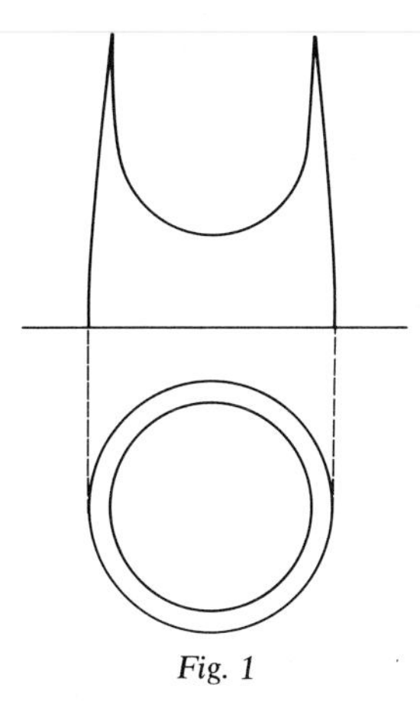

Fig. 1

thicken and contract from the edges before it breaks up. This effect can be compensated by designing the slit so as to feed more liquid into the middle of the fan. In order to produce a reasonable uniformity of sprayfall across the swath from a multi-jet boom there was at one time a recommendation to adjust the spacing, fan angle and height so that the spray patterns just met at crop height, but it is now generally recognized that one must aim for considerable overlap, preferably such that the band directly below each nozzle receives the edge of the spray pattern from the nozzles on either side and the band midway between nozzles receives a substantial contribution from two nozzles. Two factors which have not always been distinguished make this double-overlap desirable. The total sprayfall is customarily measured on a device mistakenly (for reasons explained below) called a "patternator". This is in effect a multiple raingauge. A sharply corrugated metal sheet forms a series of contiguous channels and is slightly inclined to the horizontal so that spray collected in the channels flows to one end where it drips into a series of measuring vessels. The tray must cover an area larger than the spray curtain which falls on it and the spray must be maintained during measurement for long enough to ensure no significant hold-up in the channels. The device can therefore only be used to measure the delivery, per unit horizontal area, of *very high volumes.* It is useful for testing uniformity and spread from a nozzle and particularly for detecting and demonstrating the effects of wear, partial blockage and misuse, but its relevance to reception by crop, weeds or pest of low volume sprays is limited and the nature of these limits should be appreciated.

A rainguage, situated as it should be in the open, measures the amount of rain per unit horizontal area independent of wind and therefore of angle of incidence of the drops. The patternator measures the equivalent sprayfall. For each nozzle the sprayfall is a function of distance from the centre, falling to zero on either side. We shall first accept the usefulness of this measure and examine what function is best suited to provide a nearly uniform cover from a series of nozzles.

The summation of sprayfall from various individual patterns is represented diagramatically in Fig. 2. The individual patterns are shown by faint lines: their summation by bold lines. We start with an idealized rectangular pattern. If brought exactly together, the individual rectangles would give a uniform total distribution (a, 2). If the distance between centres is 5/4 of the matching distance we have intermittent gaps (a, 1). If brought to 3/4 times the matching distance we have a total with intermittent doubling (a, 3). Uniformity is too dependent on spacing.

It will be evident that the M-type pattern given by the hollow cone nozzle will be even worse in this respect than a rectangular pattern and that approximate uniformity can be attained only by considerable overlap, the optimum spacing being 2/3 of the distance between the peaks of each individual pattern so as to put two other peaks in each valley.

An approximation to a triangular pattern is generally sought by the makers of fan nozzles. The summation of triangles is shown at (b)1, (b)2, (b)3 equivalent to the diagram for rectangles. We have taken a full overlap (2)—i.e. nozzle spacing 1/2 of the triangle base—as the desirable norm, giving uniform summation. Increase to 5/4 of the optimum distance (1) or decrease to 3/4 (3) both give a serrated summation line. Reduction to 1/2 (or, in fact, to any integral submultiple) would restore uniformity. There is therefore sensitivity to spacing but, for close spacing, the deviation from uniformity is not too serious.

In Fig. 2c we illustrate the summation of individual patterns having "normal" (6) distribution of sprayfall—i.e. proportional to e^{-Kx^2} where x is the distance from centre, K being in this case so chosen that, at the basic (2) separation, the sprayfall from each nozzle falls to 3/4 of the peak value at midway between nozzles. It will be seen that the summation is almost exactly uniform and remains so at 3/4 of the standard separation (1) or at any lower value (with no periodic serration effect). Even at 5/4 of the basic spacing (3) the deviation from uniformity is only 1·5% and at 3/2 only 6·5% where sprayfall from each nozzle drops to 1/2 at midway between nozzles: this (4) corresponds to the standard spacing in the triangle case; increase by a further 5/4 produces (5) 30% deviation. Combination of "normal" distributions thus produces a more uniform

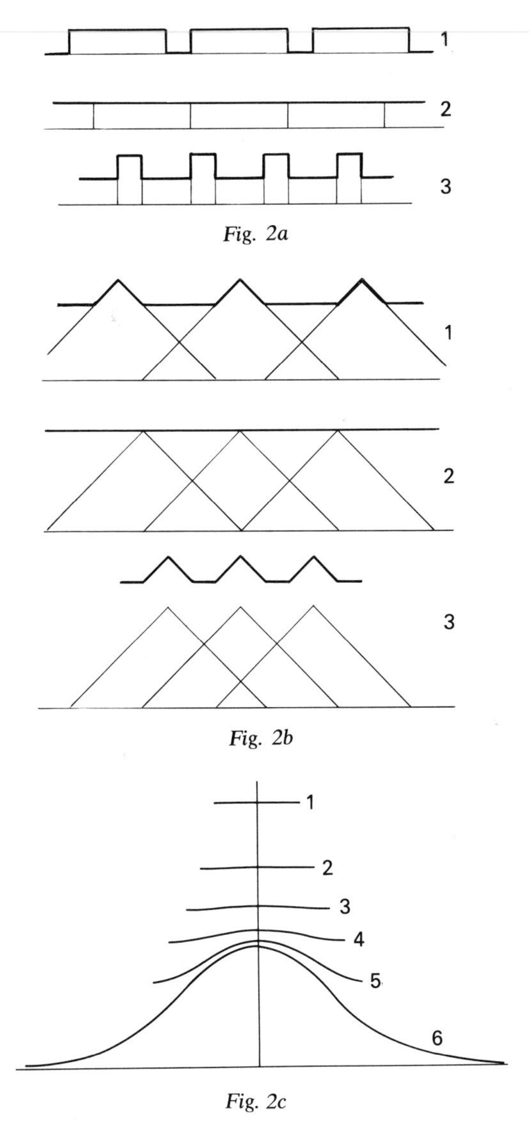

Fig. 2a

Fig. 2b

Fig. 2c

summation than the other patterns if the spacing is close enough but begins to deviate seriously when the individual sprayfalls at midway between nozzles have less than 1/2 the peak value.

We have illustrated the effect on summation of change of spacing. In practice, of course, it is height above the crop which is the main day-to-day (and even, because of boom sway, second-to-second) variable.

To a first approximation the effect is similar since each pattern spreads further the greater the height through which it falls. The pattern, however, while spreading, will also change in form because the remote part, having a higher horizontal component of velocity initially, is subject to relatively greater retardation by air friction and also takes longer to fall since it has less vertical component initially. The effect is shown diagrammatically in Fig. 3 where the "spray lines", equidistant where they intersect the horizontal level A, representing a uniform sprayfall (rectangular pattern) at this level, become relatively more crowded in the outer part as the distance of fall increases, giving the pattern indicated at level B tending to the M-shape form from a hollow-cone jet. A matching overlap at one height may therefore give a non-uniform sprayfall at another height unless the individual patterns are of the Gaussian ("normal") type arranged for multiple overlap at all probable heights. Reliable uniformity of sprayfall across the swath cannot be obtained without multiple overlap if the spray is diverging when it reaches crop level.

Before considering sprays designed to be falling vertically when they reach crop level we will examine another aspect of spray patterns and their summation. Up to about 200 ℓ. ha^{-1}, and, on many crops and weeds, even higher, there is very little run off. Spray drops may bounce but otherwise stay where they hit, spreading to a limited extent only before evaporation of the carrier. A raingauge with a well-maintained wettable funnel, or a patternator, is unreliable at less than 0·25 mm rainfall, corresponding to 2500 ℓ. ha^{-1}. At realistic sprayfalls, the soil–crop–weed system is not fairly represented as a horizontal surface. A no less valid oversimplification is to represent it as a plurality of vertical surfaces. Two

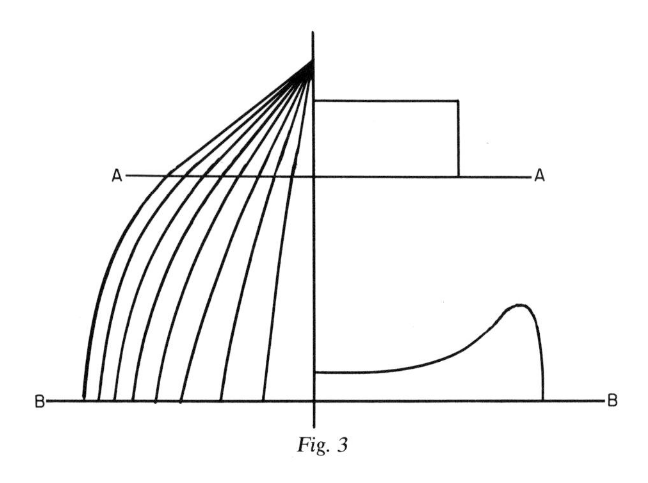

Fig. 3

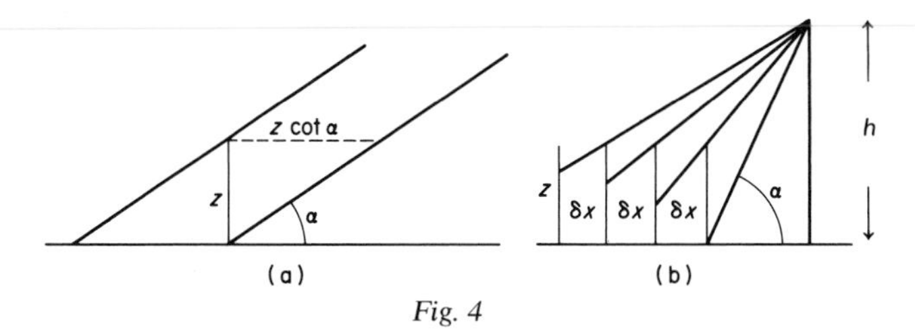

Fig. 4

extreme forms of this model, lying on each side of the real situation, will be considered. In an idealized divergent spray of the type shown above the level A in Fig. 3, the equality of horizontal spacing between the "rays" represents uniform sprayfall, i.e. a rectangular pattern. An isolated vertical "leaf" of height z (Fig. 4a) collects the spray which would otherwise fall on a horizontal area $\delta y \,.\, z \cot \alpha$, where δy is the width in the track direction. Equal *isolated* leaves therefore collect amounts proportional, under a boom of fixed height, h, to distance ($h \cot \alpha$) along the boom from a nozzle centre. Matching the nozzles to give uniform sprayfall will thus not match them for uniform reception on isolated vertical leaves. In a *dense stand* of vertical surfaces the effect will be modified by shading of one by another. At equal spacing, δx, and for $\cot \alpha > \delta x/z$ (Fig. 4b) the total reception by each surface will be the same but increasingly concentrated towards the top as distance from the nozzle centre increases. The so-called "patternator" does not measure these significant features of a real spray pattern at a realistic volume applicance but rather what the distribution of wastage would be were a very excessive volume applied.

If fan nozzles were mounted so as to send out horizontal fans from a boom sufficiently high above the crop for horizontal velocity to have died out before the crop was reached, the drops reaching the crop would all be falling vertically (or in a direction dictated by the wind) and neither the pattern nor the gross sprayfall distribution would depend on further change of boom height. Such an arrangement would, probably rightly, be condemned as creating too much drift. Where the horizontal initial dispersal is achieved by rotating discs (or, perhaps, in future, by an air assisted "anvil" nozzle) the drift problem is greatly reduced because optimum-sized drops can be formed with negligible volume of the spray in smaller ones. This principle has been employed in the experimental machine described by Cussans and Taylor (1976).

The uniform drops projected at uniform velocity from a spinning disc (axle vertical) reach the same radial distance before the horizontal component is lost and their subsequent fall is vertical. The deposit from a disc with stationary axis in still air is a sharply defined ring and, when the disc axis is moved parallel to itself, the sprayfall pattern is of an extreme M form, quite unsuitable for matching up to give a uniform distribution. This problem is surmounted by housing a series of discs on the same axle within a cylindrical shroud which restricts the spray to a limited sector issuing through a slot (providing the flat central valley of the M). The spray collected in the shroud from each disc forms the feed to the next lower disc until a final unshrouded disc dispenses the final remainder. Once spacing and disc speed have been adjusted to give uniform sprayfall across the swath, the sprayfall and pattern will remain the same from any height of the boom, provided all heights are greater than that necessary for the drop trajectories to become vertical. This applies, of course, to operation in still air. Turbulent cross wind disturbs the pattern and such disturbance increases with height.

The use of optimum-sized drops, providing a better coverage-to-drift ratio than the wide spectrum of sizes from a hydraulic nozzle, enables the CDA machine to give a satisfactory weed control at reduced volume applicance. This permits a lighter machine to service a given area and makes possible application in weather and soil conditions where a heavier machine would cause damage. It is on this economic advantage in farm practice that the Weed Research Organization in England has largely concentrated. Possibilities of improved selectivity or efficiency which may be realized with a uniform pattern of drops of selected size will doubtless be further examined.

C. Uniformity along the Swath

One important aspect of uniformity along the swath has already been considered (Section 14.IV.A) under "metering"—namely variation due to variation of speed without compensating variation of feed rate. The possibilities of linking feed rate to land speed were also briefly reviewed. Irregular variation can be produced by up-and-down motion of the boom relative to crop just as can variation across the swath when the sprayfall at crop height retains a horizontal component. An additional source of variation in the track direction comes from fore-and-aft swaying of a flexible boom. This is associated with up-and-down sway but corresponding sideways displacement is not an important factor in uniformity across the swathe. The magnitude of sway is obviously greatest at the end of a centrally-supported boom and increases rapidly with boom length. The effect has been analysed by van der Weij (1970). Various devices can be employed to reduce sway but they become very costly with increasing boom length and the effect sets a practical limit to length.

The fan jet distribution is the one most sensitive to fore-and-aft movement of the nozzles because its deposit is most localized in the track direction. The effect may be illustrated by assuming as an extreme case that the movement consists of forward jerks of 20 cm between stationary periods of 0·1 s, averaging 2 m s^{-1} general forward speed. If the fan has no spread in the track direction, there will be sharp contrast between dense lines of treatment and 20 cm gaps. If, however, the spread of the jet *in the track direction* is Gaussian with the rate of sprayfall at 20 cm in front or behind the centre equal to half that at the centre, the result of the jerking would be scarcely noticeable (Section 14.IV.B). It is for this reason that van der Weij recommends "solid cone" rather than fan jets.

D. Uniformity of Penetrated Spray

In spraying against weeds in a crop which is usually taller at the time of spraying, the weeds are partly protected by the crop: some leaf shading is inevitable. One might at first imagine that a strictly vertical spray, falling under gravity, would be less intercepted by the nearly vertical leaves of a cereal crop and so be more effective against prostrate weed seedlings. A closer analysis of the geometry of this situation (Appendix 7), however, shows that a wide-angled spray is only at a disadvantage in this respect if a significant fraction of the spray is able to strike both sides of the leaves. The leaves of a cereal crop at the usual time of herbicide spraying spread at a fairly wide angle to catch the light and this condition is not fulfilled. If only one side of each leaf is hit the leaf area effective for interception is not dependent on spray angle. The angled spray will then be more effective because the same amount of penetrated spray reaches ground level with *less sharp* shadows.

This calculation ignores the effect of leaf-flattening and flutter: the first may put the fan-jet downward spray at a disadvantage, the latter at an advantage. A down-draught is necessarily associated with hydraulic jets on a low level boom. It is greater and of a more widespread pattern when spray is delivered from a helicopter and is even considerable below fixed-wing aircraft flying low. The narrow band of spray-laden wind below a fan-jet boom is likely to have most effect in flattening the upright crop leaves to provide a transient cover for the weeds. A double-overlap spacing, desirable for general uniformity across the swath, would advantageously have the fans arranged alternately in two planes so that a different leaf disposition is induced by each series. Taylor and Merritt

(1975) compared deposits on small horizontal cards placed within a crop and on cleared areas and found no systematic difference of interception between the CDA machine giving free-falling drops and a conventional spray boom.

The calculation also ignores the effect of leaf angle on drop reflection. If the drops are large enough and of sufficient surface tension reflection will increase penetration through some crop canopies but Lake (1977) found, in laboratory experiments with controlled drops of two spray formulations falling on barley leaves, that the amount collected was not systematically dependent on leaf angle from 0° to 60° to the horizontal when the change in horizontal projection area was allowed for. At 75° there was increased reflection, particularly of solution with dye but no wetting agent having surface tension of 58 mN m^{-1}, small drops of which were reflected in high proportion at this angle. Merritt and Taylor (1978) found less reflection of commercial formulation having high surfactant concentration. They found that a drop incident on a resident one was often reflected (see p. 794 and Section 13.IV).

Uniformity of the penetrated spray on a small scale is likely to be better when the spray has a miscellany of drop sizes and incident directions. This is to some extent in conflict with the requirement for greatest uniformity across the swath at crop level. Lake emphasizes that reception will depend more on crop structure and formulation when uniform drops in free fall are applied. The name "*controlled* drop application" (CDA) was chosen so that the method would not automatically be associated with exploitation of *very* low volumes and *very* small drops. It may be desirable to spread the drop sizes deliberately, retaining the advantage of absence of very small drops and stopping far short of the great width of drop spectrum produced by hydraulic nozzles.

V. RETENTION

A. Introduction

For permanent retention of a particle on crop or other surface it must impact without bouncing or rolling and its adhesion must then become adequate to withstand subsequent wind, leaf flutter and rain. The particle can strike under its own momentum, if big enough (in practice above about 200 μm effective diameter), fall on to the surface under gravity, be attracted to it by electrostatic charge or impact it under the influence of wind. Impaction of airborne particles is considered in Chapter 6. Efficiency of "dynamic catch" is measured as the fraction of particles scoring a hit among those which would have passed through the space occupied by the object had it not been there. The efficiency increases with increase of particle size (while still small enough to be wind-borne): it increases with decrease of object size and with increase of wind speed. For particles large enough not to be carried by wind and impacting therefore by their own momentum or by fall under gravity, efficiency in this sense is 100%. If the object is there the particle will hit it. For particles small enough to show significant Brownian movement—smoke particles down to a few μm or less—this thermal motion plays a part in dynamic catch and, when we get down to molecular dispersion it is completely dominant. In this section we are concerned only with particles above 10 μm whose thermal motion is negligible.

Unless we disperse electrically charged particles which can be attracted to conducting surfaces (including all agricultural targets) from several mm distance, the only control we have over impaction as such is that of particle size and, to a limited extent for near targets, wind speed. Obviously we cannot alter the target size or shape, but we can adapt our attack in some cases to make the target less or more efficiently impacted than other objects.

Impaction of the object surface is not necessarily followed by retention. Several mechanisms can intervene. These will now be considered.

B. Reflection of Liquid Drops

It is a familiar observation that a quite heavy shower of rain will leave many leaf surfaces (e.g. pea, cabbage, banana between veins) completely dry, except where puddles have formed in hollows. If the rain has been falling long on some raised horizontal surface, e.g. a garden seat, small droplets will also be seen skating around on a water surface. Water drops can be reflected from some dry surfaces and from a free water surface. The dry surfaces which reflect water all have a micro-roughness, provided, in the case of plant cuticles, by wax crystals which give reflective leaves their characteristic "bloom". Water drops within a certain size range are reflected in their entirety, leaving no residue (Ennis *et al.*, 1952; Brunskill, 1956; Hartley and Brunskill, 1958). During all reflections of water drops some of the kinetic energy of impact is transiently stored as tension energy in the increased surface of the flattened drop which then recoils elastically becoming transiently a prolate spheroid with net outward velocity. During this process some energy is converted to heat against viscous forces. The reflection is therefore less elastic the lower the surface tension and the higher the viscosity and can be arrested altogether beyond certain limits of these properties.

During reflection of water from water there is never real contact. The oblate–prolate deformation is completed before air, trapped between flattened drop and a depression created in the larger surface, has time to escape. Some of the recoverable energy is stored in compression of this air. There is no energy of adhesion to be overcome. If some chance irregularity initiates contact at any point, coalescence of the main volumes is inevitable and often occurs so violently that some satellite droplets are transiently produced. A large drop falling on a still water surface throws small droplets upwards, some of which skate along the surface. Reflection is most reliable for small drops (Schotland, 1960), but, for very small droplets as found in mists and clouds, inefficiency of impaction is more important than inefficiency of fusion (see Woods and Mason, 1964). For drops of spray size there is a rather small range where neither splashing (large drops) nor reflection (small drops) occurs.

In reflection from a micro-rough surface there is real transient contact and work of adhesion must be overcome. This is proportional to surface area of the (flattened) drop while impact energy increases more than proportionally to drop volume, since incident velocity normally increases also. The balance is complex but it is clear in practice that drops of clean water are not reflected from perfect pea leaves if their diameter is less than 100 μm and are always reflected if it is above 150 μm. Between these limits reflection depends on velocity and angle of incidence and local quality of leaf surface.

That there is real contact cannot only be inferred from there being a lower limit of size for reflection but by two related differences in behaviour between a pea leaf and a carbon black (candle flame) deposit on glass. Reflection from the former is almost random in direction while from the latter it is very regular (compare Plates 2 and 5 of Hartley and Brunskill, 1958). Drops reflected from the leaf contain no wax particles. Those reflected from the sooty plate always carry a small cluster of black particles. Soot is weakly attached to glass and the contaminated drop leaves a smooth surface. Wax crystals are strongly attached to the leaf and the clean drop departs in a direction dependent on the shape and orientation of the last area of contact. A sufficiently large drop falling on a leaf will splash. The critical diameter is less on a "bloomed" reflective leaf than on a wettable one but in both cases it is greater than the least diameter splashing on water and greater than the normal spray drop range. Drops roll on both bloomed leaves and a water surface but it is not always easy or useful to distinguish between bouncing at a very flat angle of incidence and rolling. The difference between real contact (leaf) and air-cushioned non-contact (water) is again evident. Drops can "skate" for several metres on a smooth water surface but a finite, if small, slope is necessary to cause a droplet even with very high contact angle to roll down a leaf surface.

Water-off-water reflection can be demonstrated with spray drops from an already-sprayed wettable leaf but can only be significant in very high volume spraying, as a complicating factor in coalescence and run-off or if spraying is done during rain. None of these situations is now of practical importance but it may be noted that, under rain or overhead irrigation, it should be possible to contaminate unwettable leaves while leaving wettable ones untouched. Also spray under rain could, reflected, strike lower leaf surfaces in a way not possible under dry conditions. The significance of reflection and, particularly, splashing, of rain itself as a means of distributing pathogens and even of pesticides has been mentioned elsewhere (Section 13.IV).

Water-off-bloomed-leaf reflection is exploited to a limited extent to enhance the selectivity of some herbicide treatments where biochemical selectivity is not wholly sufficient. Thus DNBP sprayed with minimal wetting agent on to peas is mostly just not received by the crop leaves unless they have been damaged by wind, dust or some pre-emergent herbicides (Dewey *et al.*, 1956; Pfeiffer *et al.*, 1957). A careful and knowledgeable worker can even "get away with" spraying MCPA on peas but the practice is not safe enough for official recommendation. A secondary advantage of exploiting reflection by the crop is that the dose actually received by the weeds must be increased. Exploitation of reflection from the crop has the disadvantage that weeds which have reflective leaves also escape damage. It is possible under laboratory conditions to take healthy leaves of pea (*Pisum sativum*) and fat hen (*Chenopodium album*) both of which reflect perfectly water drops above 250 μm and, by bringing down size or surface tension, to have the pea still reflective but the fat hen now receptive. However, it is not practicable to apply this degree of finesse under field conditions because of variations in weather factors and between different parts of both plants.

The feeble attachment of a small water drop to a bloomed surface permits it to roll under the influence of the wind which caused the initial impaction. It therefore comes to rest on the lee side of the object struck. When many drops suffer this fate their coalescence on the lee side eventually forms a drop big enough to break away under the reduced wind pressure. If the drop has much larger diameter than the object, it is discharged again without having to wait for enlargement by coalescence and without shatter. There is thus a very clear, but broad, optimum size range for collection of droplets by filaments. A too small drop does not impact. A too large one is not retained.

The impaction, reflection and retention of spray drops must assume increasing importance with the expanding interest in ultra-low-volume and controlled droplet application. The use of much smaller drops avoids direct reflection from dry surfaces. The dispersion of drops in a narrower size range makes possible more exploitation of the differences in physical behaviour arising from difference of size.

C. Retention of Solid Particles

The impaction of particles with larger objects is not influenced by the internal properties of the particle unless a significant electrostatic charge is carried. With regard to reflection, rolling and adhesion, however, there are major differences between liquid and solid particles of the same size. Solid particles are reflected intact from a dry surface over a much greater range of size than are liquids. Reflection is much less dependent on the microscopic structure of the surface but the recoil energy of a liquid drop, in the conditions favouring reflection, can be a higher fraction of the incident energy than is that of most solids. If a liquid drop can make real contact with a smooth surface, even at high contact angle, it is more firmly attached than most solid particles of the same mass. A pendant drop of water on the convex under surface of a watch glass must reach a much greater mass before falling off than that of any dry solid object suspended from dry glass. If a water droplet impacts a solid object in the wind and makes initial contact it will be retained unless it can grow by coalescence with other droplets. A solid object is much more likely to make a series of contacts with projecting parts, none sufficient to hold it. Solids are therefore much less efficiently captured by dry objects than are liquid non-bouncing drops. Directly for this reason, as well as indirectly because of lower inter-particle adhesion, they are less likely to grow by accretion.

The factor of shape does not affect the behaviour of liquid drops, which are necessarily spherical before impact but is very important for solid particles. The spherical form is, of course, most favourable to bouncing and rolling and therefore best adapted to penetration of a leaf canopy but spherical granules finally rest on the soil in a pattern determined by the geometry of crevices. Most commercial pre-formed granules are of irregular polyhedral form, grossly spherical in outline and rendered more so when a.i. is applied as an external adherent deposit. If retention on foliage is desired, granules much less spherical would perform much better if difficulties of distribution and metering due to bad flow properties could be overcome. Core materials in short-fibre or small-platelet form could be made available.

Almost perfectly spherical particles can be produced by spraying a molten solid into cold air. The substance or mixture must be one not prone to supercooling, though a dust of crystal nuclei can reduce this tendency. The droplets may, if insoluble and unreactive, be cooled in water or on a plate covered with inert dust. The spheres produced in this modified way tend to be flattened at least on one face. This type of formulation, usually called "prill", is limited by the properties of the a.i. and its compatibility with physically suitable diluents.

Air-drying of a spray containing a volatile liquid diluent, a process widely developed for producing dried milk powder, instant coffee and other convenience foods can also produce near spherical particles but the dry particles are often of less regular shape, variously corrugated or saucer-shaped. This depends on the formation, during the drying of the surface, of a skin which is unable to contract elastically as the enclosed volume is further reduced. If the liquid to be dried consists of a suspension of fine solid without any skin-forming soluble matter, the residue, if evaporation is effectively complete while the drops are airborne, is spherical. Aerial spraying of wettable powder formulations results in drops arriving at ground level as spherical solid particles if the conditions are arid enough. At 25°C, 30% r.h. ($\Delta\mathbf{T} = 10{\cdot}5$°C) the lifetime of a 100 μm diameter water drop in free fall (Section 6.VI) is about 6 s during which, in still air, it would fall less than 1 m. In 4 s its area would have decreased to one-third of the original and the volume therefore to about one-fifth. If it contained only 12% of solids it would at this stage be mechanically dry—i.e. the residual water would be occluded entirely between the packed particles. The 100 μm liquid drop has therefore become a solid sphere of about 60 μm diameter. Such particles are very ill-adapted for retention except in crevices, mostly in the soil. Hartley and Howes (1961) observed this very clearly when spraying copper oxychloride suspensions. Petri dishes placed on the ground to collect spray revealed many small blue particles eddying into the dishes and out again.

Immediate retention of solid particles in the 20–200 μm range is very much influenced by the nature of the object surface. A hairy leaf will cushion the impact and reduce the tendency to reflect. Population density of hairs as well as their length is obviously important. Where this matches the size of the particle the latter may be retained permanently in the "mat". One may note here that a pile carpet is much the best surface on which to apply granules in testing machine performance: the granules stay where they fall but can easily be removed again with a vacuum cleaner. Dust is well-retained on tomentose (finely hairy) leaf surfaces but it must be remembered that particles caught up in a "mat" fail to make direct contact with the underlying surface.

D. Electrostatic Attraction

The pattern of initial retention of small particles (<250 μm) can be very significantly altered by giving them a high electrostatic charge (Hopkinson, 1974). This has no effect on final adhesion of each particle although more may be retained after a period of weathering because of the higher initial density. The collection pattern of such fine particles, otherwise dominated by fall under gravity (Harris, 1964) or dynamic impaction, can also be modified when the particles are charged in that deposition occurs also on lower surfaces and lee sides, but the peripheral leaves of a canopy, becoming more efficient collectors, tend to reduce gross penetration when the particles are charged.

Some confusion has arisen for those not conversant with the appropriate branch of physics between the electrostatic attraction over distances of the cm order of particles highly charged by some external process and the attraction over very short distances in water due to equilibrium potential differences. Friction potentials between dissimilar non-conductors can easily reach several thousand volts—about 3000 V is necessary to create audible sparking while combing the hair and, in very dry atmospheres, a 1 cm spark corresponding to about 30 000 V can result from walking over a nylon carpet in leather-soled soles. The ability of rubbed plastic objects to lift small bits of paper is familiar to every child. Equilibrium potentials in water, between a leaf surface, left negative by migration of mobile K^+ ions into the water film away from the chemically bound $-CO_2^-$ anions, rarely exceed 0·1 V. Even the selected materials in an electrochemical battery produce voltages rarely exceeding 2.

A highly charged particle is repelled by a similar one or attracted to an opposite charge. In most practical applications, the attraction which is evident is towards a much larger earthed, conducting surface. In the nearest part of the conductor an opposite charge collects and the force of attraction, usually called an image force, is equal to that towards an equal oppositely charged particle at twice the distance. If the particle is spherical, of radius a, and carries a charge Q when suspended in a medium of dielectric constant D, these quantities in free space, are related to its electrostatic potential ψ by

$$Q = \psi \mathrm{D} a$$

At a shortest distance, X, from a plane conducting surface, the particle is attracted to its image with a

$$\text{force} = \frac{Q^2}{\mathrm{D}(2X)^2} = \left(\frac{\psi a}{X}\right)^2 \cdot \frac{\mathrm{D}}{4}$$

One absolute electrostatic unit of potential is equal to 300 V, so, if the potential is expressed in volts and D (for air) $\simeq 1$, we find

$$\text{force} = 2{\cdot}8 \times 10^{-8} \times \left(\frac{\psi a}{X}\right)^2 \text{ mN}$$

For a particle of diameter 200 μm ($a = 10^{-2}$ cm) and $X = 1$ cm, we have

$$\text{force} = 2{\cdot}8 \times 10^{-12} \times \psi^2 \text{ mN}$$

For $\psi = 10\,000$ V, this is $2{\cdot}8 \times 10^{-4}$ mN while the weight (density assumed = 1) is only 4×10^{-5} mN. The particle would be attracted up against gravity on to an earthed ceiling from a distance 2·6 cm below. For a particle of 10 times the linear dimension this critical distance would be 0·8 cm. At a potential of 100 V these distances would be 2·6 and 0·8 μm, so there would be very little remote effect on airborne particles, while, at 10 V, the attraction would be comparable with the very short-range molecular forces and significant only at distances obscured by the micro-roughness of any but highly polished surfaces. At electrochemical potentials (<2 V) even after multiplying by 80 for the dielectric constant, the force is significant only for very small particles. The selective effect of a negatively charged leaf on positively charged small emulsion globules has been demonstrated by Haydon (1962). The attraction of a 10 000 V droplet to an earthed leaf has nothing to do with the leaf surface having a negative potential of 50 mV with respect to a water solution lying upon it.

Once the highly charged particle has been attracted to the conducting surface, the excess charge is lost to the conductor and there is left only the small electrochemical potential making its (usually minor) contribution to the adhesion. If the particle is a very good insulator and the target and atmosphere are dry it is possible for charge effects to persist; such behaviour can be demonstrated in the laboratory but is not significant for pesticides in the field. Elementary sulphur is the only pesticidal substance which is, when pure, a sufficiently good insulator, but all commercial dusts must be given "anti-static" treatment during formulation to avoid handling problems and risk of explosion.

The reason why particles can be raised to a much higher potential in air than in water is not that a much higher charge density can be acquired but that, in a conducting liquid, the fixed charge is largely neutralized by the close crowding of oppositely charged mobile ions. The charge density on a 200 μm diameter sphere necessary to raise it to 10 000 V in air is only 5×10^{11} electron units per cm^2. A close-packed paraffinic monolayer could contain 3×10^{14} end groups per cm^2, but, if all could be ionized, the potential would be drastically reduced by the close crowding of opposite ions. Even in extremely pure water, the effective thickness of the double layer (Section 2.XVIII.A) is only 0·3 μm. In other words, the capacity of the sphere instead of being equal to its radius is equal to its area divided by a thickness very much less than the radius.

There is no major difficulty in designing apparatus which can disperse liquids in the field in fine, highly charged droplets but the capital cost is, of course, higher than in orthodox sprayers and there are technical limitations. The distributing unit must be well insulated and its insulation protected because, if conducting solutions are dispersed, their precipitation on the insulation can render it useless. Small electrostatic dispensers, as used in industrial decorative painting, are convenient for local application of small amounts in the laboratory since the spray is drawn directly *to* a near target and wastage is therefore avoided as well as damage by contact. Particulate solids, if surface-conducting, can also be dispersed. Short fibres are transferred in this way to an adhesive-coated surface to give a velvet-type finish to such diverse objects as instrument and jewel cases, teddy bears and wallpaper. Electrostatic devices could perhaps assist the field dispersal of very non-spherical particles which have bad flow properties in orthodox dusters.

There have recently been promising advances in the development of equipment for the practical application of electrostatically charged liquid sprays in the field. Notable among these is the "electrodyn" system (Coffee, 1979) which is restricted to oil based formulations, but has the advantage that the formation and initial projection of the droplets are achieved electrostatically as well as their charging in a system which contains no moving parts. Capture and distribution between lower and upper leaf surfaces are greatly improved using this device. An example of the improvements in deposition obtainable using modern electrostatic systems for water based sprays is provided by the work of Arnold (1980).

E. Adhesion of Particles

The molecular forces of attraction between any two surfaces—even of very dissimilar materials such as water and paraffin—are very strong where real contact can be made, but fall off steeply with distance of separation so that solid surfaces have to be very flat for the molecular attraction to be effective. A bearing too accurately machined will seize up completely if not lubricated. A liquid does make real contact with a smooth solid, but the deformability of the liquid makes the molecular attraction seem deceptively small. If no gas is present to expand under tension at the expense of the liquid, the very adhesive *force* can be demonstrated, but the total *work* of adhesion is quite small because the

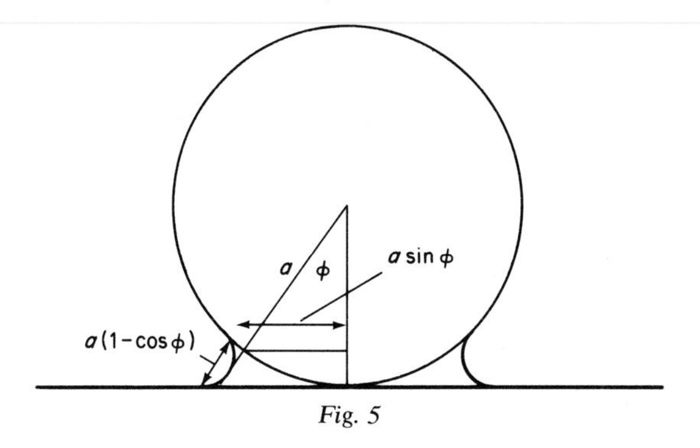

Fig. 5

distance over which the tension persists when adhesion breaks down is of molecular dimensions.

Owing to microscopic irregularities in all natural surfaces of solids, direct molecular force adhesion retains only very small particles of a few μm (see p. 468). Permanent adhesion of dry particles to leaves depends almost entirely on the presence, at some stage of the process, of a liquid film, as do practical adhesion operations in the home or workshop.

A liquid film itself produces considerable adhesion if it wets both surfaces. Its operation is usually explained in terms of a sphere resting on a solid of the same material with a small volume of liquid in between (Fig. 5). It can easily be shown that, when the maximum thickness of the liquid is much less than the radius, a, of the sphere, the force of adhesion is $4\pi a\gamma$ where γ is the surface tension of the liquid. This assumes that the liquid wets the solid perfectly. The volume of the liquid, if small, does not matter. As it decreases (e.g. by evaporation), the negative pressure in the liquid enclosed by the concave meniscus, increases, but it has a smaller area of contact on which to operate. If the contact angle, θ, is finite, it can be shown that the expression for the force must be multiplied by $\cos\theta$ and therefore becomes negative for $\theta > 90°$. A water film would thus repel a wax bead from a flat wax surface but the repulsion expresses itself in fact in a different way, the water being forced to one side rather than forming a ring around the point of contact of the solids.

Adhesion of a wetted sphere remains constant until the liquid has evaporated completely. The magnitude of this liquid film attraction should be noted. The force becomes equal to the weight of the sphere for

$$a^2 = \gamma \cos\theta/\rho g$$

for water with tension depressed by some wetting agent to, say, 40 mN m^{-1}, $\theta = 0$ and density, $\rho = 1$, a is 2 mm. A 4 mm diameter sphere will therefore be held by a water film against the lower surface of a plate until dryness.

A familiar example of this behaviour, although the particles are not spherical, is provided by clean sand and water. If there is excess water, the sand is easily plastically deformed under the water. When excess water is drained away the whole sand mass has considerable rigidity. Only when completely dry does it flow again readily. The similar behaviour of small glass spheres is more sharply defined.

Permanent adhesion results if the liquid, when it dries, leaves behind a coherent biconcave lens of involatile material, but few residues remain fully coherent during drying. Either some crystallization occurs, with reduction of volume and loss of contact, or a hard skin develops before evaporation is complete and voids are then created by further contraction. The most reliable and universal adhesives are substances which never become really hard and brittle—soluble polymers of not very high molecular weight. Most residues, however, will give some adhesion. It will be much less than necessary to hold a 4 mm diameter sphere against gravity but such strength of attachment is not required for pesticide particles in practice.

Adhesion is, like retention, very dependent on particle shape and again the sphere is the least suitable. A thin plate-like particle lying flat on the surface is much more strongly held as well as experiencing, under the same mechanical conditions, much weaker forces tending to dislodge it. It is again worth illustrating with a familiar example—the much greater difficulty experienced in swilling tea leaves, as compared with coffee grounds, along the flat bottom of the kitchen sink. The layer of water between the flat particle and the flat surface has a radius of curvature at the edge determined only by the closeness of fit. If the particle, like the wet tea leaf, is flexible, the negative pressure draws the surfaces closer together and is thus augmented. Release on drying now depends mainly on the concave-outward curling of the leaflet particle as it dries on the outside. If the tendency to curl is not too strong, the adhesive residue can, unlike that between a rigid sphere and plate, remain coherent on drying because the surfaces can be brought closer together to accommodate the decrease of volume without creation of voids or loss of contact.

Very little adhesive is necessary to fix flexible leaflets on to a flat plate, but no attempt to exploit this in a particulate pesticide formulation seems to have been recorded. The most common use of "stickers" is in spraying suspended finely ground solid pesticides, and the result achieved is not adhesion of the individual particles but rather a binding together of the composite residue with an adhesive in amount comparable with that of a.i. The smallness of the individual particles is in this case no final advantage, the only purpose served being to delay sedimentation in the spray tank. A "tea-leaf" formulation would require some special means of application as well as of fabrication.

Materials practically available as stickers are best considered after we have discussed weathering since resistance to weathering greatly restricts the choice.

F. Weathering of Deposits

Active ingredient can be lost from deposits by evaporation (Section 3.XII), by uptake into the living target (Chapters 11 and 12) by mechanical detachment, and by the action of rain. We are concerned in this section mainly with the last two processes. They are not entirely distinct since rain may not only dissolve the a.i. but its beating action can dislodge an attached residue mechanically.

Even without rain there can be mechanical detachment. If wind caused the initial impaction and there was just sufficient adhesion to hold the impacted particle, an increase of wind could dislodge it again. Moreover wind in the crop is locally very variable and made more so by movement of the leaves so that a dust particle deposited without special means of adhesion, may remain at one site for a short time, then be dislodged and collected elsewhere until it arrives eventually in some very low-level or sheltered site. Radioactive fall-out in pasture accumulates in the leaf-bases of tussocks (Possingham and Scott Russell, 1961).

Even when some means of adhesion can be called on, which is much easier with spray than with dust or micro-granules, there is some reason to believe that the natural expansion of young leaves can cause detachment. A rigid deposit cannot expand with the growth of the area underneath without cracking. Moreover the growth of epicuticular wax under the deposit, where this occurs, can be expected to loosen the attachment of the latter. It is widely believed by field workers that such detachment does occur and that, at least when the now nearly obsolete DNOC is sprayed against weeds in cereals, wind can carry off loosened deposit and create a drift problem. Definite experimental work on this possibility with pesticides seems to be lacking, but Moorby and Squire (1963) found that ^{90}Sr carbonate deposits migrated from leaves even in dry conditions. Dehiscence of wax crystals might have assisted this. The use of plastic rather than brittle adhesives seems desirable.

Gentle rain on wettable leaves will dissolve constituents of the deposit and even extract some systemic pesticides from plant tissues. It is useful to enquire into what factors limit the solution process and firstly into the behaviour of a water film on wettable surfaces.

Bikerman's (1956) verified theory of drainage from a smooth wettable rectangular plate (Section 8.VII) applied to the steady-state situation where vertical rain at a rate $\mathscr{V}_R$ (volume per unit horizontal area) falls on a surface inclined at angle α to the horizontal, gives

$$x^3 = \frac{3\mathscr{V}_R \eta}{\rho g \tan \alpha} \cdot z$$

where x is the film thickness at distance z down the slope from the upper edge. The mean thickness in the zone from 0 to z is $\bar{x} = \frac{3}{4} \cdot x$. The rate of arrival, and therefore of departure, of water per unit area of surface is $\mathscr{V}_R \cos \alpha$ and from these we can obtain the mean residence time, $\bar{t}$, of the water. Diffusive equilibrium through the thickness of the film will be practically complete if $D\bar{t}/(\bar{x})^2$ is >1. Under this condition the water leaving the zone will be nearly saturated with any substance still remaining spread uniformly on the surface in the zone.

For a reasonably typical situation, $\alpha = 45°$ and heavy summer rain $\mathscr{V}_R = 0{\cdot}00028\ \text{cm s}^{-1}$ ($1\ \text{cm h}^{-1}$); $\bar{x}$ comes to $15\ \mu\text{m}$ at $z = 1\ \text{cm}$ and $71\ \mu\text{m}$ at $z = 100\ \text{cm}$. For $D = 3 \times 10^{-6}\ \text{cm}^2\ \text{s}^{-1}$, $D\bar{t}/(\bar{x})^2$ comes to *c.* 10 and 2 for these z values. This function is proportional to $z^{-\frac{1}{3}}$ and to $\mathscr{V}^{-\frac{4}{3}}$. The lower the rainfall the more nearly saturated the drip-water will be but, of course, the less of a rather insoluble substance will be removed in a given time. For example, 1 cm of rain could dissolve $100\ \text{g ha}^{-1}$ of a substance having $1\ \mu\text{g}\ \ell.^{-1}$ solubility.

Before we turn to some important physical factors omitted from this oversimplified picture we should look at an important geometrical one. The deposit is not normally uniformly distributed. For only a fraction of the time estimated above will the film be flowing over the local areas of deposit. Lateral diffusion will be unimportant. Reduction of effective time in the above calculation by the proportion of area actually covered will not have a fully proportional effect. If $D\bar{t}/(\bar{x})^2 = 10$ is brought down to the value $0{\cdot}2$, the film will still be about 50% saturated.

The most important other physical factors are associated with this geometrical one. Most local deposits will be thicker than the mean film thickness in light rain. The film may not flow over them although some individual rain drops must do so. Surface tension effects will be of major importance for the behaviour of the descending film. When the main surface itself is wettable, an adherent spot of unwettable substance will be avoided by the film. Soluble surfactant in the residue can also reduce effective contact because the lowered tension spreads outwards over the film, thinning it down (Section 8.II). Most leaf surfaces are not smooth and flat. Often the surface is not easily wetted: much less so than the deposit itself. In the case of aqueous sprays, the deposit is left by a wetting solution which will have spread in the directions of easier wetting, e.g. along grooves over veins (Section 8.XI). Rain will therefore tend to flow preferentially along the channels where most deposit is found once the soluble surfactants have been washed out.

Very low volume application, of insecticides particularly, is often now done with pure a.i. or a concentrated solution in oil. There is much more possibility therefore of leaving a less wettable deposit. Solution and redisposition of cuticular wax can contribute to the fixation of such deposit. Only a small fraction of its area will be contacted by water except intermittently under impacting drops.

On a surface where water rests with a finite, even high, contact angle sufficiently heavy rain will produce a continuous film. The "de-wetting" of a surface needs a dry edge from which to start. Quite spectacular movement results when water is slowly drained away from above an accurately levelled unwettable plate initially submerged. The covering film thins down until some chance disturbance creates a local air-plate contact which then rapidly expands piling up the retreating water in a visible wave front. In the critical condition a falling drop of water can initiate the retreat. Film behaviour on an unwettable leaf surface is thus very dependent on surface contours and on rate of rainfall. Only on "bloomed" leaves capable of total reflection of water drops can a covering film never be established. Even when such leaves are submerged in deep water an air film is retained under the wax-crystal layer. Any deposit which spoils the microstructure invites local wetting but solution from such deposit must depend on a falling or rolling drop making direct impact. The residual wetting agent from most sprays makes the leaf locally wettable until washed clean, but a volatile wetting agent, e.g. octyloxyethanol (Section 8.VI.A) can spread water on a pea leaf and evaporate to restore reflective behaviour.

Heavy rain not only maintains complete (and agitated) cover over any non-reflective leaves but its purely mechanical impact can dislodge coherent residues *en masse.* This was clearly observed by Hartley and Howes (1962) applying very heavy artificial rain on residues of amine stearate formulations. The number, not size, of the conspicuous white residues decreased. Each departing lens deposit left a "ghost" ring which indicated its earlier presence. Prolonged exposure to gentle rain had much less effect on the amine stearate residues. There was fair agreement between chemical estimate and visual number estimate as found by Courshee and Iresan (1961) for a wider variety of formulation residues.

Most laboratory testing of formulations has been done with very heavy artificial rain because it appears to give quicker results and is much easier to produce than a uniform precipitation of small drops. The impact effects of the short-period, excessively heavy, laboratory rain can obviously exceed those of natural rain but the laboratory rain may be much less effective than prolonged drizzle in dissolving a component out of a permanent residue or even in dislodging the residue if the dried adhesive requires inward diffusion of water to soften it. It is very easy to show that a 5-minute heavy shower leaves dried gelatin residues on glass but knocks off wax deposits, but that immersion in still water for several hours removes the gelatin but leaves the wax.

Byass (1969) described a more elaborate laboratory system in which a pattern of hypodermic needles is sealed into the bottom of a shallow tank at 5 cm spacing each projecting through holes in an underlying air chamber so that a downward draught tears off the drops at less than their normal size. This can produce drops rather less than 2 mm diameter with a precipitation rate down to 8 mm h^{-1}. Fine hydraulic or air-assisted spray jets can, of course, produce much smaller drops but the rainfall rate can be reduced only by making it intermittent. A rail- or pendulum-based traverse can be arranged and 0·1 s bursts at 10 s intervals over a small target is probably a reasonable imitation of continuous drizzle.

There seems to have been little attempt to exploit the fairly obvious differences between residues or the differences arising from leaf morphology. It should be possible to adapt formulations more efficiently to specific tasks or to different climatic regions.

G. Selective Adhesion

We have so far assumed that the spray drop residue behaves as a totality. This is not always even approximately true but becomes more nearly so the smaller the drops and the smaller the total volume applicance, both these trends reducing spreading and coalescence.

Consider applicance 1 kg a.i. (assumed density 1 g cm^{-3}) per ha in 100 ℓ. water ha^{-1}, a fairly typical low volume crop spray operation. If it consisted entirely of 200 μm diameter drops and these fell uniformly on a perfectly flat area, their mean centre to centre distance would be 695 μm. If leaf area were 4 times horizontal projection area their distance apart on leaves would be twice this value. At 20° contact angle each drop would cover 6·12 times the cross-sectional area of the sphere. Its maximum height when spread would be about 120 μm and mean height about 40 μm. A fraction 0·13 of the leaf area would be covered by these spread drops. If the a.i. (1% by volume) were entirely in particles of volume equivalent to 2 μm diameter spheres, each drop would carry 10^4 particles. If deposited uniformly within the area of spread they would lie

8·2 μm centre-to-centre apart. Each particle could therefore be separately fixed to the underlying leaf surface: a very small relative volume of good adhesive would be necessary. One would need to raise the volume concentration to nearly 20% or reduce the particle size to about 0·18 μm before a uniform layer of particles would have to be in contact.

The deposit, however, will never be uniform, even on a perfectly plane surface. There are several important factors leading to a non-random distribution and, of course, a random distribution is not uniform. First there is flocculation in the suspension—a tendency for particles to cluster loosely together under the influence of attractive forces.

Secondly, if the a.i. has high specific gravity as in the case of copper fungicides (*c.* 4 g cm^{-3}), sedimentation can occur within the drop as shown in Fig. 17 of Section 8.XIII. A 2 μm diameter particle of specific gravity 4 falls at about 6 μm s^{-1} in water. This is about 2 cm h^{-1}, passing most standard tests of suspendibility and not important in an agitated spray tank. However, at this rate only 10 s is needed to settle to the pole of a suspended 200 μm diameter drop making 20° contact angle. About 80 s is needed to clear the diameter of such drop on a vertical surface. Flocculation, of course, assists sedimentation. Thirdly, and opposing sedimentation, there will usually be a cyclic convection within the drop as it evaporates. This arises because evaporation is relatively faster at the edge than the centre: this usually produces a lowering of surface tension due to increase of concentration of the surfactants present. The lower tension spreads in the surface of the drop towards the centre. Particles cannot reach the extreme limit of liquid spread because here the drop is too thin. There is a tendency therefore for particles to be rolled along towards the edge where they will pile up against others already stranded as the liquid volume decreases. Marginal accumulation is helped by dissolved substances crystallizing out in the most concentrated solution. It accounts for the almost invariable presence of at least a "ghost" ring marking the extremity of spread before drying. This ring may be only a ghost, if most of the a.i. and/or fillers have sedimented, mainly in the centre, before drying, or it may, particularly when the involatile content was entirely dissolved, contain most of the deposit (see Fig. 17, Section 8.XIII; Fig. 12, Section 8.V.C; Fig. 18, Section 8.XIII). In examining deposits of wettable powder formulations it must be remembered that the "filler" may be present in considerable excess over the a.i. and that it is generally more visible so that the visible distribution of density may not correspond to that of the a.i.

Where the deposit is thin enough for the particles to settle separately there can be considerable difference in the firmness with which particles of different kinds adhere to the leaf surface. Generally one can expect the a.i. to have more affinity for the leaf surface, since it is usually much less polar than water or the mineral fillers. Indeed the principal function of the mineral filler in a w.p. formulation is to prevent the adhesion of the crystals of organic a.i. and it can therefore be expected to exert the same, no longer useful, function between the a.i. and the cuticle if the conditions of formulation and drying bring the particles together. Usually, any adhesive is present in amount sufficient to cement the particles in an agglomerate.

In high volume spraying it is generally desirable to have the a.i. selectively precipitated on the leaf surface so that the carrier water only runs off. This can be achieved on most leaf and stem surfaces, including those of winter-dormant fruit trees, by using an unstable emulsion with the a.i. dissolved in the oil phase. There is obviously some conflict with the operator's requirement for a stable emulsion in his machine. With specially designed machines it would be possible to spray an unstabilized emulsion using good agitation to keep the mixture grossly homogeneous and relying on the nozzles to disperse the oil phase more finely. A more practical answer is an emulsion which is stable in the tank but becomes unstable on exposure. Loss of a constituent by evaporation or increase of acidity by absorption of CO_2 are the changes most easily exploited on passing from the closed spraytank to a fully exposed spray. Tar oil washes can be marketed as self-emulsifying oils using an alkali or ethanolamine soap as stabilizer, assisted in the emulsification process by the content of phenolic substances. Such emulsion is not stable long when CO_2 has access and the fatty anions retreat into the oil phase as the pH decreases. The process can be accelerated by using a volatile amine to form the soap in the initial solution. Trimethylamine is most rapidly lost but emulsions stabilized by ammonium soaps break more quickly than those stabilized by alkali metal soaps. The emulsion must not break too quickly because spreading is more rapid if accomplished initially by the high volume phase.

The possibility of selective adhesion of oil globules without breaking of the emulsion has been explored by Furmidge (1962c) and Haydon (1962) with some success in laboratory experiments. The run-off from over-sprayed leaves had a lower oil content than the initial spray when the emulsions systems were carefully chosen. Haydon showed that it was possible, using a basic protein as fixed stabilizer of the globules in a pre-formed emulsion to have the globules positively, but the leaf surface still negatively, charged. Such finesse is not practicable under field conditions using a variety of untested natural waters and both high volume spraying and stock emulsions are, for reasons already explained, of very limited use. Practical development along these lines seems unlikely. There is more future in fine droplet spraying of undiluted oils.

H. Adhesive Formulations

The most reliable and versatile adhesives are substances which never really harden. They have the advantage that molecular-force incompatibility does not manifest itself in non-adhesion until the brittle stage is reached. This phenomenon is very strikingly and usefully illustrated by the behaviour of cellulose acetate films on glass. If the acetone "dope" from which the film is cast contains adequate involatile plasticizer the dried film adheres fairly well to the glass but can be peeled off without damage. If the cellulose acetate is completely unplasticized the drying (and contracting) film goes through a stage where its adhesion is great enough to tear chonchoidal chips out of the glass but, when, finally dry, the brittle film releases them, the final state being clean glass chips and separated brittle film.

To resist erosion by rain it is desirable that the adhesive be water-insoluble and, since most leaf surfaces are hydrophobic, this will also favour adhesion. Polyisobutene would be ideal but is soluble only in paraffins which are poor solvents for most pesticides. Other more polar resins are more practical, but there is still a formulation problem in that they increase the viscosity of the self-emulsifying oil. Polymers of lower molecular weight than are useful in the plastics field are the compounds of choice. The need for a very high viscosity when the solvent has evaporated but a low one when it is in solution is one reason for the very successful development of emulsion glues in construction trades. If the very viscous polymer can be stabilized in emulsion in a predominantly aqueous solvent, the viscosity of the whole is quite low as long as the dispersed phase volume does not much exceed 50%. When the water, which thus facilitates quick application and spreading, evaporates, the globules are forced into distorted contact and eventually coalesce. Softening by a solvent of intermediate volatility is usually called on to assist this process. Many paints, as well as glues, are made on this principle and enable a water-based fluid to leave a very water-resistant end-product. A second advantage in emulsion paints is that, since much less volatile solvent need be used, there is less toxicity problem. Suitable polymers are also more easily prepared in emulsion than in massive form and in fact much of the polymer for moulded plastic articles is produced initially in emulsion form. The most widely used polymer for paints and glues is poly(vinylacetate) but there is a bewilderingly wide range of modifications introducing other monomers along with vinylacetate in the polymerization process to secure special properties.

An unpigmented emulsion paint "vehicle" would seem to offer great advantages for pesticides applied in water and intended to leave a water-resistant residue. Somers (1956) demonstrated that such formulations could be efficient without damaging leaves. There are, however, practical difficulties. The emulsion paint base must be supplied in the emulsion form. If the pesticide is to be provided in the same container it must not only not interfere with the stability of the emulsion on storage but must also not itself be vulnerable to hydrolysis. The latter requirement is severely restrictive now that the environmental hazards of very stable substances such as the chlorinated hydrocarbons have led to severe restrictions in their use. The requirement for physical stability of the mixture is also restrictive. The polymer latex is an extremely fine emulsion with globules mostly $<0{\cdot}5\ \mu m$ diameter and is prone to coagulation by shear when the composition is allowed to get outside limits established by careful trial. Paints are applied as packaged and their non-polymer variable content is made up of inert pigments. Pesticides must usually be diluted with water having uncontrolled impurities. Different pesticides are often applied together after mixing in the tank. Coagulation in the spray tank could lead to very intractable filter- and nozzle-blockage.

Oil-soluble resinous adhesives cannot be used with active ingredients which, because of low solubility, can only be formulated as wettable powders without employing some indirect method. An obvious answer is the supply of a.i. and formulant in separate packages for tank mixing, but this form of supply has never become acceptable either to users or to the technical sales force because of the possibility of poor performance and complaint if one component is omitted or wrong proportions used. It should be remembered, however, that the earliest and very widely used fungicide—Bordeaux mixture—involved a pre-tank mix of two separately packaged chemicals. The trend has always been away from this, into single-pack "fixed coppers", but Bordeaux is still preferred by many orchardists and vine growers because of its combined activity and tenacity. The alleged prejudice against the two-pack issue could certainly be overcome and could be assisted by novel forms of packaging such as a two-compartment polyethylene bag from which the contents would escape together when the top was cut off and which would provide, of course, a standard area applicance. The formulator could give the user much more help for particular uses of environmentally acceptable pesticides if dual packaging were popularized.

J. Selective Availability

Ideally a persistent residue should not be available to weathering processes, particularly rain, but should be available to the target organism. These requirements are bound to be in conflict and some compromise is inevitable. It would be possible in many cases so to occlude the a.i. that it is not available at all. The logical extreme would be to leave it sealed in the supply package, as many opponents of pesticides would wish. Unless protected by some specialized formulation such as micro-encapsulation it is free to evaporate as soon as it is dispersed over the enormous area of the crop and will probably be visited by far more rain drops than insects or fungus spores. In a very light shower of 1 mm "depth" composed only of 1 mm diameter drops any point on an exposed leaf will most probably receive 5 or 6 impacts, even assuming the drop spreads to no greater contact area than its own cross section. The chance of no impact is 1 in 4000. In a slightly heavier shower of 5 mm depth composed of 2 mm diameter drops, the corresponding figures are 14–15 and 1 in 3 million. This is childishly obvious but is too often forgotten in the pursuit of visually and mechanically acceptable all-purpose formulations. If the residue is made to be unaffected by rain it is inevitably less available to other contacting objects.

The factors determining what compromise is to be sought are discussed elsewhere in this book. They are complex and often inadequately investigated. It is in this field that there is particular need for close collaboration of biologist and formulator. The habits of the target pest, physical properties of the pesticide, means of transfer available and prevalent climate are all important. Four fairly distinct situations exist. (1) Systemic action in the host plant into which quick entry is required, preferably before rain is likely (Section 11.V.D). (2) A residual deposit to attack leaf-eating insects for which it is possible to provide some form of rain protection which does not, however, resist mastication or digestion (Section 13.III.F). (3) A residual deposit to attack pests in a nonphytophagous phase. The most common use is on walls against flies, mosquitoes etc. where rain-fastness is not necessary. However, there are related field situations where the problem of making the deposit available by contact but not to rain is a considerable challenge (Section 13.III.B). (4) Prophylactic attack on incident fungus spores where surface water must provide the means of transfer and is often a major vector of the spores (Section 13.IV).

Chemical formulation (Section 4.IV.B), where the chemistry permits, can help in situations (1) and (2) by supplying an oily derivative which can be reconverted to a more hydrophilic active compound in plant or pest. Stickers or microcapsules which are selectively degraded by the digestive processes of the insect should have useful application in situation (2). Adhesive additions to a formulation of a.i. of low solubility can help in situation (4) and it should be borne in mind that the optimum solubility of, e.g. a copper salt may be altered in this process—too often this sort of "lateral thinking" is omitted in compound selection where an optimum is chosen in orthodox activity tests and then further formulation steps are taken to improve performance. Situation (3) is clearly the most difficult from the point of view of formulation improvement. One can change viscosity: one can change fillers: one can change dilution and add adhesives but the effect on *relative* availability to the tarsi of an insect (a particular insect) and to raindrops requires a degree of experimental attention which no organization appears willing to finance.

The related problem of obtaining selective availability to harmful organisms rather than beneficial species is considered in Section 13.III.F.

VI. CONTROL OF DRIFT

A. General

The transport of particles and vapour in air and their reception on objects have been considered in Section 6.VII.B. Drift with the wind is deliberately exploited in some pesticide operations but in this section we shall be concerned only with how undesirable drift can be reduced.

Measurement of drift in the field is subject to large experimental error and there are also problems in interpretation. Fall-out on flat sampling plates distributed on flat areas as used by Yates and Akesson (1973) is not simply related to catchment on multiple surfaces in a crop. As Chamberlain showed (p. 373), this catchment may be much greater than anticipated from gravitational fall-out and this could shorten the whole drift cloud when it passes over a vegetated area as compared with an airfield runway. Nordby and Skuterud (1974) used thin vertical sampling rods but it is not clear how the recorded amounts were related to horizontal area or to the fraction of the total discharged spray which drifted.

When the record shows that more tracer is collected at, say, 20 m downwind than at 50 m, it cannot be assumed that the difference is due to deposit on the intermediate ground or crop. It will be part due to upward dilution of the drifting cloud. A very long tail to any deposit–distance plot is to be expected (p. 530). The exponential decrease postulated by Nordby and Skuterud is not justified either by theory or by their results. The use of test plants in this investigation, however, gave a direct estimate of the near downwind damage. This damage was found to be greater, in relation to amount of spray caught on the sampling rods, when a pressure of 10 bars than when one of 2·5 bars was used with the same nozzles, presumably due to the increased drift being mainly of smaller drops.

In ground machines using orthodox booms and fanjet nozzles, the drift hazard can be reduced by keeping the boom low and avoiding strong winds, particularly winds across the track direction. In field tests, using rod samplers to collect fluorescent tracer and barley seedlings to register damage by aminotriazole, Nordby and Skuterud concluded that the total *amount* of spray escaping from the swath, as well as its downwind spread, increased with both height and wind speed. They consider that the downwind created by the jets is weakened by lateral displacement in a crosswind and the impaction on crop surfaces therefore made less efficient. This factor is, of course, less important in spraying from fixed-wing aircraft.

The use of a box, with flexible curtain sides, enclosing the boom has been proposed and had limited success in the field. This was shown by Edwards and Ripper (1953) greatly to reduce damaging herbicidal drift. Much of the fine-drop spray retained in the box must be smeared over the crop by the trailing curtain and could lead to increased crop damage. Courshee (1959) had some success with a deflector exerting part of the function of the closed box in a more practical manner.

Drift from aircraft is necessarily greater than from land machines delivering the same drop-size spectrum. It can be reduced by designing the machine to avoid, as far as possible, introducing spray into the wing-tip vortices. Total amount drifting will be reduced by operation only in the inversion condition of late evening or early morning, but this, while reducing remote drift, may increase direct damage over a short distance downwind. The use of very small drops in ULV-spraying necessarily increases drift and is acceptable only in the treatment of large areas or where deposit of insecticide or fungicide off the target area can be tolerated. The problems of drift are, of course, usually most serious when herbicides are sprayed. A narrow drop spectrum would enable a smaller mean size to be used with less drift creation but the spinning disc devices, so effective for this control on slow ground machines, are not easily adapted to deliver drops over 200 μm diameter when exposed to the aircraft slip stream.

B. Avoidance of Small Drops

Avoidance of undesirably small drops is an obvious objective of nozzle design, but some compromise must be made between the requirement for efficient control of the pest on target and the requirement for a low level of off-target contamination. In our various previous discussions of drop-size effects (Sections 6.VI, 6.VII, 8.VIII, 8.XI, 8.XIII, 14.IV.A) we have given rather different limits. This partly reflects serious gaps of knowledge, but also a difference of emphasis in different situations. Under all practical conditions 500 μm drops are aimable and will arrive on target. They will obviously be more precisely aimable by a hand sprayer than by aircraft but the target size is also different. One does not expect to carry out spot-spraying of local weed patches from the air. At the other extreme, 50 μm droplets, evaporating in flight to a much smaller residue, are not aimable. If one uses them to secure better kill of small insects in flight one must weigh this advantage against the increase of significant effects downwind. The decrease of effect with distance downwind is dependent on wind speed, turbulence, filtering action of the crop and dosage-time relationships of toxic effect. For ground-based sprayers one can bring the limit of aimability down from 500 to 300 μm and even to 200 μm, except in extremely dry conditions or where the target itself must be defined to a few cm. On the other hand, in dry and turbulent air and if the main concern is remote drift, the limit of smallest acceptable size would have to be raised from 50 to 100 μm or more.

The recent development of rotating-disc sprayers for field use greatly increases the scope for control and study of drift. We emphasize again that the important advance is not in the production of very small drops but of drops of almost any desired size in a very close size range. For this reason it is desirable to dissociate the development from ultra-low volume (ULV) spraying and refer instead to controlled drop application (CDA). Not only will the development make possible the use of a more desirable mean size without the associated production of much smaller drops but it can greatly increase our experimental knowledge of the relationship of drop size and drift.

To illustrate the advantage of a close drop spectrum we consider two very arbitrary but not unrealistic examples. The output of a hydraulic nozzle operated so as to give a useful volume of liquid in 100–200 μm drops could be represented crudely as a mixture of 100, 200, 300, 400 μm diameter in numerical ratio 1:3:3:1 and therefore in volume ratio 1:24:81:64, each eight drops having a total volume 170 times that of one 100 μm drop, volume average diameter 277 μm. We will assume that one 200 μm or eight 100 μm drops are sufficient for certain kill of the pest or weed contacted. Each 300 or 400 μm drop will be no more useful than one 200 μm drop so that, with the best possible distribution, the effective volume of the eight drops will be $1+24+24+8=57$, i.e. about one-third of the total. Compare this with a spray in which all the drops are 200 μm and therefore the whole volume, on the same assumptions, useful.

If drift is wholly associated with the 100 μm droplets, the uniform size spray has unambiguous advantage. Suppose, however, that the distinction is less sharp, say a 50% escape of 100 μm droplets and a 1% escape of 200 μm, all larger sizes remaining on target. From the volume equivalent to 170 100 μm droplets there is a contribution of $1/2+24/100=0{\cdot}24$ to drift. From the same volume in the form of 200 μm drops only the drift contribution would be 1·70, but we would not need to use the same total volume since the hydraulic spray is only one-third as efficient on the target. We therefore save two-thirds on chemical usage and cut down drift risk by a quarter.

The CDA approach may therefore make it possible to eliminate drift by avoiding completely the drift-generating size and even where this is not possible it can reduce drift by avoiding waste in "overkill" drops without dispensing unnecessarily small drops.

The CDA machine as developed by the W.R.O. (p. 789) puts out the spray horizontally at such height that the horizontal component of drop movement has died out before the crop is reached and the drops are falling vertically without downward wind. The mechanism postulated by Nordby and Skuterud for reducing drift by immediate impaction in the crop is thus not operative. Reduction of drift in the W.R.O. machine depends on non-generation of driftable drops. Nordby and Skuterud do not give a drop size analysis of their spray and therefore no indication whether crosswind causes finer drop formation. It is possible that uniform drop size could be produced, and perhaps more economically, by air blast over a toothed edge of a long sheet on which the liquid is uniformly spread—an extension of the anvil-type nozzle combined with air blast. If this is found practical, the uniform drop curtain could now be directed downwards into the crop, securing both advantages.

C. Spray Thickeners

Hydraulic nozzles on aircraft produce smaller drops than on land machines because of the much greater air speed. This can be compensated to some extent by facing the jets backwards so as to reduce the relative speed of the liquid sheet and air. Attention has also been given to increasing drop size by change of formulation. The principle is to increase viscosity sufficiently.

Ford and Furmidge (1969) showed that the mean drop size from a hydraulic nozzle is independent of viscosity over a considerable range above that of water. Size then *de*creases with increase of viscosity until the trend is reversed and the drops become much larger at viscosities above 30 times that of water. To increase the viscosity enough without using expensive materials "invert emulsions" have been widely used.

Emulsions having a fractional volume of the dispersed (globular) phase less than about 50% usually have a viscosity not much greater than that of the continuous phase. (Dairy cream is exceptional.) When the fractional volume rises above about 60%, the viscosity increases markedly because the close-packed globules interfere with each other's motion and must be distorted before they can pass. The viscosity of the globular phase then becomes important: if it is very high, as in emulsion paints, the system becomes virtually solid at about 70% fractional volume, but, when the globular phase is mobile, volume fractions well over 90% may be attained and the emulsion, although showing some rigidity against feeble forces, behaves as a viscous liquid at high rates of shear. An invert emulsion, containing 80% or more dispersed water, provides a cheap way of attaining high viscosity and resistance to break-up at the spray jet. Securing high viscosity by dispersing water in oil is not only more economic than dissolving some high molecular mass additive but the product has the further advantage that, when the water evaporates, the oil residue is mobile and spreads on the target surface.

Soluble high molecular mass thickeners have been used to reduce small drop formation by ground sprayers with good effect, but, in this case, where high air velocity is not involved, the use of lower pressure nozzles offers a more economic solution, unless the thickener eliminates very small "satellite" droplet formation without increasing the size of the main drops correspondingly. In order to eliminate satellites it is probably essential for the flow properties to be non-Newtonian and for the fluid to have appreciable elasticity. These two properties are not equivalent. A Newtonian liquid is defined as one in which rate of shear, under non-turbulent conditions, is proportional to stress—i.e. the viscosity is independent of shear rate. The most common deviation is that viscosity decreases with rate of shear. Ford and Furmidge (1969) found that the limiting value of viscosity for very high shear rate is the value determining drop size in emulsions.

Frequently this deviation from Newtonian behaviour is accompanied by others, some dependent on time. Many stiff suspensions are liquefied by stirring (thixotropy) and take varying times at rest to recover their stiffness. Other solutions may show appreciable elastic behaviour, tending partly to recover the form they possessed before shear was imposed. In some degree this property is universal in solutions of molecules of very high molecular mass and is very strongly developed in some fluids of living tissues. Very high viscosity can lead to a long thread between a falling drop and the liquid still remaining on the object from which the drop falls. This thread can break up into a large number of very small satellites but, in an elastic liquid, the two parts recoil from the first place of rupture without satellite formation.

The use of non-simple liquids to reduce small drop formation creates other problems. A very viscous liquid is difficult to stir and therefore the thickening agent tends to remain for a long time incompletely swollen—in culinary terms the sauce tends to be lumpy. Moreover the viscous spray liquid needs more energy in pump and pipe lines. The invert emulsion provides a way out of these difficulties because its two component liquids are, separately, quite mobile and develop the high viscosity only when the greater volume of water is dispersed in the lesser volume of oil. If appropriate emulsifying agents are chosen, the desired phase type is produced by simple, violent agitation. A twin-tank, twin-pipe system which brings the liquids together only at the nozzle can serve the purpose of an emulsifying mill in addition to means of dispersal. This advantage can be gained to some extent in a normal sprayer since the single tank can be charged with a coarse emulsion which becomes much more viscous as globule size is reduced in the swirl chamber of a hollow cone nozzle (Ford and Furmidge, 1969).

Quick solution of a thickener in a spray tank or its supply tanker can be achieved by products of emulsion polymer technology. Ethylenic polymers having free carboxylic groups are insoluble in water until the acidic group is ionized by raising the pH. Admixture of maleic anhydride or acrylic esters in hydrophobic monomers before polymerization yields polymers which have this property after mild hydrolysis. A fine emulsion can be supplied which is stirred into water. Alkali is then added to give, quickly, a smooth, highly viscous solution.

Both these methods of reducing small drop content by manipulating flow properties of the spray involve the use, in some form, of a two-product supply to the operator (Section 14.V.H, where we consider the use of emulsion polymers to secure adhesion).

D. Deposit Stickers

If, as many field workers believe, a deposit can be directly blown off and contribute to delayed particulate drift from the crop, it would be desirable as considered above to employ a highly viscous or plastic adhesive. The problem is only likely to arise where a powder deposit is left in a brittle condition unable to accommodate expansion of young leaves.

E. Reduction of Vaporization

The origin of drift in vapour form has been considered in Chapter 6 where it was shown, for pesticides of sufficiently low volatility to be applied as spray, that evaporation from deposits on leaves or soil is much the most important source. In addition to selection of the pesticide applied, drift can be reduced to some extent by methods used to conserve the dosage necessary on the target area—i.e. use of crevice-penetrating granules rather than absorbed spray or, when the pesticide is applied pre-planting, incorporation in soil. In application to foliage, adjuvants which assist penetration of the leaf cuticle could, in theory, reduce vapour drift and formulations which leave a skin of low permeability on the surfaces of deposits exposed to air could also assist, but no definite evidence of success by either device is available.

F. Reduction of Reception

It is probable that crop-damaging drift in the near downwind area could be reduced by reducing the availability of the a.i. from particles which have arrived by drift. We are not aware of a deliberate attempt to exploit this possibility, but it is worthy of attention. If the spray contains a considerable proportion of inert filler, a 100 μm droplet drifting a dangerous distance down a dry wind becomes a rather smaller solid sphere. Such particles, as was shown by Hartley and Howes (1961) have very little adhesion and cascade through a leaf canopy on to the ground. Droplets arriving directly on the target do so as liquids and adhere, and the diffusion of a.i. in the pores of the eventually dry residue need not be seriously impeded. Addition, for example, of 10% of an inert powder and 1% of glycerol to a spray containing 2% of ethanolamine 2,4-D would leave a deposit nearly as effective as that of the herbicide alone on a target leaf on which the drops arrived in the liquid state. If drifting on to a neighbouring crop it would arrive as solid, non-adherent spherical particles.

It could well be possible to augment this effect, or replace it, by some destructive chemical reaction not operative when the drops make early contact with leaves. A photochemical sensitizer, inactive in a deep layer of liquid in a dark tank, could destroy the pesticide when small drops are fully exposed to light as they drift. The sensitizer itself could react with native compounds on the leaf surface and therefore not remain effective in drops which have spread on leaves. The search for suitable reactions would be a worthy challenge to chemical ingenuity.

G. Drift of Dust and Micro-granules

Dusts are very cheap and convenient for hand application in domestic gardens and peasant agriculture and also for wide and rapid distribution including deliberate drift into canopies. Distributing machinery ranges from a hand shaker, for which an old stocking toe is very suitable, though a small plastic bottle puffer to large multi-outlet air-blast machines. Even in very local application one relies on air movement to reduce leaf shadowing. Some uncontrollable drift must be accepted and dusting is generally restricted to fungicides or insecticides where drift on to a neighbouring crop is unimportant.

Traditional dusts have a wide and poorly specified range of properties. Many are produced locally by diluting a manufacturer's powder—often, unnecessarily, a *wettable* powder—with some local inert mineral dust. The desideratum is "dustiness", i.e. fineness of particles without any adhesion which would cause them to clump together, so that the product is easily dispersed, settles slowly and is responsive to slight air currents.

The object of dilution—from the 20–80% range of the wettable powder to around 1 or 2% a.i.—is to bulk out the product, making it easier to distribute at low a.i. rate from simple devices and rendering the deposit visible. This visibility, necessary in small scale use, can be misleading unless the dust is prepared with greater knowledge and supervision than is usually the case, since it is the 98% filler which is seen. The a.i. particles may not follow the same pattern.

Dusts, as referred to in various technologies, cover a very wide range of size. Present in most acceptably clean air the particles may number several million to the cubic metre but be less than 1 μm across. The range from 2 to 10 μm, particularly of silica and some silicates, is most dangerous to health as it penetrates the nasal defences and settles in the lungs. Typical dust fillers used for pesticides have dispersible units in the 10–30 μm range which may be aggregates of smaller particles. Much finer particles (individual clay mineral particles are usually less than 10 μm) are not easily dispersed as such in ordinary equipment. Adhesion becomes more effective the smaller the ultimate particles and when a wide spectrum of size is present. Coarser dusts are distributed more "dustily" than much finer, but "claggy", powders.

The micro-granules now being produced by several pesticide formulators are intermediate between ordinary granules and dusts. They offer some prospect of covering the target more uniformly than granules while avoiding the high drift risk of dusts. They can be prepared, with some formulations, in as narrow a size range as drops from a spinning disc, one method of manufacture being to use a spinning disc drying tower of the type which produces dried milk or instant coffee. As the particles are in the 100–200 μm order, it is necessary that they be fairly uniform in size with absence of much finer particles in order to secure good flow properties. Micro-granules of such narrow size-range could offer some of the advantages and be subject to the same problems as CDA. They are not, however, so easily distributed as spray. The only machinery at present developed for wide boom application distributes the product in air flow. Micro-granules are also not so easily collected by crop surfaces as are liquid drops of low surface tension and they are not so immediately adherent.

The importance of shape cannot be too strongly emphasized. When we depart from the spherical simplicity of liquid drops, we open up, literally, two new dimensions. Dried drops can be spherical when formed from an insoluble powder suspended in a simple liquid. They can be variously corrugated or flattened if the liquid solution forms a skin at an early stage of drying and the resulting bag must collapse as its volume decreases: an extreme form is a saucer-shaped particle. Platelet particles can be produced by fragmentation of a brittle film and elongated particles by chopping of continuous fibres as is done in the production of artificial velvet. Spherical granules have poor receptibility, platelets transiently stick to most surfaces and short fibres are easily arrested on hairy surfaces.

To define micro-granules only in terms of size—usually the diameter of a sphere which would have the same volume—is a gross oversimplification. It will prevent the full exploitation of the potential of micro-granules in providing some new compromise solutions to the conflicting requirements of cover, penetration, pick-up and drift hazard.

VII. APPLICATION OF PESTICIDES TO SOIL

A. Introduction

Pesticides are frequently applied to soil as granular formulations. Often this is for reasons of convenience already discussed (Section 14.II.F), but in addition pesticides have frequently been found to persist longer when applied in this way compared with sprays. It is often claimed that granules release their contents more slowly, an advantage in some situations, and special slow-release granules have been prepared to enhance this property. The benefits of slow release have been unambiguously demonstrated and exploited in application to water, described later. The evidence for its usefulness with soil applications is often less convincing because of confusion about the process of release and its significance. Comparisons have been mainly between fine spray and coarse granules when differences may be more attributable to the much larger population of sources in the case of sprays than to difference in the physical form of these sources. As long as granules were always coarser than spray drops, such errors in interpretation could be regarded as relatively unimportant from the standpoint of practical use, but there is now much interest in microgranules with sizes well down into the range produced by conventional hydraulic sprays, while dribble applicators have been developed which deliberately employ much coarser drops. At the other extreme, rotary atomizers make it possible to apply a very small mean drop size by drift. This implies very low volume applicance. Granules and liquid formulations now overlap in size and concentration, therefore, and comparisons must be more carefully considered.

B. Release from Initial Source and its Significance

Whatever method of application is used, the pesticide is introduced into or on to the soil in particulate form ranging from several cm^3 of a very voltaile liquid fumigant at injection sites many centimetres apart to a fine spray of solution, or a fine dispersion. Each soil crumb which has absorbed a spray drop, each injection site or granule represents a discrete source from which the active chemical spreads by diffusion or water flow and dispersion according to the principles considered in Chapter 3, and by the activities of soil-inhabiting organisms which we shall not discuss here. We shall first restrict consideration to molecular diffusion, in the air space or water content of the soil (Section 3.IV.E).

If the a.i. is very soluble in water (e.g. sodium TCA) or very volatile (e.g. methyl bromide) it very quickly becomes molecularly dispersed as vapour, aqueous solution or adsorbed in the immediate environment of the source. If molecular dispersion is the criterion of release, the a.i. can be regarded as released as soon as applied. It is subsequently diluted, progressively more slowly in the surrounding soil until eventually the "territories" of the individual sources begin to overlap and a region of nearly uniform concentration is then further dispersed only by slower downward spread of the whole zone, upward loss by evaporation and degradation. Whether the progress from separate regions of very high concentration near the sources towards uniform dilution in the general zone of application is to be classed as further release depends on biological factors.

If the a.i. is a herbicide which has no direct effect on root growth and which is taken up passively by a well-ramified root system it may matter little whether a large portion of this system is exposed to a low concentration or a much smaller portion is exposed to a correspondingly high concentration. The same will be true of a small animal making contact during random and rapid wandering. If, however, the herbicide is one which destroys root function or inhibits growth, its confinement to local volumes results in only localized damage which the plant can survive by drawing on the remaining undamaged roots as was demonstrated by Lyndsay and Hartley (1963) in the case of pea, linseed and MCPA. Similarly, localized treatments might be ineffective against relatively immobile and widely dispersed pests, because the probability of contact with a treated zone would be low. Such effects could extend to mobile insects where the localized insecticide had a repellent action. Conversely, the lethal effects of a localized formulation having attractant properties would be reduced by diffusive spread. Enhancement of herbicidal action when a herbicide is incorporated with root-stimulating fertilizer has been reported.

Although therefore the processes of departure from the original source can be identified and analysed, "release" of a soluble or volatile pesticide cannot properly be defined precisely in purely physical terms: it depends on the physiological and behavioural responses to the chemical of the organisms affected. Release of a chemical of very low solubility or volatility has a more definite physical component. It is generally assumed that the chemical must be dispersed in molecular state (in solution or vapour) before it can be taken up. The amount so dispersed will increase slowly with time and there is therefore an increasing amount, however distributed, available for uptake—increasing until all has dissolved. This increase we can describe as a physical process, but its significance is still greatly dependent on biological factors. Moreover, the assumption that solution in soil water or evaporation into soil air must precede uptake is invariably valid. Particularly in the case of insects and if ingestion is involved, a microcrystal may deliver up its contents to some oily surface more rapidly if direct contact is made although the actual area of contact may be much smaller than that with dilute aqueous solution. Such behaviour can be important when the chemical owes its low solubility in water to weak molecular attraction to water (and therefore high oil/water partition) rather than to high melting point.

Having emphasized the significance of biological factors we will now consider concentration distribution as a physical process and later examine how the process may be modified.

C. Patterns of Distribution

To get some idea of the distances over which diffusion must occur and the volumes of soil which each source must supply we will consider two representative situations and assume that the granules (or spray drops) are all of the same size and are distributed perfectly uniformly. In one situation we assume the sources to lie on a perfect surface so that each is equidistant from its six nearest neighbours. In the other, the sources are mixed uniformly into a limited depth, X, of soil so that each lies equidistant in space from 12 nearest neighbours.

Each source in the first of these idealized distributions lies at the centre of a regular hexagonal cell, the whole surface being occupied by a close-packed "honeycomb" of such cells. The inscribed radius of each cell is half the distance, ℓ_0, between nearest-neighbour sources. Diffusion from each source would be unaltered if each cell were bounded by an impervious wall. A more convenient model for calculation is a circular cell with impervious wall. To account for the total area, the radius, b_0, of the "circle of influence" is $(2\sqrt{3}/\pi)^{\frac{1}{2}} = 1{\cdot}05 \times$ the inscribed radius of the hexagon. The equivalent circles therefore overlap slightly, compensated by gaps of equal area at the points of the hexagon. The difference, for the very crude approximations we must make for other reasons, is quite trivial. In the second case we replace the space-filling dodecahedra by spheres of equal volume. Their radius, b_X, is $(3\sqrt{2}/\pi)^{\frac{1}{3}} = 1{\cdot}11 \times$ the half distance between nearest neighbour sources.

Let the diameters of the sources as applied be d μm (more precisely this is the diameter they would have if their fixed volume were spherical). Let the applicance be $\mathscr{A}$ kg a.i. ha^{-1} and therefore $\mathscr{A}/\rho\ell$ of liquid or solid ha^{-1} where ρ is the mass a.i. per total volume of formulation in g cm^{-3}. The radii of the circle and sphere of influence are

$$b_0 = (\rho d^3/60\mathscr{A})^{\frac{1}{2}} \times 10^{-3} \tag{1}$$

and

$$b_X = (\rho d^3/80\mathscr{A})^{\frac{1}{3}} \times 10^{-2} \tag{2}$$

We tabulate these quantities below (Table 1) for two values of ρ, namely

Table 1
Radii of territory (cm); applicance 2 kg a.i. ha^{-1}

Granule or drop diameter (μm)		Spray drops				
		normal granules		micro-granules		powder
		800	400	200	100	20
ρ	Depth X (cm)					
0·11	0	0·69	0·24	0·086	0·030	0·003
1·1	0	2·18	0·76	0·272	0·095	0·008
0·11	10	1·52	0·76	0·38	0·19	0·04
1·1	10	3·27	1·64	0·82	0·40	0·08

1·1 representing an average pure pesticide and 0·11 representing an 11% aqueous suspension or solution or a 5% w/w granule on a mineral of density 2·2. We assume an applicance of 2 kg ha^{-1} and therefore a volume of total material applied of 1·8 or 18 ℓ. ha^{-1}. We have exemplified granule (or spray drop) diameters covering the normal and micro range and also the coarse end of the powder range, although powder grains will not of course usually reach the soil separately.

Equations (1) and (2) indicate how these figures change for other values of $\mathscr{A}/\rho$, d and X. The treatment is, of course, idealized. Not only is the distribution in practice at best random rather than uniform but it will deviate from simple randomness because of the non-uniform soil structure and imperfections of machinery. We calculate these values only to make order-of-magnitude estimates of times taken to reach nearly uniform concentration and to indicate how these may be influenced. Deliberately non-uniform application has been considered in respect of "depth-protection" in Section 5.IX and in respect of localized seed protection in Section 13.II.B. Formulation for seed treatment and delayed release is considered below (Sections 14.VIII.A, 14.IX.A).

D. Diffusion from Sources Distributed in Limited Volume

Diffusion downwards from sources on the surface of a penetrable medium is initially on a pattern of expanding hemispheres. When these begin to interfere a complex stage is entered but eventually spread becomes limited only by general downward advance from what has become a combined plane source. If upward diffusive loss into air is important this will usually remain complex as the thickness of the stagnant layer is comparable with the distance between sources, a situation considered (Section 6.V) in respect of evaporation of residues from leaves. If water-phase diffusion is the major transport mechanism in soil, redistribution is necessarily very variable since the sources lie on a surface usually dry. It is quite impractical to consider downward diffusion from surface sources independently of leaching by rain. If air-space diffusion is important, the variable process of evaporation must also be involved. This important situation is discussed qualitatively in a later section. Here we consider diffusive spread from sources buried in a supposedly uniform soil.

For simplicity we consider sources of two distinct types. The first supplies a substance so soluble (or volatile) that it can be assumed wholly in solution even in the small volume in which it is applied. The second supplies a substance with so low a solubility that depletion of the source can be neglected over the period of interest: it is assumed to maintain a constant, saturated, concentration at its outer surface, radius a. For convenience, these sources will be called "dissolved" and "saturating" respectively. Many sources have, of course, intermediate character. Initially they have saturating behaviour but lose their content of saturating substance before their "spheres of influence" are saturated.

In practice we are always concerned with spread much beyond the initial radius of the source, when the cumbersome exact equation for diffusion from a dissolved source of finite radius is indistinguishable from the much simpler expression for a point source (starting with the mathematical fiction of an infinite concentration in zero volume). The point source equations, for diffusion without limit, are

$$C_r = \tfrac{1}{8}\mathbf{M}(\pi Dt)^{-\frac{3}{2}} \exp(-r^2/4Dt) \tag{3}$$

$$\mathbf{M}_r^\infty = \mathbf{M}\left(\operatorname{erfc}(r/2\sqrt{Dt}) - \frac{r}{\sqrt{\pi Dt}} \exp\left(-\frac{r^2}{4Dt}\right)\right) \tag{4}$$

where $\mathbf{M}$ is the initial content of source, C_r is the concentration at radius r and time t and $\mathbf{M}_r^\infty$ is the amount which has escaped beyond r in this time. The rapidly decreasing concentration at the centre is given by eq. (3) with $r = 0$, so that

$$C_0 = \tfrac{1}{8}\mathbf{M}(\pi Dt)^{-\frac{3}{2}} \tag{5}$$

For the saturating source and diffusion without limit we have the exact equation

$$C_r = C_{\text{sat}} \times \frac{a}{r} \operatorname{erfc}\left(\frac{r-a}{2\sqrt{Dt}}\right) \tag{6}$$

in which, with sufficient accuracy, we may replace $r-a$ by r. From this it follows that the total amount dissolved in time t is

$$\mathbf{M}_a^\infty = C_{\text{sat}}(4a^2\sqrt{4\pi Dt} + a\,.\,4\pi Dt) \tag{7}$$

Evaluation of the corresponding expression for amount escaped beyond r shows that, for times and distances of practical interest, we may simplify to

$$\mathbf{M}_r^\infty = \mathbf{M}_a^\infty[4i^2 \operatorname{erfc}(r/2\sqrt{Dt}) + (2r/\sqrt{Dt}) \operatorname{ierfc}(r/2\sqrt{Dt})] \tag{8}$$

for which the integral error functions can be obtained from statistical tables.

We may note at this stage that, while the size of a dissolved source is unimportant, the size of a saturating source has direct control of the rate of release (eq. (7)) but no effect on the relative distribution outside the source of what has been released.

The effect of neighbouring sources is equivalent to that of an impervious barrier formed around each source passing through the mid points between it and its neighbours. Without significant error this dodecahedral surface can be replaced by a spherical one of equal volume, as calculated in the previous section. At the earliest stage of overlap, when the concentrations at the distance of the barrier but in its absence would be very small, according to eqs (3) and (6), the effect is to double the concentration at the barrier, the extra concentration coming from the neighbour but represented as a reflection. In linear diffusion, the reflection method can be applied indefinitely but, at a spherical surface it soon becomes inapplicable. Also, overlap becomes important very early in the spherical case because the rapid increase of volume with radius makes the *amount* escaping disproportionately great.

For the 800 μm diameter buried granules of pure a.i. (Table 1, bottom left), i.e. $a = 0{\cdot}04$ cm, $b = 3{\cdot}27$ cm, we have plotted, in Fig. 6, concentration *vs* r for $\sqrt{Dt} = 2$ cm from eqs (3) and (6)—i.e. for isolated granules in a limitless environment but with the value of $\mathbf{M}$ for the dissolved source chosen to be equal to the amount of the saturating solute which would be contained in the 3·27 cm radius sphere. The eventual state for both solutes in the restricted volume is therefore the same, represented by the rectangle. It is at once evident that, for the time illustrated, the other sources, when present, must have had a major effect even on the central concentration in the dissolved case but that the amount of adjustment needed is, overall, greater in the saturating case. To illustrate better the effect of increasing volume the data are replotted in Fig. 7 as Cr^2 *vs* r so

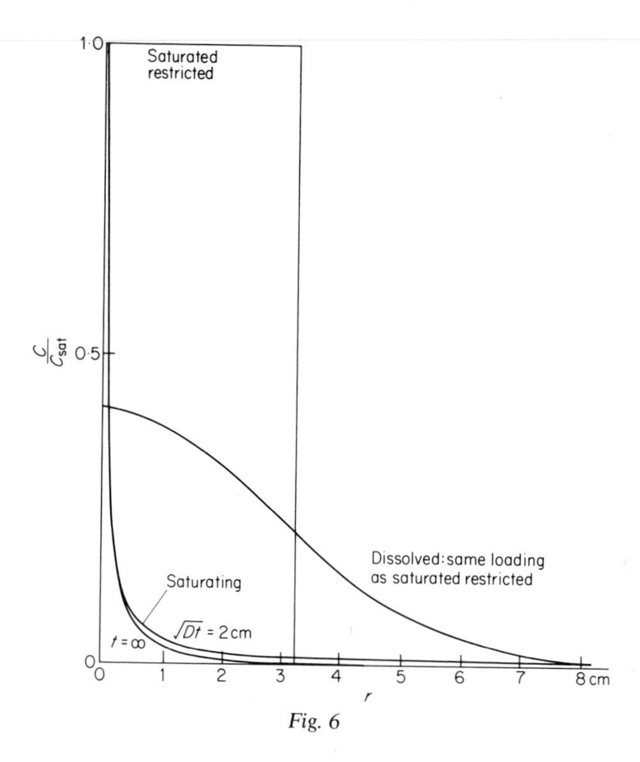

Fig. 6

that the area under the curves represents amount. We have plotted only the $t=\infty$ version for the saturating case but, for the dissolved case, we have included curves for $\sqrt{Dt}=1$ and $0{\cdot}5$ as well as that for 2 shown in Fig. 6. These are all for the case in which there is no restriction. The common curve for $t=\infty$ in the restricting shell at $r=3{\cdot}27$ is a parabolic rise to this limit. The sharp drop at $r=3{\cdot}27$ is a consequence of the barrier: to be more realistic we should plot a mirror-image curve descending to the next source at $r=6{\cdot}54$ cm and, to allow for the approximation of representing the symmetrical (but not regular) dodecahedron as a sphere, we should then round off the sharp peak.

Exact treatments of the diffusive filling of a sphere from an internal source do not seem to be available. Approximate estimates can be made on the basis that the approach to the final state tends to an exponential form in the later stages. In this crude model crescent state diffusion in the whole volume is replaced by pseudo steady-state diffusion from a uniform concentration in one part (saturated or decreasing) to an increasing uniform concentration in the other through half the thickness or radial distance. For the dissolved source the hypothetical boundary between the

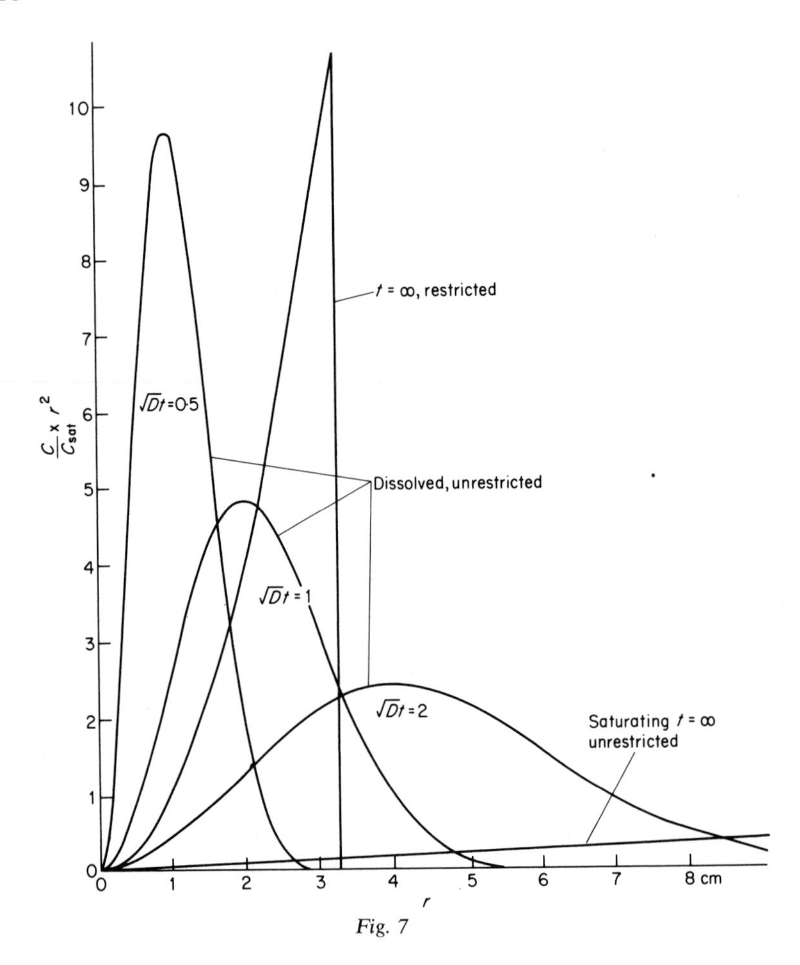

Fig. 7

volumes is at the half distance. For the saturating source the real boundary is at the saturating surface.

From a dissolved source, initially in very small volume, the condition of 90% homogeneity, i.e. maximum concentration difference = 10% of final uniform concentration, will be reached in about $0{\cdot}45b^2/D$. The corresponding numerical factors for 80 or 95% will be about 0·08 less or more than for 90%. The $0{\cdot}45b^2/D$ compares with $0{\cdot}37X^2/D$ for a parallel sided block of thickness X, a dissolved source being placed on one face. For a cylinder approached from the centre, a model of lateral spread from columns of leachate descending under rain, the result is therefore likely to be of the same order. One can assume that $0{\cdot}4b^2/D$ will give the 90%

time with a confidence of ±50% where b is the mean half-distance between neighbours.

For the saturating source the time for 90% homogeneity for the block, saturated on one face, is $1{\cdot}04X^2/D$ but that for the granule is $0{\cdot}7b^3/Da$, dependent on radius of the source as well as on distance apart. The form of dependence of time on a and b is more certain than the numerical factor, provided the solubility is so low that a does not change significantly while the volume within b is being filled. Two further comments must be made. If the incorporation depth is kept the same and also the total volume applied is the same, b is proportional to a and therefore the times for dissolved and saturating granules, both proportional to a^2. Since $a \ll b$ in any practical case the time to reach homogeneity is much greater in the saturating than in the dissolved case.

The importance, for the saturating case, of the small radius a has significant implications in relation to the influence of the nature of the granule surface and its discontinuous contact with soil water. This matter is taken up later (Section 14.VII.H). Even assuming the granule and soil to be homogeneous and continuous, allowance must generally be made for decrease of size of the dissolving granule. At the simplest, during the period when the speed of accommodation in the region of important resistance is much greater than the speed of retreat, we can equate (for pure a.i.) the rate of decrease of mass of the granule and the rate of escape from it by diffusion into a limitless medium (eq. (7) Section 3.II)

$$4\pi a^2 \rho\left(-\frac{\mathrm{d}a}{\mathrm{d}t}\right) = 4\pi a D C_{\mathrm{sat}} \tag{9}$$

or

$$-\frac{\mathrm{d}a^2}{\mathrm{d}t} = 2DC_{\mathrm{sat}}/\rho$$

The extension of this argument to the behaviour of a granule inside a restricted sphere of influence would be very complex. It is evident, however, that decrease of a will lead to times based on constant a being underestimates. Once there is no longer excess a.i. at the site of the granule, however, the time to achieve a given degree of homogeneity will be described by the dissolved case equations and indeed will be overestimated because some of the process has already occurred. The full time for the dissolved case is much shorter than the constant time for the saturated case so that disappearance of the granule by solution appears to accelerate diffusion. The explanation of the apparent contradiction lies in a change in the nature of the final states. When the granule content is more than enough to saturate the available environment we measure completeness by the attainment everywhere of a concentration of more than, say, $0{\cdot}9C_{sat}$. If we have a much smaller granule containing only enough a.i. to bring the whole volume to, say, $0{\cdot}1C_{sat}$, the initial slower rate of supply will be faster relative to the final ceiling of one-tenth the level demanded of the larger granule. That the concentrations C_a and C_b close up much more rapidly after the supply is exhausted is due to the major contribution being rapid fall of C_a from C_{sat} to near $0{\cdot}1C_{sat}$.

To enable the reader to appreciate the significance of the various predicted times in the practical range, a matter of major importance when formulations for the control of rate are considered, we list in Table 2 estimates corresponding to the situations of Table 1. In addition to the "90% complete" estimates, $Dt = 0{\cdot}7b^3/a$ and $Dt = 0{\cdot}4b^2$, we have included values of Dt for $C_b/C_a = 0{\cdot}2\%$ (i.e. 0·1% in eqs (5) and (6) remembering that neighbouring granules will double the concentration when this is very small) and also the times for complete solution of isolated granules.

D ranges, in fertile soil, from about $10^{-8}\ cm^2\ s^{-1}$ to $10^{-2}\ cm^2\ s^{-1}$ (p.

Table 2
Applicance 2 kg a.i. ha^{-1} incorporated to depth 10 cm. Final concentration if all dissolved/adsorbed in this zone 2×10^{-6} g cm^{-3}

Diameter of source (μm)		800	400	200	100	20
ρ = g.a.i. cm^{-3} in source	code	Product Dt in cm^2				
0·11	Sat. 0·2%	0·4	0·09	0·02	0·0045	0·00015
	Sat. 90%	60	16	4	1	0·04
	Soln. 10^{-5}	9	2	0·5	0·1	0·005
1·1	Sat. 0·2%	2·2	0·46	0·10	0·02	0·0008
	Sat. 90%	600	160	40	10	0·4
	Soln. 10^{-5}	88	22	5	1	0·05
0·11	Dis. 0·2%	0·08	0·02	0·005	0·001	0·00005
	Dis. 90%	1·3	0·3	0·08	0·02	0·0007
1·1	Dis. 0·2%	0·39	0·10	0·02	0·005	0·0002
	Dis. 90%	6	1·5	0·4	0·1	0·004

Sat. = saturating source of constant size (i.e. solubility $\ll 2 \times 10^{-6}$) Dis. = wholly dissolved source. Soln. 10^{-5} = complete solution of a.i. if solubility 10^{-5} g cm^{-3} and no other sources present 0·2% or 90% is the ratio of lowest to highest concentration at the estimated time.

264) according to moisture status, volatility and adsorption. C_{sat} refers to the overall concentration, g.a.i. dissolved, evaporated and adsorbed, per cm^3 of whole soil. At the highest D value 800 μm diameter sources would need only about 1 day to bring the zone to 90% homogeneity even if saturating (an unlikely condition for a very volatile chemical). If the a.i. were dissolved at the source 1 h would suffice. At the lowest D value even 20 μm diameter particles of pure pesticide, if distributed uniformly (but of course in practice they would be clustered in drop residues), would take 3 days to bring about 90% homogeneity, while for 800 μm sources it would require 11 years for a significant concentration of diffusing pesticide to be reached at mid-points. Taking a moderate D value of 10^{-7} cm^2 s^{-1} for an effectively involatile, extensively adsorbed pesticide, the predicted time for 90% homogeneity from 800 μm granules of pure pesticide persisting as such is about 240 years. C_{sat} (dissolved/adsorbed) for this condition must be well below 10^{-6} g cm^{-3}. For the concentration we have exemplified, $C_{sat} = 10^{-5}$, only 1/20th saturating the 10 cm zone eventually, the granules will be effectively isolated while dissolving and this process will take about 29 years. The time for 90% homogeneity if the solubility were not limiting would be about 1 year, adding very little to the time of solution at $C_{sat} = 10^{-5}$. If the solubility were 100 times as great, the 1 year would add on to 0·29 years, a very different matter.

The time of solution is inversely proportional to the solubility and, for a given a.i. it can be reduced only by decreasing the granule content. The most economical way to do this is to decrease the granule size. Greater acceleration could be obtained by reducing the content and maintaining the size if the a.i. is situated on the outside of a non-porous core, but this involves increasing the total mass applied. It will be seen that there may be need to delay release from buried granules only if the a.i. is very soluble. The effect of leaching is considered in a later section.

E. Effect of Chemical Decay

Chemical decay will not only reduce the mean availance resulting from any given applicance but will accentuate the non-uniformity arising from application at discrete sources throughout the life of the chemical. These effects are simply analysed if the decay can be assumed first order. The steady-state analogy (Section 3.V.C) enables us to calculate the ultimate availance ($\int_0^\infty C\,dt$) which builds up in the soil surrounding the source. It bears to the released dose the same relationship as does the steady state concentration to the steady rate of supply at the source which is necessary to maintain it.

If the a.i. could be applied homogeneously throughout the volume of soil treated, the steady supply necessary to maintain a concentration C is necessarily equal to the rate of decay, $kC\mathbf{V}$. Therefore the uniform availance is $1/k \times$ amount per unit volume (= applicance/depth)

$$\bar{I} = \mathcal{A}/Xk \tag{10}$$

The relevant steady state around a source in a uniform distribution of sources is that obtaining between the surface of the source, assumed to provide a steady supply, $\mathbf{F}_a$, and the boundary of the effective sphere of influence, radius b_X (we shall omit the suffix in the equations below), where $\mathrm{d}C/\mathrm{d}r = 0$ because there is no net flux. The general equation applicable to the annular space is

$$r\frac{\partial C}{\partial t} = D\frac{\partial^2(Cr)}{\partial r^2} - kCr \tag{11}$$

At the steady state $\partial C/\partial t = 0$ and the solution is

$$Cr = A\mathrm{e}^{\alpha r} + B\mathrm{e}^{-\alpha r} \tag{12}$$

where $\alpha^2 = k/D$ and A and B are constants determined by the boundary conditions already specified and given by:

$$A = (C_b/2\alpha)(1 + \alpha b)\mathrm{e}^{-\alpha b} \tag{13}$$

$$B = -(C_b/2\alpha)(1 - \alpha b)\mathrm{e}^{\alpha b} \tag{14}$$

The flux at $r = a$ and the concentration at r are obtained from these equations. The volume supplied is $\frac{4}{3}\pi(b^3 - a^3)$. We find that the ratio of availance at r to availance if uniform is

$$\frac{I_r}{\bar{I}} = \frac{b^3 - a^3}{3} \cdot \frac{\alpha^2}{r} \cdot \frac{(1 + \alpha b)\mathrm{e}^{-\alpha(b-r)} - (1 - \alpha b)\mathrm{e}^{\alpha(b-r)}}{(1 + \alpha b)(1 - \alpha a)\mathrm{e}^{-\alpha(b-a)} - (1 - \alpha b)(1 + \alpha a)\mathrm{e}^{\alpha(b-a)}} \tag{15}$$

This fraction is a function of $\alpha^2 = k/D$ and the geometry. It does not depend on solubility, but, of course, the time scale of the $C-t$ plot does. The availance ratio from (15) sets the nearest approach to homogeneity that α^2 and the geometry permit. This may prevent the concentration in some parts of the regions between sources ever reaching a level sufficient to kill a small organism while, close to the source, there may be a substantial "overkill" potential. This might not be undesirable. If the source contained an attractant for the target species to which beneficial species were indifferent, the non-uniformity could be exploited to give better control.

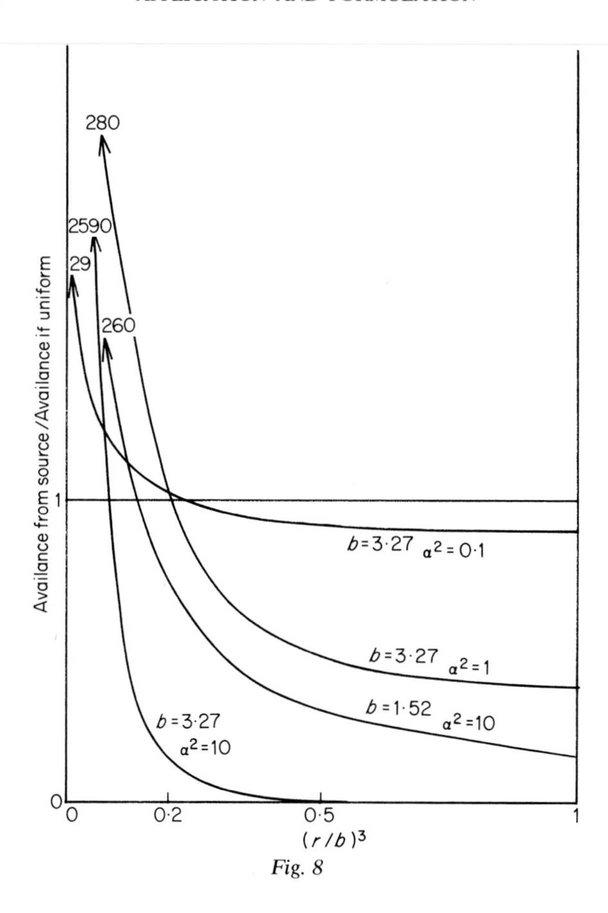

Fig. 8

The function in eq. (15) is plotted in Fig. 8 for three values of α^2, 0·1, 1·0 and 10 cm^{-1} and for $b = 3{\cdot}27$ cm (see Table 1). Source diameters of 800 μm are assumed but this has little influence on the availance as long as $\alpha a \ll 1$. It is 0·4 for the highest value of α assumed. The availance ratio is here plotted against $(r/b)^3$ instead of its product by $(r/b)^2$ against r/b because we are now interested mainly in values in the outer part of a limited volume. The areas below in both cases measure volume. The limiting values of the function at the surface of the source, which would be inseparable from the ordinate line on the graph, are indicated near the arrows. In interpreting the significance of these plots it should be remembered that α^2 is proportional to k at any one value of D and that the availance if uniform would be inversely proportional to k. The absolute availance values (for the same applicance) are little altered by k at the source surface and the values in the regions remote from the source have differences much greater than appears in the plot of the ratio. For example, the availance, if uniform, for $\alpha^2 = 0{\cdot}1$ would be $0{\cdot}01$ times that at $\alpha^2 = 10$. The non-uniformity effect brings this down at $(r/b)^3 = 0{\cdot}4$ to 1/4700. The mean availance over the outer half of the volume at $\alpha^2 = 10$ is about 1/20 000 of that at $\alpha^2 = 0{\cdot}1$.

We have included in the graph the function for $b = 1{\cdot}52$ cm and $\alpha^2 = 10\ \text{cm}^{-2}$ to illustrate the effect of increasing the population density of sources 10-fold. No change of availance if uniform is involved but there is a *c.* 40-fold increase in availance in the least contaminated half of the available volume.

To put these predictions in perspective we note that $k = 10^{-6}\ \text{s}^{-1}$ corresponds to a half-life of about 8 days and, combined with $D = 10^{-7}\ \text{cm}^2\ \text{s}^{-1}$ gives $\alpha^2 = 10\ \text{cm}^{-2}$. To realize $\alpha^2 = 0{\cdot}1\ \text{cm}^{-2}$ the half-life must be *c.* 800 days or D be $10^{-5}\ \text{cm}^2\ \text{s}^{-1}$, the latter possible only with considerable vapour-phase contribution. The half-lives chosen represent approximately the limits of the too ephemeral and too persistent. Over this range there will be profound differences in non-uniformity of availance if large granules are employed at a reasonable applicance. The microgranule range would result in fairly uniform availance unless the a.i. were very short-lived.

The availance calculated here is the full integral $\int_0^\infty C\,dt$. It is only this function which is readily calculated. Generally an effect is required over a time of exposure during which only a small fraction of this availance is created. An availance made up of a very low concentration over a very long period could be quite useless. The calculated values, however, are the highest possible at the distribution and applicance specified: improved distribution of smaller granules will usually be much more effective and economic than increased applicance. Comparison of the times necessary for distribution in the absence of decay (Table 2) will give some idea of the duration of a pulse in the region between sources. The predictions of course are interrelated. The time for 90% homogeneity from dissolved sources $3{\cdot}27$ cm apart (mean radius, b_X) is found from Table 2 to be $4{\cdot}8/D$ s. This is about 550 days for $D = 10^{-7}\ \text{cm}^2\ \text{s}^{-1}$. If the decay sets a half life of 8 days very little pesticide will move undecomposed far from the source, consistent with the availance being $<10\%$ of the mean over 80% of the volume ($<1\%$ over 50% of the volume). Where a much longer time is estimated because of very low solubility, the effect of decay will be to hold the concentration between sources at a very small fraction of the concentration it would attain if the a.i. were stable.

F. Leaching from Discrete Sources

We have so far considered release by diffusion only and thus have been concerned only with an application incorporated into a depth of soil where retained moisture usually permits diffusion in the water phase as well as the soil air. A surface application leaves sources generally dry. If vapour diffusion into the soil is important, so also is loss into the atmosphere, considered in a later section. Downward release of involatile pesticide depends almost entirely on periodic rain or overhead irrigation.

Furmidge *et al.* (1966) assumed it to be entirely so and investigated the leaching from granules lying separately on a fine wire gauze under artificial rain. They expected proportionality between initial rate of leaching and the fraction of horizontal area covered by granules (i.e. $\pi a^2 n$ where n is the population density). If this fraction of the rainfall were fully saturated with the test chemical the initial rates of leaching would be about one-third of the lowest observed. That the "rain" drops were probably larger than the separated granules (mean diameter in most experiments *c.* 700 μm) would ensure a contact area much greater than the actual projection area. An individual drop contacting a smaller granule lying on a fine gauze would remain in enveloping contact until other drops increased its size to the fall-through condition. The conditions of experiment make it unlikely that this wet-waiting period would be more than a few seconds and saturation by diffusion of a volume of water three times that of the granule would not occur in so short a time. It is probable that removal of microcrystals of a.i. under the impact of rain on the granules, which were prepared by agglomeration, was responsible for greater loss than simple contact and solubility theory predicts: certainly this explanation was put forward as the only possible one in cases where a much less soluble compound was leached more rapidly than a more soluble one. Tests were made of "wet strength" of granules by agitation in water and re-sieving, and there was, for a given a.i., a correlation between tendency to disintegrate and rate of leaching when different fillers and binders were compared. These tests would not record attrition of fine particles from the surface only. In a later paper (1968) these authors examine further the influence of different fillers and binders, the latter having much more influence.

These experiments were designed to represent conditions on a soil surface under rain. Disintegration and attrition, as well as diffusion in solution, operate in both cases. Granules incorporated below the soil surface will be largely protected from mechanical impact of rain and the flow past the surface of a buried granule will be slow and steady during infiltration. There seems to be neither experimental data nor sound theory of such leaching. One might expect that, subject to qualification below, the granule could saturate a column of its own diameter so that rate of solution would be $=\pi a^2 v$, v being the mean downward water velocity. There must, however, always be a dependence on diffusion coefficient since no removal by solution could occur without diffusion. In an investigation into the solution of beads of hydroxy benzoic acids in flowing water, Mullin and Cook (1965) obtained results which fitted well an equation in which $\pi a^2 v C$ is multiplied by a factor $3{\cdot}2(D/\nu)^{\frac{2}{3}}$, ν being the kinematic viscosity. For free water this factor is about 0·02. While the extension of this equation to water percolating slowly in a porous bed would not be permissible, the work suggests that $\pi a^2 v C$ is probably a gross overestimate.

In non-turbulent percolating water and in the absence of air space (see Section 14.VII.H), there is no reason to expect dependence of rate on any parameters other than solubility, radius, diffusion coefficient and velocity in any given structure of porous material. Obviously, solubility must appear as the first power so that the combined dimensions of the others must be L^3T^{-1}. There is no unique choice but, if velocity appears as the first or higher power, diffusion coefficient must appear to zero or negative power. The most rational choice is $a^{\frac{3}{2}}D^{\frac{1}{2}}v^{\frac{1}{2}}$. If we replaced uniform movement with velocity v by a series of jumps through distance δ with intervals during which eq. (7) would apply we should find a mean rate of solution

$$-\frac{\mathrm{d}\mathbf{M}}{\mathrm{d}t}=C_s(8\sqrt{\pi}a^2(Dv)^{\frac{1}{2}}\delta^{-\frac{1}{2}}+4\pi aD) \tag{16}$$

The second term corresponds with the second term of the rate of diffusive loss into a static environment. To be at all realistic, δ must be of the order of the source diameter, $2a$. If it were much shorter each period of static diffusion would be reduced by the residual concentration from the last. Putting $\delta=2a$ we find

$$-\frac{\mathrm{d}\mathbf{M}}{\mathrm{d}t}=C_s(4\sqrt{2\pi}a^{3/2}(Dv)^{\frac{1}{2}}+4\pi aD) \tag{17}$$

For the largest of our tabulated sources, $a=0{\cdot}04$ cm and taking $D=10^{-7}$ and $v=10^{-4}$ cm s^{-1} (corresponding to mean linear flow in soil of 20% v/v water capacity under 1·7 cm rain in 24 h) the two terms in the bracket are $2{\cdot}5\times10^{-7}$ and $0{\cdot}5\times10^{-7}$ cm^3 s^{-1} respectively while $\pi a^2 v$ is 5×10^{-7}. While the numerical value for the Dv term is dubious there is no doubt about the dimensional argument. The relative excess of the flow and diffusion term over the static diffusion term is less the smaller the source and its relative deficit below the simple flow term is smaller the larger the source. Its value increases as $\sqrt{v}$ and therefore real intermittent flow, as exemplified by the $\delta^{-\frac{1}{2}}$ factor in eq. (16) will leach less solute than steady flow of the same total volume.

This last effect would make leaching from a surface deposit less effective under intermittent rain than leaching from buried sources where the flow is less discontinuous. Directly opposing this is the greater vulnerability of surface sources to disintegration and attrition under the mechanical impact of rain drops.

G. Accelerated Solution of Active Ingredient

We have so far assumed that the rate of solution of a.i. from a granule is determined only by the saturation concentration (solubility) of the a.i. in soil water plus air and the combined external diffusion coefficient. We consider later deliberate retardation of this rate by composition of the granule matrix. It is sometimes asserted, in the case of drugs, particularly aspirin, that the rate of solution even of the pure substance is determined by additional factors but a study of the literature suggests that the assertion is based on inadequate definition.

An essential requirement in experiments to relate rate of solution to physical properties of the solute, is to standardize the geometry and dynamics of the system. Usually a standard area is exposed to a solvent stirred in a standard manner. Preferably the area exposed should be part of a plane face of a perfect crystal, but not many substances can be obtained in this form. A highly compressed tablet of standard shape is a more practical alternative, but the surface should not disintegrate, releasing fragments which greatly extend the area of solid–liquid contact as was shown by Furmidge *et al.* (1966) to occur with some pesticide granules.

Working with single crystals of two forms of aspirin, Tawashi (1968) found that the form obtained by slow crystallization from hexane solution, having a m.p. of 124°C, dissolved 50% faster in water under comparable conditions than the normal stable form, m.p. 143°C, obtained by crystallization from ethanol solution. This is entirely to be expected from the greater equilibrium solubility of the metastable form. Recrystallization as the stable form during the solution process would not alter the rate since the new nuclei would form only in the super-saturated diffusion layer from which they would be swept into the solvent. The result supports the classical observations (Noyes and Whitney, 1897) that rate of solution depends only on solubility, diffusion coefficient and dynamics in the surface layer.

Hamlin *et al.* (1965) found that the rates of solution in water under standard conditions of 50 drugs formed into tablets under very high pressure were proportional to solubility, even over the wide range of 10^5 covered. The proportionality was within a factor of 2, mostly much closer. Calvet *et al.* (1975) measured the rate of solution of three *s*-triazine herbicides by the compressed tablet technique. The amounts dissolved after standard times were closely proportional to solubility for atrazine and propazine, the former six times as soluble, but simazine, slightly more soluble than propazine, appeared to dissolve about 30% less rapidly after the first 100 minutes. These authors considered that the solution of simazine followed a second order equation with regard to the external concentration. We can envisage no mechanism for such an improbable relationship and it seems more likely that surface disintegration contributed in all cases but less with simazine after an initial period.

Water-soluble binding agents such as gelatine or a non-crystalline sugar are incorporated in many pharmaceutical formulations intended for ingestion. Solution of the binding agent facilitates disintegration of the granule which in turn results in accelerated solution of the dispersed contents. Without a soluble binder, compressed tablets will dissolve fastest when made up from particles of a.i. in an optimum size range since adhesion between particles becomes more effective with decreasing size but the rate of solution of individual particles when released decreases with increasing size. Acid-base reactions can also be exploited to accelerate the overall solution process where the chemistry permits, as in dry-mixing of magnesium or calcium carbonate or magnesium oxide with the acidic aspirin. These complications in the solution process, which still leave the rate-determining step external to the solid (unlike the exceptional behaviour of anhydrous lithium salts, see p. 187), are described by Finholt *et al.* (1966).

Disintegration of the granule, as in some of the compositions examined by Furmidge *et al.* (1966), can certainly render the contents of granules on the soil surface more rapidly and widely available under the influence of rain, since the impact of rain drops mechanically disperses the granule fabric which has been weakened by wetting.

There is clearly less scope for mechanical dispersal of buried granules even if they are greatly weakened by wetting, but, in an open-structured soil, fine particles released from granules could be carried downwards by percolating water. The extent to which this could occur must depend on the structure and compression of the soil and also on the adhesion of the particles of a.i. to the surfaces of soil crumbs. Such movement would directly accelerate downward transport of dissolved a.i. and would indirectly assist lateral diffusion also because the three-dimensional pattern of "point" sources would tend towards a two-dimensional pattern of vertical line sources. Not only would the lines lie closer together (see Table 1, Section 14.VII.C) but the diffusion from them would be less divergent.

We have, in the above discussion, assumed the a.i. to be a powdered solid. Pre-formed porous granules can also serve to carry liquids. Another mechanism of rapid release then becomes possible—water may displace the active liquid from the pores if it wets the walls preferentially. Graham-Bryce *et al.* (1972) found that disulfoton in granular pumice was released much more rapidly from a single layer of granules on gauze when subjected to short showers of artificial rain than could be accounted for by solubility. The first 150 ml of water took with it about 70 mg of a.i. from 4 g of granules containing initially 300 mg. Only 2·3 mg could have been dissolved. Disintegration of nearly one-quarter of the granule bulk clearly did not occur. Moreover, phorate absorbed in pumice and both insecticides absorbed in fuller's earth were released in amount corresponding only to saturated solution.

This initial rapid surge flattened off when only about one-third of the a.i. was released. Such behaviour is common with immiscible liquid displacement, which has great economic importance in extraction of oil from porous strata. The water advances more rapidly in the larger pores and oil by-passed in narrower pores cannot then escape. Displacement will only occur when the water can make a finite contact angle on a surface previously wetted by oil. It is therefore a rather critical process. Displacement may not occur at all. If it does occur, some, but not all, the oil is easily displaced. Surfactants present in the formulation could influence the behaviour, as the authors suggest, and in this case their removal with the first leachate could hasten the arrest of the displacement. Evaporation rate from *dry* pumice granules was greater than from dry fuller's earth granules for *both* insecticides. Whether a liquid a.i. of low water solubility displaced from the granules becomes more available to the pest depends on physical interaction of the liquid with moist soil solids as well as on the processes of uptake. In this investigation, although the pumice formulations gave a higher aphis mortality when applied to field beans than fuller's earth formulations, the difference was relatively smaller than that between the rates of release from granules (in the absence of soil).

Surfactants can complicate contact of soil-water with the granule surface where water has a lower surface tension, as considered in the next section. There is evidence that there may be some effect of surfactants even when a granule is completely enveloped by water. This could be due to "solubilization" (Section 2.IX.C). If the increased solubility is allowed for, however, the rate of solution may appear anomalously low because the extra solute, present in surfactant micelles, diffuses more slowly than that dissolved in the water itself. In pharmaceutical products when rapid disintegration is required, surfactants may accelerate this process. The kinetics of solution and diffusion in micelles has been reviewed by Mysels (1969) and Chan *et al.* (1976). The significance for solution of cholesterol gallstones is examined by Sehlin *et al.* (1975) and Carey and Small (1978). "Solubilization" is unlikely to be important in application to soil. When all concentrations are low, the surfactant carries less than its own volume of oily solute into solution. In soil it will itself be adsorbed on organic matter. Although it would be possible to demonstrate effects in soil, quite uneconomic amounts of surfactants would normally be needed.

H. Granule–Soil-water Contact

The radius of a saturating source is a primary determining factor in the rate of release. This radius may be comparable with the dimensions of adjacent soil crumbs and particularly with the interstitial concave-sided water volumes. While the influence of soil structural details on the diffusion coefficient can be treated as a combined effect when we are concerned with diffusion over distances much greater than the structural dimensions, detail of contact cannot be ignored when all the released pesticide must transfer across contacts between a comparatively small number of foreign bodies and the neighbouring native particles. Specific contact effects are not to be expected where vapour diffusion is the chief means of transfer. Even with rather involatile materials, vapour diffusion could reduce specific contact effects by providing the means of transfer from a granule surface over a distance of a few μm into what can then be considered "bulk" soil.

However, when volatility is so low that vapour transfer is negligible, the detail of surface contact could be critical. Where a smooth unwettable surface makes contact with points and edges of wettable particles there will be no aqueous contact until the soil is flooded. Diffusive transfer from the smooth surface to the concave water units is likely to be much more, and critically, dependent on moisture content than is diffusion in the bulk of the soil itself. Spherical wax-based granules could in this respect behave very differently from irregular mineral or vegetable particles on to the outside of which the a.i. is fixed by some hydrophilic adhesive. The dependence on moisture content and therefore on flow rate could be even greater in leaching than in static diffusion. This subject merits much more experimental investigation. The technique used by Collier *et al.* (1978) to charge granules with a radioactive a.i., dry the soil after a period of moist diffusion, fix it with a setting resin and then take sections and radioautographs would provide a direct means of examining detail of non-uniform diffusion but would be very laborious.

Some unpublished observations by Hartley indicate the reality of contact effects. Solidified drops of molten ^{14}C stearic acid were used as model granules. These were buried in sand and leached with solvent. To examine the effect of solubility, the solvent was varied, using mixtures of water, methanol and ethanol. Results were highly erratic and the granules, on exhumation, were found to be of very irregular stellate shape and frequently fragmented. This did not happen when they were partly dissolved in agitated, free solvent. Around the stearic acid granules surface tension changes would promote local stirring effects; these can also occur or be deliberately introduced in real granules. Surface behaviour is worth more study. (See also Sections 8.V.C, 14.VIII.D.)

J. Internal and External Delay

The desirability and functioning of slow release formulations are more important in other applications than to soil and are therefore considered in a separate section (Section 14.IX). There is always some delay in dispersion in the immediate environment of the special source and this is particularly important in soils. We deal here therefore with the relationship of internal and external delay which has some, though less, relevance to dispersal into free air or water. The simplest type of slow release granule is one in which the solid a.i. is incorporated into some porous or molecularly permeable matrix which remains *in situ* offering high resistance to escape while the core of the a.i. is slowly retreating. At time t, let the undissolved core have retreated from the original radius, a_0 (which is still the radius of the restrictive matrix) to the value a. We will assume steady-state radial diffusion to operate and no significant interference by neighbour granules over the times considered. Let the concentration at equilibrium within the granule matrix be $\mathcal{P}$ times the overall concentration in the soil and let the apparent diffusion coefficient (p. 122) within the granule be D_g and D_a that in the soil. Retaining C_{sat} to denote the saturation concentration in the soil, that at the surface of the retreating core is $\mathcal{P}C_{sat}$. Diffusion occurs with coefficient D_a from radius a to radius a_0 at which the concentration jumps from $\mathcal{P}C_1$ just within the matrix to C_1 just within the soil. Equating the diffusive flux through the annulus to the outward flux in the soil and to the rate of decrease of core volume $\times\rho$, we have

$$\rho \,.\, 4\pi a^2\left(-\frac{\mathrm{d}a}{\mathrm{d}t}\right) = \mathcal{P}(C_{\mathrm{sat}} - C_a) \,.\, 4\pi D_{\mathrm{g}}\left(\frac{1}{a} - \frac{1}{a_0}\right)^{-1} = C_1 \,.\, 4\pi D_{\mathrm{a}} \,.\, a_0 \tag{18}$$

Eliminating C_1 and integrating from initial radius a_0 (at $t = 0$) to the value a_1 at t, we obtain

$$\frac{C_{\mathrm{sat}} D_{\mathrm{a}} t}{\rho} = \frac{a_0^{\,3} - a_1^{\,3}}{3a_0}\left(1 - \frac{D_{\mathrm{a}}}{D_{\mathrm{g}}\mathcal{P}}\right) + \frac{(a_0^{\,2} - a_1^{\,2})}{2D_{\mathrm{g}}\mathcal{P}}\, d_{\mathrm{a}} \tag{19}$$

Putting $a_1^{\,3}/a_0^{\,3} = f$, so that f is the fraction of solid a.i. core still not dissolved, we obtain

$$\frac{C_{\mathrm{sat}} D_{\mathrm{a}} t}{\rho a_0^{\,2}} = \left(\frac{1}{6} + \frac{f}{3} - \frac{f^{\frac{2}{3}}}{2}\right)\frac{D_{\mathrm{a}}}{D_{\mathrm{g}}\mathcal{P}} + \frac{1 - f}{3} \tag{20}$$

If $f = 0$, t becomes the time for complete solution and the R.H.S. of eq. (20) simplifies to $\frac{1}{3} + D_{\mathrm{a}}/6D_{\mathrm{g}}\mathcal{P}$. It will be noted that $D_{\mathrm{g}}\mathcal{P}$ appears throughout as the product. $D_{\mathrm{g}}\mathcal{P}$ is the permeability (Section 9.I.B) of the matrix. The time t in these equations applies to the retreat of the solid core of a.i. from a_0 to a and is thus only applicable where the a.i. *dissolved within* the granule is negligible. If the a.i. were initially wholly dissolved within the granule matrix, increase of $\mathcal{P}$ would increase the time since the interface soil concentration would not be saturated and C_{sat} would not occur in the relationship to be increased by $\mathcal{P}$. The a.i. would therefore tend to be more retained within the matrix.

Continuing to assume that only a negligible proportion of the a.i. in the granule is not solid we may obtain from eq. (20) an expression for time of release when the processes inside the granule are so fast that only external resistance matters. This is

$$\frac{C_{\mathrm{sat}} D_{\mathrm{a}} t}{\rho a_0^{\,2}} = \frac{1 - f}{3} \tag{21}$$

which could be obtained from the equation for steady-state radial diffusion where the environment is kept at the saturation concentration at the surface, a_0. The granule behaves, in the limit, as though its a.i. were wholly on the surface. Correspondingly, we can obtain from eq. (20) the time of release into violently agitated water by the mathematical device of letting D_{a} become very large while keeping $D_{\mathrm{g}} \times \mathcal{P}$ constant, or from the basic equation by putting $C=0$ at a_0, giving

$$\frac{C_{sat}D_a t}{\rho a_0^2}=\left(\frac{1}{6}+\frac{f}{3}-\frac{f^{\frac{2}{3}}}{2}\right)\frac{D_a}{D_g\mathscr{P}} \tag{22}$$

It will be seen that the times are additive, i.e. the time for a given fraction of the a.i. to be dissolved from an initially uniform granule into a stagnant environment is the sum of the times for the same fraction to be dissolved into water and for the same fraction of an external coating to be dissolved into the stagnant environment.

It is important to appreciate this additive relationship because rate of solution into water is easily and quickly measured in laboratory tests, and it is sometimes assumed that if one formulation takes, say five times as long to release its contents to water as another, it is going to provide significant delay in the field; 50% loss from either formulation might take 30 days by diffusion into soil: whether the same loss into water takes $\frac{1}{2}$ h or $2\frac{1}{2}$ h is unimportant. If the formulation is to delay release significantly into soil, the time for release into water must be increased to many days. Otherwise expressed, $D_g\mathscr{P}/D_a$ in eqs (20) and (22) must be considerably less than 1 for the internal solution time (eq. (22)) to be significant compared to the external solution time (eq. (21)). The larger f the greater the influence of internal resistance. To illustrate these points we have plotted in Fig. 9 time against $D_g\mathscr{P}/D_a$ on a log scale for $f=0{\cdot}5$ (50% solution) and $f=1{\cdot}0$ (100% solution). For the freely washed granules, the plots are indefinite straight lines and almost 10 times as long is needed to dissolve all as to dissolve 50%. There is not such a long tailing effect where the resistance is mainly external. To demonstrate this we have drawn the curves back into the $D_g\mathscr{P}/D_a>1$ region where they are asymptotic to a linear rate of solution where time for complete solution is twice as long as for 50%. Increasing internal resistance (decrease of $D_g\mathscr{P}/D_a$) has an earlier and greater effect on complete solution than on 50% solution. In this respect, this type of granule (of initially uniform composition) does not produce a satisfactory release sequence. In Fig. 10 is plotted the fraction released *vs* time for two values of $D_g\mathscr{P}/D_a$. The conclusion that differences in the performance of granules having widely different rates of loss in laboratory tests may prove disappointing, is borne out by practical observations. For example, Graham-Bryce *et al.* (1972) found that the effectiveness of the systemic insecticides disulfoton and phorate against aphids on beans was influenced remarkably little when adsorptive granules having the wide range of properties discussed earlier (p. 358) were compared, probably because transport and uptake

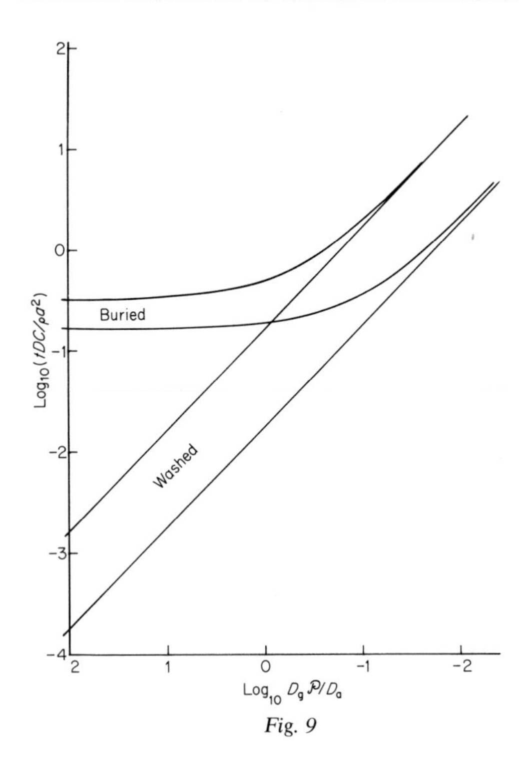

Fig. 9

were limited largely by properties of the environment over which there was little control.

Different mechanisms of internal resistance—e.g. an envelope of high-resistance material around the active core (Section 14.IX.B)—will give different release sequences, but the approximate additivity of times will still apply. Seaman and Warrington (1972) report on some pirimicarb-in-pumice granules, coated and uncoated, tested for rate of release both into water with intermittent agitation and into soil exposed to rain. They claim a broad correlation between the rates although the time scales were very different. Half the contents were released into water, from the most thickly coated granules, in a few hours but release then became very slow, indicating that much of the a.i. had migrated into the coating during processing. The uncoated and singly-coated granules released over 90% into water in 2 h and 24 h. The earliest observation in soil was after 1 month and the last-mentioned forms showed substantial difference, over 90% release from the uncoated (mostly over 98%) and between 80 and 85% for the singly coated in most of the tests (less in sand than loam

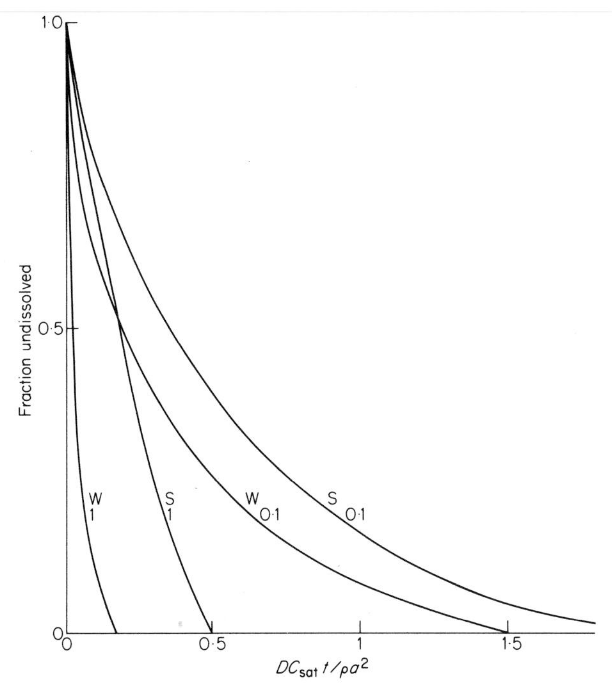

Fig. 10. W = surface washed (water); S = environment stagnant (soil). Numbers are values of $D_g\mathscr{P}/D_a$.

soils). The thickly coated granules had released only 70–90% of their contents even after 3 months. The comparison of uncoated and thinly coated granules provides an apparently clear exception to the additivity rule inasmuch as coating causes a delay of less than 24 h in release into water but of more than 1 month in release into soil. The soil test must introduce some new factor of delay which is not operative in the water test and therefore makes the latter of limited value. The most probable such factor is reduced contact with water as considered in the previous section. Not only must the water make contact over all the outer surface of the granule but it must penetrate the pores also (unless vapour phase transfer is of major importance which is questionable for pirimicarb, $\mathscr{P}^{\text{water}}_{\text{air}}$ being about 10^7). The pumice had a 30% pore volume but carried only 5% a.i.

Free water is more likely to penetrate the pores than is water held in the capillary spaces of soil. If water can bridge across the entrance of a 1 μm diameter pore the walls of which it can wet, it will exert more than 1 atm pressure on the air and so remove it by solution and diffusion if it cannot displace it. An unwettable outer surface could stop the water spreading across the mouth of the pore. This could be an important general difference between free water and soil water but, in the experiments reported, the results were probably also influenced considerably by the way the granules were presented to the soil: 4 g of granules (*c.* 300 μm diameter) were placed in a 4×4 cm tea bag which was itself laid across soil in a *c.* 10 cm diameter pot on which further soil was placed and pressed. The granules must have formed more than a single close packed layer within the bag. One effect of this congestion would be to make diffusion away into the soil slower than from isolated granules, but this would reduce even further the importance of delay within the granules. The segregation of the granules from direct contact with smaller particles of soil would make wettability more important. Rainfall percolating through the layer could do so as a film over the whole surface of the uncoated granules but only in discrete and intermittent channels between some of the coated granules.

The very rough parallel, on different time-scales, between release into free water and release from a dense layer of granules through a fabric may thus be fortuitous; more generally these considerations emphasize the complexities of devising and interpreting tests to evaluate granules.

K. Loss from Surface Application

It is widely believed that surface application of an appreciably volatile pesticide in granular form results in less loss and more persistent effect than application by spray. Obviously evaporation will be less when a uniform surface application is replaced by discrete sources, but the availability to the soil-borne target will also in general be decreased. For there to be overall advantage from application of discrete sources there must be some mechanism of favourable selectivity between these processes. Analysis of possible selective mechanisms could help their better exploitation. Three mechanisms can operate.

The first is essentially geometrical. It depends on the change of diffusion patterns when a uniform application is replaced by discrete sources. The steady-state analogy (Section 3.V.C) is invoked to predict the effect on efficiency of dosage transfer. In Fig. 11 is shown a highly schematic diagram of the passage of a volatile pesticide applied, on the left, uniformly to the soil surface represented by the horizontal straight solid line. The vertical lines represent the diffusion equivalent of magnetic lines of force, rate of transfer across an area normal to them being proportional to their density. A steady state is assumed to be established by a maintained supply. Upward diffusion terminates at the wavy line above

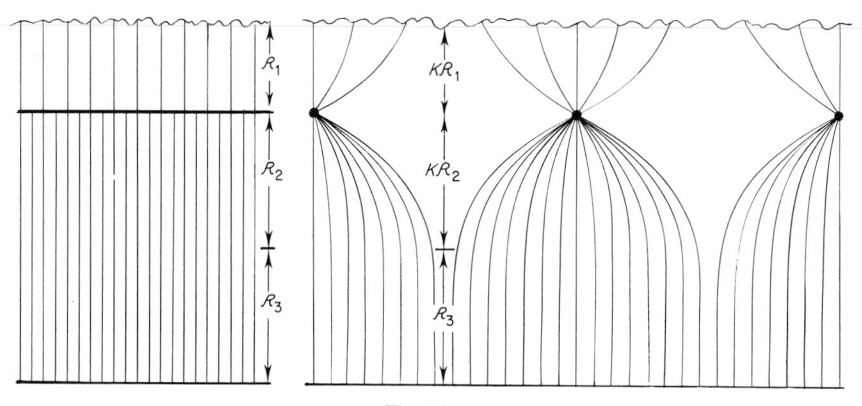

Fig. 11

which turbulence effectively sweeps away the escaping pesticide. Downward diffusion carries the rest of the supplied pesticide towards a buried target. We have indicated this target in an exaggerated manner as consuming all the pesticide reaching it and doing so at a definite level where the "lines of force" terminate. The argument below would remain valid, but become more complex, if we had attempted to show a more realistic diffuse target accepting the pesticide inefficiently.

On the right-hand part of the diagram we have replaced the uniform application by three discrete sources. Each feeds an area of the target determined by the population density of sources and the middle source is therefore typical. The "lines of force" must diverge from each source. Upwards they are cut off by the boundary layer surface before becoming parallel. Downwards they curve over to a vertical course before they reach the target. They are less dense than under uniform application because the contraction to discrete sources necessarily cuts down all the rates of delivery. The increased resistance causing this decrease of rate is geometrical in origin and confined to the regions of divergent diffusion. This geometrical effect is shown as extending through the air boundary layer (a few mm in thickness) but affecting only the upper part of the much greater depth of soil which must be penetrated. We may roughly divide the resistance in the soil layer at the level of the short line. If $\mathcal{R}_1$ represents the air resistance in the uniform case and $\mathcal{R}_2$, $\mathcal{R}_3$ the upper and lower resistances in the soil then, approximately, the change from uniform to discrete application multiplies $\mathcal{R}_1$ and $\mathcal{R}_2$ by the same factor of geometrical origin and leaves $\mathcal{R}_3$ unchanged. The ratio of rate of consumption at the target to total rate of release is therefore $\mathcal{R}_1/(\mathcal{R}_1+\mathcal{R}_2+\mathcal{R}_3)$ in the uniform case and $K\mathcal{R}_1/(K\mathcal{R}_1+K\mathcal{R}_2+\mathcal{R}_3)$ in the discrete case where K is a (>1) geometrical factor. The second is greater than the first.

Provided therefore the diffusion path to the target is considerably longer than the diffusion path to the air, application in discrete sources will increase the efficiency of uptake. It will also delay both processes and whether there is net advantage depends on concentration–time relationships in the biological response.

The second mechanism is less general. It depends on there being economic advantage in delayed release. To a first approximation evaporation of the dissolved pesticide, like chemical decay, proceeds at a rate proportional to concentration. Evaporative loss, under given geometrical and physical conditions, therefore increases, additively, the effective decay constant. Suitably delayed release of a chemical subject to decay can reduce the applicance necessary compared with instantaneous release (Section 14.IX.A). The advantage to be gained increases with the decay rate. It therefore increases further when evaporative loss is added to decay. The loss of course will always be a loss, and must be paid for, but it will be relatively less under optimum delay conditions (which can only be secured by granulation) than when a freely dispersed formulation is used.

The third mechanism is purely physical, restricted to granules, and is probably the most important, but seems to have been overlooked. The surface of well-tilled soil is never perfectly flat on a mm scale and on many seed beds in heavy soil is a mosaic of small clods 10 mm or more in diameter and crevices several mm deep. The "stagnant layer" thickness over an exposed surface controls the rate of evaporation (Section 6.III.B) and is rarely more than 10 mm except during night-time inversion. Spray drops will usually be absorbed quickly where they first contact the soil. The pesticide they carry is therefore deposited mainly on the tops of the small clods and on impervious stones when present. Small granules however, particularly if smoothly rounded, bounce and roll off summits and come to rest in crevices. The granule deposit starts therefore at a significantly lower level and in air protected from wind. It therefore evaporates at a significantly lower rate even if there is no retardation in the granule itself. For the same reason it is better situated for downward diffusion into the soil. The surface geometry of the soil is therefore likely to be an important factor and a fairly rough seed bed may be a positive advantage when a granular formulation of a rather volatile pesticide is used, unless the associated reduction of uniformity of deposit has a greater, adverse effect.

The mean depth of penetration of granules applied to a cloddy soil surface will depend not only on the roughness of the soil structure but also on the nature and size of the granules. A granule cannot penetrate a crevice too small for it and this effect would lead to deeper penetration the smaller the granule but, opposing this, is the catchment of very small granules by micro-ridges in small soil clods over which larger granules would roll into deeper valleys. Roundness of the granules will always assist penetration, probably during incorporation as well as when surface applied.

VIII. SEED TREATMENT

A. Introduction

We have frequently referred to the restricted mobility of pesticides in soil and the consequent difficulties of delivering effective quantities of toxicant to soil-borne target organisms. In many cases some form of localized treatment which brings toxicant and intended recipient together is essential. The advantages of seed treatments in this connection are obvious. They provide an economical means of placing the chemical in the critical region to protect the germinating seed and emerging seedling during their most vulnerable stages. The principles governing the outward spread of toxicant into the soil from such localized sources are discussed mainly in Section 3.II. For controlling mobile insect larvae, seed treatments exploit the action of the pest itself as it moves into the zone of soil containing toxicant during its search for the host plant. Some implications of this movement in relation to optimum specifications for seed treatment formulations were considered in Section 13.II.B: here we are concerned essentially with the processes of application.

In introducing this method, we should also recall that it can have substantial advantages for applying systemic pesticides in addition to protectant insecticides and fungicides. A reservoir for extended uptake can be provided conveniently sited close to the absorbing organs from the earliest stages of growth. This use is well-established for products such as the systemic fungicide ethirimol, applied as a seed treatment to barley for protection against powdery mildew. Seed loadings for systemic compounds must generally be substantially higher than for traditional protectant pesticides.

Nor are the potential advantages of seed treatment confined to protection against pests and diseases. The method could be exploited for weed control by protecting selected plants against herbicide action. The great attraction of this approach is that the crop is protected amongst vulnerable weed seedlings because it has been treated, not because of anatomical or biochemical differences. This permits closer (and chosen) selectivity to be obtained (for example, oats could be freed of wild oats, sorghum of shattercane or tomatoes of *Solanum nigrum*). Various attempts have been made to utilize the very powerful general adsorptive properties of active charcoal for this purpose. Limited success has been obtained by dusting the roots of transplanted seedlings. Band treatment over the seed is effective (Ripper and Scott, 1957; Burr *et al.*, 1972) in certain cases but is barely economic and tends to protect also those weed seedlings which are most damaging because they are in the most competitive positions.

Coating seeds with active charcoal has been generally unsuccessful because absorption of herbicide by the plant is mainly through radicals or shoots which quickly grow out of the protected zone. The recent advent of chemical "antidotes", "safeners" or "protectants", however, offers new prospects for controlled selectivity because the agents are diffusible and possibly systemic so that protection can spread from a seed treatment which is clearly not possible with charcoal. The compounds available at present are antagonistic to only a limited range of herbicides, chiefly the thiolcarbamate type, but other examples which will widen the scope of the approach can be expected in future. So far the most successful example of the method, a mixture of the antidote diallyldichloroacetamide with the herbicide EPTC (Chang *et al.*, 1973) has been conventionally sprayed over seed beds planted with maize. This use is based on the fact that the compounds are antagonistic on maize seedlings but not significantly so on some troublesome weed grasses. Seed treatment with such antagonists could be more selective, effective on more crops and more rewarding to the farmer.

With most pesticide treatments the final application is out of the hands of the manufacturer and his formulations must be sufficiently safe, robust and unsophisticated to allow for wide variations in operator skill and conditions of use. In principle these restrictions are very much less for seed treatment. The seed is the one stage of the crop which can be pretreated under the complete control of the manufacturer or merchant: seed treatment should therefore be more reliable and more amenable to sophisticated operations demanding precision, technical skill or factory mechanization. It can be reasonably claimed that these advantages have not been fully recognized and that there is much scope for further advance, particularly if more complex operations such as pelleting or, for example, sandwiching seeds between wafers of larger area are considered.

B. Methods of Applying Pesticides to Seeds

The detailed procedures for applying pesticides to seeds will depend on the crop species, but the broad principles are of general application. The following discussion is based mainly on methods used for cereals: in many countries protectant fungicides are applied routinely to cereal seed and insecticides are also commonly applied for protection against soil-borne insect larvae. A full account of methods, equipment and other aspects of seed treed treatment is given in the monograph edited by Jeffs (1979).

A satisfactory seed treatment process must fulfill several different types of requirement. First it must clearly achieve its intended purpose of placing the appropriate amount of chemical on each seed without damaging it and in such a way that the applied dose adheres during subsequent handling and drilling. Application should be done without hazard to operator and without releasing the formulation unnecessarily into the neighbouring environment. For practical purposes the seed treatment machinery should be capable of rapid throughput (for example, to process rapidly the large amounts of seed to be treated between harvest and autumn sowing) and should be versatile so that it can handle a range of seed species and formulations without elaborate modification. These formulations include both liquids and solids, powder treatments probably being still most extensively applied.

No existing commercial process fully meets these idealized requirements, shortcomings being mainly attributable to unsatisfactory adhesion which may be either insufficient, so that the formulation is not retained, or excessive so that seed-to-seed distribution is uneven. We shall give examples later but we may note here that adhesion to heterogeneous surfaces such as those of seeds is complex and there appears to have been little attempt to analyse the process in any detail. A range of factors are involved from short-range molecular forces to the entrainment of particles in surface irregularities and hairs. With powders, molecular forces are likely to be of limited importance because the area of close contact between rough seed surface and irregularly shaped particle is normally small. Contact area can be increased by extending the period or vigour of mixing so that the pesticide particles or seed surface are distorted or the particles have more opportunity to find their way to more stable locations (Jeffs, 1976). Alternatively adhesion can be improved by adding adhesives, but both approaches are likely to increase costs and therefore be unpopular. Adhesion also depends on particle size; probably because there is an optimum size for physical entrainment in the surface irregularities (Jeffs and Tuppen, 1979). Molecular forces are likely to be more important with liquid treatments which leave a film of pesticide in close contact with the surface and may effect some penetration of the seed coat.

The majority of seed treatment machines employ one of three basic principles with numerous variations of detail and different metering and dispensing devices.

(1) *Revolving drum mixing.* An advantage of these machines is the simplicity of their operating principle. Seed is admitted to one end of a cylindrical drum rotating about its longitudinal axis which is slightly inclined to the horizontal. The formulation is applied to the seed as it passes through the drum, possibly by way of tines which dip into the flowing seed and assist mixing. The treated seed leaves the drum at the lower end and is collected in holding bins or directly into bags. These machines are probably used more for applying liquids than solids, but versions capable of handling powder treatments are available.

(2) *Auger mixing.* Seed and formulation are introduced into one end of a mixing chamber which may be horizontal or inclined, and are mixed while being conveyed to the opposite outlet end by a rotating auger. In some inclined versions, there is sufficient clearance between the blades of the auger and the walls of the mixing chamber to allow seed to fall back down the conveyor so that remixing can occur. Machines working on the auger mixing principle are available for applying both powder and liquid formulations.

Residence times within the mixing chamber of modern auger machines are surprisingly short: typically seed takes less than 10 seconds to pass through the auger. The resulting mixing is remarkably effective but the rapid throughput tends to limit adhesion and substantial amounts of loose powder may be lost.

(3) *Spinning disk application.* This approach has proved popular for applying liquid formulations. The liquid is supplied to a rapidly spinning disk from which it is thrown by centrifugal action as a mist of fine droplets: the principle is essentially the same as that employed in the rotary atomizers used for crop spraying (Section 14.VI.B). The seed to be treated passes over a cone above the spinning disk which distributes it as a uniform cylindrical curtain that falls through the droplets, picking up the treatment. It is important that the curtain is not more than two or three seeds thick if those on the outside are to be treated adequately.

In addition to these three basic approaches, there are various machines working on miscellaneous other principles. One of the most interesting and important is the Rotostat (Ellsworth and Harris, 1973). The mixing chamber consists of a dish-shaped rotor with vertical axis which fits closely inside a stationary cylindrical sleeve. A weighed batch of seed is transferred directly into the mixing chamber. The rapid rotation of the rotor accelerates the seed up the side of the dish by centrifugal action until it meets the stationary sleeve, loses momentum and falls back onto the rotor. The seed batch is thus in constant rapid movement, forming a "toroidal doughnut" shape into which the formulation (either liquid or solid) can be admitted. After a brief period of mixing the treated seeds are discharged centrifugally through a trap directly into a bag. The machine is particularly adaptable: since it operates on a batch process, changing to a different formulation causes little difficulty. Analytical results indicate that mixing may be more satisfactory than with traditional machines (Middleton, 1973).

C. Evaluation of Seed Treatment Processes

For the purposes of protecting the crop, seed treatment machines must be assessed in terms of the average loading of formulation on the seed, the seed-to-seed distribution of formulation and the subsequent retention of the treatment during storage and handling. While traditional machines achieve an impressive throughput, their performance on the basis of these criteria may be less satisfactory.

A survey of seed treated commercially in Britain (Lord *et al.*, 1971b), undertaken in collaboration with manufacturers and merchants, indicated that both liquid and powder treatments can have serious shortcomings. With powders, where the performance of three types of machine was examined, average loadings were usually well below the target. Table 3 shows that only five out of the 33 sets of samples treated with insecticide by different merchants had an average loading greater than 50% of the target and the maximum loading was 70%.

Table 3
Relative amounts of insecticide on seeds treated by different merchants (summarized from Lord et al., 1971*b*). *Figures in each column give the number of sets of samples with the stated percentage of the target dose.*

	Percentage of target dose						
Formulation	0–10	10–30	30–50	50–70	70–100	>100	Total
Powder	12	7	9	5	0	0	33
Liquid	0	1	6	5	3	2	17

With liquid treatments analysis of bulk samples indicated that on average loadings were closer to the target (Table 3). However, there were very large differences in the amounts on individual seeds, the distribution being much less even than with powders. Some seeds carried as much as 10 times the average loading, 5–10% had more than twice the average and 50–60% carried 25–75% of the average. The seed examined had been treated by either rotating drum or spinning disk types of machine. Each type gave a characteristic pattern of seed-to-seed distribution, the spinning disk giving slightly better results.

Because the powder treatments gave such unsatisfactory loadings, commercial methods of application were investigated (Jeffs *et al.*, 1972). This examination showed that powder and seed were generally mixed in the correct ratio in the mixing chamber but the powder did not adhere adequately and so separated from the seed after leaving the chamber, in holding bins, bags and during subsequent handling and drilling. In seeking possible methods of improvement Jeffs (1973) showed that adhesion of insecticidal powders can be greatly enhanced by pretreating the seeds with adhesives such as polybutenes, vegetable oils or gum arabic.

D. Effects of Seed Treatment Formulation on Biological Activity

While control will obviously be unsatisfactory if most of the applied chemical falls off before sowing, it cannot be automatically assumed that strong adhesion will give a correspondingly favourable performance and it is necessary to have a good understanding of the biological requirements to suggest the most suitable formulations. Some of the factors to be considered in the control of soil borne insect larvae are discussed in Section 13.II.C.

The composition of the seed treatment formulation can also markedly affect the nature of the protective zone by influencing the release of toxicant to water percolating down the soil profile. It might be expected that the beneficial effects of adhesives, discussed above, would be at some expense of subsequent effectiveness in keeping with the frustrating and widely applicable maxim that biological availability and resistance to physical loss are often inversely related. However, investigations by Jeffs (personal communication) show that this is not inevitable. Leaching of insecticide from seeds treated with equal loadings (0·3 g per 100 g seed) of lindane powder alone or after various pretreatments and then embedded in moist sand was investigated. Surprisingly, most insecticide was released from seeds pretreated with gum arabic, which increases adhesion, and least from those pretreated with surfactant, intended to facilitate release (Table 4).

The results shown in Table 4 can probably be explained in terms of the detailed nature of the contact between the water network in the soil and the different formulations on the seed surface. Water in soil pores is subject to capillary suction resulting from surface tension effects at air–water interfaces. When water in the capillary network reaches the

Table 4
Release of insecticide from different formulations applied to wheat seeds (results obtained by K. A. Jeffs)

Treatment	Amount released to 100 ml leachate by 100 seeds (mg)
lindane alone	22
Pretreatment with surfactant (Myrj 52)	31
Pretreatment with gum arabic	68

surfactant on the seed surfaces, there will be an immediate reduction in surface tension, causing local withdrawal of water away from the seed and retreat along the pores. Such an effect can easily be demonstrated by applying a small source of surfactant to the end of a capillary part-filled with water. The surfactant formulation could thus cause a "drying" action around the seed which could at least partly offset the normally expected wetting action. On the other hand when the gum arabic coating is wetted, it swells, exposing a greater area to the leaching solution. Such considerations demonstrate that formulation can influence biological activity in ways which may not be obvious at first sight and which can only be interpreted by detailed analysis; they also emphasize that seed treatment offers much scope for manipulating performance.

Considerations of biological activity must also include possible harmful effects, notably phytotoxicity. Powder treatments can have advantages in this respect: for example, potentially phytotoxic pesticides such as lindane can be safely applied as powders at rates which would be damaging to the seed if applied as liquids. The phytotoxic effects of liquid formulations may be attributed to penetration through the seed coat, particularly via the scutellum (Jeffs and Griffiths, 1973). We may also note that it is in general possible to apply much larger loadings of pesticides with powder treatments if adhesives are used and to apply both insecticides and fungicides together more readily as combined dual-purpose formulations.

E. Pelleting and Coating

For high-value crops or where there are specialized requirements, more elaborate seed treatments may be justified. Seeds can be *pelleted* by alternately spraying them with adhesive in a revolving drum and then applying clay and fibre mixtures which adhere to the sticky surface. Pesticides, nutrients or other materials (for example, herbicide antagonists or growth regulators) can be incorporated into the mixture and the operations repeated until a coat of the desired thickness has accumulated.

The advantages of the method are therefore that it makes possible the application of higher loadings of more complex materials and produces a seed of more uniform and often more convenient shape than the original. This last factor may be of considerable benefit with crops having irregularly shaped seed which must be sowed with precision, as in the case of sugar beet for which the method is now widely used. For satisfactory performance it is essential that water can penetrate the pellet readily after sowing to allow germination and emergence of shoot and radical. Some delay in germination may occur compared with untreated seed, but the method allows the incorporation of fungicide to protect against any undesirable consequence of such delay. Provided the individual seeds are of reasonably similar size (which cannot always be safely assumed) and the final pellets are sufficiently uniform, there should be far less seed-to-seed variation in loading than with traditional methods of treatment. Evidence in support of this expectation was obtained by Germains, UK (see Jeffs and Tuppen, 1979) when applying treatments to wheat seed giving 20% lindane in the coat. Amounts of lindane on individual pelleted seeds ranged from 21–37 μg with a mean of 27 μg.

At present, despite its advantages, the use of pelleting is limited by its relatively high cost and slow throughput compared with conventional treatments. Jeffs and Tuppen quote figures of 2 to 5 tons per hour for pelleting processes, whereas most modern seed treatment machines would be expected to handle at least 10 tons. However, if the full potential of pelleting for applying more sophisticated treatments was exploited, the method could become economically viable even for crops such as cereals, particularly if precision sowing methods are introduced which allow lower seeding rates and if costs of cereal seed continue to rise.

In seed *coating*, the seed receives one or more complete coats of material but the process is less elaborate than pelleting and the seed retains essentially its original shape. The simplest method of coating is to dip the seeds into suspensions of coating material. Such steeping is widely employed with sugar beet and certain vegetable seeds. An example of a more complex procedure is the Wurster Air Suspension Technique (Wurster, 1959), devised for coating drug particles in the pharmaceutical industry and employed by Schreiber and LaCroix (1967) for applying coatings to delay germination. Seeds are supported on a column of warm air, the velocity distribution of which varies so that the seeds ascend to regions of low velocity and then fall back. Coatings of different composition incorporating the chemical agents required can be applied as aerosols at the lower end of the column. The seeds then dry very rapidly as they ascend. As with pelleting, such techniques would seem capable of much further development.

IX. CONTROLLED RELEASE

A. The Advantages

The chief potential advantage of controlled release is the maintenance of an effective concentration for longer than when the same applicance is released all at once and therefore subject to maximum loss through chemical decay, evaporation and leaching. This advantage is clearly explained by Neogi and Allan (1974, p. 196). McFarlane (1976) envisages even greater economy, and therefore less environmental impact, if the pesticide could be released in a broad pulse form matching the future progress of the pest, as far as this can be anticipated. To secure this advantage, the formulation might have to provide a time-lag before release commenced as well as (or in place of) reducing the rate of later release. The expression "controlled release" is now widely used, but it must be appreciated that at present only the simplest type of control is practicable in an agricultural field situation.

There has been more progress in developing controlled release for applying pesticides to flowing water and volatile pheromones to air for two important associated reasons. Chemicals released into moving air or water are further dispersed relatively rapidly. Little delay occurs outside the units of the special formulation. Delay can be useful on a shorter time-scale than is required in soil application and the sources can be further apart and therefore larger and of more elaborate structure. They can therefore, economically, be better adapted to perform a simpler task.

Although the potential advantages in soil application are more difficult to realize, in practice they could be substantial. Delayed (as distinct from slow) release could enable soil treatment to be laid down at the time of sowing a crop, thus avoiding a separate passage of machinery and permitting incorporation which is impracticable after crop emergence.

The argument for economy of active chemical is illustrated in the schematic concentration–time diagram (Fig. 12) where the scales are arbitrary. The a.i. is assumed to decay exponentially at such a rate that an initial concentration of 8 units is necessary if 20 time units must elapse before it falls below 1 unit, as shown by curve 3. This requires the exponential factor to be 0·104 reciprocal time units. The biological requirement is assumed to be a concentration in excess of 1 unit over the time period 10–20 i.e. a curve falling outside the rectangle 1, 1, 2. The minimum consumption of a.i. to achieve this, assuming consumption in the toxic process to be negligible, would follow introduction of 1 unit at time 10, then a steady supply at rate $10 \times 0{\cdot}104$ up to time 20 and finally withdrawal of the "capital" of 1 unit—net consumption 1·04 units. Withdrawal is not a possible operation in practice so, at best, the 1 unit present at time 20 must be allowed to decay—i.e. pathway 1, 1, 3, consuming 2·04 units. If a controlled release formulation supplied a.i. at a constant rate from time zero so that a concentration of 1 unit was reached at time 10, the rate would have to be 0·169 and the supply continue to time 17·3 when the residue would be left to decay—pathway 4, 4, 3, consuming 2·78 units.

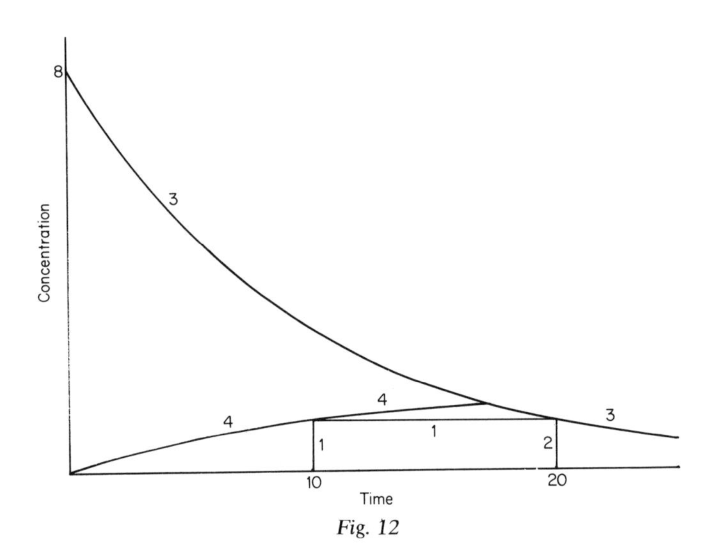

Fig. 12

The more rapid the decay the more a.i. is necessary by any route and the greater is the relative advantage of controlled release. If a single application giving 100 concentration units at time 0 was necessary the lowest practicable consumption (1, 1, 3) would be 3·30 and that necessary for constant release (4, 4, 3) would be 4·99.

Economic advantage would still be gained if no initial lag were required. Had the demand in our schematic diagram been for maintenance of concentration 1 from time 0 to 20, 8 units of the compound with the decay rate first assumed would still be required for a single application and 100 units of the second compound. Below we put figures for the second compound in brackets. The minimum practicable consumption for an initial flood of 1 unit followed by maintenance would be 3·08 (5·60). Linear release would not meet the requirement: instead of lag, one would need an initial surge—i.e. an extreme of the form of release most easily obtained. A formulation meeting this requirement would satisfy the first (C above 1 for t between 10 and 20) as well and the additional economy from introducing delay for the first requirement would be only 33% (41%) compared with economies over the single applicance without decay of 62% (94%).

Allan *et al.* (1972) report another advantage for slow release in soil. 2,4-DB is not seriously toxic to Douglas Fir but its β-oxidation product, 2,4-D when formed in the soil from an application of the free butyl compound, is severely damaging. A slow-release formulation of 2,4-DB gave good control of weeds in a Douglas Fir plantation without damage to the young trees. As McFarlane (1976) points out, once a compound is released into the soil it is subject to the same environmental hazards however released. There is no simple physico-chemical mechanism which would yield a lower *proportion* of the derivative when the parent compound is released slowly and it would be unwise to generalize this interesting effect. It must arise from a lower biological response to a long-period low-level presentation of 2,4-D. Other unpredictable effects of slow release may appear.

In one respect, McFarlane's view could be an understatement. Many compounds, among which one may cite particularly the hormone herbicides, are degraded by soil microorganisms and an adaptive acceleration of this process is well established (e.g. Audus, 1964). Slow release of pesticide from separate sources provides conditions very favourable for such adaptation. Somewhere in the region of decreasing concentration, an adapted population is likely to develop giving a situation in which each source can be envisaged as surrounded at an appropriate distance by microorganisms waiting for a food source to arrive. This is a potential limitation of slow-release technique that should be taken into account.

An important, specialized advantage of a *delayed* release formulation lies outside the scope of our text but should be mentioned for completeness: namely its potential for reducing the toxic hazard to the user. This applies particularly to volatile toxic compounds encapsulated in coating material impermeable in the dry state but becoming permeable when swollen or disintegrated by uptake of water when distributed in a moist environment. Such formulations are not absolutely safe. The interior of a gumboot or glove is a moist environment. The encapsulated formulation can improve the performance and safety of the intelligent operator but, like many mechanical safety devices, over-reliance on its properties can have the reverse effect.

B. Retardation of Release—Physical

In the simplest types of slow-release granule (1) a solid a.i. is initially uniformly distributed in a slowly permeable matrix or (2) it is dissolved in a polymer matrix. These are extremes since some of the a.i. may be dissolved but excess exist as discrete particles. The form of the release–time curve is conveniently plotted as fraction of initial content released *vs* time under conditions of free escape outside—i.e. when concentration in the ambient air or water is kept effectively zero. The curves differ somewhat between these two extreme forms although in both the rate decreases steadily from its initial high value. If in the first case the amount of a.i. dissolved within the granule was negligible compared with the amount in the retreating zone containing solid, the "ceiling" of total release would be reached at a finite time. This is, of course, artificial since there must always be a dissolved content (without which there could be no diffusive escape) and this will lead to an asymptotic approach to total release in the final stages. Otherwise expressed, the first type of behaviour must at some stage pass into the second but perhaps so late as to be insignificant. Two other release forms although less easily realized are conveniently compared. These have envelopes of negligible capacity and low and constant permeance. In one case (3) the core is supposed to remain of constant volume while a dissolved a.i. (of negligible volume) diffuses out. In the other (4) the core of a.i. is assumed to contract without altering the permeance of the envelope, a condition which could be approached in practice by the envelope becoming flattened or wrinkled rather than elastically contracting in area and increasing in thickness. These forms, for spherical geometry (although strictly inconsistent with the last case) are compared in Fig. 13 with the controlling quantities, D,

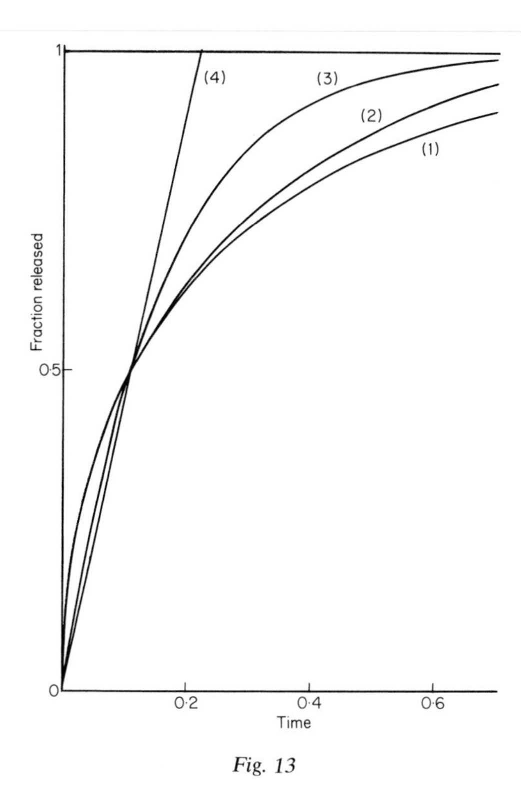

Fig. 13

$\mathscr{P}$, a and, where relevant, C_{sat}, adjusted to give 50% release at the same arbitrary time.

Curve (4) is linear and reaches complete release at a finite time for the same artificial reason as curve (1) (where the time is beyond the limit of the diagram). Curve (3) is simply exponential. Curve (2) is taken from the computation of Crank (1956, p. 90) and curve (1) from eq. (28). (1) and (2) are indistinguishable below 50% release. The exponential (3) is, over the first 70% of release, more nearly linear than either (1) or (2) which, initially, have amount released proportional to $\sqrt{\text{time}}$.

None of these curves shows a lag period and none settles to a linear slope after an initial surge. McFarlane (1976) proposed introducing the lag by the time taken to build up a gradient in an initially "empty" envelope (Section 3.V.F). The envelope on composite granules would, however, become saturated during storage if not during manufacture so that an initial surge rather than a lag would result unless the envelope material had neither capacity for nor permeability to the a.i. in the dry state but acquired both when swollen by water.

Formulations for controlled release, many more complex than the basic types exemplified above, have been the subject of much research in the pharmaceutical field. The basic principles were established by Higuchi (1963 and earlier references therein). Several aspects of later developments have been reviewed by Ritschel (1973), Katz (1973) and Poulsen (1973). The discussion by Flynn (1974) is particularly relevant. Baker and Lonsdale (1975) have reviewed the pharmaceutical methods in relation to possible pesticide applications. While a study of the pharmaceutical developments will assist consideration of possible agricultural formulations, some very important differences must be borne in mind. First, the time-scale of desirable drug release in medicine is shorter and better defined. Secondly, medical products are used in smaller amount in a higher-priced market—a course of multi-granule pills may contain a thousand units: a hectare of crop needs 100 million. Thirdly, and more important, the ingested drug formulation liberates its a.i. into a fairly free, mobile aqueous environment of well-regulated pH and remarkably constant temperature: granules in the field must be subject to very variable and much less predictable future humidity, pH and temperature.

Controlled release of pesticides into air or water has received considerable attention at several recent conferences. Ashare *et al.* (1975) describe the "Conrel" device which contains volatile agents in plastic microtubes laid on adhesive tape and sealed across at intervals. Release is initiated by cutting across the assembly and then proceeds at a diminishing rate, determined by combined cross section and vapour pressure, until the exposed section is exhausted. Multiple-layer plastic strips are described by Kydonieus *et al.* (1975). In these the pesticide must diffuse from a source layer through a "protective barrier" on to a surface where a deposit forms which can be picked up by walking insects, particularly roaches.

In this last system the gradient of thermodynamic potential driving the pesticide from interior to surface exists because crystallization is inhibited within the polymer network but occurs, with associated lowering of potential, on the surface. The method is an elaboration of that, almost as old as DDT itself, of dissolving the insecticide in wall paints on the surface of which it slowly forms a microcrystalline bloom, often assisted by transfer of nuclei by walking insects. Migration to the surface will also be maintained if the pesticide is lost by evaporation. Control of evaporation of dichlorvos into room-space by plastic strips has been exploited for several years (Slomka, 1970; Gillett *et al.*, 1972). Kydonieus *et al.* claim that their multilayer strips can develop a deposit of sub-micron crystals and also that the product can be reduced to confetti-type granules. The technique can greatly extend the effective life of one treatment with a pesticide which has a much more limited life without special formulation, but evidence is not disclosed about what fraction of the applied dose can eventually be released. Economy of application time and labour rather than of a.i. is the principal object.

Retardation of release by physical mixture of soluble a.i. and insoluble filler to form a highly compressed tablet of uniform composition has very limited potential because, unless the a.i. is present in only a small proportion, the residual filler must be left with high porosity. The pores of a compressed granular structure are freely permeable. Some special minerals, such as some types of pumice, have an almost sealed-bubble structure but if this structure is very slow to empty it is also difficult to fill. McFarlane (1976) showed that granules formed by solidification of paraffin-wax containing an oil soluble dye in the molten state released the dye with an apparent diffusion coefficient as low as $3\times 10^{-14}\ \mathrm{cm^2\ s^{-1}}$ but the concentration of dye is not stated. Compressed wax, particularly with addition of a viscous additive such as polyisobutene or a water-swellable additive such as starch, can provide a barrier of very low permeability but it must be used as an outer layer in a composite granule so that a useful loading of a.i. can be contained in the core.

The coating of granules is, of course, a well-established process in the pharmaceutical trade and even extends to fertilizer granules for the high-price horticultural market. The basic technique is to tumble the core granules in an oblique rotating drum (concrete mixer) while alternately adding filler dust and sprayed adhesive solution. Under condition of good temperature control molten waxes can build up in the solid state by this process.

For homogeneous granules certainly, and for coated granules preferably, the resistance layer has a high content of some polymer without which the necessary very low permeability cannot reliably be achieved. Neogi and Allan (1974) report that a condensation polymer of ethylene diamine and the cyclic dimer of linoleic acid is a useful slow-diffusion solid solvent for many pesticides. As with all solutions in polymers, the diffusion coefficient increases rapidly with concentration. Using hexamethylorthophosphoramide as a model pesticide in their system, Neogi and Allan found $D \simeq 4{\cdot}9\times 10^{-10} \exp(21C)\ \mathrm{cm^2\ s^{-1}}$ where C is the concentration in $\mathrm{g\ cm^{-3}}$. D increases 19-fold from $C=0$ to $C=0{\cdot}14$. While the amount released from a semi-infinite block initially at $C=0{\cdot}14$, proportional to $\sqrt{\text{time}}$, is not increased so much (*c.* 2·5 times above the $C \rightarrow 0$ rate), the rate of release from a limited volume would decrease more rapidly than were D constant, giving greater wastage in an undesirably long "tail". Also, the bulk of the granule is material as costly as the a.i. although some of the volume could be replaced by mineral filler.

The polymeric coating, therefore, like a particulate one, is more efficiently used as an outer layer rather than in a homogeneous granule. Something must then be done to stop the a.i. spreading from core to coating during manufacture. There are three possible ways, alone or in combination. The coating polymer may be cross-linked, thus limiting its swelling capacity: the a.i. may be a crystalline solid, thus limiting its solubility or the polymer may be so chosen that the a.i., if liquid, swells it to a limited extent only. A modification of the last method alters the a.i. rather than the polymer by dissolving the a.i. in a liquid which is a non-solvent for the polymer. This, however, would reintroduce the tailing effect: at best the release would be exponential, because the concentration of diffusible a.i. in the core decreases, but the slowing-off would be exaggerated because of diminishing swelling, and therefore permeance, of the envelope. The best system for an approximately steady rate of slow release is to have a crystalline a.i. encapsulated in a cross-linked polymer. Slow-release granules for soil application are therefore expensive to produce and require considerable research effort in selection, and perhaps fabrication, of the coating material. An aspect we have not so far mentioned, but one of which the professional formulator must be well aware, is that the combination must retain during storage the properties essential for the desired performance. Since only a small fraction of the chemical bonds on swollen polymers, particularly when cross-linked, need react to influence their physical properties disproportionately, storage stability may not be an easy requirement to meet.

C. Retardation of Release—Chemical

As we have just indicated, chemical reaction may be a limiting factor in the performance of granules retarding release by a physical mechanism. However, it can also itself be exploited as a means of retarding release. Examples of compounds which function as progenitors of the active ingredient have been considered in Section 4.IV. If the progenitor is an insoluble polymer such "chemical formulation" can provide the retardation of release. The method is limited by the necessity for the active pesticide to have a residual functional group which can be liberated by a reaction initiated on exposure. The hormone (carboxylic acid) herbicides are the most obvious examples. Esters are widely used as free progenitors. If the active acid is bound by ester linkage to an –OH containing polymer it can be converted to an insoluble, granular, progenitor. Cellulose can be esterified with many carboxylic acids, most easily by displacement of acetic acid from the acetate. The 2,4-D ester of cellulose, or a cellulose acetate with many acetic groups replaced by dichlorophenoxyacetic, is a water-insoluble solid which will slowly release free 2,4-D.

Cellulose-acetate-2,4-D certainly behaves in this way. The link through an O atom of a chain C atom and the CO group of the active acid is the most easily attacked (by hydrolysis) link in the polymer. More complex cases may undergo undesired reactions. For example, an insecticidal carbamate, of general formula RO–CO–NH–CH_3 could be linked through the N atom to a C atom in a polymer chain carrying a reactive "pendant" atom or group such as Br, CO_2H, CO_2Cl but the linked insecticide could decompose by hydrolysis at linkage 1 or 2, or, with assistance of oxidation, even at 3, before the desired liberation occurs at linkage 4. The method is a potentially powerful one but much chemical preparation and testing may be necessary for each active compound.

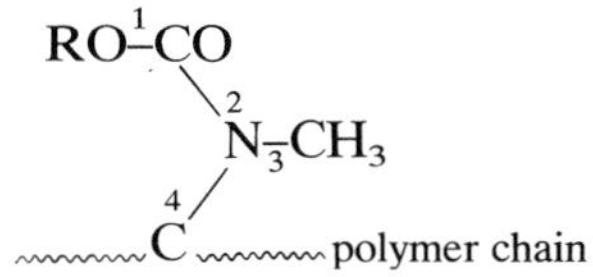

```
RO–¹–CO
       \2
        N–₃–CH₃
      4/
~~~~~C~~~~~ polymer chain
```

Not only must the right reaction be more facile than inactivating reactions but it must also proceed at the right rate. Hydrolytic reactions are often strongly pH-dependent so that ester-fixed 2,4-D could be liberated too rapidly in an alkaline soil and too slowly in an acid one. In the solid polymer the ester groups are largely protected because water, and especially the ions dissolved in it, do not have free access. This is already evident in cellulose acetate which is unaffected by immersion in water under conditions where ethylacetate or glycol diacetate would be extensively hydrolysed. The glucose units in cellulose itself are linked by ether groups involving the acetal or diether group –O–C–O– for one of every 6 C atoms. Acetal itself is easily hydrolysed, yet cellulose in the absence of destructive bacteria or fungi, is remarkably durable. Cellulose is degraded by inanimate chemistry only under very vigorous conditions (hot, strong acids) and then breaks down completely to glucose. Once the protection of the organized crystalline structure is lost, the linkages in the water-soluble oligosaccharides are rapidly broken.

This "topochemical" behaviour is important for slow release because it introduces other controlling factors—crystallinity, orientation and swelling by "plasticizers" can influence the rate of release. It also enables the course of release, particularly from fragments of oriented sheet to be more nearly linear than exponential. The possibilities of release by chemical reaction in polymers have been discussed more fully by Baker and Lonsdale (1975) and Feld *et al.* (1975). Mehltretter *et al.* (1974) used 2,4-D esters of starch and Neogi and Allan have described successful use of 2,4-D ester-linked on to crude cellulosic materials.

Beasley and Collins (1970) showed that herbicidal carboxylic acids could be bound into resins by precipitating ferric salts at suitable pH and temperature in the presence of aldehydes. The products are not complex crystals like many basic salts. They are soluble in acetone but not in water. Some other metals can substitute for iron. It was claimed that these products could give slow release into soil but this use does not seem to have been developed. Since the release is bound to be pH dependent and decay of the released herbicides is dependent on soil fertility and microflora one cannot expect the method to give a performance which can be reliably forecast.

D. Delayed Release

Most simple slow-release formulations will commence release as soon as they are distributed in an environment which can accept the active compound and release will then continue at a declining rate. In some situations a long lag period before release would be desirable—if its length could be matched to the biological requirement. Even if the subsequent release is not retarded, a very short lag period can serve a useful purpose in reducing toxic hazard or loss of a valuable volatile compound during packaging and distribution in the dry state, release being initiated when the formulation absorbs moisture.

For effective breakdown of an outer protective coating or for release of a chemically bound a.i., swelling, solution or hydrolysis by water is the obvious means to exploit when the formulation is distributed in the open. Evaporation of a volatile constituent of the formulation is not likely to lead directly to increased permeability but change of pH consequent on evaporation of an amine or acid could perhaps be made to break some linkage or at least assist hydrolysis. Coatings which would break down on oxidation could be devised but they would create more difficulties in manufacture than those sensitive to water and moreover oxidation reactions would be more susceptible to other uncontrolled factors in the environment such as access of transition metal ions. Water-sensitivity is likely to remain the most easily and reliably exploited property.

Increase of permeability by reduction, if a mechanism could be devised, would have a function which hydrolysis alone could not fulfil. Heavy granules are used to carry herbicides into the bottom mud in lakes or slow rivers to attack bottom-rooting weeds. Such mud has a reducing potential and if the granule released its content only after reduction, there would be less wastage into the free water.

In pharmaceutical practice some so-called enteric coatings are used which continue to occlude a drug in the acid environment of the stomach but liberate it in the more alkaline intestine contents. Cellulose acetate in which some acetyl groups have been replaced by dibasic phthalic acid, only one acid group being esterified, is the favoured coating material. The acid form is only slightly swollen by water but it dissolves, or becomes so extensively swollen that it disintegrates, when the free $-CO_2H$ groups are ionized. Although the pH variation in fertile soils is not great enough to make so large a change of permeability, the rate at which cellulose-acetate-phthalate would disintegrate in soil would depend on cation exchange capacity. There seems, however, no obvious use for such variation.

In some operations a delay of many days may be desirable after the granules (or other formulated units) have come to rest in the new environment which initiates release. There is no major difficulty in arranging the shorter delays usually required within warm blooded animals in an environment of constant very high humidity and constant temperature. Water diffuses more rapidly than organic solvents into all macromolecular substances which it can swell appreciably. According to the extensive data reviewed by Barrie (1968) few diffusion coefficients for water at ordinary temperature fall below $10^{-9}\ cm^2\ s^{-1}$. At this value a thickness of 100 μm will be penetrated near to equilibrium in *c.* 24 h. Small granules could not economically be protected by a much thicker layer (100 μm thickness around a 500 μm diameter core adds 170% to the volume). There seems little probability of attaining delays of several days with a simply slowly-swellable coating.

A swellable inner coating protected by a thin outer coating of water-resistant material could induce much longer delay until the swelling of the inner layer bursts the outer one but such mechanism would probably be difficult to make reliable and might be beyond the acceptable price range. Another mechanism of delay could be provided by a water-impermeable material internally loaded with closely-packed microcrystals of some substance slowly soluble in water. Gypsum ($CaSO_42H_2O$) would be an obvious choice for the latter. It is soluble to about 1 in 10^3 by volume and solubility depends very little on temperature. A slab of pure gypsum immersed in completely stagnant water would lose about 8 μm of thickness in 1 day, 80 μm in 100 days. Solution rate in moist soil would, of course, be slower according to the principles considered in Section 5.IV, and would be accelerated by flow under rainfall or irrigation. Delay produced by imbibition of water into swellable polymer coatings followed by disintegration would be little influenced by moisture content of soil, unless very dry, because water uptake would be from nearly saturated vapour. Temperature effect, however, would be great. The gypsum solution method, which could probably be provided by a slurry of ground gypsum in molten paraffin with added polybutene, would be little influenced by temperature but greatly by moisture conditions.

It will be seen that there are considerable technical problems to be solved if granules with protracted delay are to find a useful place in application to soil rather than to water, air or the tissues of animals. The biological demand as well as the physical implementation is not reliably predictable. One general requirement is to have the pesticide released at the time when the crop seeds, or seeds of some troublesome weed with a dormant behaviour, germinate. Could seed of the weed species be combined in the granule so that germination of the seed disrupts the coating of the pesticide? If swelling of the seed were the factor mainly dependent on weather-history, this idea should be practicable but, in many crop seeds, imbibition of water is not greatly dependent on temperature but lag period before emergence of radicle is. In experiments with radish seeds enclosed in selected straws, Hartley (unpublished results) found that a tightly surrounding straw was split by swelling of the seed, but that, if the straw were of large enough bore to accommodate the swollen seed but not the expanding cotyledons, it was the seedling that suffered, the hypocotyl emerging and leaving the detached cotyledons behind. While there are problems, therefore, the method is perhaps worth more research.

E. Encapsulation

Coating of particles individually by physical methods is probably limited by cost factors to pharmaceutical products, although high-speed methods have been devised (see review by Goodwin and Somerville, 1974). In order to apply coating procedures to particles in the range of a few μm diameter it is necessary to use chemical engineering processes which can be carried out in bulk. These processes are referred to as micro-encapsulation but many can be applied also to globules in the mm range.

These processes involve the spontaneous formation of a coherent envelope around particles dispersed in liquid, usually an aqueous solution. The particles to be coated are usually initially oil globules although these may contain suspended solid. Direct coating of aciculate crystals tends to leave points exposed.

The envelope material is almost necessarily one of high molecular mass. The polymer may be a pre-formed one held initially in solution in the continuous phase and slowly precipitated by change of temperature or addition of salts or non-solvents. Suitable envelope substances under suitable conditions form around existing particles, at first as a tenuous gel, characteristically thicker at opposite poles under the influence of necessary agitation and becoming more compact as the conditions of precipitation are slowly made more extreme. After further contraction by the action of added tanning agents, the coated particles may be washed, filtered off and dried.

This is the coacervation process extensively exploited by the National Cash Register Co. to produce a clean substitute for carbon paper in business stationery. An encapsulated chromogenic substance is released by the pressure of a pencil from the coating on one surface and reacts to produce a colour with a coating on the adjacent surface. The coacervate coating in this case was a complex of gum arabic and gelatin brought out of an aqueous formulation by change of temperature and forming around dispersed oil globules. A di-aldehyde such as glutaraldehyde is used as tanning agent.

There are many other uses for materials encapsulated by this method and many other, most more expensive, soluble polymers which can be coacervated by similar procedures.

A different method is to form the polymer coating material by chemical reaction *in situ*. A nylon-type polymer is formed in the interface when a di-acid chloride dissolved in oil is brought into contact with a diamine dissolved in water, a method first announced by Morgan and Kuclek (1959). The polymer forms around emulsified oil, encapsulating the contents. Much thinner envelopes can be produced than by the coacervation method and these were first exploited by Chang (1964, see also 1972) to form artificial biological cells which could hold enzymes while permitting exchange of small molecule substrates.

Other agents than acid chlorides to react with diamines have since been exploited including diisocyanates to produce polyureas, e.g.

$$\text{OCN}-C_6H_3(CH_3)-\text{NCO} \xrightarrow{H_2N(CH_2)_nNH_2} -\text{NH}-C_6H_3(CH_3)-\text{NH-CO-NH-}(CH_2)_n-$$

The chief technical problem is to avoid reaction of these very chemically active substances with the physiologically active constituents of the oil phase. If this problem can be solved the drug or pesticide may be held in a more stable condition by exclusion of oxygen and water. By addition of triamines to the amine reagent, cross-linking can be introduced and a coating of very low permeability obtained.

So far micro-encapsulated pesticides are mainly materials of promise rather than established performance. They can play a part in reduction of handling risks and delayed release. Phillips *et al.* (1976) examined their use to delay the rapid action of synthetic pyrethroids in baits for the leafcutting ant so that the poison was carried to the nests (Section 13.V.A) and de Savigny and Ivy (1974) claim that micro-encapsulated methyl parathion had improved and more persistent residual contact action as an insecticide while its toxicity to mammals was reduced.

Micro-encapsulation has many technological uses in higher-price markets than agriculture (Vandegaer, 1974) and one may expect further development in the pesticide field when several difficulties, including economics, have been surmounted. It should be remembered that the techniques are not necessarily "micro" and it is fitting to add, in conclusion, Cardarelli's (1979) conception of "in flight" encapsulation by spraying solutions which will leave a coherent envelope on drying as the spray floats to its target. Other aspects of this solidification we have already (Section 14.V.C) mentioned.

Appendix 1. Coefficients of Molecular Diffusion

I. GASES

Molecules in a gas spend most of their time in free motion. Their mean kinetic energy is a measure of absolute temperature and is independent of mass or size, the relationship being

$$\overline{v^2} = 3\mathrm{RT}/M \qquad (1)$$

The proportionality of mean velocity (strictly, of the square root of the mean of the squares of velocities) to $1/\sqrt{M}$ and the concept that molecular motion directly produces diffusion lead to the expectation that $D\sqrt{M}$ should be the same for different diffusing molecules. This is frequently quoted as true for gases and even for liquids but is in fact true, and that only approximately, for moderately heavy vapours in air.

The molecules in two volumes of gas placed in contact advance into one another's territory much more slowly than at their mean velocity because collisions, though of short duration, are very frequent and result in complete chaos with regard to direction.

Collision frequency depends on the size of the molecules and the operation of short-range forces between them. In a gas containing molecules of one kind only and assumed to collide as elastic spheres, the mean free path between collisions is

$$\mathscr{L} = (8\sqrt{\pi} n a^2)^{-1} \qquad (2)$$

where a is the molecular radius and n the number per unit volume. This leads to the prediction that D for "self-diffusion" (measurable with isotopes) at given temperature and pressure should satisfy the equation

$$Da^2\sqrt{M} = \text{constant} \qquad (3)$$

The expression for the mean free path in a gas mixture is much more complex as is its relation to diffusion and it is necessary to distinguish collision between like and unlike molecules. The kinetic theory of diffusion was elaborated by Chapman and Cowling (1952) and their simplest prediction, assuming elastic spheres and no interference before impact, is

$$D = \frac{3}{8n}\sqrt{\frac{\mathbf{kT}}{2\pi}}(a_A + a_B)^{-2}\left(\frac{1}{M_A} + \frac{1}{M_B}\right)^{\frac{1}{2}} \tag{4}$$

for the (single at constant pressure) interdiffusion coefficient of two gases only, A and B. The pressure is defined by the temperature and the number, n, of molecules per unit volume. If one gas, A, is the same in a series of diffusion processes and the ratio of molecular masses, B to A, is m and of radii is r, we can write this equation, for constant pressure and temperature, in the form

$$D = \text{const.} \times (1 + r)^{-2}\left(1 + \frac{1}{m}\right)^{\frac{1}{2}} \tag{5}$$

To a rough approximation, the radii will be proportional to the $\frac{1}{3}$ power of the volume in the liquid state so that if ρ is ratio of density of B to that of A in this state

$$D = \text{const.} \times \left(1 + \left(\frac{m}{\rho}\right)^{\frac{1}{3}}\right)^{-2}\left(1 + \frac{1}{m}\right)^{\frac{1}{2}} \tag{6}$$

For m between 1 and 30 and ρ between 1 and 2 the product of the function by $(m/\rho)^{\frac{1}{2}}$ ranges only from 0·306 to 0·354. Most organic vapours in air would come into this range. Air can be regarded, accurately enough, as a single gas of molecular mass 29 with liquid density extrapolating to 0·7.

Many approximations are made in this theory. Collision radii should be derived from other data rather than assumed to be those of spheres which would, *in toto*, comprise the whole volume of the liquid. For benzene in air, eq. (4) with the radii derived as in eq. (6) gives $D = 0{\cdot}05\ \mathrm{cm^2\ s^{-1}}$ instead of the experimental 0·09, indicating that the space-filling spheres over-estimate the collision radii by about 35%. There have been further refinements of theory (see Chapman and Cowling) and several attempts to fit more elaborate equations to experimental data (see e.g. Fuller *et al.*, 1966). None are fully satisfactory. Fortunately values within ±30% are as accurate as other uncertainties in the application of interest in this text will allow to be significant. For vapours in air at normal pressure and at temperatures in the agricultural range, the following equation is quite good enough

$$D = \left(\frac{\mathbf{T}}{293}\right)^{1\cdot5} \times 0{\cdot}79\left(\frac{\rho_L}{M}\right)^{0\cdot5} \tag{7}$$

Values from this equation are compared with experimental data, corrected where necessary to 20°C in Table 1 below. Its predictions are within 10% for ordinary organic vapours, but more seriously low for water and high for iodine and mercury. For the important case of water the special equation

$$D=\left(\frac{\mathbf{T}}{293}\right)^{1\cdot5}\times 0\cdot24 \tag{8}$$

should be used.

In Table 1 the product of the experimental D value by $\sqrt{M}$ is compared with its product by $\sqrt{M/\rho}$. The experimental Ds cover a 5-fold range (2-fold if we omit water, mercury and iodine), $D\sqrt{M}$ covers a 3·8-fold range (2-fold with the same omissions). For $D\sqrt{M/\rho}$ the ranges are 2-fold and 1·4-fold. It would in fact be adequate to take $D=0\cdot07$ cm^2 s^{-1} for all organic vapours in the pesticide range in air under normal conditions. The object of this discussion is simply to show the origin and limitations of the $D\sqrt{M}$ constant assumption.

Table 1
Diffusion in air

Vapour	Exp D [a]	Exp D [b]	D from eq. (7)	$D_{exp}\sqrt{M}$	$D_{exp}\sqrt{\frac{M}{\rho_L}}$
Water	0·24	0·25	0·186	1·039	1·039
EtOH	0·109		0·103	0·739	0·827
iPrOH	0·095		0·090	0·697	0·789
nBuOH	0·083		0·082	0·714	0·794
Et_2O	0·088		0·077	0·757	0·898
Benzene	0·089	0·095	0·084	0·812	0·865
EtOAc	0·084	0·084	0·080	0·788	0·831
Cl benzene	0·071	0·070	0·078	0·736	0·698
$C_3H_4Cl_2$		0·077[c]		0·811	0·739
CCl_4	0·071			0·881	0·729
$C_2H_4Br_2$		0·081[d]	0·085	1·111	0·754
Cl_3CNO_2		0·086	0·079	1·101	0·857
diphenyl		0·065	0·068	0·918	0·863
n-decane		0·058	0·057	0·691	0·809
tri Me heptane		0·046	0·056	0·548	0·646
Mercury	0·132		0·202	1·899	0·515
Iodine	0·078		0·110	1·243	0·560

[a] From the compilation of Jost (1953).
[b] From the compilation of Fuller *et al.* (1966).
[c] From Leistra (1971).
[d] From Call (1957a).

II. EQUILIBRIUM GRADIENTS

The extension of kinetic analysis to liquids, where molecules are all the time under the powerful influence of intermolecular forces, is prohibitively difficult. A more general relationship to quantities other than diffusion coefficient has, however, some utility for liquids although none for gases. If molecules of one kind in a binary mixture are subjected to a steady unidirectional force a gradient of composition is built up until the selective movement created by this force is balanced by the randomizing effect of diffusion. A familiar practical example, where molecules of one kind are very large and heavy, is the sedimentation equilibrium in a gravitational—or, more usually, a much larger centrifugal—field, but the principle is valid whatever the nature of the force and whether it produces a measurable gradient or not. In dilute solution an exact thermodynamic relation exists at equilibrium. It is

$$\mathbf{RT}.\frac{\mathrm{d}\ln C}{\mathrm{d}x} = G \qquad (9)$$

where G is the force per unit of solute and $\mathbf{R}$ the gas constant in consistent units.

The gradient and the force are in equilibrium. No transfer of solute across any section is occurring. We now assume that the transfer produced by diffusion down the concentration gradient if the force were removed would be equal and opposite to that produced by the force in a uniform solution. This must be valid if the gradient is very small and must remain valid at higher gradients if Fick's law is applicable.

In Einstein's (1905) original use of this principle, the concentration gradient was envisaged as creating a force. The principle does not give us the diffusion coefficient directly but it relates it exactly to the sedimentation velocity

$$D = \mathbf{RT}/\mathbf{N}\boldsymbol{\phi} \qquad (10)$$

where $\boldsymbol{\phi}$ is the friction coefficient of a single molecule, i.e. the unidirectional component of velocity multiplied by $\boldsymbol{\phi}$ is equal to the force. If we are dealing with spherical colloidal particles, we can use Stokes' equation, for spheres moving in a fluid of viscosity η, to replace $\boldsymbol{\phi}$ by $6\pi\eta a$, a being the radius of the spheres. Einstein (1908) made the bold assumption that ordinary molecules can be treated in the same way, leading to the Stokes–Einstein equation

$$D = \frac{\mathbf{RT}}{6\pi\mathbf{N}\eta a} \qquad (11)$$

This equation predicts values of the right order. It does not involve molecular mass, only molecular size. It is important to appreciate how this apparently simpler equation applies to complex liquids and not simple gases. Equation (10) is exact and universal but it is no use to us in the case of gases because the sedimentation velocity is not simply related to force. The Stokes equation is applicable to small spheres moving slowly in a viscous *continuum*. Compact colloidal particles are a reasonable approximation. Where the diffusing molecules are smaller than the patterns of movement or structure in which they diffuse, one cannot expect the effect of the solvent to be represented simply by its viscosity. Sucrose diffusing in water is a system to which one might expect eq. (11) to apply. It is right within 10% for sucrose in water. It fails in gases because the free paths are much greater than the molecular radii. It predicts D for benzene vapour in air to be 1/200 of the experimental value. It fails for small molecules diffusing in a swollen polymer because the rigidity of the latter (its macroscopic viscosity) is created by a network structure on a coarser scale than the diffusing molecules.

III. ELECTROLYTES

In the case of ions in a strongly dissociating solvent we have another measure of the friction coefficient since the velocity of the ions in an electric field (under an exactly calculable force) is exactly related to their contribution to electrical conductance. By this reasoning, Nernst (1888) obtained the well-known equation which carries his name. It is exact for a pure salt in dilute solution. Extended to multivalent salts by Haskell (1908) it can be put most simply in the form

$$D = \frac{\mathbf{RT}}{F^2}\left(\frac{1}{Z^+}+\frac{1}{Z^-}\right)\left(\frac{1}{\mathbf{u}^+}+\frac{1}{\mathbf{u}^-}\right)^{-1} \tag{12}$$

where F is the Faraday equivalent of electricity in appropriate units, Zs are the charges on the ions in electron units but without regard to sign (i.e. 1 for K^+, 2 for $SO_4^=$ etc.) and **u**s are the equivalent conductances of the separate ions. For a simple uni-uni-valent salt, (12) assumes the more familiar form

$$D = \frac{\mathbf{RT}}{F^2}\frac{2\mathbf{u}^+\mathbf{u}^-}{\mathbf{u}^+ + \mathbf{u}^-} \tag{13}$$

(when **u**s are expressed as usual in $\text{ohm}^{-1}\,\text{cm}^2\,(\text{g eq.})^{-1}$ and D in $\text{cm}^2\,\text{s}^{-1}$, the numerical value of $\mathbf{RT}/F^2$ at 20° is $2{\cdot}62 \times 10^{-7}$).

Since large colloidal ions are multivalent they have equivalent conductances not greatly less than those of simple ions and if the simple ions associated with the large one are univalent, eq. (12) predicts diffusion of a pure colloidal electrolyte to be around half that of a simple salt. The value is high because of the potential gradient created by the tendency of the small ions to advance more rapidly. This effect is eliminated if there is a large excess of other simple salts present (Hartley and Robinson, 1931), in which case the diffusion of a single ion becomes

$$D = \frac{\mathbf{RT}}{F^2} \frac{\mathbf{u}}{Z} \tag{14}$$

As with all intensive properties of electrolytes, the values for infinite dilution fall rapidly with increase of concentration due to inter-ionic attraction effects. At the simplest, eqs (12) and (13) require a factor $\left(1 + \frac{\mathrm{d} \log f}{\mathrm{d} \log C}\right)$, where f is the activity coefficient. The effects are greater the higher the values of Z, but the most important of them (the activity factor) disappears when an ion diffuses down a small gradient in the presence of a large excess of other ions at uniform concentration. This factor therefore is unnecessary in eq. (14). Diffusion rates of ionic species, particularly of large or colloidal ions, are therefore best measured, because less sensitive to impurity and more easily interpreted, in the presence of a "swamping excess" of a simple salt.

IV. DIFFUSION OF NEUTRAL MOLECULES IN LIQUIDS

A. Variation with Size and Shape of Molecules

The average volume occupied by a molecule in the pure state is easily calculated from the molecular weight, density and Avogadro number. The simplest estimate of the radius, a, to be put into the Stokes–Einstein equation is made by assuming this volume to be that of a sphere of radius,

$$a = \left(\frac{3V}{4\pi\mathbf{N}}\right)^{\frac{1}{3}} \tag{15}$$

$$V = \frac{\mathbf{M}}{\rho} \tag{16}$$

Inserting numerical values and substituting eq. (15) in eq. (11), we obtain

$$D = 9{\cdot}96 \times 10^{-10} \cdot \frac{\mathbf{T}}{\eta} \cdot V^{-\frac{1}{3}} \tag{17}$$

where $\mathbf{T}$ is in °K, η in poise, V in $cm^3\ (g\,mol)^{-1}$ and D in $cm^2\,s^{-1}$.

Diffusion coefficients calculated from eq. (17) are found in general to be considerably lower than the experimental values in the case of small molecules (e.g. D for methanol in water at 25° $9{\cdot}9 \times 10^{-6}$ from eq. (17) compared with experimental 17×10^{-6} (Gary-Bobo and Weber, 1969)) and appreciably higher for large molecules (e.g. $4{\cdot}9 \times 10^{-6}$ for raffinose, $C_{18}H_{32}O_{16}$, cf. $4{\cdot}3 \times 10^{-6}$ (Longsworth, 1953)).

Equation (17) applied to toluene diffusing in hexane predicts $D = 2{\cdot}1 \times 10^{-6}\ cm^2\,s^{-1}$ at 25°C while the experimental value (Wilke and Chang, 1955) is $4{\cdot}2 \times 10^{-6}$. In tetradecane the discrepancy is greater still ($0{\cdot}30$ and $1{\cdot}02 \times 10^{-6}$).

Several workers have proposed empirical equations to provide a better fit. Longsworth used the partial molal volume in water solution rather than in the pure state in his investigation of the diffusion of aminoacids and sugars in water. This, however, makes inadequate improvement and he had to subtract a constant length from $V^{\frac{1}{3}}$ and use a lower factor in eq. (17). Scheibel (1954) proposed an equation which has rather similar effect but includes a dependence on the molecular volume of the solvent other than that attributable to viscosity. He found it to fit the data assembled by Wilke (1949), but Wilke and Chang (1955) considered a rather simpler equation to give a better fit with all data if special adjustment is made for associated solvents. Lusis and Ratcliff (1968) have reconsidered the data and propose an equation which, when expressed in the same form as Scheibel's, differs only in the numerical coefficients: they claim it to give the best fit with organic solvents. These three equations may be expressed

$$\frac{D\eta}{T} \times 10^{10} = 8{\cdot}2\,V_A^{-\frac{1}{3}} + 12{\cdot}2\,V_B^{\frac{2}{3}}\,V_A^{-1} \quad \text{(Scheibel)} \tag{18}$$

$$= 11{\cdot}9\,V_A^{-\frac{1}{3}} + 8{\cdot}5\,V_B^{\frac{2}{3}}\,V_A^{-1} \quad \text{(Lusis and Ratcliff)} \tag{19}$$

$$= 7{\cdot}4\,M_B^{0\cdot5}\,V_A^{-0\cdot6} \quad \text{(Wilke and Chang)} \tag{20}$$

where A refers to (dilute) solute, B to solvent, Vs are molal volumes (Longsworth would use partial volumes in solution) and M is molecular weight. It may be noted that $\sqrt{M}$ has reappeared in the Wilke and Chang equation but refers to the solvent and is in the numerator. The case for using M rather than V is not strong: the only solvent for which data are available and which is significantly denser than the rest is carbon tetrachloride and eq. (20) overestimates all data for this solvent.

When $V_B \ll V_A$, i.e. the case of large molecules diffusing among very small molecules, eqs (18) and (19) simplify to the form of the Stokes–Einstein equation, (17), but with different numerical coefficients. Equation (20) never assumes this form. If V_A and V_B are put equal, corresponding to self-diffusion (measurable by isotope methods) (18) and (19) become equal and predict

$$\frac{D_{\text{self}}\eta}{\mathbf{T}} = 20{\cdot}4\,V_A^{-\frac{1}{3}} \tag{21}$$

in which the numerical coefficient is just over twice that from the Stokes–Einstein equation. Equation (20) predicts

$$\frac{D_{\text{self}}\eta}{\mathbf{T}} = 7{\cdot}4\rho^{\frac{1}{2}}V_A^{-0{\cdot}1} \tag{22}$$

indicating very little dependence on V_A for compounds of the same density. This seems rather nonsensical and eq. (22) is certainly not true of a comparison of D_{self} for water ($24\times10^{-6}\ cm^2\ s^{-1}$ according to Wang *et al.*, 1953) and carbon tetrachloride ($1{\cdot}4\times10^{-6}\ cm^2\ s^{-1}$ according to Wilke and Chang, 1955).

It will be seen that an exact relationship of diffusion of non-electrolytes to other measurable properties is still elusive but it must be emphasized that, for compounds in the pesticide range of molecular size, the rates do not cover a very wide range. Longsworth's (1953) data and those of Gary-Bobo and Weber (1969) cover water-soluble non-electrolytes ranging in size from methanol to raffinose. D at 25°C goes from 17×10^{-6} to $4{\cdot}3\times10^{-6}\ cm^2\ s^{-1}$ only. Of non-ionic pesticides few have a molecular weight lower than 100, except for those used as fumigants, and few higher than 350, a range approximately equivalent to that between proline and raffinose, among the measured compounds, i.e. $8{\cdot}8-4{\cdot}3\times10^{-6}\ cm^2\ s^{-1}$. The Lusis and Ratcliff eq. (19) would provide an estimate within 15%. Even to use a standard value of $6{\cdot}5\times10^{-6}\ cm^2\ s^{-1}$ for all ordinary pesticides in water at 25°C would not be more than 30% in error. In all practical problems involving diffusion of pesticides other factors, soil structure, convection etc. will introduce greater uncertainty.

One should mention the effect of shape even if only to dismiss it as, in the context, unimportant. The axial ratio for an ellipsoidal particle has to be very far from unity before the average resistance differs greatly from that of a sphere of equal volume. This subject has been considered particularly in sedimentation and diffusion measurements on proteins, where the monodispersity and structural rigidity make interpretation more certain. The interested reader is referred to Alexander and Johnson (1949, p. 260 *et seq.*) for discussion.

B. Variation with Concentration

Diffusion in free liquids (or gases) is interdiffusion at necessarily equivalent rates. Only when some selective septum (a semi-permeable membrane in liquids or a finely porous plug in gases) allows volume or pressure changes to occur can we separate the diffusing tendencies of two molecular species. When we are concerned with diffusion of large molecules among smaller we cannot expect the function $D\eta/T$, even when non-ideal thermodynamics are allowed for, to remain constant throughout the whole range of composition because it is mainly determined by the molecular size of one species when it is in dilute solution and by the size of the second species when the first is nearly pure. For this reason, diffusion coefficients in aqueous solution (except in the case of dilute salts) increase with increase of concentration. The subject is outside our present range of interest except as an important source of error in many measurements. Key papers for the interested reader are by Hartley and Crank (1949), Miller and Carman (1959), Anderson and Babb (1961), Bearman (1961), Bidlack and Anderson (1964) and McCall and Douglas (1967). The effect takes on an extreme form, and one of great importance for permeation as considered in the main text (Section 9.II.F), when one species is a macromolecular one.

C. Solute–Solvent Interaction

In all but very dilute solutions, deviation from thermodynamic ideality can seriously affect observed diffusion rates. An interesting and extreme form of this effect has been shown by Khazanova and Kalsina (1961) in triethylamine-water solutions. The diffusion coefficient goes through zero near the critical miscibility condition, where the thermodynamic potentials are on a plateau.

There are also interaction effects not predictable from thermodynamic properties. The real volume of the diffusing entity can only plausibly be equated to the average volume in the pure state or in solution if these are equal. When the partial molal volume in solution is less than the molal volume in the pure state the compression must result from forces of attraction which create at least transient associations of molecules. A dilute solution of magnesium sulphate provides an interesting extreme example. The solution actually occupies slightly less volume than the constituents separately, i.e. the partial volume is negative, but one should not conclude that diffusion will be backwards. The immediately surrounding water molecules, oriented into a "hydration sheath" have a higher density than free water, but they are at least transiently attached to the hydrated ions and therefore increase its volume and resistance to movement. One could say that, in general, where the partial volume in solution is less than that in the pure state, the volume to be put in eq. (15) should be greater, but one cannot say by how much.

Another intermolecular force effect is apparent in aqueous solutions. The behaviour of monatomic argon in comparison with "isotopic" potassium chloride clearly illustrates this. Smith *et al.* (1955) found D for Ar in water at 25°C to be $14{\cdot}6 \times 10^{-6}\ \mathrm{cm^2\ s^{-1}}$. Muller and Stokes (1957) pointed out that this is significantly less than D for KCl $(20{\cdot}0 \times 10^{-6})$ despite the complete inertness of Ar and the significant hydration of K^+, Cl^-. Maharajh and Walkley (1973), using a different method of measurement, confirmed the low value for argon and found the temperature coefficient to be higher than for simple salts. Stokes (1964) points out that the naked radii of K^+ and Cl^- in a medium of high dielectric constant will be, as in the crystal, less than in the vapour and, at 0·133 and 0·181 nm respectively, are both less than the radius of Ar, 0·192 nm, but he also considers that the hydration layer must add an annulus of thickness 0·28 nm, making it necessary to conclude that the water just outside the hydration layer is "destructured". Muller and Stokes found the mobility of the singly negative citrate ion to be significantly less than that of the undissociated citric acid molecule. Gary-Bobo and Weber come to a similar conclusion to explain that the diffusion coefficients for lower amides and alcohols in water are somewhat greater than those of the hydrocarbons, measured by Witherspoon and Saraf (1965)—a typical comparison is (equating, C, O, N for volume) *n*-butane 9·6, *n*-propanol 11·5, acetamide $13{\cdot}2 \times 10^{-6}\ \mathrm{cm^2\ s^{-1}}$ at 25°C. The alkanes, like argon, are "structure promoters". The alcohols, and especially the even more hydrophilic (Section 2.X.C) amides, are "structure breakers". The larger hydrated molecules (or ions) are surrounded by a zone of water much less viscous than the bulk.

Again, these effects are small in relation to the errors involved in application of diffusion equations to pesticide problems. They are considered here in order to dispel any idea that rapid transport in some biological systems can be due to special physical factors rather than to metabolically-powered carrier mechanisms, the simplest of which may be cyclic stirring processes.

Water itself, as a diffusing substance, presents some rather special anomalies. When its transport is measured, very conveniently, by use of the radioactive hydrogen isotope, tritium, it is in fact the diffusion of this atom which is being measured. Hydrogen-bonded water molecules form chains in which unique attachments between H (or T) and O are not maintained. The T atom could progress by exchange of partners as well as by diffusion of a TOH molecule as such. It is this mechanism which gives the hydrogen ion in water a much higher velocity in the electric field than any other ion. Wang *et al.* (1953) measured the self-diffusion of water independently labelled with deuterium, tritium and ^{18}O and found that it was the latter which gave a slightly higher value than either of the others, but only by 14%. The exchange effect is therefore a small one. The authors attribute this to most of the water being in structured units and diffusion rate being determined by rate of escape from these units. The rapid mobility of ionized hydrogen is reflected in the rapid diffusion of strong acids in agreement with the Nernst equation (12).

V. DIFFUSION IN SOLIDS

Diffusion can occur for particular substances in crystalline solids—e.g. interdiffusion in metals and of ions in isomorphous crystalline solutions. In our present context, it is negligible.

Diffusion in the amorphous part of macromolecular solids such as "plastics", plant cuticles etc. is significant and is discussed in the main text. The value of D cannot usefully be predicted. Commonly it is several powers of ten lower than in liquids, but is much higher than would be expected from the Stokes–Einstein equation, the apparent value of η from this equation being of the order of a viscous oil whereas the material may be completely solid judged by its dimensional stability.

The mechanism, of course, is that small molecules can exchange places with segments of flexible chains even when the chains themselves cannot change places and the material is therefore, in bulk, rigid. Simple salts diffuse almost as rapidly in a "rigid" table jelly as they do in water. They diffuse through the spaces between the rigidifying gelatin network. A dense polymer is intermediate because most of the matter in it is part of the network structure. Diffusion is therefore much faster than the Stokes–Einstein equation would predict but much slower than in simple liquids and it is *much more dependent on molecular size and shape.* This property tends to be overlooked in comparison with partition effects when permeability of skins is considered (see Section 9.II.F).

VI. VARIATION WITH TEMPERATURE

For diffusion in air at ordinary temperature and pressure, the diffusion coefficient is approximately proportional to $\mathbf{T}^{\frac{3}{2}}$, i.e. the increase is about 0·5%/°C.

For diffusion in liquids, the Stokes–Einstein equation predicts proportionality to $\mathbf{T}/\eta$ and this part of the prediction has not been seriously questioned. For water at 20°C, $1/\eta$ increases by about 2·7%/°C, so that D increases by about 3%/°C, much more than in air.

The vapour pressure of all substances increases with temperature. For substances in the pesticide range the increase is of the order of 5%/°C. When diffusion in both air and water is important (Sections 3.IV.E and 5.II.C), vapour phase diffusion become relatively more important the higher the temperature.

Diffusion in macromolecular materials generally increases much more rapidly with temperature than in simple liquids; 10–20%/°C is common for simple "solvents" such as ethyl acetate or benzene diffusing in such materials as poly(vinylacetate) or rubber.

Appendix 2. Vapour Pressure—General

McGowan (1965) gives a useful empirical equation for non-associated liquids which is an improvement on the simplest combination of Trouton's rule and the Clapeyron–Clausius equation and approximately valid over a wider range.

The equation is

$$\log_{10} \mathbf{p} = 5{\cdot}580 - 2{\cdot}7(\mathbf{T}_b/\mathbf{T})^{1{\cdot}7} \qquad (1)$$

where $\mathbf{p}$ is the saturation vapour pressure in mmHg at temperature $\mathbf{T}$ (absolute, centrigrade) and $\mathbf{T}_b$ is the boiling point at 760 mmHg.

This equation can be extended to the case where the boiling point, $\mathbf{T}_1$, is known only at a lower pressure, $\mathbf{p}_1$. To predict the vapour pressure, $\mathbf{p}_2$, at temperature, $\mathbf{T}_2$, we eliminate $\mathbf{T}_b$ from the two substitutions in equation (1) and obtain

$$\log_{10} p_2 = \left(\frac{\mathbf{T}_1}{\mathbf{T}_2}\right)^{1{\cdot}7} \log_{10} \mathbf{p}_1 - 5{\cdot}580\left[\left(\frac{\mathbf{T}_1}{\mathbf{T}_2}\right)^{1{\cdot}7} - 1\right] \qquad (2)$$

To facilitate the application of this equation to estimate vapour pressures of liquids at 20°C, the values have been tabulated (Table 1), for $\mathbf{T}_1$ in °C over the useful range at 10°C intervals,

of $(\mathbf{T}_1/\mathbf{T}_2)^{1{\cdot}7}$ (Col. 2)

of the complete second term (Col. 3), and

of the final sum for the case of $\mathbf{p}_1 = 760$ mm, i.e. of $\log_{10}\mathbf{p}_{20°C}$, when the boiling point expressed in the first column is that at 760 mm.

To obtain $\log_{10} \mathbf{p}_{20°C}$ when the boiling point expressed in the first column refers to another pressure, $\mathbf{p}_1$ mm, multiply the second column figure by $\log_{10} \mathbf{p}_1$ and subtract the third column figure from the product.

There is, of course, exact relationship between the saturation vapour concentration (SVC) and vapour pressure

$$\log_{10} \text{SVC}(\text{g}\ \ell.^{-1}) = K + \log_{10} \mathbf{p}(\text{mmHg}) + \log_{10} (\text{Mol. mass})$$

Table 1

Tabulation of McGowan's vapour pressure equation

			For $\mathbf{p}_1 = 760$ mm	
b.p. (°C)	$\left(\frac{\mathbf{T}_1}{\mathbf{T}_2}\right)^{1\cdot 7}$ Value and diff. for 1°C	Second term eq. (2) Value and diff. for 1°C	$\log_{10} \mathbf{p}_{20}$ and diff. for 1°C	SVC in g $\ell.^{-1}$ for mol. mass = 200
60	1·243	1·355	2·226	1·840
	0·0064	0·0358	0·0175	0·0610
70	1·307	1·713	2·051	1·230
	0·0065	0·0365	0·0177	0·0377
80	1·372	2·078	1·874	0·853
	0·0067	0·0372	0·0179	0·0311
90	1·439	2·450	1·695	0·542
	0·0068	0·0379	0·0182	0·0186
100	1·507	2·829	1·513	0·356
	0·0069	0·0386	0·0187	0·0124
110	1·576	3·215	1·326	0·232
	0·0071	0·0394	0·0190	0·0082
120	1·647	3·609	1·136	0·150
	0·0072	0·0401	0·0194	0·0054
130	1·719	4·010	0·942	0·096
	0·0073	0·0408	0·0198	0·0035
140	1·792	4·418	0·744	0·061
	0·0074	0·0415	0·0200	0·0023
150	1·866	4·833	0·544	0·038
	0·0076	0·0422	0·0203	0·0014
160	1·942	5·255	0·341	0·024
	0·0077	0·0429	0·0208	0·0008
170	2·019	5·684	0·133	0·016
	0·0078	0·0435	0·0210	0·00068
180	2·097	6·119	$\bar{1}$·923	0·0092
	0·0079	0·442	0·0215	0·00036
190	2·176	6·561	$\bar{1}$·708	0·0056
	0·0080	0·0449	0·0219	0·00022
200	2·256	7·010	$\bar{1}$·489	0·0034
	0·0081	0·0456	0·0222	0·00014
210	2·337	7·466	$\bar{1}$·267	0·0020
	0·0083	0·0462	0·0223	0·00008
220	2·420	7·928	$\bar{1}$·044	0·0012
	0·0085	0·0469	0·0225	0·000048
230	2·505	8·397	$\bar{2}$·819	0·00072
	0·0085	0·0476	0·0230	0·000038
240	2·590	8·873	$\bar{2}$·589	0·00034
	0·0086	0·0482	0·0234	0·000009
250	2·676	9·355	$\bar{2}$·355	0·00025
	0·0088	0·0488	0·0236	0·000008
260	2·764	9·843	$\bar{2}$·119	0·00017

Values of K are at

15°C	20°C	25°C	30°C
$\bar{5}$·7459	$\bar{5}$·7384	$\bar{5}$·7311	$\bar{5}$·7239

In the fifth column of Table 1 are given values of SVC in g ℓ^{-1} corresponding to the 20°C vapour pressure in Col. 4 for a hypothetical compound of mol. mass 200.

The measurement of low vapour pressures is not easy and many discrepant values appear in the literature. Direct measures are often erroneously high because of impurities or decomposition products more volatile than the "target" compound. Vapour pressures, however, are not capriciously variable. No substituted paraffin is more volatile than the unsubstituted compound (except for some polyfluoro compounds). If a dichloro compound is reported more volatile than its parent, the report should be discounted. Most linkages in organic compounds other than through C atoms confer some polarity which reduces vapour pressure. If N and O are reckoned as C, Cl reckoned as 1·5 × C and S and P as 2 × C we arrive at an "equivalent paraffin" and this will certainly be *more* volatile than the compound under enquiry. Thus the ethyl ester of 2,4-D has the empirical formula $C_{10}H_{10}O_3Cl_2$ and the "equivalent paraffin" would be $C_{16}H_{34}$ with a vapour pressure at 20°C of $5{\cdot}8 \times 10^{-3}$ mmHg. A report of a higher pressure for the ester would be ludicrous. A lower one is to be expected because of the polarizability of the benzene ring and the

Table 2
Vapour pressures of n-paraffins over liquid state at 20°C

no. C atoms	mmHg
10	1·7
11	0·61
12	0·25
13	0·089
14	0·035
15	0·014
16	0·0058
17	0·0025
18	0·0011
19	0·0005
20	0·0002
21	0·0001
22	0·00005

polarity of the ester group. The value of 10^{-4} mm (p. 363) is acceptable. Comparison with the vapour pressure of the "equivalent paraffin" will not give a reliable approximation for highly polar compounds but will guard against acceptance of erroneously high values. In Table 2 are values for the saturated paraffins in the range of chief interest. These were obtained from the compilation of G. G. Schlessinger (1972) summarized in the "Handbook of Physics and Chemistry" (52nd edn) (Chem. Rubber Co.). One cannot expect the extrapolation to 20°C to be very accurate. The values probably range in accuracy from ±5% at C_{10} to ±30% at C_{20}.

Appendix 3. Ionization Constants—General

The ionization of weak acid is classically represented as $HA \leftrightharpoons H^+ + A^-$ equilibrating according to the equation

$$K_A[HA] = [H^+] \times [A^-] \quad (1)$$

the square brackets denoting concentrations and K_A being the ionization (or ionic dissociation) constant.

In reality the hydrogen ion does not exist as a naked proton but is closely associated with molecules of solvent. In water, at the simplest it exists as an oxonium ion H_3O^+, but this is of no consequence for the validity of eq. (1) if the concentration, $[H^+]$, is understood to include the various forms of association with the solvent in an equilibrium which will be constant in dilute solution. The different ability of solvents to accept protons is part of the mechanism of variation of K_A with solvent. K_A is much higher in water than in most organic solvents.

A weak base was classically represented as dissociating into cation and hydroxyl ion

$$K_{base}[MOH] = [M^+] \times [OH^-] \quad (2)$$

but, the hydroxyl ion being exclusive to water, it is now more usual to express the equilibrium as

$$K_A \times [MH^+] = [M] \times [H^+] \quad (3)$$

i.e. as the protonation of the base. In water, since all species will be associated with water molecules in a way unimportant for the formal equations, we could write [MOH] in eq. (2) as $[MH_2O]$ and $[M^+]$ as $[MH^+]$ and identify $[MH_2O]$ with [M] in eq. (3), e.g. it is immaterial whether we call ammonia, in solution in water, ammonium hydroxide, NH_4OH, in eq. (2) or NH_3 in eq. (3).

Since water itself ionizes, according to

$$K_w = [H^+] \times [OH^-] \quad (4)$$

it follows that, for a base and the conjugated acid,

$$K_{\text{base}} = K_w / K_A \tag{5}$$

Ionization constants are frequently expressed in logarithmic form. Since pH is defined as $-\log_{10}[H^+]$, pK is similarly defined as $-\log_{10} K_A$. On this convention eq. (1) may be rewritten

$$\log_{10}\left(\frac{[A^-]}{[AH]}\right) = pH - pK$$

Note that a low pK denotes a strong acid or a weak base. From this we see that the ionization stage considered is half complete when pH = pK, only 9% complete at 1 pH unit more acid, only 1% at 2 pH units more acid: correspondingly 91% complete at 1 pH unit more alkaline and 99% at 2 pH units more alkaline. Otherwise expressed, 82% of the amount of acid necessary to change the A content of a solution from all A^- to AH, produces only 2 units change of pH. pH changes most slowly with added strong acid when pH = pK. This is the basis of the most widely used method for measuring ionization constants—potentiometric titration. pH is plotted against quantity of added acid or alkali. Steps in the curve correspond to pK values and several pKs may be measured in one solution if they are not too close together.

Other methods are spectroscopic, when a difference in light absorption between ionized and unionized molecules enables relative proportions to be measured, and conductimetric. The latter is, however, easily applied only to pure solutions of single acids or bases and is not accurate if they are very weak.

Ionization constants are moderately temperature-dependent, very dependent on solvent. They are somewhat increased in water by the presence of neutral salts, because of compression of the ionic atmospheres (Section 2.XVIII.A). Acetic acid, for example, is about 20% more ionized in a 1 N sodium chloride solution than in pure water. An immiscible second solvent does not directly affect the dissociation constant but shifts the equilibrium by removing undissociated acid. The important interplay of ionization constant and partition coefficient is considered in Section 2.XVII. Solubility again has no direct effect on ionization constant but, by setting a maximum value to [HA] in eq. (1) or [MOH] in eq. (2), limits the ionic concentrations that can be obtained.

The ionization constant is, of course, dependent on chemical structure. The effects of substituents follow some fairly clear trends and enable reasonable estimates to be made where direct measurement is not available, as is the case with many pesticides.

With very few, but important, exceptions, acidic pesticides owe this property to ionization of a carboxyl, $-CO_2H$, group or a phenolic –OH group and basic pesticides to protonation of an amino, $-NH_2$, group. We now consider the effect of substituents on the ionization of these important chemical groups. Values for pesticides are listed in Appendix 4.

The simplest series is the saturated fatty acids. pKs at 20°C of the first five, i.e. formic to *n*-valeric, are 3·75, 4·76, 4·87, 4·81, 4·84. As is frequent in such homologous series, the first member is anomalous, formic being ten times as strong as any other. There is then a small, but significant, further decrease in strength (increase of pK) from acetic to propionic and consistently, pivalic acid (trimethylacetic) has a significantly higher pK (5·1) than straight chain acids. Beyond propionic, there is no important change. Alkyl substitution has, therefore, a weakening effect which is small when it has to pass through a $-CH_2-$ group, and negligible when it has to pass through two $-CH_2-$ groups.

The straight chain ($\alpha\omega$) dibasic acids introduce some important principles. Two ionizations are possible, and therefore 2 pK values, the second necessarily higher than the first. If the intrinsic strengths of the two acid groups were equal, as we might expect them to be if symmetrically placed and sufficiently separated by $-CH_2-$ groups, the first K_A would be twice that for a corresponding unibasic acid for purely statistical reasons. The undissociated dibasic acid can lose a proton equally easily from either group to give indistinguishable univalent ions, but each univalent ion can accept a proton in one place only. Correspondingly the second dissociation constant would be half that of the unibasic acid. From propionic acid on in the unibasic series, pK is close to 4·83 and we should expect the values for a dibasic acid to be $\log_{10}2$ different, i.e. 4·53 and 5·13. The actual values for suberic acid, in which 6 $-CH_2-$ groups separate the carboxyls, are 4·52 and 5·41—the first as expected, the second significantly weaker.

There are two possible mechanisms for physical interference between the two acidic groups. First is the so-called "inductive" effect due to disturbance of mean electron distribution within the molecule. The second is an external effect through the solvent. When both groups are ionized, the two ionic charges on one molecule are much closer together than the mean distances between separate ions in the dilute solution. Each anion increases the mean concentration of cations, including H^+ ions, around it and therefore each anion finds itself in a rather greater H^+ ion concentration than if the other were not present. The dissociation eq. (1) refers to mean bulk concentration and the second dissociation constant calculated from it is therefore less than it would be in the absence of this "ionic atmosphere" (Section 2.XVIII.A) effect. The behaviour of suberic acid is consistent with negligible inductive effect through 6 $-CH_2-$ groups but appreciable ionic atmosphere effect.

For oxalic acid, the pKs are 1·27 and 4·27 indicating a great acid strengthening inductive effect greater even for the second dissociation than the ionic atmosphere effect. For succinic, pKs 4·22, 5·66, the inductive effect on K_1 is still significant but that on K_2 is less than the atmosphere effect, the latter greater than in suberic because of the less distance between the charges.

The acid-strengthening inductive effect is due to electron withdrawal by the substituent group against which a few saturated C atoms provide insulation. Unsaturation facilitates electron displacement and we find accordingly than when the $-CH_2-CH_2-$ of succinic is replaced by $-CH{=}CH-$ pK_1 is reduced, to 1·9 in fumaric (*trans*) and 3·0 in maleic (*cis*).

Another illustration of the insulating effect of saturated C atoms can be seen in the chloro substituted fatty acids. For monochloroacetic, the pK is 2·82 (1·9 more acid than acetic) and for α chloropropionic, 2·87 (2·0 more acid than propionic), while for β chloropropionic it is 4·0 (0·87 more acid than propionic). Malonic acid has $pK_1 = 2{\cdot}85$ which we must correct, for the statistical factor (above), to 3·15, rather weaker than chloroacetic, indicating that $-CO_2H$ has less acid-strengthening effect than –Cl. We cannot quote for chloroformic in comparison with oxalic because the former has only microsecond stability. The greatest effect for a single substitution in acetic acid is for the –CN group—cyanacetic acid has pK 2·46, a displacement of 2·3 units more acid than acetic. Multiplication of acid-strengthening groups will usually increase the effect. Trichloroacetic acid has pK 0·66, 4·1 units more acid than acetic.

Some compounds of pesticidal interest, either as formulants or as metabolites of active compounds, have phosphonic, phosphoric, sulphonic or sulphuric acids, in part esterified so as to leave at least one acid function free. The first, or only, ionization of such acids can be considered strong.

Phenol itself is a very weak acid, pK 10·0 but is much more influenced by substituents than are the aliphatic acids. The resonating "aromatic" structure of the benzene ring has more ability to accommodate electrons and allow displacement to substituent groups. Methyl is slightly weakening; chloro, cyano, nitro groups are greatly strengthening and their effect is greatest in the ortho and para positions. The pKs for *o*-, *m*- and *p*-nitrophenols are 7·2, 8·4, 7·2. For 2,4-dinitro and 2,6-dinitro they are 4·1 and 3·7. While acetic acid is nearly one million times as strong as *o*-cresol, it is weaker than DNOC.

With benzoic acid, substitution in the benzene ring has much less, but still significant, effect and strengthening increases in the order $p < m < o$. Typical are the monochlorobenzoics with pKs 4·0, 3·8, 2·9 compared with 4·2 for no substitution. When the $-CO_2H$ is spaced off from the ring by $-CH_2-$ or $-O-CH_2-$, the effect is much less. Phenylacetic, pK 4·3, is reduced only to 3·7 by 2,3,6-trichloro substitution, which reduces benzoic from 4·2 to 1·3. Phenoxyacetic at 3·0 is stronger but this is mainly the effect of the ether group, ethoxyacetic having pK 3·8. The effect of ring substituents in phenoxyacetic is minor.

The relationship of basic behaviour of tervalent nitrogen to chemical structure is more complex than in the case of the phenolic and carboxylic acids. We begin the discussion with ammonia, NH_3, having a pK in water of 9.25. At this pH, half the NH_3 content is protonated to NH_4^+. The base constant (eq. (2)) is $1{\cdot}8 \times 10^{-5}$ close to that of acetic as an acid. Alkyl substitution, which weakens acids, strengthens nitrogen bases, the pKs of ethylamine and methylamine being 10·75 and 10·30 (note: increase of pK indicates a weaker acid but a stronger base). For triethylamine and trimethylamine the values are 10·80 and 9·87. Trimethylamine, as often with the first member of a series, is anomalous. Higher trialkylamines are close to triethyl.

The alkylamines exhibit a behaviour which has no parallel in the acids. A fourth attachment of N to C is possible but this enforces a positive charge so that a "quaternary nitrogen" is necessarily a cation, the hypothetical base being strong. Paraquat and diquat, of course, have this cationic structure.

Most other substituents weaken the basic property, parallel with their effect in increasing acidity. Fairly typical is OH which strengthens acetic acid by 1 pH unit to glycollic and reduces ethylamine by about 1 pH unit when ethylamine goes to ethanolamine. Ethanolamine, pK 9·5, is still just stronger than ammonia, but triethanolamine, pK 7·9, much used in formulation of herbicidal acids, is weaker.

The phenyl group is greatly weakening, pK of aniline being 4·6 (6 units more acid than ethylamine) and acid-strengthening substituents in the benzene ring weaken the base further (*o*-, *m*-, *p*-nitro anilines having pKs of −0·3, 2·5 and 1·0). Basicity is virtually absent in the 2,6-dinitroaniline herbicides (trifluralin etc.). As in the acids, a $-CH_2-$ group partly insulates the basic N against substituent effects. Benzylamine (pK 9·3) is much stronger than aniline.

The amino group itself is base-weakening. Ethylene diamine, which, for statistical reasons, could be expected to have pK 0·3 units greater than ethylamine has pK_1 of 10·07 (i.e. 0·7 units less). pK_2 is 7·0. The effect is greater without $-CH_2-$ insulation, hydrazine having pK_1 of 5·5.

An imino group on the same C atom as amino produces considerable strengthening. Guanidine (I) is the strongest known non-quaternary nitrogen base, pK 13·6. Methylation in this case slightly weakens the base (13·2). The *N*-phenyl compound is weaker still, as expected, but much stronger (at 10·7) than aniline. A long alkyl will probably have less effect than methyl, so that the fungicide dodine (II) is a very strong base consistent with strong alkali being necessary to liberate it from the standard acetate formulation. The fungicide glyodin (III), 2-heptadecyl imidazoline, also has the amine-imine combination. We can find no record for this compound itself but pK for the 2-ethyl compound is reported as 11·0 and is not likely to be much different.

$$H_2N\text{–}C(NH_2)\text{=}NH \quad \text{(I)} \qquad C_{12}H_{25}NH\text{–}C(NH_2)\text{=}NH \quad \text{(II)} \qquad C_{17}H_{35}\text{–}C{<}(\text{=N–CH}_2\text{–CH}_2\text{–NH–}) \quad \text{(III)}$$

When the N atoms are included in a ring having at least two double bonds in the classical structural formula and therefore classed as "aromatic" (the real structure having a distributed charge in the resonance hydrid) the basic function is drastically weakened and acidic property may even result. The contrast between pyridine (IV, pK 5·3) and the corresponding saturated piperidine (V, pK 11·3) serves to illustrate the trend

$$\text{(IV) pyridine: } CH{=}CH\text{–}CH{=}CH\text{–}N{=}CH \text{ (ring)} \qquad \text{(V) piperidine: } CH_2\text{–}CH_2\text{–}CH_2\text{–}CH_2\text{–}CH_2\text{–}NH \text{ (ring)}$$

Before considering more complex ring compounds, which include many important pesticides, we should consider the significance of the "zwitter-ion" structure.

n-Propylamine has pK 10·68, propionic acid has pK 4·87. At pH midway between, each will be only about 0·1% *un*ionized, whether alone or mixed. We should not except a major change in acid or base strength if we replace a mixture of $O_2^-CCH_2CH_3$ and $CH_3CH_3CH_2NH_3^+$ by the covalently linked 6 amino hexanoic acid. This will exist in ionized form, each molecule carrying a + charge at one end and a − charge at the other. The overall charge is zero. Only because we expect the individual pKs to be not much altered do we know that this "zwitterion" and not the neutral molecule will be the species most abundantly present in solution. We might expect that the enforced proximity of the charges will slightly increase the tendency to ionize thus placing the two pK values a little further apart. The experimental values are 4·4 and 10·8, indicating that some slight internal weakening effect masks the external strengthening effect in the case of the base function.

Contrast now aniline, pK 4·6, and phenol, pK 9·9. In this case, at pH midway between, each will be only about 0·25% *ion*ized. If we replace by a covalent single molecule having both functions, e.g. 4-amino, 4-hydroxy diphenylmethane, we should know that the "neutral" species will really be unionized, only a very small proportion of zwitterion being present.

These are extreme cases, and simple ones. In glycine, $H_2NCH_2CO_2H$, the acid and base functions are so close that inductive effects may be large. So will the atmosphere effect and this certainly in the direction of making the acid pK lower than the 4·4 of 6 amino hexanoic acid and the base pK higher than the 10·8. Potentiometrically determined pKs in this case are 2·35 and 9·78. We know from other evidence in this well known case that the major neutral species is still the zwitterion. If the decrease from 4·4 to 2·35 represents the atmosphere effect then the internal one must greatly weaken the base function, since its pK also decreases.

Potentiometric methods will give us the pK values associated with change of ionic charge. If we come down the pH scale by adding acid to a strongly alkaline solution the successive pK values indicate decreasing unit steps of negative charge but, without other evidence, give us no information on whether an anion is suppressed or a cation created.

A good example of the uncertainty is illustrated by the herbicide glyphosate. The measurements of Sprankle *et al.* (1975), confirmed by Wauchope (1976) list the following pK values and the ionic changes which they are believed to represent:

pK	predominant species
	$HOCOCH_2\overset{+}{N}H_2CH_2PO_3H_2$
>2	
	$HOCOCH_2\overset{+}{N}H_2CH_2PO_3H^-$
2·6	
	$^-OCOCH_2\overset{+}{N}H_2CH_2PO_3H^-$
5·6	
	$^-OCOCH_2\overset{+}{N}H_2CH_2PO_3^=$
10·6	
	$^-OCOCH_2NHCH_2PO_3^=$

There can be no doubt about the first two and last two species, but it could perhaps be argued that the middle species should have two charges on the phosphonic group and none on the carboxyl.

The closer the ionizable groups are the more difficult it is to allocate a correct structure because the interaction, both atmosphere and inductive, increases. Interposition of at least one $-CH_2-$ group cuts down the inductive interference, but when we have the ionizable groups part of an "aromatic" ring structure the uncertainty of structure greatly increases. In the aromatic rings, alternative "canonical" structures become freely interchangeable and a resonance hybrid results having two important effects on ionization. Firstly a generalized affinity for electrons makes negative ionization more probable—i.e. it has an acid-strengthening effect, which may even give an apparent acid function to an =NH group. Secondly the anionic charge is a mobile property of the ring. It is not localized on one particular atom and distinction between a zwitterion and a highly polar molecule largely disappears.

We have already commented on the weakness of aromatic pyridine as a base compared with the hydrogenated, piperidine. Despite this weakness, and a significantly higher b.p. (115°C cf. 106°C), both are miscible with water. The weaker base is thus more self-attracting and more water attracting (Section 2.X). The difference is more evident if the ring methyl compounds are compared, all methyl pyridines being water miscible while two of the methyl piperidines have limited, through high, solubility. One could have expected the stronger base to be more water-attracting even unionized. The anomaly could be explained by postulating a zwitterion form of "unionized" pyridine behaving only as a strong dipole in non-polar solvents but with separation of charge increasing in an ionizing solvent (water) because of the mobility of the electrons in the resonating ring system.

The three compounds where two –CH– groups of benzene are replaced by –N– are even weaker bases than pyridine and symmetrical triazine is weaker still. They are components of many pesticide molecules, usually with an amino group attached to C. Some of these compounds are particularly interesting with regard to base strength since they contain (around the top C atom in formulae (VI), (VII), (VIII)) the grouping characteristic of the very strong base, guanidine.

Nevertheless 2-amino pyrimidine (VII, R = H) has pK only 3·4 and even melamine (VI with R = H and the other Hs replaced by NH_2) has pK only 5·1. 3-amino, 1,2,4-triazole, the herbicide amitrol (VIII) also contains the guanidine structure but has pK only 4 (cf. 13 for guanidine). It is nevertheless very water-soluble (limited by crystallinity) and is

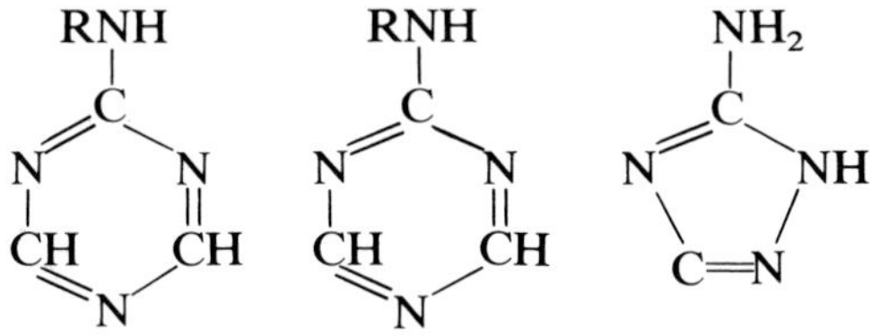

(VI) (VII) (VIII)

another example of the anomaly of pyridine solubility referred to above. The crystalline compound has higher solubility in water than in most solvents and very low solubility in non-polar solvents. It behaves therefore very much as an organic salt but is apparently unionized except in acid solution.

Maleic hydrazide (IX) is apparently an example of acidic properties in a nitrogen containing ring. Its acidic property is unquestioned. (Albert and Phillips, 1964, give $pK_1 = 5 \cdot 7$ and $pK_2 = 13$. The value for the first dissociation is confirmed by Barlin (1974) but that for the second disputed, as also by Cookson and Cheeseman (1972) who find protonation in solutions more acid than 10 N.) The structure can be put in the benzenoid form (X) and the acidity then appears as a "phenolic" –OH dissociation, much stronger however than in phenol.

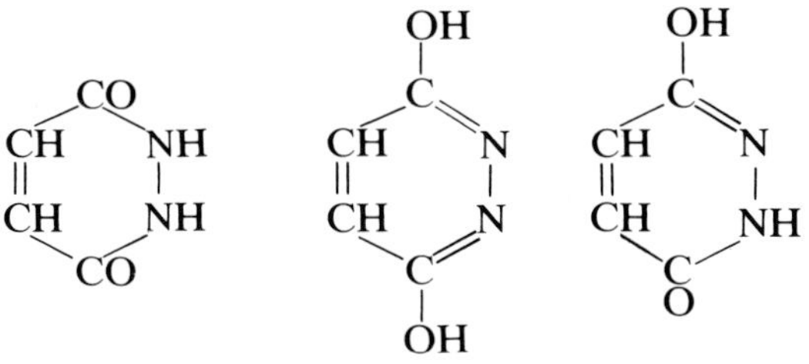

(IX) (X) (XI)

The structure is frequently written in the form (XI), but this practice rather misses the significance of its acid-strengthening resonance hybridization.

The herbicide asulam (XII) is sufficiently acidic to form a sodium salt by solution in bicarbonate, and is formulated as a stable solution of the salt. The acidic group is held to be the –NH– between $-SO_2-$ and –CO–, but, if the –N= in this form is doubly bound to S or C, the − charge can be located on these atoms. The ion is a resonance hybrid.

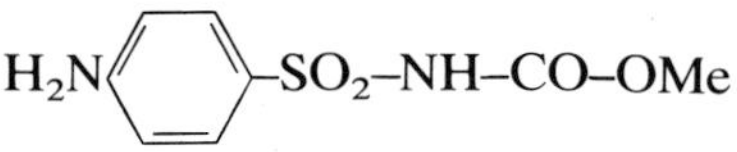

$H_2N-C_6H_4-SO_2-NH-CO-OMe$

(XII)

Appendix 4. Physical Properties of Pesticides

Notes on Entries and Abbreviations

The accepted common names of pesticide as listed in the British Crop Protection Council's "Pesticide Manual" (6th edn) edited by C. R. Worthing are given.

Only compounds which have agreed common names or short chemical names and for which at least two entries are available (see "Sources") are listed. We have also omitted compounds of indefinite composition.

Column 1. Molecular mass to nearest whole number.

Column 2. Melting point. For many compounds melting below room temperature no figure is available and is replaced by "ℓ". If a m.p. with a short range (<5°C) is given in the source, only the upper figure is listed. Larger ranges are quoted in full, the limits being separated by a dash. If two or more figures are separated by a comma the values refer to distinct crystal forms.

Columns 3, 4, 5. Solubility is quoted for room temperatures (normally 20–25°C) in g ℓ^{-1} but, in the water column (3), an index ^{3} or ^{6} indicates mg or μg in place of g. "V" means very low, "V. Sol" very soluble. In column 5, referring to paraffinic hydrocarbon solvents (PHS), "V" means very low, "L" means low and "M" (moderate) indicates solubility not greatly less than in a "good" solvent. Column 4 refers to a solvent or solvents in which the compound concerned is much more soluble than in the extremes of water or paraffin, but it cannot be assumed that the best possible solvent is listed. Where no figure is given in this column the solvent range most probably useful is indicated.

Abbreviations have the following meanings: MISC = the liquid pesticide is miscible with most organic solvents: where a dash extends into the water or paraffin columns, this indicates that miscibility is so extended; WSS = the acid or basic pesticide forms water soluble salts;

PO = polar organic solvents; ALC = alcohol, i.e. pesticide most soluble in hydroxylic solvents; AHC = aromatic hydrocarbons; KET = ketones; ACET = acetone; ALK = alkaline (water); DMF = dimethyl formamide; DMSO = dimethyl sulphoxide; ISOPH = isophorone; BENZ = benzene; TOL = toluene; XYL = xylene; ANIL = aniline; GEN = generally soluble in organic solvents but more specific information not available. Other specific solvents are indicated by chemical formula or accepted abbreviation.

Column 6. Vapour pressure in torr (mmHg). An index x is an abbreviation for 10^{-x}—i.e. $9{\cdot}1^{7} = 9{\cdot}1 \times 10^{-7}$. Vapour pressure has been corrected to a standard temperature of 20°C. Where a 25°C value was obtained, the correction was made by McGowan's equation (Appendix 2) and no indication is made. Where similar extrapolation was made from a temperature between 25°C and 50°C this is indicated by "a" below the figure: if from temperature above 50°C, by "b"; "c" indicates a large extrapolation stated by the source. Most values quoted will, or course, have involved an unspecified extrapolation. neg. = negligible, appearing in this or under water solubility indicates use of this word by the source or some such phrase as "non-volatile" or "insoluble in water". It does *not* mean that the authors of this book endorse the "negligibility". In many cases the value, if measured, would be higher than values listed for some other compounds.

Column 7 lists the saturation vapour concentration calculated from the molecular mass and the column 6 entry. Index has the same significance as in column 6. When b.p. is <20°C, the density at standard atmospheric pressure is listed. Most accurate methods of measuring low vapour pressures in fact measure directly the SVC, but only the calculated v.p. is usually reported. As the SVC is the more useful quantity we would urge that it be directly reported.

Column 8 is the partition coefficient water/air calculated as ratio of water solubility to SVC. In this column the index x is abbreviation for 10^{+x}, i.e. $7{\cdot}1^{10} = 7{\cdot}1 \times 10^{10}$.

Column 9 is the pK as defined in Appendix 3 with indication of ionic states, transition between which is governed by the value quoted. U indicates instability above or below quoted pH, where an ionic change might be expected. We have attempted no systematic notes on pH-controlled hydrolysis. For other qualifications in this column, see "Sources".

Sources

Most of the data are taken from the British Crop Protection Council's "Pesticide Manual" (6th edn), where full structural formulae, analytical methods, proposed uses, etc. are also listed. Permission of the Council and the help of the editor, Dr C. R. Worthing, are gratefully acknowledged. Data from *other* sources are indicated by letters as follows:

d in $\mathcal{P}_A^w$ column, for the compilation of Goring (1967);
e in pK column, from the compilation of Kortüm *et al.* (1961);
f in pK column, from the compilation of Perrin (1965, 1972);
g in pK column, from Matell and Lindenfors (1957);
h in pK column, from Wershaw *et al.* (1967);
j in pK column, from Nelson and Faust (1969);
k in pK column, from Sargent *et al.* (1969);
l in pK column, from Robinson and Bates (1966);
m in pK column, from Barlin (1974);
n in pK column, from Albert and Phillips (1956);
p in pK column, from Cookson and Cheeseman (1972);
q in pK column, from Weber;
r in pK column, from Massini (1958);
(note, some data from g to r inclusive are quoted in e and f)
s in pK column, from Graham-Bryce and Coutts (1971);
t in pK column, pK of 2,4-D, has been measured many times. Values in addition to that quoted from the "Pesticide Manual" (6th edn) are 2·90 from g, 2·93 from h, 2·73 from j, 2·8 from k;
u in pK column, from T. Drapala, *Chem. Abs.* 1969, **70,** 10945.

Notes on Reliability

( ) a bracket round the pK value indicates an approximate value estimated by the authors by methods of Appendix 3 from data on closely related compounds.

H (column 4): vapour pressure listed is too high by comparison with the "equivalent paraffin" (Appendix 2)

HH (column 4): Camphechlor. Vapour pressure is *much* too high on this basis or any comparison. Presumably a total pressure was measured over an impure specimen containing much volatile impurities.

HH (column 4): chlorthal dimethyl. This vapour pressure is much too high. The value for hexachlorobenzene is only 10^{-5} of that quoted for chlorthal dimethyl and is in the range expected from other comparisons. $-CO_2Me$ normally has a greater depressing effect on v.p. than $-Cl$.

L (column 4): dichloran has a m.p. 150°C higher than trifluralin, but this is unlikely to outweigh the effect of heavier substitution in trifluralin (2 propyls in place of 2 H, NO_2 and CF_3 in place of 2 Cl). We suspect v.p. listed for dichloran is too low.

The value for dichlorophen is much lower than many in the literature for other bis-phenols.

LL (column 4): difenoxuron has listed v.p. only 1/500th of that of chloroxuron, although having slightly lower m.p. and $-OCH_3$ in place of –Cl. This change of substituents in other related compounds (cf. chlortoluron, diuron, metoxuron) makes, as expected, much less difference. Value for difenoxuron probably erroneously low.

The value for quinalphos is too low on any comparison.

Less explicit doubts could be raised about other v.p. values. Most of these must involve long extrapolation and some from measurements—such as "vacuum boiling points"—of doubtful value. The values for most of the higher mol. mass pesticides at 20°C must be regarded as indicative only to a few powers of 2. More data and more publication of methods of measurement and calculation would be desirable.

Table 1

Name	Mol. Mass	m.p. (°C)	Solubility: Water	Solubility: Good solv.	Solubility: PHS	v.p.	SVC	$\mathscr{P}^{W}_{A}$	pK
Acephate	183	82–9	650	Water	V	$9{\cdot}1^{7}$	$9{\cdot}1^{9}$	$7{\cdot}1^{10}$	
Acrolein	56	−88	208	MISC	M	220	0·64	325	
Alachlor	270	41	240^{3}	PO	V	—	—	—	
Aldicarb	190	100	6	PO	V	$5{\cdot}3^{5}$	$5{\cdot}5^{7}$	$1{\cdot}1^{7}$	
Aldoxycarb	222	140	9	75 MeCN	V	$4{\cdot}7^{5}$	$5{\cdot}7^{7}$	$1{\cdot}9^{7}$	
Aldrin	365	104	27^{6}	AHC	M	$7{\cdot}5^{5}$	$1{\cdot}5^{6}$	18	
Allethrin	302	ℓ	V	MISC		$6{\cdot}1^{7}$	$1{\cdot}0^{8}$	—	
Allyl Alcohol	58	ℓ	MISC	MISC		a 17·3	·055		
Ametryne	227	85	190^{3}	PO	L	$3{\cdot}8^{7}$	$4{\cdot}7^{9}$	$4{\cdot}0^{7}$	$_{+}(4{\cdot}0)_{0}$
Aminotriazole	84	158	280	ALC	V				$_{+}4{\cdot}0_{0}$
Amitraz	293	86	*c.* 1^{3}	300 ACET		—			e U<7
Ancymidol	256	110	650^{3}	PO	L	$8{\cdot}2^{9}$	$1{\cdot}1^{10}$	$6{\cdot}0^{9}$	
Anthraquinone	208	285	V	9 $CHCl_3$	L				

Asulam	230	143	4	300 ACET WSS	L	—	—	—	$_{0}4{\cdot}8_{-}$
Atrazine	216	175	30^{3}	PO	L	$3{\cdot}0^{7}$	$3{\cdot}5^{9}$	$8{\cdot}6^{6}$	$_{+}1{\cdot}3_{0}$ $_{+}(1{\cdot}8)_{0}$
Azinphos Ethyl	345	53	neg.	PO	V	$2{\cdot}2^{7}$	$4{\cdot}1^{9}$	—	
Azinphos Methyl	317	73	33^{3}	PO	V	neg.	—	—	
Aziprotryne	225	95	55^{3}	PO	L	$2{\cdot}0^{8}$	$2{\cdot}5^{10}$	$2{\cdot}2^{8}$	$_{+}(4)_{0}$
Barban	258	75	11^{3}	PO		—	—	—	
Benazolin	244	193	600^{3}	PO WSS		neg.			$_{0}(3)_{-}$
Bendiocarb	223	129	40^{3}	200 ACET	L	$2{\cdot}4^{6}$	$2{\cdot}9^{8}$	$1{\cdot}4^{6}$	
Benfluralin	335	66	$<1^{3}$	PO	M	$1{\cdot}2^{4}$	$2{\cdot}1^{6}$	$<5^{2}$	
Benodanil	323	137	20^{3}	120 EtOH	L	neg.			
Benomyl	290	dec	$3{\cdot}8^{3}$ pH 7	130 $CHCl_3$	L	neg.			
Bensulide	397	34	25^{3}	ACET	L	—			
Bentazone	240	137	500^{3}	1200 ACET	L	—			
Benzthiazuron	207	287 dec	12^{3}	5–10 ACET, AHC	L	$2{\cdot}2^{10}$ a	$2{\cdot}5^{12}$	$4{\cdot}8^{9}$	
Benzoylprop-ethyl	366	70	20^{3}	PO	M	$3{\cdot}5^{8}$	$7{\cdot}0^{10}$	$2{\cdot}9^{7}$	
Bifenox	342	85	$0{\cdot}3^{3}$	400 ACET	L	$5{\cdot}2^{7}$	$8{\cdot}3^{9}$	$3{\cdot}6^{4}$	
Binapacryl	322	69	neg.	780 ACET	L	$4{\cdot}7^{7}$	$8{\cdot}3^{9}$	—	

Table 1 (*contd.*)

Name	Mol. Mass	m.p. (°C)	Solubility: Water	Solubility: Good solv.	Solubility: PHS	v.p.	SVC	$\mathscr{P}_A^W$	pK
Bioresmethrin	338	30–5	neg.	PO	M	4·8^{11} a	8·9^{13}	—	
Biphenyl	154	70		43 BENZ	M	3·7^{2} a, u	3·1^{4}		
Bromacil	261	158	815^{3}	ACET WSS	L	3·2^{8} a	4·6^{10}	1·8^{9}	
Bromofenoxim	461	196	0·1^{3}	PO	L	<1^{8}	<2·5^{10}	>4^{5}	$_0$(4)$_-$
Bromophos	366	53	40^{3}	PO	L	1·3^{4}	2·6^{6}	1·5^{4}	
Bromophos-ethyl	394	ℓ	2^{3}	MISC		1·2^{5}b 2·1^{9}c			
Bromoxynil	277	194	130^{3}	WSS	L	2·2^{6} a	3·3^{8}	4·0^{6}	$_0$4·1$_-$
Bromoxynil Octanoate	403	45	neg.	MISC		1·3^{4} a	2·8^{6}	—	
Brompropylate	428	77	<5^{3}	PO		5·1^{8}	1·2^{9}	<4^{6}	
Bronopol	200	130	250	300 ACET	V	1·3^{5}	1·5^{7}	1·7^{9}	
Bufencarb	221	29	<50^{3}	PO	L	7·6^{6}	9·2^{8}	<5^{5}	
Bupirimate	316	50	22^{3}	PO	V	5·0^{7}	8·6^{9}	2·6^{6}	
Butachlor	312	ℓ	20^{3}	MISC		3·3^{8} a	5·7^{10}	3·5^{7}	

Butamifos	332	ℓ	neg.	PO	L				
Buthidazole	256	133	3·4	208 DMF	L	—	—		
Buthiobate	373	33	neg.	GEN		$4·5^{7}$	$9·2^{9}$		
Butocarboxim	190	37	L	PO	L	$8·0^{5}$	$8·3^{7}$		
Butralin	295	61	$1·0^{3}$	PO	L	$1·3^{5}$	$1·0^{7}$		
Buturon	237	145	30^{3}	279 ACET	L	—	—		
sec-Butylamine	73	ℓ	MISC	MISC		135	0·54		$_{+}(10·5)_{0}$
Butylate	217	ℓ	45^{3}	MISC ——		$7·2^{4}$	$8·6^{6}$		
Camphechlor	414	70–95	3^{3}	V. Sol	M	0·2 HH			
Captafol	349	160	$1·4^{3}$	PO		neg.			
Captan	301	178	$3·3^{3}$	100 $CHCl_3$	V	$5·0^{6}$	$8·2^{7}$	$4·0^{3}$	
Carbaryl	201	142	120^{3}	PO	L	$2·1^{5}$	$2·3^{7}$	$5·2^{5}$	
Carbendazim	191	300 dec	$5·8^{3}$	300^{3} EtOH	V	—	—		$_{+}4·5_{0}$
Carbetamide	236	119	3·5	ALC, KET	L	neg.			
Carbofuran	221	150	700^{3}	270 DMF	V	$6·5^{6}$	$7·8^{8}$	$9·0^{6}$	
Carbon disulphide	76	ℓ	2·2	MISC		291 298d	1·0	2·2 1·8d	
Carbon tetrachloride	154	ℓ	280^{3}	MISC		90	0·76	0·37	

Table 1 (*contd.*)

Name	Mol. Mass	m.p. (°C)	Solubility: Water	Solubility: Good solv.	Solubility: PHS	v.p.	SVC	$\mathscr{P}_A^W$	pK
Carbophenothion	343	ℓ	$<40^{3}$	MISC	M	$3{\cdot}0^{7}$	$5{\cdot}6^{9}$	$<7^{6}$	
Carboxin	235	92	170^{3}	600 ACET	M	—	—		
Cartap	237		200 HCl	PO WSS		—	—		
Chloralose	309	187	4·4	L	V	—	—		
Chloramben	206	200	700^{3}	PO WSS	L	$8{\cdot}4^{7}$ a	$9{\cdot}4^{9}$		$_{0}(4)_{-}$
Chlorbromuron	293	98	35^{3}	PO		$4{\cdot}0^{7}$	$6{\cdot}4^{9}$	$5{\cdot}5^{6}$	
Chlorbufam	224	45	540^{3}	286 MeOH		1·2 H	$1{\cdot}5^{2}$	$3{\cdot}6^{3}$	
Chlordane	410	ℓ	neg.	MISC		5^{6}	$1{\cdot}1^{7}$		
Chlordimeform	197	32	250^{3}	200 ACET	M	$3{\cdot}6^{4}$	$3{\cdot}9^{6}$	$6{\cdot}4^{4}$	
Chlorfenac	239	156	200^{3}	PO WSS	L	$1{\cdot}1^{6}$ a	$1{\cdot}5^{8}$	$1{\cdot}3^{7}$	$_{0}3{\cdot}7_{-}$ j
Chlorfenprop-methyl	233	ℓ	40^{3}	MISC		$2{\cdot}8^{4}$ b	$3{\cdot}6^{6}$	$1{\cdot}1^{4}$	
Chlorfenson	303	86	neg.	ACET	M	—	—		

Chlorflurecol-methyl	273	152	18	200 ACET	L				
Chloridazon	222	205	400^{3}	PO		$1{\cdot}2^{2}$ H	$1{\cdot}4^{4}$	$2{\cdot}8^{3}$	
Chlormephos	235 Cl^{-}	ℓ	60^{3}	PO		$2{\cdot}3^{2}$ H	$3{\cdot}0^{4}$	2^{2}	
Chlormequat	158	245 dec	> 1000	V HCs	V	—	—		
Chloroacetic acid	94	56, 63	V. Sol	ALC WSS	L	3^{1} u	$1{\cdot}5^{3}$		$_{0}2{\cdot}8_{-}$ e
Chlorobenzilate	325	36	neg.	PO	L	$6{\cdot}8^{6}$	$1{\cdot}2^{7}$		
Chloromethiuron	229	175	50^{3}	PO	L	8^{9}	1^{10}	5^{8}	
Chloroneb	207	133	8^{3}	133 CH_2Cl_2	M	$1{\cdot}7^{3}$	$1{\cdot}9^{5}$	4^{2}	
Chloropicrin	164	ℓ	2·27	MISC		17·9	0·16	14 11d	
Chloropropylate	339	73	$<10^{3}$	PO		$1{\cdot}8^{7}$	$3{\cdot}3^{8}$		
Chlorothalonil	266	250	$0{\cdot}6^{3}$	89 XYL	M	$1{\cdot}2^{3}$	$1{\cdot}8^{5}$	33	
Chloroxuron	290	151	$3{\cdot}7^{3}$	PO	L	$1{\cdot}8^{9}$	$2{\cdot}0^{11}$	$1{\cdot}8^{8}$	
Chlorphoxim	333	66	$1{\cdot}7^{3}$	600 TOL	L	$7{\cdot}5^{6}$	$1{\cdot}4^{7}$	$1{\cdot}2^{4}$	
Chlorpropham	214	41	89^{3}	PO		—	—		
Chlorpyrifos	351	43	2^{3}	790 OCTANE		9^{6}	$1{\cdot}7^{7}$	$1{\cdot}2^{4}$	
Chlorpyrifos-methyl	322	46	5^{3}	PO		$2{\cdot}0^{5}$	$3{\cdot}5^{7}$	$1{\cdot}4^{4}$	
Chlorthiamid	206	151	950^{3}	PO		$1{\cdot}0^{6}$	$1{\cdot}1^{8}$	$8{\cdot}6^{7}$	

Table 1 (*contd.*)

Name	Mol. Mass	m.p. (°C)	Solubility			v.p.	SVC	$\mathscr{P}^{W}_{A}$	pK
			Water	Good solv.	PHS				
Chlorthal-dimethyl	332	156	$0{\cdot}5^{3}$	300 BENZ		0·1 HH			
Chlorthiophos	361	ℓ	neg.	MISC		$1{\cdot}1^{7}$ a	$2{\cdot}2^{9}$		
Chlortoluron	213	147	70^{3}	PO	L	$3{\cdot}6^{8}$	$4{\cdot}2^{10}$	$1{\cdot}7^{8}$	
Chlofop-isobutyl	349	39	180^{3}	AHC		—	—		
Crimidine	172	87	neg.	ACET		$3{\cdot}5^{4}$	$3{\cdot}3^{6}$		
Crotoxyphos	314	ℓ	1	MISC AHC	L	$1{\cdot}4^{5}$	$2{\cdot}4^{7}$	4^{6}	
Crufomate	292	60	neg.	ACET	V	—	—		
Cyanazine	241	167	17^{3}	PO		$1{\cdot}6^{9}$	$2{\cdot}1^{11}$	$8{\cdot}1^{8}$	$_{+}(1{\cdot}5)_{0}$
Cyanide (Hydrogen)	27	−14	MISC	MISC WSS		b.p. 26°	0·7		$_{0}9{\cdot}3_{-}$ e .
Cyanofenphos	303	83	$0{\cdot}6^{3}$	KET, AHC		$6{\cdot}6^{6}$	$1{\cdot}1^{7}$	$5{\cdot}5^{3}$	
Cyanophos	243	14	46^{3}	MISC		—	—		
Cycloate	215	ℓ	85^{3}	MISC		$3{\cdot}6^{3}$	$4{\cdot}3^{5}$	$2{\cdot}0^{3}$	
Cycloheximide	281	117	21	$CHCl_3$, PrOH		—	—		

Cypermethrin	416	c. 60	$<0{\cdot}2^{3}$	450 ACET	M	$3{\cdot}9^{12}$	$8{\cdot}8^{14}$	$<3^{10}$	
2,4-D	221	140	620^{3}	ACET WSS		a $9{\cdot}4^{7}$ a	$1{\cdot}1^{8}$	$5{\cdot}6^{7}$	$_{0}2{\cdot}64_{-}$ t
Dalapon	143	186	500	ALC	L				$_{0}1{\cdot}8_{-}$ j
2,4-DB	249	117	46^{3}	ALC WSS					$_{0}4{\cdot}8_{-}$
Daminozide	160	155	100	WATER	V	—	—		
Dazomet	162	99	3	391 $CHCl_3$					
DDT (pp)	354	108	$1{\cdot}2^{6}$ d	250 AHC	M	$1{\cdot}9^{7}$	$3{\cdot}8^{9}$	$3{\cdot}2^{2}$	
Dehydroaceticacid	168	110	<1	300 ACET	L	$3{\cdot}9^{3}$	$3{\cdot}6^{5}$	$<3^{4}$	
Demeton-S (P = O)	258	ℓ	2	MISC	L	a $2{\cdot}6^{4}$	$3{\cdot}7^{6}$	$5{\cdot}4^{5}$	
Demeton (P = S)	258	ℓ	60^{3}	MISC	L	$2{\cdot}8^{4}$	$4{\cdot}0^{6}$	$1{\cdot}5^{4}$	
Demeton-S-methyl	230	ℓ	3·3	MISC	L	$3{\cdot}6^{4}$	$4{\cdot}5^{6}$	$7{\cdot}3^{5}$	
Demeton-S-methyl sulphone	262	60	SOL	ALC	V	$5{\cdot}0^{6}$	$7{\cdot}2^{8}$		
Desmedipham	300	120	7^{3}	400 ACET		$1{\cdot}2^{9}$	$1{\cdot}9^{11}$	$3{\cdot}7^{8}$	
Desmetryne	213	85	600^{3}	PO		$1{\cdot}0^{6}$	$1{\cdot}2^{8}$	$5{\cdot}0^{7}$	$_{+}(4)_{0}$
Diallate	270	ℓ	14^{3}	MISC		$8{\cdot}8^{4}$	$1{\cdot}3^{5}$	$1{\cdot}1^{3}$	
Diazinon	304	ℓ	40^{3}	MISC	M	$1{\cdot}4^{4}$	$2{\cdot}3^{6}$	$1{\cdot}7^{4}$	

Table 1 (*contd.*)

Name	Mol. Mass	m.p. (°C)	Solubility: Water	Solubility: Good solv.	Solubility: PHS	v.p.	SVC	$\mathscr{P}^{W}_{A}$	pK
1,2-Dibromo-3-chloropropane	236	ℓ	1	MISC	—	7^{1}	9^{3}	1^{2}	
Dicamba	221	115	4·5	ALC WSS		$3{\cdot}7^{3}$	$4{\cdot}5^{5}$	1^{5}	$_{0}1{\cdot}8_{-}$ k
Dichlobenil	172	145	18^{3}	PO		$2{\cdot}8^{4}$	$2{\cdot}7^{6}$	$6{\cdot}7^{3}$	
Dichlofluanid	333	105	neg.	70 XYL		1^{6}	$1{\cdot}8^{8}$		
Dichloran	207	195	neg.			$1{\cdot}2^{6}$	$1{\cdot}4^{8}$		
Dichlorfenthion	315	ℓ	$0{\cdot}2^{3}$	MISC	—	$1{\cdot}7^{5}$ a	$3{\cdot}0^{7}$	$6{\cdot}8^{2}$	
p-Dichlorobenzene	147	53	80^{3}	AHC		$6{\cdot}9^{1}$	$5{\cdot}5^{3}$	14	
Dichlorophen	269	178	30^{3}	ALC WSS		$3{\cdot}5^{11}$ L	$5{\cdot}2^{13}$	$5{\cdot}8^{10}$	$_{0}(9{\cdot}7)_{-}$
Dichloropicolinic acid	192	151	1	ALC WSS		$6{\cdot}0^{6}$	$6{\cdot}3^{8}$	$1{\cdot}6^{7}$	$_{0}(3)_{-}$
1,2-Dichloropropane	113	ℓ	2·7	MISC	—				
1,3-Dichloropropene	111	ℓ	1	MISC	—				
Dichlorprop	235	117	350^{3}	ALC WSS		neg.			$_{0}(3)_{-}$
Dichlorvos	221	ℓ	10	MISC		$1{\cdot}2^{2}$	$1{\cdot}4^{4}$	$7{\cdot}1^{4}$	

Dieldrin	381	175	$0{\cdot}19^{3}$	AHC	V	$3{\cdot}1^{6}$	$6{\cdot}5^{8}$	$3{\cdot}0^{3}$	
Dienochlor	475	122	neg.	AHC	L	$4{\cdot}9^{6}$	$1{\cdot}3^{7}$		
Diethatyl-ethyl	312	49	105^{3}	AHC		—	—		
Diethyl-*m*-toluamide	191	ℓ	neg.	MISC ———		$5{\cdot}6^{4}$	$5{\cdot}8^{6}$		
Difenoxuron	286	138	20^{3}	156 CH_2Cl_2		a $9{\cdot}3^{12}$ LL	$1{\cdot}4^{13}$	$1{\cdot}4^{11}$	
Difenzoquat (+)	214			WSS					
Diflubenzuron	311	231	$0{\cdot}2^{3}$	PO	V	—	—		
Dimefox	154	ℓ	MISC	MISC	L	$2{\cdot}4^{1}$	$2{\cdot}0^{3}$		
Dimethachlor	256	47	2·1	PO	L	$1{\cdot}6^{5}$	$2{\cdot}2^{7}$	$1{\cdot}0^{7}$	
Dimethametryn	255	65	50^{3}	PO	L	$1{\cdot}4^{6}$	$1{\cdot}9^{8}$	$2{\cdot}6^{6}$	$_{+}(4)_{0}$
Dimethirimol	209	102	1·2	1200 $CHCl_3$	V	$2{\cdot}6^{6}$	$3{\cdot}0^{8}$	$4{\cdot}0^{7}$	$_{+}4{\cdot}8_{0}$ s
Dimethoate	229	51	25	KET	V	$4{\cdot}1^{6}$	$5{\cdot}1^{8}$	$4{\cdot}9^{8}$	
Dimethylarsinic acid	138	193	2000	WSS					$_{0}6{\cdot}3_{-}$ e
Dimetilan	240	69	SOL	$CHCl_3$	L	$1{\cdot}0^{4}$	$1{\cdot}3^{6}$		
Dimethyl phthalate	194	ℓ	4·3	MISC	M	$1{\cdot}1^{4}$	$1{\cdot}2^{6}$	$3{\cdot}7^{6}$	
Dinitramine	322	98	$1{\cdot}1^{3}$	400 ACET		$1{\cdot}7^{6}$	$3{\cdot}0^{8}$	$3{\cdot}7^{4}$	

Table 1 (*contd.*)

Name	Mol. Mass	m.p. (°C)	Solubility: Water	Solubility: Good solv.	Solubility: PHS	v.p.	SVC	$\mathscr{P}_A^W$	pK
Dinoseb	240	30–40	100^{3}	ALC WSS	L	—	—		$_{0}4{\cdot}6$
Dioxacarb	223	114	6	550 DMF	L	$3{\cdot}0^{7}$	$3{\cdot}7^{9}$	$1{\cdot}6^{9}$	
Diphenamid	239	135	270^{3}	PO		neg.			
Dipropetryn	255	105	16^{3}	PO		$7{\cdot}3^{7}$	$1{\cdot}0^{8}$	$1{\cdot}6^{6}$	$_{+}(4)_{0}$
Diquat (+)	184		700 Br-	WSS					
Disulfoton	274	ℓ	25^{3}	MISC		$1{\cdot}8^{4}$	$2{\cdot}7^{6}$	$9{\cdot}3^{3}$	
Ditalimfos	299	83	133^{3}	AHC	M	$6{\cdot}7^{7}$	$1{\cdot}1^{8}$	$1{\cdot}2^{7}$	
Diuron	233	158	42^{3}	53 ACET	L	$3{\cdot}1^{8}$	$4{\cdot}0^{10}$	$1{\cdot}0^{8}$	
DNOC	198	86	130^{3}	KET WSS		b $5{\cdot}5^{5}$	$6{\cdot}0^{7}$	$2{\cdot}2^{5}$	$_{0}(4{\cdot}0)_{-}$
Dodemporph	281	ℓ		MISC					$_{+}(9)_{0}$
Dodine	227		136 acetate	ALC WSS					$_{+}(13{\cdot}2)_{0}$
Drazoxolon	238	167	neg.	100 $CHCl_3$ ALK		$8{\cdot}9^{7}$	$1{\cdot}2^{8}$		?
Eglinazine ethyl	260	228	300^{3}	GEN					

Endrin	381	226	neg.	AHC		$8{\cdot}8^{8}$	$1{\cdot}8^{9}$		
EPN	323	36	neg.	PO		$7{\cdot}3^{9}$	$1{\cdot}3^{10}$		
Epoxyethane	44	ℓ b·p· 11°C	MISC	———		a 1095			
EPTC	189	ℓ	365^{3}	MISC		$8{\cdot}0^{3}$	$8{\cdot}3^{5}$	$4{\cdot}4^{3}$	
Etacelasil	317	ℓ	25^{3}	GEN		b 20^{4}	$3{\cdot}5^{6}$	$7{\cdot}1^{3}$	
Ethalfluralin	333	55	$0{\cdot}2^{3}$	AHC		$4{\cdot}2^{5}$	$7{\cdot}6^{7}$	$2{\cdot}6^{2}$	
Ethiofencarb	225	33	1·82	>600 TOL		$2{\cdot}7^{5}$	$3{\cdot}3^{7}$	$5{\cdot}5^{6}$	
Ethirimol	209	159		WSS		b $9{\cdot}4^{7}$	$1{\cdot}1^{8}$		$_{+}5{\cdot}3_{0}$ s
Ethofumesate	286	70	110^{3}	400 ACET	L	$2{\cdot}8^{7}$	$4{\cdot}4^{9}$	$2{\cdot}5^{7}$	
Ethohexadiol	146	ℓ	6	MISC	L	3^{2}	$2{\cdot}4^{4}$	$2{\cdot}5^{4}$	
Ethoprophos	242	ℓ	750^{3}	MISC		a $1{\cdot}7^{4}$	$2{\cdot}2^{6}$	$3{\cdot}4^{5}$	
Ethylene dibromide	188	9·3	4·3	MISC	———	b 8·1	$8{\cdot}4^{2}$	51	
Ethylene dichloride	99	−6	4·3	MISC	———	78	4·2	10	
Etridiazole	247	ℓ	50^{3}	MISC		1^{4}	$1{\cdot}3^{6}$	4^{4}	
Etrimfos	292	ℓ	40^{3}	MISC		$4{\cdot}9^{5}$	$7{\cdot}8^{7}$	$5{\cdot}1^{4}$	
Fenamiphos	303	49	700^{3}	PO					

Table 1 (*contd.*)

Name	Mol. Mass	m.p. (°C)	Solubility: Water	Solubility: Good solv.	Solubility: PHS	v.p.	SVC	$\mathscr{P}_A^W$	pK
Fenarimol	331	118	$13\cdot7^{3}$ pH 7	PO		$1\cdot0^{7}$	$1\cdot8^{9}$	$7\cdot6^{6}$	
Fenbutin oxide	1053	139	neg.	380 CH_2Cl_2	L				
Fenchlorphos	321	40	40^{3}	AHC	M	$4\cdot4^{4}$	$7\cdot8^{6}$	$5\cdot1^{3}$	
Fenfuram	201	109	100^{3}	340 CYH		$1\cdot5^{7}$ C	$1\cdot6^{9}$	$6\cdot1^{7}$	
Fenitrothion	277	ℓ	neg.	AHC	L	$6\cdot0^{6}$	$9\cdot1^{8}$		
Fenoprop	269	180	140^{3}	WSS		neg.			
Fensulfothion	308	ℓ	$1\cdot54$	PO		$9\cdot9^{9}$	$1\cdot7^{10}$	$9\cdot1^{9}$	
Fenthion	278	ℓ	55^{3}	AHC	M	$3\cdot0^{5}$	$4\cdot6^{7}$	$1\cdot2^{5}$	
Fentin acetate	409	120	28^{3}	L	V	$2\cdot8^{7}$	$6\cdot3^{9}$	$4\cdot4^{6}$	
Fenuron	164	133	$3\cdot85$	ALC		$8\cdot4^{7}$	$7\cdot6^{9}$	$5\cdot1^{8}$	
Fenvalerate	420	ℓ	$<1^{3}$	KET, AHC		$<1^{5}$			
Flamprop-isopropyl (R)	364	70	10^{3}	677 CYH		—	—		
Flamprop-methyl	336	86	35^{3}	414 CYH		$3\cdot5^{7}$	$6\cdot4^{9}$	$5\cdot5^{6}$	

Fluorodifen	328	94	2^{3}	ACET 750 ACET	L	7^{8}	$1{\cdot}2^{9}$	$1{\cdot}7^{6}$	
Fluotrimazole	379	132	$1{\cdot}5^{6}$	500 BENZ 400 CH_2Cl_2					
Flurecol-butyl	282	71	36^{3}	1450 ACET	M	—	—		
Fluridone	329	151	12^{3}	PO	V	$4{\cdot}3^{8}$	$7{\cdot}7^{10}$	$1{\cdot}6^{7}$	
Folpet	297	177	1^{3}	AHC		$<1^{5}$	$<1{\cdot}6^{7}$	$>6^{3}$	
Formetanate	221	103	<1	200 MeOH		neg.			
Formothion	257	26	2·6	MISC		6^{6}	$8{\cdot}4^{8}$	3^{7}	
Fosthietan	241	ℓ	50	PO	L	$3{\cdot}1^{6}$	$4{\cdot}1^{8}$	$1{\cdot}2^{9}$	
Fuberidazole	184	286	78^{3}	HClaq.					
Furalaxyl	301	70, 84	230^{3}	GEN		$5{\cdot}3^{7}$	$8{\cdot}7^{9}$	$2{\cdot}6^{7}$	
Glyphosate	122			Water and alkali					see App. 3,
Heptachlor	373	95	56^{6}	189 Kerosine		$1{\cdot}6^{4}$	$3{\cdot}3^{6}$	17	
Heptenophos	251	ℓ	2·2	MISC		$7{\cdot}5^{4}$	$1{\cdot}0^{5}$	$2{\cdot}2^{5}$	
Hexachlorobenzene	285	226	neg.	PO	M	$1{\cdot}1^{5}$	$1{\cdot}7^{7}$		
Hexazinone	252	116	33	3880 $CHCl_3$ 2650 MeOH	M	$6{\cdot}0^{9}$ a	$8{\cdot}3^{11}$	$4{\cdot}0^{11}$	

Table 1 (*contd.*)

Name	Mol. Mass	m.p. (°C)	Solubility			v.p.	SVC	$\mathscr{P}_A^W$	pK
			Water	Good solv.	PHS				
Imazalil	297	ℓ	L	MISC		7^8	$1{\cdot}1^9$		
Indol-3-butyric acid	203	123	neg.	ALC WSS					$_0(4{\cdot}5)_-$
Iodofenphos	413	76	$<2^3$	600 BENZ 860 CH_2Cl_2	M	8^7	$1{\cdot}8^8$	$>1^5$	
Ioxynil	371	209	50^3	PO WSS		$6{\cdot}2^7$	$1{\cdot}2^8$	$4{\cdot}2^6$	$_04{\cdot}0_-$
Iprodione	330	136	13^3	500 CH_2Cl_2		$<1^6$	$<1{\cdot}8^6$	$>7^5$	
Isocarbamid	185	96	1·3			$<6^3$	$<6^5$	$>2^4$	
Isofenphos	345	ℓ	24^3	600 CH_2Cl_2		a 4^6	$7{\cdot}5^8$	$3{\cdot}2^5$	
Isomethiozin	268	159	10^3			$3{\cdot}5^7$	$5{\cdot}1^9$	$2{\cdot}0^6$	
Isoproturon	206	156	55^3	AHC		$2{\cdot}5^8$	$3{\cdot}5^{10}$	$1{\cdot}6^8$	
Lenacil	234	316	6^3	<10 PO	L				
Lindane	291	113	10^3	KET, AHC	L	$9{\cdot}4^6$	$1{\cdot}5^7$	$6{\cdot}6^4$	
Linuron	249	94	75^3	KET		$8{\cdot}6^6$	$1{\cdot}2^7$	$6{\cdot}2^5$	
Leptophos	412	70	$2{\cdot}4^3$	1300 BENZ		—	—		

MCPA	201	118	825^{3}	WSS					m,n,p $_{0}3{\cdot}1_{j}$
MCPB	229	99	44^{3}	WSS					$_{0}4{\cdot}8_{j}$
Mecarbam	329	ℓ	<1	MISC	L	neg.			
Mecoprop	215	94	620^{3}	ALC WSS					(3)
Mefluidide	310	185	180^{3}	350 ACET		$<5^{5}$	$<8^{7}$	$>2^{5}$	
Menazon	281	160	240^{3}	200 EtOH	L	$4{\cdot}6^{7}$	$7{\cdot}1^{9}$	$3{\cdot}4^{7}$	
Mephosfolan	269	ℓ	57	ALC, KET		$2{\cdot}8^{9}$ a	$4{\cdot}1^{11}$	$1{\cdot}4^{12}$	
Mercuric chloride	271	277	69	Water, ALC		6^{3}	9^{5}		
Mercurous chloride	472	>400	2^{3}						
Metamitron	202	167	1·8	30 CH_2Cl_2					
Metalaxyl	279	72	7·1	GEN		$2{\cdot}2^{6}$	$3{\cdot}3^{8}$	$2{\cdot}2^{8}$	
Metaldehyde	176	246	200^{3}	AHC, $CHCl_3$					
Methabenzthiazuron	221	119	59^{3}	116 ACET		1^{6}	$1{\cdot}2^{8}$	5^{6}	
Methamidophos	141	44	V. Sol	ALC	L	$8{\cdot}7^{5}$ b	$6{\cdot}7^{7}$		
Metham-sodium	129		720	ALC	V	*See*	Methyl	isothiocyanate	
Methazole	261	124	$1{\cdot}5^{3}$	550 XYL KET	L	—	—		

Table 1 (*contd.*)

Name	Mol. Mass	m.p. (°C)	Solubility: Water	Solubility: Good solv.	Solubility: PHS	v.p.	SVC	$\mathcal{P}_A^W$	pK
Methidathion	302	39	240^{3}	AHC, KET		1^{6}	$1{\cdot}6^{8}$	$1{\cdot}5^{7}$	
Methomyl	162	78	58	800 MeOH		$2{\cdot}6^{5}$	$2{\cdot}3^{7}$	$2{\cdot}5^{8}$	
Methoprotryne	271	69	320^{3}	PO		$2{\cdot}8^{7}$	$4{\cdot}2^{9}$	$7{\cdot}0^{7}$	$_{+}(4)_{0}$
Methoxychlor (pp)	346	89	neg.	AHC					
Methyl bromide	95	−93 b.p. 4·5	13·4	MISC					
Methyl isothiocyanate	73	35	7·6	V. Sol	M	20·7	·083	91	
Metobromuron	259	175	230^{3}	ALC, KET		$2{\cdot}3^{7}$	$3{\cdot}2^{9}$	$7{\cdot}2^{7}$	
Metolachlor	284	ℓ	530^{3}	GEN		$1{\cdot}3^{5}$	$2{\cdot}0^{7}$	$2{\cdot}6^{6}$	
Metoxuron	229	126	678^{3}	KET	V	$3{\cdot}2^{8}$	$4{\cdot}0^{10}$	$1{\cdot}7^{9}$	
Metribuzin	214	125	1·2	450 MeOH	L	$1{\cdot}1^{6}$ a	$1{\cdot}3^{8}$	$9{\cdot}2^{7}$	
Mevinphos	224	ℓ	MISC	KET, ALC	L	—	—		
Molinate	187	ℓ	800^{3}	MISC		$3{\cdot}3^{3}$	$3{\cdot}4^{5}$		
Monalide	240	88	23^{3}	KET, AHC		—	—		

Monuron	199	175	230^{3}	KET	V	$2{\cdot}3^{7}$	$2{\cdot}5^{9}$	$9{\cdot}2^{7}$	
Naled	381	ℓ	neg.	AHC	L	2^{3}	4^{5}		
Naphthalene	129	80	30^{3}	285 BENZ	M	$4{\cdot}9^{2}$	$3{\cdot}5^{4}$	86	
1-Naphthylacetic acid	186	135	420^{3}	ALC WSS					$_{0}4{\cdot}2_{-}$ e
2-Naphthoxyacetic acid	202	156	V	ALC WSS					$_{0}(3{\cdot}2)_{-}$
Napropamide	271	75	73^{3}	PO					
Naptalam	291	185	200^{3}	WSS					$_{0}(3)_{-}$
Neburon	275	102	$4{\cdot}8^{3}$	KET	V				
Niclosamide	327	230	5^{3}	WSS					
Nicotine	162	−80	MISC	MISC	M	$2{\cdot}6^{2}$	$2{\cdot}3^{4}$		$_{+}8{\cdot}0_{0}$ f
Nitralin	345	152	$0{\cdot}6^{3}$	360 ACET					
Nitrapyrin	231	62	neg.	PO		$2{\cdot}0^{3}$	$2{\cdot}5^{5}$		
Nitrilacarb	197	125	Sol	ALC	V	—	—		
Nitrofen	284	71	1^{3}	PO	L	$4{\cdot}3^{7}$ b	$6{\cdot}7^{9}$	$1{\cdot}6^{6}$	
Norflurazon	304	180	28^{3}	ACET		$2{\cdot}0^{8}$	$3{\cdot}3^{10}$	$8{\cdot}5^{7}$	
Nuarimol	315	126	26^{3}	PO	L	$8{\cdot}3^{9}$	$1{\cdot}4^{10}$	$1{\cdot}9^{8}$	

Table 1 (*contd.*)

Name	Mol. Mass	m.p. (°C)	Solubility			v.p.	SVC	$\mathscr{P}_A^W$	pK
			Water	Good solv.	PHS				
Omethoate	213	ℓ	MISC	KET	L	$2\cdot5^{5}$	$2\cdot9^{7}$		
Oryzalin	346	141	$2\cdot4^{3}$	ALC, KET	V				
Oxadiazon	345	90	$0\cdot7^{3}$	1000 BENZ 1000 $CHCl_3$		$<1^{6}$	$<2^{8}$	$>1^{5}$	
Oxamyl	219	110	280	Water 10 TOL	V	$1\cdot2^{4}$	$1\cdot5^{6}$	$1\cdot9^{8}$	
Oxycarboxin	267	128	1	360 ACET 2200 DMSO	V	—	—		
Oxydemeton methyl	246	−10	MISC ———		L				
Paraquat (++)	214			WSS					
Parathion	291	ℓ	24^{3}	MISC	L	$3\cdot8^{5}$	$6\cdot0^{7}$	$4\cdot0^{4}$	
Parathion-methyl	263	35	55^{3}	PO	L	$9\cdot7^{6}$	$1\cdot4^{7}$	$3\cdot9^{5}$	
Pebulate	203	ℓ	60^{3}	MISC ———		$2\cdot7^{2}$	$3\cdot0^{4}$	$2\cdot0^{2}$	
Pendimethalin	281	57	$0\cdot3^{3}$	AHC		$1\cdot5^{5}$	$2\cdot4^{7}$	$1\cdot2^{3}$	
Pentachlorphenol	266	191	20^{3}	ALC WSS	L	$6\cdot1^{5}$ a	$8\cdot8^{7}$	$2\cdot3^{4}$	$_{0}4\cdot8\_$ 1
Pentanochlor	294	85	8^{3}	550 ISOPH		—	—		

Phenmedipham	300	144	3^{3}	EtOH, XYL 200 ACET		—	—		
Phenothrin	350	ℓ	2^{3}	MISC		3^{7} a	$5{\cdot}7^{9}$	$3{\cdot}5^{5}$	
Phenthoate	320	18	11^{3}	MISC		$2{\cdot}6^{6}$ a	$4{\cdot}6^{8}$	$2{\cdot}4^{5}$	
Phenylmercury acetate	337	150	4·4	150 MeOH					
Phenylmercury nitrate	634	188 dec	600^{3}	9 ANIL					
2-Phenylphenol	170	57	0·7	PO					$_{0}11{\cdot}2\_$ u
Phorate	260	ℓ	50^{3}	MISC		$8{\cdot}4^{4}$	$1{\cdot}2^{5}$	$4{\cdot}2^{3}$	
Phosalone	368	48	10^{3}	KET	V	neg.			
Phosfolan	255	37–45	Sol	ALC, KET	V	$5{\cdot}7^{9}$	$7{\cdot}9^{11}$		
Phosmet	317	72	25^{3}	> 100 PO		$2{\cdot}9^{5}$	$5{\cdot}0^{7}$	$5{\cdot}0^{4}$	
Phosphamidon	300	ℓ	Sol	KET	L	$2{\cdot}5^{5}$	$4{\cdot}1^{7}$		
Phoxim	298	6	7^{3}	MISC		—	—		
Picloram	241	215 dec	430^{3}	WSS		$5{\cdot}5^{8}$ a	$7{\cdot}3^{10}$	$5{\cdot}9^{8}$	$_{0}(1{\cdot}5)\_$
Pindone	230	110	18^{3}	PO ALK		—	—		?
Piperophos	353	ℓ	25^{3}	MISC	M	—	—		
Piproctanyl (+)	266		V. Sol. Br^{-}	WSS					

Table 1 (*contd.*)

Name	Mol. Mass	m.p. (°C)	Solubility: Water	Solubility: Good solv.	Solubility: PHS	v.p.	SVC	$\mathscr{P}^{W}_{A}$	pK
Pirimicarb	238	90	$2\cdot7$	PO		$7\cdot6^{6}$	$9\cdot9^{8}$	$2\cdot7^{7}$	
Pirimiphos-ethyl	333	ℓ	20^{3}	PO		$2\cdot9^{4}$	$5\cdot3^{6}$	$3\cdot8^{4}$	
Pirimiphos-methyl	305	ℓ	5^{3}	PO		$2\cdot7^{5}$ a	$4\cdot5^{7}$	$1\cdot1^{4}$	
Prochloraz	377	41	47^{3}	3500 ACET		$6\cdot0^{10}$	$1\cdot2^{11}$	$4\cdot0^{8}$	
Procymidone	284	166	neg.	GEN		$7\cdot0^{5}$	$1\cdot1^{6}$		
Profenfos	374	ℓ	20^{3}	MISC		1^{5}	2^{7}	1^{3}	
Promecarb	207	88	92^{3}	200–400 KET	L	$1\cdot5^{5}$	$1\cdot7^{7}$	$5\cdot4^{5}$	
Prometone	225	92	620^{3}	PO	L	$2\cdot3^{6}$	$2\cdot8^{8}$	$2\cdot2^{7}$	$_{+}4\cdot3_{0}$ q
Prometryne	241	120	40^{3}	PO	L	$1\cdot6^{6}$	$1\cdot3^{8}$	$3\cdot1^{6}$	$_{+}4\cdot0_{0}$ q
Propachlor	212	67–76	700^{3}	PO	L	$2\cdot4^{6}$	$2\cdot8^{8}$	$2\cdot5^{7}$	
Propanil	218	93	225^{3}	600 ISOPH		$4\cdot1^{7}$	$4\cdot9^{9}$	$4\cdot6^{7}$	
Propazine	230	214	5^{3}	KET	V	$2\cdot9^{8}$	$3\cdot6^{10}$	$1\cdot4^{7}$	$_{+}1\cdot8_{0}$ q
Propetamphos	281	ℓ	110^{3}	MISC		$2\cdot2^{6}$ a	$3\cdot4^{8}$	$3\cdot2^{6}$	

Prothate	285	28	2·5	V.Sol	L	$7{\cdot}2^{6}$	$1{\cdot}1^{7}$	$2{\cdot}3^{7}$	
Prothiophos	345	ℓ	1·7	MISC		$<7{\cdot}5^{6}$	$<1{\cdot}4^{7}$	$>1{\cdot}2^{7}$	
Pyracarbolid	217	111	0·6	ALC, KET	L	$7{\cdot}7^{3}$	$9{\cdot}1^{5}$	$6{\cdot}6^{3}$	
Pyrazophos	373	51	$4{\cdot}2^{3}$	PO		$1{\cdot}5^{8}$	$3{\cdot}1^{10}$	$1{\cdot}3^{7}$	
Quinalphos	298	32	22^{3}	ALC, KET	L	a $3{\cdot}9^{12}$	$6{\cdot}4^{14}$	$3{\cdot}4^{11}$	
Quinmethionate	234	170	neg.	18 CYH		LL $2{\cdot}0^{7}$	$2{\cdot}6^{9}$		
Quinonamid	318	213	3^{3}	GEN		$8{\cdot}4^{8}$	$1{\cdot}5^{9}$	$2{\cdot}0^{6}$	
Quintozene	295	146	neg.	AHC		$8{\cdot}3^{3}$	$1{\cdot}3^{4}$		
Resmethrin	338	48	neg.	AHC					
Rotenone	394	163	15^{3}	PO	L				
Schradan	286	20	100° MISC	MISC	L	$5{\cdot}6^{4}$	$8{\cdot}8^{6}$		
Sesamex	298	ℓ	—	MISC		$4{\cdot}1^{7}$	$6{\cdot}7^{9}$		
Secbumeton	225	88	600^{3}	PO		$7{\cdot}3^{6}$	$9{\cdot}0^{8}$	$6{\cdot}6^{6}$	
Siduron	232	138	18^{3}	100 DMF, ISOPH					
Simazine	202	227	$3{\cdot}5^{3}$	PO	L	$6{\cdot}1^{9}$	$7{\cdot}1^{11}$	$4{\cdot}9^{7}$	$_{+}(1{\cdot}8)_{0}$
Simetryne	213	83	450^{3}	PO	L	$7{\cdot}1^{7}$	$8{\cdot}3^{9}$	$5{\cdot}4^{7}$	$_{+}(4{\cdot}0)_{0}$

Table 1 (*contd.*)

Name	Mol. Mass	m.p. (°C)	Solubility: Water	Good solv.	PHS	v.p.	SVC	$\mathscr{P}_A^W$	pK
Strychnine	334	270–80	143^{3}	HCl WSS					$_{+}8{\cdot}3_{0}$ f
Sulfallate	224	ℓ	92^{3}	PO		$2{\cdot}2^{3}$	$2{\cdot}7^{5}$	$3{\cdot}4^{3}$	
Sulfotep	322	ℓ	25^{3}	PO		$1{\cdot}7^{4}$	$3{\cdot}0^{6}$	$8{\cdot}3^{3}$	
Sulfoxide	324	ℓ	neg.	PO	M				
Sulphuryl Fluoride	102	gas (b.p. – 55)	750^{3}	MISC MeBr				0·18	
2,4,5-T	255	154	238^{3}	ALC WSS					$_{0}2{\cdot}85_{-}$ g,j
2,3,6-TBA	225	125	7	ALC WSS					$_{0}1{\cdot}3_{-}$ k
TCA	163	56	10000	ALC	L	0·1	9^{3}		$_{0}0{\cdot}6_{-}$
Tebuthiuron	228	162	2·5 1200 Na	70 ACET WSS	L	$9{\cdot}4^{7}$	$1{\cdot}2^{8}$	$2{\cdot}1^{8}$	 e
Tecnazene	261	99	neg.	AHC					
TEPP	290	ℓ	MISC	MISC	L	$1{\cdot}5^{4}$	$2{\cdot}5^{6}$		
Terbacil	217	176	710^{3}	KET					
Terbufos	288	ℓ	$10–15^{3}$	MISC		$7{\cdot}5^{5}$	$1{\cdot}2^{6}$	1^{4}	

	[illegible]	[illegible]	[illegible]	100 DMF	L	$1\cdot1^{6}$	$1\cdot4^{8}$	3^{5}	
Terbutryne	241	105	25^{3}	PO		$9\cdot6^{7}$	$1\cdot3^{8}$	$2\cdot0^{6}$	(4·0)
Tetrachlorvinphos (Z)	366	97	11^{3}	400 $CHCl_3$		$4\cdot2^{8}$	$8\cdot4^{10}$	$1\cdot3^{7}$	
Tetradifon	356	149	200^{3} 50°	AHC		—	—		
Tetramethrin	331	65–80	—	400 ACET		$3\cdot5^{8}$	$6\cdot3^{10}$		
Tetrasul	324	87	L	AHC		$7\cdot5^{7}$	$1\cdot3^{8}$		
Thiabendazole	201	305	10	80 DMSO	V	neg.			
Thiazfluron	240	137	2·1	600 DMF	V	$2\cdot0^{6}$	$2\cdot6^{8}$		
Thiobencarb	258	3	30^{3}	MISC		$3\cdot6^{8}$	$5\cdot0^{10}$	$6\cdot0^{7}$	
Thiocyclam	271	128	84	Water	V				
Thiofanox	218	57	5·2	AHC	L	$9\cdot1^{5}$	$1\cdot1^{6}$	$4\cdot7^{6}$	
Thiometon	246	ℓ	200^{3}	PO	L	$2\cdot0^{4}$	$2\cdot7^{6}$	$7\cdot4^{4}$	
Thionazin	248	ℓ	1·14	MISC		1^{3} b	$1\cdot4^{5}$	8^{4}	
Thiram	240	155	30^{3}	ACET $CHCl_3$					
Tolyfluanid	165	76	2·6	570 BENZ					
Triadimefon	294	82	260^{3}	1200 CH_2Cl_2		$<7\cdot5^{10}$	$<1\cdot2^{11}$	$>2^{10}$	
Triallate	305	30	4^{3}	MISC		—			

Table 1 (*contd.*)

Name	Mol. Mass	m.p. (°C)	Solubility: Water	Solubility: Good solv.	Solubility: PHS	v.p.	SVC	$\mathscr{P}_A^W$	pK
Triamiphos	294	168	250^3	PO		—			
Triazophos	313	0–5	39^3	AHC		$6{\cdot}4^7$	$1{\cdot}1^8$	$3{\cdot}5^6$	
Tributyltinoxide	596	ℓ	100^3	MISC		$3{\cdot}2^6$	$1{\cdot}1^7$	9^5	
Trichloronate	334	ℓ	50^3	MISC		a $5{\cdot}3^7$	$9{\cdot}8^9$	$5{\cdot}1^6$	
Trichlorphon	257	83	154	AHC	L	a $7{\cdot}8^6$	$1{\cdot}1^7$	$1{\cdot}4^9$	
Triclopyr	256	149	440^3	ALC WSS		$5{\cdot}9^7$	$8{\cdot}2^9$	$5{\cdot}3^7$	$_0 2{\cdot}7_-$

Trietazine	230	101	20^{3}	500 $CHCl_3$		—			$_{+}(1{\cdot}8)_0$
Trifenmorph	329	176, 187	20^{6}	300 CCl_4		$1{\cdot}4^{7}$	$2{\cdot}5^{9}$	$1{\cdot}6^{4}$	
Trifluralin	335	49	$<1^{3}$	500 XYL		$5{\cdot}4^{5}$	$9{\cdot}9^{7}$	$<1^{3}$	
Triforine	435	155	27^{3}	DMF, DMSO		$8{\cdot}8^{8}$	$2{\cdot}1^{7}$	$1{\cdot}3^{5}$	
Undecan-2-one	170	13	neg.	MISC	——				
Vamidothion	287	48	4000	PO	V	neg.			
Vernolate	203	ℓ	107^{3}	MISC	——	$6{\cdot}3^{3}$	$7{\cdot}0^{5}$	$1{\cdot}5^{3}$	
Vinclozolin	286	108	1	Sol. acid 435 ACET					

Appendix 5. Meteorological Factors Influencing Pesticide Behaviour

I. ATMOSPHERIC COMPOSITION

The total mass of the earth's atmosphere in a vertical column of 1 cm^2 area is, conveniently, near 1 kg. This is 100 ktonnes ha^{-1}, made up of about 76 ktonnes of nitrogen, 23 of oxygen and 1 of the inert argon. Of great importance for reducing u.v. intensity at sea level is ozone, totalling only 70 kg ha^{-1} but almost confined to a deep belt at very low pressure in the upper layers. Other relatively constant constituents are methane and nitrous oxide at about 250 and 150 kg ha^{-1}, generally distributed. More variable is sulphur dioxide at about 40 kg ha^{-1}. Most of these three minor constituents is of biological origin in balance with chemical destruction. A fluctuation of the order of $\pm 30\%$ in the sulphur dioxide content is mainly due to volcanic activity. Annual production from coal and oil burning may produce per year a few per cent of the reservior amount.

Of major biological importance is carbon dioxide at about 45 tonnes ha^{-1}, the fifth most abundant constituent and inevitably a rather variable one. The fourth most abundant—water vapour—is still more variable. It averages around 200 tonnes ha^{-1}, making the world total as vapour about 10^{13} tonnes, compared with 10^{18} tonnes in the oceans.

The variability of water content is of course due to the SVC being frequently exceeded locally so that there is erratic interchange between vaporized and condensed water. The saturation concentrations are tabulated in Table 1.

II. SOLAR RADIATION

(Based largely on Parry Moon, *J. Franklin Inst.* 1940, **230,** 583 and Cobblentz *Bull. Amer. Meteor Soc.* 1949, **30,** 204).

Table 1
Saturated water vapour

Temperature (°C)	Over ice pressure (mmHg)	Over ice ($g\,m^{-3}$)	Over liquid water pressure (mmHg)	Over liquid water ($g\,m^{-3}$)
−30	0·286	0·34	—	—
−25	0·476	0·55	—	—
−20	0·776	0·89	—	—
−15	1·241	1·39	1·436	1·61
−10	1·950	2·14	2·149	2·36
−5	3·013	3·26	3·163	3·42
0	4·579	4·85	4·579	4·85
5	—		6·543	6·80
10	—		9·209	9·41
15	—		12·79	12.83
20	—		17·53	17·30
25	—		23·76	23·04
30	—		31·82	30.35
35	—		42·17	39·6
40	—		55·32	51.1

The total energy incident on a plane area normal to the sun's rays and outside the earth's atmosphere is 1·90 g cal cm^{-2} min^{-1} which is 1322 W m^{-2} or 13·3 MW per hectare. This is nearly constant, there being a small periodic fluctuation (±3·9%) because of the variation in the radius of the earth's elliptical orbit (minimum radius, maximum energy in January) and a spasmodic variation in the far u.v. and X-radiation due to sunspots.

On a horizontal surface there is, of course, diurnal variation and this, owing to the inclination of the earth's axis to the plane of its orbit, has a seasonal component. The mean value on a horizontal area at the equator and at the equinox is π^{-1} of the maximum since the sun illuminates a diameter and this, in 24 h, is spread over a circumference.

The seasonal-cum-latitude variation of the 24 h total (purely trigonometrical ignoring atmospheric effects) is surprisingly complex. At the equinox it falls steadily from π^{-1} of the full value at the equator to zero at the poles, but at the solstice it increases from the winter arctic circle (zero at latitude above this) through the equator and the summer tropic to a rather flat maximum at about 42° latitude, then falls to a minimum around 60° and increases again to a highest value at the summer pole. Expressed as a multiple of the equatorial (π^{-1}) average, the maxima are 1·156 and 1·253 and the minimum (60°) 1·138. Note that at 42° latitude more energy is received in 24 h than at the equator at any time. The variation of climate with latitude depends much more on winter than on summer levels. Minimum 24 h levels as a fraction of maximum are 0·92, 0·44, 0·14 at latitudes 0°, 30° and 52°.

The earth's atmosphere cuts down and scatters radiation. The effect is complex because scattering depends on fine dust and cloud in addition to having a molecular component, and absorption (mainly in the ultraviolet and infrared) is very dependent on wavelength. At some of the absorption peaks the energy is reduced almost to zero whatever the inclination of the sun while, on the fringes of the absorption bands, the intensity depends greatly on the inclination—i.e. on the depth of atmosphere traversed.

It is the u.v. content of solar radiation which is particularly reduced by the atmosphere. There is no detectable radiation below a wavelength of 290 nm and very little below 300 nm. Table 2 is adapted from Parry Moon (1940). The values are W m^{-2} on a surface normal to the sun's rays. The third column must therefore be reduced by half to apply to a horizontal surface.

Although the intensity in the far u.v. is very small relatively, it is still enough to have important potential for photochemical decomposition. At a wavelength of 310 nm one quantum of radiant energy is $6{\cdot}41\times10^{-12}$ erg, equivalent to $9{\cdot}22\times10^{4}$ g cal^{-1} mol^{-1}. The energy at <320 nm incident

Table 2
Screening of solar radiation by earth's atmosphere

	Wavelength band (nm)	Outside atmosphere	Sea level, sun above horizon by 90°	Sea level, sun above horizon by 30°
u.v.	<300	3·2		0·0008
	300–310	5·4		0·03
	310–320	6·7		0·3
	320–350	24·7		3·8
	350–400	54·6		15·7
	<400	94·6	40·1	19·8
Visible	400–700	540	420	328
i.r.	700–1100	365	309	267
	110–1500	162	95	70
	1500–1900	73	51	45
	>1900	87	13	9
	>700	687	468	391

at 30° to horizontal on 1 ha is 1500 W, which is equal to $2 \cdot 17 \times 10^4$ g cal min^{-1}, containing therefore enough quanta to decompose 0·23 mol min^{-1} at unit efficiency. Even if the "u.v.-day" is effectively only 200 min this is of the order of 10 kg ha^{-1} day^{-1}. Few quantum efficiencies, except for energetic chain reactions, reach unity even in solution or vapour state but the calculation shows that there is abundant energy "in hand" even below 320 nm wavelength. It should be noted that erythema under human skin is produced by a rather narrow band about 316 nm wavelength, near to the limit where absorption in the atmosphere is rapidly changing. It is also of interest that the molecular light scattering responsible for the blue of the sky increases very much with decrease of wavelength. Could we but see it, the sky would be strongly ultraviolet. About half the radiation below 350 nm comes diffusely from a clear sky and half directly from the sun.

If all the energy in the spectrum could be harnessed to the reduction of carbon dioxide to carbohydrate, about 13 tonnes of the former would be consumed per ha on a temperate midsummer day to give 9 tonnes of dry matter. Only exceptional crop production exceeds 2% of this.

III. PRECIPITATION

Some of the water content of the atmosphere is frequently, locally, in particulate form (liquid or ice) as mist or cloud. Only a few mg in *particulate* form per m^3 can severely restrict vision and as there is nearly 5 g m^{-3} vaporized water in air saturated even at 0°C it is evident that fog can persist only in constant or falling temperature (or in severely contaminated atmosphere). In dense clouds there is inevitable coagulation of particles and consequent precipitation as rain or snow.

Descending light rain may evaporate in warmer air before reaching the ground and, since smaller drops evaporate more rapidly, the average drop size increases towards ground level. Very large drops, formed from melting hail, break up in falling rapidly and this process sets an upper limit (about 5 mm diameter) to ground-level rain drops. There is therefore a broad positive correlation between mean drop size and intensity of rainfall. Mason (1971) considers the many factors leading to variation and spread of drop sizes in rainstorms but gives (his appendix B) the following rough guide

mean drop mass in μg $= 180 \times (\text{precipitation in mm h}^{-1})^{\frac{3}{4}}$

Corresponding mean diameter in mm $= 0 \cdot 7 \times (\text{precipitation in mm h}^{-1})^{\frac{1}{4}}$

$$\frac{\text{mg of liquid water in}}{\text{m}^3 \text{ of traversed air}} = 70 \times (\text{precipitation in mm h}^{-1})^{0\cdot85}$$

Evaluation yields the following table covering the extreme range.

Precip. rate (mm h^{-1})	Diam. of drop of mean mass (mm)	g liquid per m^3 air
0·1	0·4	0·01
1·0	0·7	0·07
10	1·2	0·49
100	2·2	3·5
1000	3·9	25

It is noteworthy that only the last figure, corresponding to exceptionally heavy rain encountered only locally and rarely, exceeds the amount of water normally present as vapour (17 g m^{-3} at 20°C, 5 g m^{-3} at 0°C).

The frequency and duration of rainfall of given intensity is difficult to classify. The area over which a given value obtains must also be recorded. Very heavy rain is characteristically localized and of short duration. According to the data assembled by Foster (1949) in the USA total rainfall exceeds 1 inch in 24 h in the New England States on about 7–10 days per year. This rate of course averages about 1 mm h^{-1} but is normally made up of heavier, shorter showers. Twice this figure is exceeded on average once per year and 3 inches in 24 h only once in 5–10 years. One inch in 10 min (1500 mm h^{-1}) is attained in Concorde, N.H., about once in 8 years but the shortness of duration of such excessive rainfall is indicated by 2 inches in 100 min (300 mm h^{-1}) being rather less frequent while for 4 inches to be reached with this frequency requires the time of continuous collection to be extended to 1500 min.

An empirical relationship much used for relating frequency, duration and amount of rainfall is that derived by Bilham (1935) as follows:

$$n = \frac{1\cdot25t}{(r+0\cdot1)^{3\cdot55}}$$

where n = frequency per decade; t = duration in hours; r = rainfall amount in inches.

This relationship, although overestimating the frequency of very heavy showers, has been found to apply reasonably well throughout the British Isles for periods of up to a few hours.

The low rate rains are more uniformly spread over a larger area, but with persistent "Scotch mist" or precipitation from drifting fog and cloud a different scale of non-uniformity emerges. The droplets are largely collected by impaction (see Section 6. VII.B) and a forest—or crop—floor receives water on a pattern determined by the geometry of the canopy.

Appendix 6. Drop Geometry and Population Density

I. RELATIONS OF DIMENSIONS OF FREE AND SPREAD DROP

We assume that a drop of pure liquid of radius a in the free state spreads on a smooth surface to contact a circular area of radius r. We also assume, without appreciable error for drops of normal spray size, that gravity is unimportant so that the curved surface of the drop is spherical, radius R. h is the maximum height of the spread drop. These measures are shown in Fig. 7, Section 8.V.B.

Simple trigonometry gives

$$r = R \sin \theta$$

$$h = R(1 - \cos \theta)$$

whence

$$\frac{h}{r} = \frac{(1 - \cos \theta)}{\sin \theta} \tag{1}$$

The volume of the spread drop (shaded area) is given by the more complex but still exact function

$$\text{Volume} = \tfrac{4}{3}\pi r^3 \left(\frac{2 - 3\cos\theta + \cos^3\theta}{4\sin^3\theta} \right)$$

We will rewrite this

$$\text{Volume} = \tfrac{4}{3}\pi r^3 K_\theta^{-\frac{3}{2}}$$

where

$$K_\theta = \left(\frac{4\sin^3\theta}{2 - 3\cos\theta + \cos^3\theta} \right)^{\frac{2}{3}}$$

Since different forms are used in different texts, note the trigonometrical identities

$$\frac{2-3\cos\theta+\cos^3\theta}{4\sin^3\theta}=\frac{(1-\cos\theta)^2(2+\cos\theta)}{4\sin^3\theta}=\frac{(1-\cos\theta)(2+\cos\theta)}{4\sin\theta(1+\cos\theta)}$$

This volume is necessarily equal to that, $\frac{4}{3}\pi a^3$, of the free drop, whence

$$\frac{r^2}{a^2}=K_\theta \tag{2}$$

$$\frac{r}{a}=K_\theta^{\frac{1}{2}} \tag{3}$$

for

$$\theta\to 0 \qquad \frac{r}{a}\to\left(\frac{16}{3\theta}\right)^{\frac{1}{3}}, \quad \theta \text{ in radians}$$

The mean thickness of the spread drop, $\bar{h}$, is equal to volume divided by contact area, so that

$$\frac{\bar{h}}{a}=\frac{4}{3}K_\theta^{-1} \tag{4}$$

for

$$\theta\to 0 \qquad 2\bar{h}\to h\to\tfrac{1}{2}r\theta=a\,.\left(\frac{2\theta^2}{3}\right)^{\frac{1}{3}}$$

Table 1 gives values of the ratios in eqs (1), (2), (3) and (4) in columns 2, 3, 4, 5 for values of θ in column 1 in the range of interest.

We may note further that

$$\mathbf{A}_c=\text{area of curved surface of spread drop}=2\pi Rh$$

and

$$\mathbf{A}_p=\text{area of contact with plane}=\pi r^2$$

and it can be shown that

$$\left(\frac{\partial\mathbf{A}_c}{\partial\mathbf{A}_p}\right)_{\text{vol. const.}}=\cos\theta \tag{5}$$

Since the surface energy change consequent on a small displacement

$$=\gamma_L\delta\mathbf{A}_c+(\gamma_{SL}-\gamma_S)\delta\mathbf{A}_p \tag{6}$$

and this must be zero at the equilibrium position, it follows from eq. (5) that

$$\gamma_{SL} + \gamma_L \cos\theta = \gamma_S \tag{7}$$

identical with the balance of the hypothetical forces at the edge of the drop but with the advantage that result can be obtained by substitution of the real works of adhesion in eq. (6).

Table 1

$\theta°$	$\frac{1-\cos\theta}{\sin\theta}=\frac{h}{r}$	$K_\theta=\frac{r^2}{a^2}$	$K_\theta^{\frac{1}{2}}=\frac{r}{a}$	$\frac{4}{3}K_\theta^{-1}=\frac{h}{a}$
2·5	0·222	24·62	4·96	0·054
5	0·044	15·49	3·94	0·086
7·5	0·066	11·82	3·44	0·113
10	0·087	9·74	3·12	0·137
15	0·132	7·40	2·72	0·180
20	0·176	6·07	2·46	0·220
30	0·268	4·55	2·13	0·294
40	0·364	3·66	1·91	0·364
50	0·466	3·05	1·75	0·438
60	0·577	2·58	1·61	0·516
70	0·700	2·20	1·48	0·606
80	0·839	1·89	1·37	0·706
90	1·000	1·59	1·26	0·838
100	1·192	1·32	1·15	1·010
110	1·428	1·07	1·04	1·246
120	1·732	0·840	0·92	1·584
130	2·145	0·622	0·79	2·146

II. SEPARATION OF DROPS IN UNIFORM APPLICANCE

No application is ever uniform. Random incidence and shadowing make it impossible. It is useful, however, to know what distance would separate drops of various sizes at different rates of spraying if uniformity could be approached. We calculate therefore for a spray of equal sized drops delivered on to a flat target in the perfect face-centred honeycomb pattern where every drop is equidistant from six nearest neighbours and as far from them as is possible.

For this configuration the area per drop $=(\sqrt{3}/2)\ell^2$ where ℓ is the distance between nearest centres. This is, of course, the inverse of the number of drops per unit area which we can calculate from the drop size and the total volume per unit area. While we could approximate to the calculated figures if spraying a billiard table, the density on leaves in a crop will necessarily be more variable because of shadowing and smaller because the leaf area is usually greater than the projection area of the whole plant. The ratio is called the "leaf area index" and should be used to divide the total volume per unit field area as a basis for calculation of ideal distribution on leaves. We will denote by $\mathscr{V}$ the volume per unit

Table 2

Separation of drops in ideal uniform distribution

$2a$ \ $\mathscr{V}$	10	15	20	25	30	40	50	60	80	100
50	275	224	194	174	159	137	123	112	97	87
75	505	412	357	319	292	253	226	206	179	160
100	778	636	550	491	449	389	348	317	274	246
125	1090	887	768	687	628	543	486	444	384	344
150	1430	1160	1010	903	825	714	639	583	505	452

$2a$ \ $\mathscr{V}$	100	150	200	250	300	400	500	600	800	1000
200	695	570	491	439	401	348	311	284	245	220
250	970	797	687	614	562	486	435	398	344	308
300	128	104	903	807	738	639	571	522	452	404
350	161	131	114	102	933	806	720	659	570	510
400	197	161	139	124	113	975	880	804	693	622
450	235	183	166	148	136	117	105	957	830	744
500	275	224	194	174	159	137	123	112	972	870
600	362	295	256	228	208	181	161	148	128	114
800	556	456	393	351	321	278	249	227	196	176
1000	778	636	550	492	449	389	348	317	274	246
1·2	10·2	8·4	7·2	6·5	5·9	5·1	4·6	4·2	3·6	3·2
1·4	12·9	10·5	9·1	8·1	7·4	6·4	5·8	5·2	4·5	4·1
1·6	15·7	12·9	11·1	10·0	9·1	7·9	7·0	6·4	5·6	5·0
1·8	18·8	15·4	13·3	11·9	10·8	9·4	8·4	7·7	6·6	5·9
2·0	22·0	18·0	15·6	13·9	12·7	11·0	9·8	9·0	7·7	7·0

$\mathscr{V}$ = litres ha^{-1} area on which drops fall; $2a$ = drop diam., in μm above dotted line, in mm below; ℓ = distance between nearest centres, in μm above stepped line; ×10 for μm below stepped line, above dotted line; mm below broken line.

area so corrected and expressed in litres per hectare (= litres per field hectare + the leaf area index). If ℓ is measured for convenience in μm, as is the drop diameter $2a$, the relationship is

$$\ell^2 = 6{\cdot}05(2a)^3/\mathcal{V} \tag{8}$$

There is a further latent assumption in the application of this equation that there is no interference and fusion of drops.

Table 2 lists values of ℓ over a practical range of values of $\mathcal{V}$ and $2a$. Note that if ℓ is required for values of $\mathcal{V}$ or $2a$ outside the range listed, ℓ is proportional to $a^{\frac{3}{2}}$ and to $\mathcal{V}^{-\frac{1}{2}}$. If, for example, we multiply $2a$ by 4 we should have to multiply $\mathcal{V}$ by 64 to give the same population density.

Appendix 7. Interception of Non-vertical Spray by Non-horizontal Leaves

In this appendix we consider the interception of spray by a fully exposed leaf in relation to angle of incidence of the drops (θ to the vertical) and the inclination of the leaf (α to the horizontal). In practice, of course, with incompletely matched spreading jets from fan or hollow-cone nozzles, spray arrives at different angles according to position of the leaf in relation to the nozzles. The general principle can however be demonstrated by considering uniform spray incidence, i.e. we assume that all angles are similarly distributed at all locations under the boom in the absence of crop.

The geometrical situation is therefore as shown in Fig. 1 where the spray arriving on the leaf at angle θ is shown by parallel lines. The first simplification we can introduce from our assumption that all spray angles are fairly represented is that there will be the same intensity of spray falling in the $-\theta$ direction as in the $+\theta$ direction.

The area of the leaf relative to its true, assumed plane, area, apparent to an observer looking in the spray direction is the sine of the angle between spray direction and leaf plane. This angle is $90-(\alpha+\theta)$ for the $+\theta$ spray and $90-(\alpha-\theta)$ for the $-\theta$ spray. The amount of spray intercepted by the leaf is therefore proportional to the intensity of spray in these directions times the sines of these angles and therefore, for the two directions together

$$= \text{intensity} \times (\cos(\alpha+\theta) + \cos(\alpha-\theta)) \tag{1}$$

$$= \text{intensity} \times 2\cos\alpha\cos\theta \tag{2}$$

It will be seen that the contributions of α and θ are simple and independent, so, however intensity varies with θ, the integration over all values of θ does not depend on α. Let the intensity be represented as a

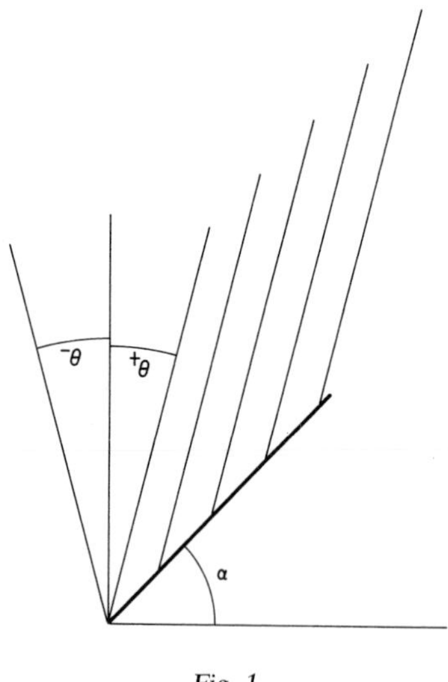

Fig. 1

general function f(θ) of θ, then the total interception per unit plane leaf area is

$$\int_0^{\theta\,\max} 2\cos\alpha\cos\theta\,\mathrm{f}(\theta)\,\mathrm{d}\theta \tag{3}$$

$$= 2\cos\alpha\int_0^{\theta}\cos\theta\,\mathrm{f}(\theta)\,\mathrm{d}\theta \tag{4}$$

Now for $\alpha = 0$ this becomes the "sprayfall" or applicance per unit gross field area, so that

$$\mathscr{A} = 2\int_0^{\theta\,\max}\cos\theta\,\mathrm{f}(\theta)\,\mathrm{d}\theta \tag{5}$$

and interception by the leaf at angle α to the horizontal is simply

$$\mathscr{A}\cos\alpha \tag{6}$$

independent of θ. The fraction of spray received is therefore independent of angle of spray, while the sharpness of shadow cast is of course greatly blurred by angular spread of the spray.

This deduction is mathematically general as long as eq. (1) is valid, but, if the spray is flatter than the leaf, i.e. $\alpha + \theta > 90°$, so that the reverse side of the leaf is hit by $+\theta$ spray, $\cos(\alpha + \theta)$ in eq. (1) is negative—i.e. we are subtracting the spray on the reverse side. In practice, of course, we must add it and therefore, for all $\theta > 90° - \alpha$, we must replace (1) by

$$\text{intensity} \times (\cos(\alpha - \theta) - \cos(\alpha + \theta)) \tag{7}$$

$$= \text{intensity} \times 2 \sin\alpha \sin\theta \tag{8}$$

While (5) remains valid, since $\theta + \alpha < 90°$ for $\alpha = 0$, (4) must be replaced by

$$2 \cos\alpha \int_0^{\alpha} \cos\theta \, f(\theta) \, d\theta + 2 \sin\alpha \int_{\alpha}^{\theta\,\max} \sin\theta \, f(\theta) \, d\theta \tag{9}$$

The simplicity of eq. (6) is lost and the complex function replacing $\cos\alpha$ therein will be dependent on α, $\theta_{\max}$ and $f(\theta)$.

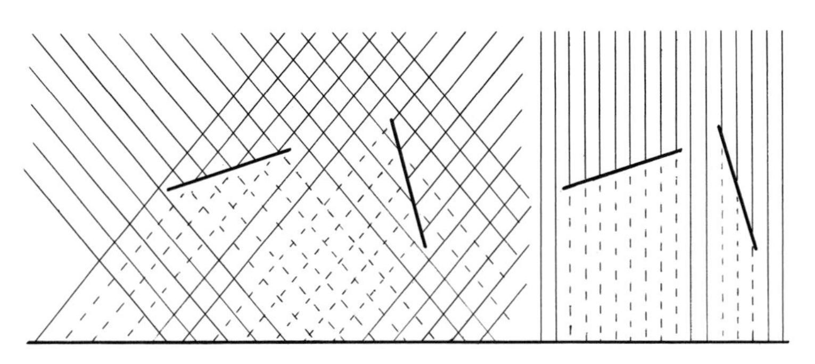

Fig. 2. On the left is shown schematically a symmetrical two-angled spray giving, when not interrupted, the same total "sprayfall" on the ground as the vertical spray on the right. A flat-angled plane leaf interrupts the same number of spray lines in each case $(3+5=8)$. A steep-angled leaf, receiving spray on both sides from the angled spray interrupts more $(5+2)$ than from the vertical spray (3). The latter value is in fact the difference $(5-2)$, not the sum. The angled spray casts broader but more diffuse shadows.

A qualitative simple conclusion is, however, general. Since (6) would be valid if spray on the reverse side of the leaf were counted negative, the true value of spray intercepted, when this spray is counted positive, must be greater than indicated by (6). Therefore when leaves receive spray on both sides, interception is greater for the spread spray than for a vertical spray although if all spray falls on one side only of each leaf the total leaf interception is independent of spray angle (if symmetrical from $-\theta$ to $+\theta$).

Figure 2 illustrates these points schematically.

Appendix 8. Some Dimensions of Organs and Pests

I. INTRODUCTION

Sizes of pests in various stages of development and of the accessible organs of the host can have important influence on pesticide efficiency and selectivity and choice of size of pesticide and carrier particles. Sizes however are often not recorded in the relevant biological literature and scales not included in photographs. They may be too familiar, at least intuitively, to the biological worker and he may feel that they are too variable to specify. The result is that the worker in pesticide physics or formulation is left without even order-of-magnitude information on unfamiliar objects, and may make guesses which are highly erroneous. In this appendix we have assembled some data illustrative of the relevant biological sizes in the immediate environment of the airborne or deposited pesticide.

Most of the data are very approximate, but at least a figure x can be assumed to lie between $\frac{1}{2}x$ and $2x$ which is better than no figure at all. Obviously, in the detailed study of any particular crop–pest situation, measurement will be necessary, but we hope this list will be of value in providing a general background.

The schematic picture (Fig. 1) of part of an aphis on part of the thickness of a cabbage leaf displaying some sizes approximately to scale will serve as an introduction and will probably contain some surprises for most readers.

II. AERIAL SYSTEM OF THE CROP AND WEEDS

Crop and weed leaves vary greatly in shape and area, but the total functional area in the crop relative to area of the gross land surface, the Leaf Area Index, is much less variable being, at maturity, in the range of 5–10.

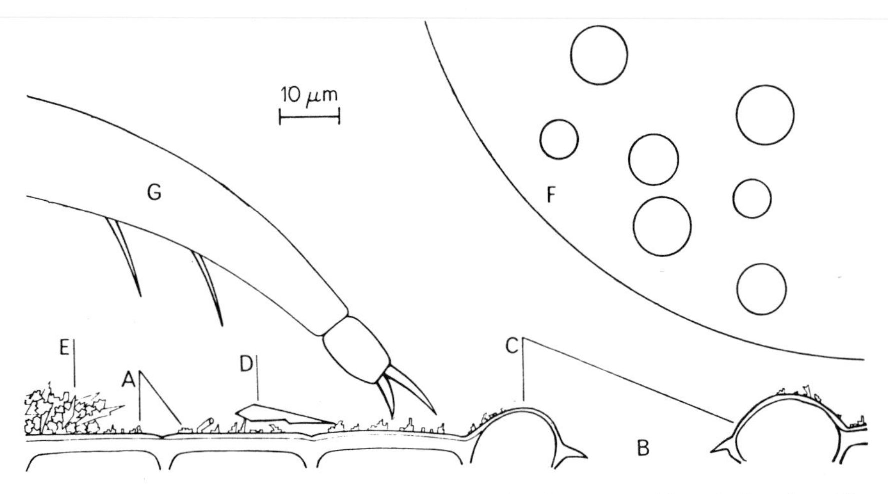

Fig. 1. Diagrammatic representation of some relative sizes on a cabbage leaf surface. The upper walls of the epidermal cells are shown in section, with the continuous cuticle shown as two layers, the outer one of cutin on which rest the castillated wax platelets, A. At B is an open stoma flanked by guard cells, C. D represents a small aciculated insecticide crystal typical of dust deposit and E the edge of a dried deposit from a wettable powder drop composed of insecticide crystals and filler. At F is shown part of a falling spray drop of medium-small size (150 μm diameter) containing in this case typical emulsion droplets. If 1 lb per acre of insecticide could spread as a continuous layer over the leaves of a cabbage crop, the layer formed would be about as thick as the line showing the outer surface of cutin. At G is shown one half of a typical aphis tarsus: the complete leg is about 20 times this length. The hook-fringed padded proleg of a third instar *Pieris brassica* caterpillar would cover more than the whole diagram.

The size of large plain leaves is not of much significance in the present context but that of small leaves or the finer branches of divided leaves can be. The smallest receiving organ of a crop is not a true leaf but the very numerous green cladodes of asparagus which are about 0·35 mm diameter and 15–20 mm long. The terminal pinnae of the much-divided carrot leaf are 2–3 mm wide on the greater part of their 5–10 mm length and no part of the leaf is more than 5 mm across.

The thickness of the leaves of most crops and weeds of temperate agriculture is much less variable. Their very variable rigidity is deceptive, being more dependent on other features of internal structure and on the support of thick veins. Desert succulents and maritime species have much thicker tissues and many widely distributed species develop thicker leaves when growing on maritime sands.

The succulent pineapple has the thickest leaves of common crops, the section being nearly triangular with thickness tapering from around 8 mm in the centre to near 0·1 mm at the green tissue nearest to the sharp edge.

In most mesophytes the thickness increases sharply close to the veins, most of the interveinal lamina having fairly constant thickness. Only the Iris among the species measured by us (directly in fresh transverse section under 25× magnification) and listed below showed a smooth "catenary type" transition from one vein to the next and an average thickness is thus given.

A mean thickness of 0·25 mm corresponds, when air space is allowed for, to about 20 mg wet mass cm^{-2} (actual measurement on garden pea, cv Onward). Assuming a modest L.A.I. of 5 this is equivalent to 10 tonnes ha^{-1}. The actual mass is often greater because stems and seeds are not allowed for.

We have tested one extremely hairy leaf in Table 1 to indicate the extent which hairiness can reach, but the size, shape and frequency of hairs vary too widely for any but a comprehensive list to be useful.

Table 1
Leaf thickness

Species	Mean thickness (mm)	Species	Mean thickness (mm)
Brassicas		*Allium porrum* (leek)	0·3–0·4[b]
Cauliflower near tip	0·25–0·4	*Stachys palustris*	0·15
Cauliflower near base	0·3–0·4	*Stachys lanata*	0·1[c]
Cabbage	0·5–0·6	*Beta* (fodder)	0·2–0·35
Scotch (curly) kale	0·2–0·4	*Senecio vulgaris*	0·25
Phaseolus multiflorus	0·2	*Stellaria media*	0·15
Pisum sativum	0·25	*Hedera helix*	0·25–0·3
Rose (hybrid tea)	0·25	*Ilex aquifolium*	0·25–0·4
Pear	0·2[a]	*Escalonia macrantha*	0·25
Euphorbia lathyrus	0·3–0·4	*Rhododendron ponticum*	0·4–0·5
Iris	0·6	*Taxus baccata*	0.5[d]
Zea mays	0·15	*Cedrus atlantica*	0·8[d]
		Pinus sylvestris	0.7[d]
		Ulex europaeus	1·4[e]

[a] A very stiff leaf despite thinness. The only leaf in the series which did not float in water.
[b] Two layers of this thickness on either side of near-rectangular cavities 0·5 mm deep.
[c] The very hairy leaf of the garden "rabbits' ears"; 1·3–1·5 mm depth of dense hair-cushion on the abaxial surface; 1–1·4 mm depth on adaxial surface with hairs, upright near leaf surface, bending over to form a dense canopy.
[d] The needles of *Cedrus* are almost cylindrical, of *Taxus* flat and of *Pinus* flattened triangle, the greatest depth of which is indicated.
[e] The leaves of *Ulex* are near cylindrical but with considerable air space.

Cuticle thickness is covered in the text. Wax covering is also dealt with in the text. Stomata are mouth-shaped and have slit length usually between 10 and 20 μm. They may be absent on the adaxial surface or number, on either surface, from a few to 100 or more per mm^2.

III. SEEDS OF HIGHER PLANTS

Seed sizes vary greatly and are not closely correlated with the size of the mature plant or the longevity or dormancy of the seed. Many crop seeds are larger than most weed seeds. This is often due to the seed forming the desired harvest and the plant breeder having selected cultivars for, *inter alia*, seed size. One must not, however, assume this to be a general rule, tobacco for instance having one of the smallest seeds known and many weeds having larger seeds than that of lettuce. Table 2 gives the sizes of a reasonable selection of crop and weed seeds, measured directly under low magnification. The size is of course an "envelope" size measured to the outside of any small ribs or protrusions present and not allowing for concavities. Seeds are generally of approximately ellipsoidal form.

The dimensions are listed as xy^2, x being the diameter of the unique axis of revolution, the shape being prolate for $x > y$ and oblate for $x < y$. For a true ellipsoid the volume is $(\pi/6)xy^2$. An idealized cylindrical rod or lozenge would have volume $(\pi/4)xy^2$. If three "diameters" are needed, as in cucumber seeds, all are given. For many seeds, mass is better recorded and in such case the diameter of a sphere which would have this mass at a density of 1 is indicated in brackets. Density of the seed solids often exceeds 1 but there may be cavities and frequently corrugations of the surface in which case the "envelope" size is greater than that estimated from mass (e.g. *Sagina apetala*).

IV. RATE OF ROOT GROWTH

Even those seeds, including those of most crops, which do not have a protracted dormancy, requiring some special environmental sequence to break it, may take many days inbibition before the radicle breaks through the testa. Extension of the primary root is then very rapid, 1–3 cm day^{-1} being usual, and a length of 10–15 cm is frequently attained before the cotyledons emerge and photosynthesis can begin. Even lateral branch roots may begin to develop before photosynthesis can contribute and,

Table 2
Seed sizes

Weed species	Axes (mm)	Crop species	Axes (mm)
Arctium lappa	$7·5 \times 3 \times 2$	Bean (*Phaseolus multiflorus*)	$20 \pm 2 \times 11 \pm 2 \times 7·5 \pm 0·5$
Sonchus oleraceus	$4 \times 0·9^2$	Bean (*P. vulgaris*)	$15 \pm 2 \times 7 \pm 1 \times 5–6$
Euphorbia peplus	$1·6 \times 1^2$	Bean (*Vicia faba*, garden)	$22 \pm 3 \times 15 \pm 2 \times 7 \pm 1$
Stellaria media	$0·7 \times 1^2$	Bean (*Vicia faba*, field)	$0·6 \pm 1$ g (11)
Rumex acetosella	$0·9 \times 1·1^2$	Bean (*Glycine max*) Soy	$0·1 \pm 0·02$ g (6)
Papaver rhoeas	$0·8 \times 0·7 \times 0·5$	Pea (*Pisum sativum*)	$0·25 \pm 0·05$ g (8)
Capsella bursa-pastoris	$0·8 \times 0·3^2$	Cotton (*Gossypium hirsutum*)	$12 \times 5 \times 4$; 0·12 g (7)
Digitalis purpurea	$0·7 \times 0·5^2$	Cucumber (*cucumis sativus*)	$10–14 \times 5 \times 1·5$
Epilobium montanum	$0·7 \times 0·3^2$	Maize, corn (*Zea mays*)	$0·3 \pm 0·05$ g (8·4)
Juncus communis	1×10^{-5} g (0·27 mm)	Wheat (*Triticum aestivum*)	$7 \pm 1·5 \times 3·6 \pm 0·05 \times 3·5 \pm 0·6$; $0·05 \pm 0·015$ g (4·6)
Sagina apetala	$0·4 \times 0·3^2$ $7·5 \times 10^{-6}$ g (0·25 mm)	Barley (*Hordeum vulgare*)	0·046 g (4·5)
		Sanfoin (*Onobrychis viciaefolia*)	$4·5 \times 3^2$
		Lucerne, alfalfa (*Medicago sativa*)	$3 \times 1·5^2$
		White Clover (*Trifolium repens*)	$1 \times 0·6^2$
		Brassicas	
		Mustard	$2·2^3$
		Rape	$2·0^3$
		Cabbage	$1·8^3$
		Turnip	$1·6^3$
		Grasses	
		Italian Rye (*Lolium italicum*)	2×10^{-3} g (1·6)
		Timothy (*Phleum pratense*)	3×10^{-4} g (0·8)
		Agrostis stolonifera	7×10^{-5} g (0·5)

when it is fully established, the extension rate of lateral roots may exceed that of the primary root.

Weaver (1926) and Weaver and Bruner (1927) record careful examination of roots of many crop plants revealed on the walls of freshly dug trenches. A striking and frequent picture was the rapid near-horizontal spread of many side roots before they turned downwards. Roots of the pea, for example, at 6 weeks after germination had reached a depth of 50–60 cm and a lateral spread of 45 cm. The most rapid extension observed by these workers was 6 cm day^{-1} in lateral roots of potato.

Pavlychenko (1937) carefully exhumed wheat plants, some growing in weed-free isolation in a 10 m^2 area others growing at normal crop density. At maturity (80 days), the isolated plants had a total root length (excluding root hairs) of 8×10^6 cm. Those at normal density were restricted to 6×10^4 cm, but development of this, if uniform over the period of growth, needs 25 cm of new root length per cm^2 ground per day.

Tree roots appear to have a smaller potential for extension than the field crops discussed above. Rogers (1939) who studied apple root growth in relation to rootstock, soil, seasonal and climatic factors quotes maximum rates of approximately 1 cm day^{-1}. Similar maximum rates have been found for cherries, quince and grape.

V. SIZES OF INSECTS

A very wide range is found among insects and mites. The extreme species quoted by Imms (1947) are both beetles, the African Goliath having a body length of 100 mm and width of 50 mm, while one of the fungus feeders has a length of only 0·2 mm. The volume ratio must be about 20 000 000 to 1. Of insects, the smallest of pest status in agriculture are species of whitefly (Aleurodes), *c.* $1{\cdot}5 \times 1$ mm body, and thrips *c.* 1–2 mm $\times$ 0·25 mm. The greenhouse red spider is even smaller as are some non-pest insects such as certain chalcid wasp hyperparasites.

Eggs cover a narrower range than adults. The eggs laid by the large yellow underwing (*Triphena pronuba*) have each about 1/70000th the volume of the adult female, while those of the whitefly are about 1/10th. There is here, however, a broad correlation of sizes not evident among plants and their seeds—the red spider could not lay an egg 20 times her own volume!

The newly hatched larva or nymph has a larger volume than the egg from which it emerges. This is achieved by air-inflation stretching the elastic and corrugated new skin to the full capacity of which the tissues then grow until the next moult. In the case of the emperor moth, *Saturnia pavona*, there are five caterpillar instars before pupation and the volume ratio of mature larva to egg is about 4500. If each stage involves the same *relative* increase, the volumes at moult going up in geometric progression (Dyer's law), the factor for each stage would be 5·4. For *Triphena pronuba* the same calculation gives a factor of about 8.

Rate of growth depends on food supply and its nutritive value and on temperature. Under favourable conditions the maggot of the flesh fly spends only 71 h in the larval states, increasing in mass 450 times while the tree-boring caterpillar of the Goat Moth (*Cossus ligniperda*) takes 3 years to increase 72 000 times to maturity. If these increases are exponential, the respective times for 10-fold increase are 1·1 days and 220 days. With freely available food the caterpillar of the small cabbage white (*Pieris rapae*) takes 8½ days from egg to pupa at 30°C but 55 days at 10°C.

Table 3
Insect sizes (mm)

Species	Egg	Mature larva	Adult (body)
Pieris brassicae	$1{\cdot}2\times0{\cdot}65^2$	40×5^2	$22\times2{\cdot}5^2$
Pieris rapae	$1{\cdot}0\times0{\cdot}5^2$	30×4^2	18×2^2
Mamestra brassicae	$0{\cdot}5\times0{\cdot}65^2$	35×5^2	17×5^2
Smerinthus ocellata	$1{\cdot}5\times2^2$	60×10^2	28×8^2
Tiphena pronuba	$0{\cdot}4\times0{\cdot}5^2$	50×7^2	21×6^2
Saturnia pavona	$1{\cdot}5\times1{\cdot}3^2$	130×10^2	15×5^2♂
Saturnia pavona			25×7^2♀
Locusta migratoria	$6{\cdot}5\times1{\cdot}5^2$		46×7^2
Schistocerca	18 mg		1·4–1·8 g
Aleurodes brassicae	$0{\cdot}26\times0{\cdot}1^2$		$0{\cdot}6\times0{\cdot}2^2$
Thrips		$2\times0{\cdot}5^2$	$1\text{–}2\times0{\cdot}25^2$
			(vivip. ♀)
Aphis fabae	$0{\cdot}5^3$		$2{\cdot}5\times1{\cdot}6^2$
G'ho. Red spider[a]	$0{\cdot}15^3$		$0{\cdot}29\times0.18^2$♂
Tetranychus urticae			$0{\cdot}46\times0{\cdot}29^2$♀
Syrphid aphis predators			
Scaeva pyrasti	$1\times0{\cdot}4^2$	18×6^2	15×6^2
Platychines sp.	$0{\cdot}9\times0{\cdot}4^2$	$7\times1{\cdot}5^2$	10×2^2
Braconid aphis parasite			
Aphidius sp.	$0{\cdot}06\times0{\cdot}02^2$	$1{\cdot}4\times0{\cdot}6^2$	$2\times0{\cdot}7^2$

Note: For index significance, see Table 2. Body sizes, especially of larvae are variable by stretching, but the conjugate sizes quoted should give approximate volume.

[a] From measurements at all stages by Gasser (1950).

VI. PLANT PATHOGEN SPORE SIZES

The largest spores such as those of some vesicular–arbuscular mycorrhizae found in soil may approach 1 mm in diameter. Most spores are, of course very much smaller but there is a very wide range of size among plant pathogens as shown in Table 4 which gives some representative examples.

Table 4
Plant pathogen spore sizes (μm)

Alternaria solani	145–370 × 16–18
Leptosphaeria maculans	3·5–5·0 × 1·5–2·0
Actinomyces scabies	0·8–0·9 × 1·3–1·5
Erwinia amylovora	0·9–1·5 × 0·7–1·0

Appendix 9. Structural Formulae of the Pesticides listed in Appendix 4

Knowledge of structural formulae is necessary for prediction or interpretation of properties but two-dimensional formulae are space-demanding and expensive to print. The Wisswesser line formulae were developed for computer input but are difficult for inexperienced readers to interpret. We have used below orthodox in-line formulae where possible, extending their use by some additional abbreviations and conventions which leaves us with about 140 out of 390 compounds for which two-dimensional representation, at least in part, is necessary.

Standard abbreviations for very frequently occurring groups are used: Me for $-CH_3$, $-Et$ for C_2H_5, Pr for $-CH_2CH_2CH_3$, ^{i}Pr for $-CH(CH_3)_2$, Bu for $-CH_2CH_2CH_2CH_3$, ^{i}Bu for $-CH_2CH(CH_3)_2$, ^{s}Bu for $-CH(CH_3)$ Et and ^{t}Bu for $-CMe_3$.

∅ represents a benzene ring with one valence free, i.e. a phenyl group and, where the ring carries other substituents these are indicated within brackets before or after the ∅ as convenient.

Where several groups are attached to a multivalent one we reduce this to in-line arrangement by writing the univalent radicals with an unattached valence dash to the right and then the multivalent group with several unattached valence dashes to the left. This method is unambiguous only where the plural valences are equivalent, as in ≡PO, ≡PS and ≡sT the last symbol standing for the symmetrical triazine ring. Extension to multivalent asymmetric ring structures has not been attempted.

In non-ring compounds connecting links in the main chain are indicated by . or, where greater clarity is needed by – or, for a double band by =, but these symbols are usually omitted in a chain of CH_2 groups. Where . or – is omitted within a grouping, the links indicated by .. outside are both assumed attached to the first atom nominated in the group. This will usually also be clear from the valences of the atoms. Where a more complex "side chain" is attached it will, for clarity, be enclosed by brackets.

Exemplifying these conventions we give the two-dimensional and in-line formulae of one hypothetical compound.

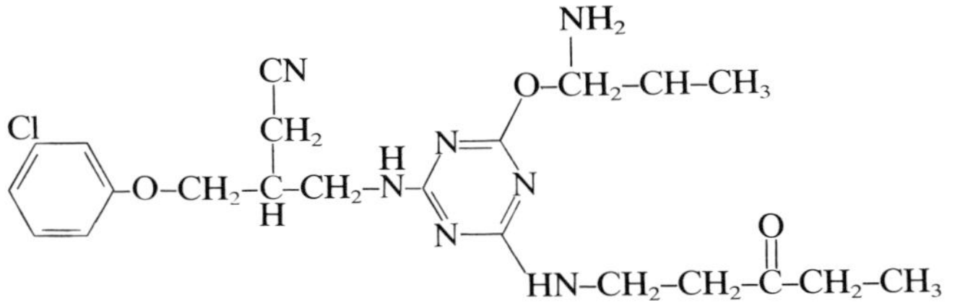

(3Cl)∅.O.CH_2.CH(CH_2CN).CH_2NH– Me.$CHNH_2$.CH_2.O–
Et.CO.CH_2CH_2NH– ≡sT

CHEMICAL STRUCTURES OF PESTICIDES

Acephate	MeS– MeO– Me.CO.NH– ≡PO
Acrolein	CH_2=CH–CH.O
Alachlor	MeO.CH_2– ClCH_2.CO– =N∅(2,6 diCl)
Aldicarb	MeS.CMe_2.CH=N–O–CO.NH.Me
Aldoxycarb	Me.SO_2.CMe_2.CH=N–O–CO.NHMe

Aldrin (HHDN)

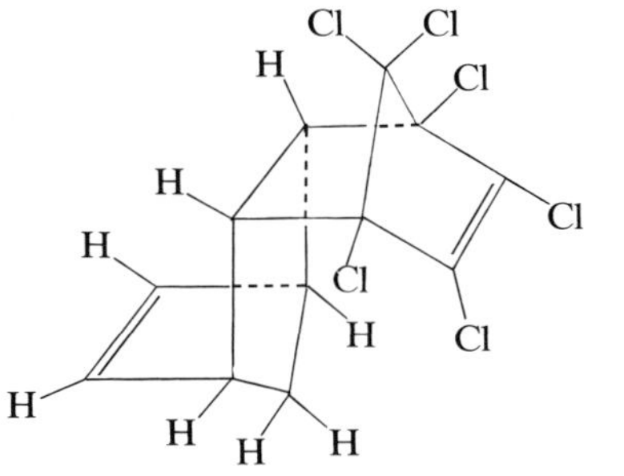

Allethrin

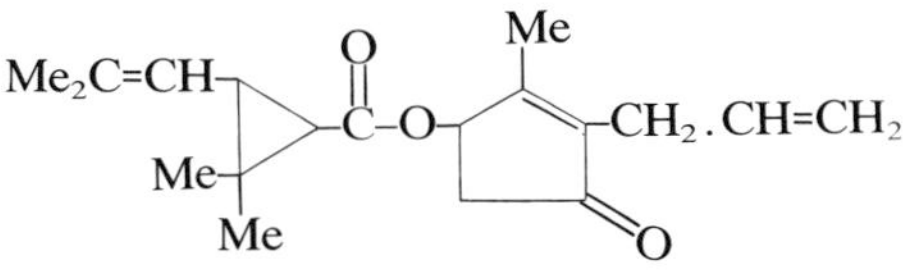

Allyl alcohol CH_2=CH–CH_2OH

Ametryne MeS– EtNH– ^{i}PrNH– ≡sT

Aminotriazole (Amitrole)

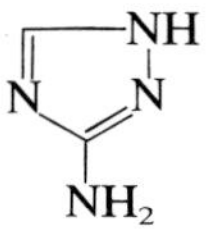

Amitraz (2,4 diMe)∅–N=CH–NMe–CH=N–∅(2,4 di Me)

Ancymidol

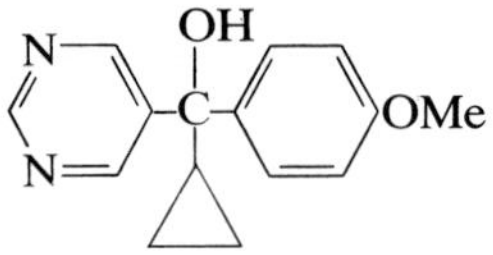

Anthraqinone

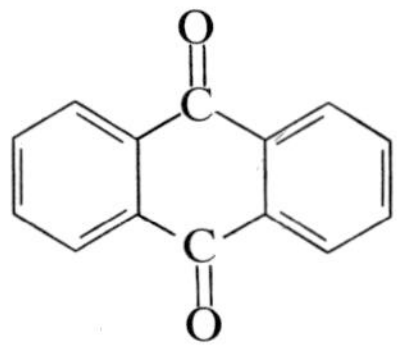

Antu

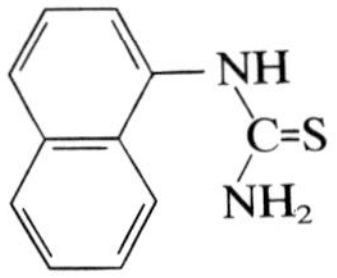

Asulam MeO.CO.NH.SO_2.∅(4 NH_2)

Atrazine Cl– EtNH– ^{i}PrNH ≡sT

Azinphos-ethyl

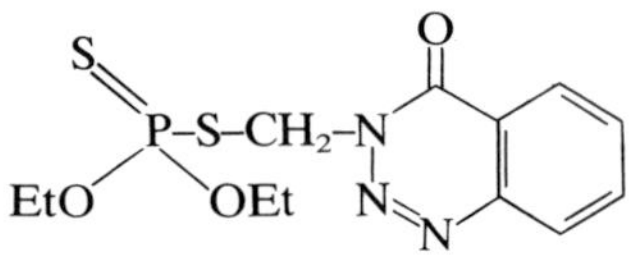

Azinphos-methyl as above with Me in place of Et

Aziprotryne MeS– N_3– ^{i}PrNH– ≡sT

Barban (3 Cl)∅–NH.CO.O.CH_2.C≡C.CH_2Cl

Benazolin

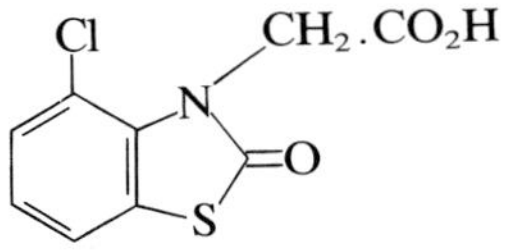

Bendiocarb

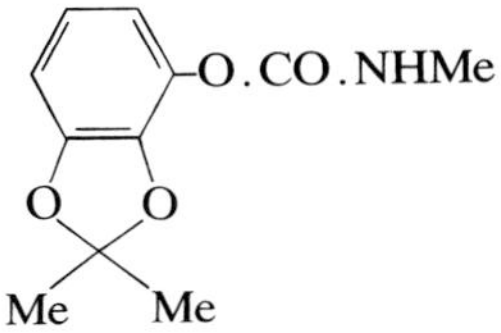

Benfluralin Et– Bu– =N∅(2,6 di NO_2, 4 CF_3)

Benodanil (2 I)∅–CO–NH–∅

Benomyl

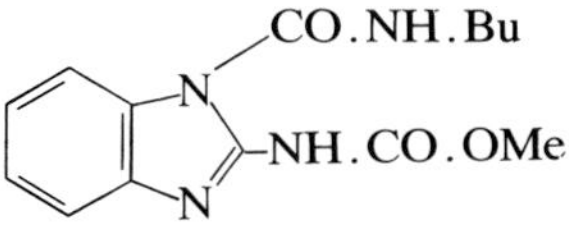

Bensulide $Me_2CH.O–$ $Me_2CH.O–$ ∅$SO_2.NH.CH_2.CH_2.S–$ ≡PS

Bentazone

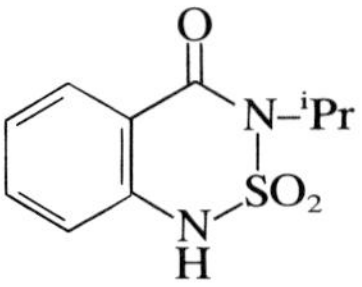

Benzoylprop-ethyl (3,4 diCl)∅–N(CHMe.CO.OEt)–CO–∅

Benzthiazuron

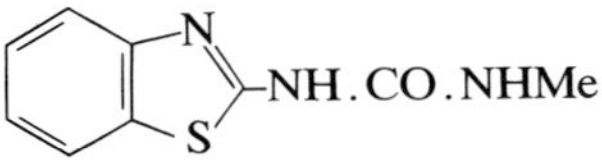

Bifenox (3 CO.OMe, 4 NO_2)∅–O–∅(2,4 diCl)

Binapacryl (2,4 di NO_2, 6 ^{s}Bu)∅–O–CO–CH=CMe_2

Bioresmethrin

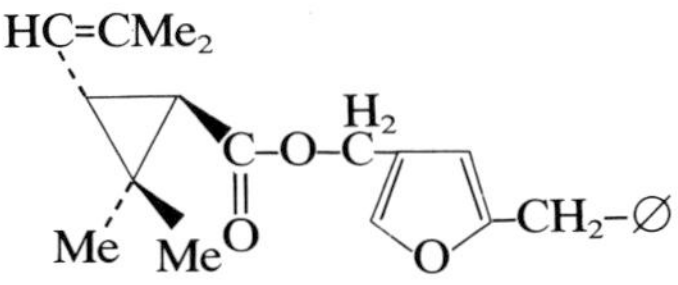

Biphenyl ∅–∅

Bromacil

Bromophenoxim (3,5 di Br, 4 OH)∅–CH=N–O–∅(2,4 di NO_2)

Bromophos (2,5 di Cl, 4 Br)∅O– MeO– MeO– ≡PS

Bromophos-ethyl as above with Et in place of Me

Bromoxynil (3,5 di Br, 4 OH)∅CN

Bromoxynil octanoate as above with O.CO.nC_7H_{15} in place of OH

Brompropylate (4 Br)∅–C(OH, CO.O^{i}Pr)–∅(4 Br)

Bronopol HO.H_2C.C(Br, NO_2)–CH_2.OH

Bufencarb (3 CH.MePr *or* 3 CH$(Et)_2$)∅–O.CO.NHMe

Bupirimate

Butachlor Bu.O.CH_2– $ClCH_2$.CO– =N∅(2,6 di Et)

Butam Me_3C.CO– Me_2CH– =N.CH_2.∅

Butamifos (2 NO_2, 5 Me)∅O– ˢBuNH– EtO– ≡PS

Buthidazole

Buthiobate

Butoxycarboxim MeS.CHMe.CMe=N–O–CO.NHMe

Butralin	(2,6 di NO_2, 4 CMe_3)∅–NH.^{s}Bu
Buturon	(4 Cl)∅–NH.CO.NMe.CHMe.C≡CH
^{s}Butylamine	$^{s}BuNH_2$
Butylate	EtS.CO.N(isoBu)$_2$
Camphechlor	approx. $C_{10}H_{10}Cl_8$
Captafol	O, O, N–S.CCl_2.$CHCl_2$
Captan	O, O, N–$SCCl_3$
Carbaryl	–O–CO–NHMe
Carbendazim	N, N, –NH.CO.OMe
Carbetamide	∅–NH.CO.O.CHMe.CO.NHEt
Carbofuran	–O.CO.NHMe, O, Me, Me
Carbon disulphide	CS_2
Carbon tetrachloride	CCl_4
Carbophenothion	(4 Cl)∅–S–CH_2S— EtO– EtO– ≡PS

Carboxin	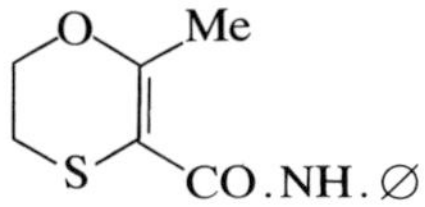 O, Me, S, $CO.NH.\varnothing$
Cartap	$Me_2N.CH(CH_2.S.CO.NH_2)_2$
Chloralose	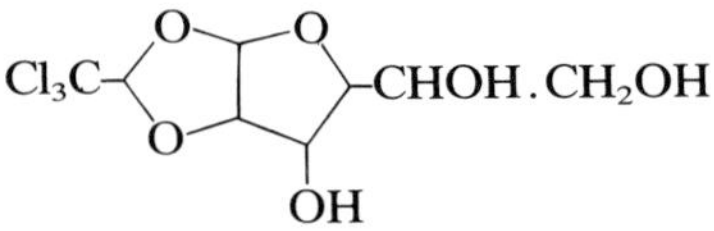 Cl_3C, O, O, O, $CHOH.CH_2OH$, OH
Chloramben	$(2,5\ di\ Cl,\ 3\ NH_2)\varnothing CO_2H$
Chlorbromuron	$(3Cl, 4Br)\varnothing.NH.CO.NMe.OMe$
Chlorbufam	$(3\ Cl)\varnothing.NH.CO.O.CHMe.C{\equiv}CH$
Chlordane	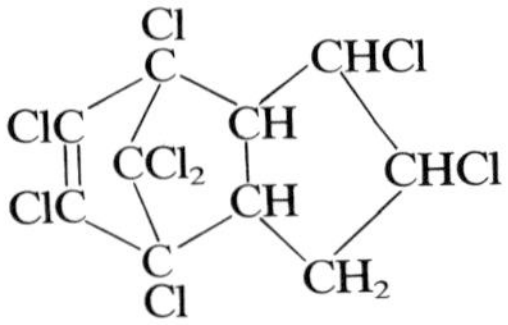 Cl, C, CHCl, ClC, CH, CCl_2, CHCl, ClC, CH, C, CH_2, Cl
Chlordimeform	$(2\ Me,\ 4\ Cl)\varnothing{-}N{=}CH{-}NMe_2$
Chlorfenac	$(2,3,6\ tri\ Cl)\varnothing.CH_2.CO_2H$
Chlorfenprop-methyl	$(4\ Cl)\varnothing{-}CH_2{-}CHCl{-}CO.OMe$
Chlorfenson	$(4\ Cl)\varnothing.SO_2.O.\varnothing(4\ Cl)$
Chlorfenvinphos	$(2,4\ di\ Cl)\varnothing.C({=}CHCl).O{-}$ EtO– EtO– ≡PO
Chlorflurecol-methyl	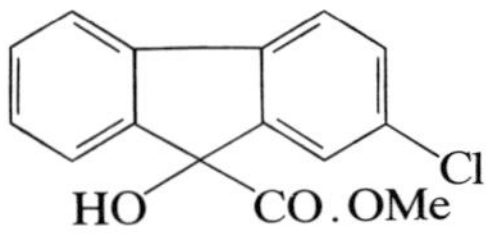 HO, CO.OMe, Cl
Chloridazon	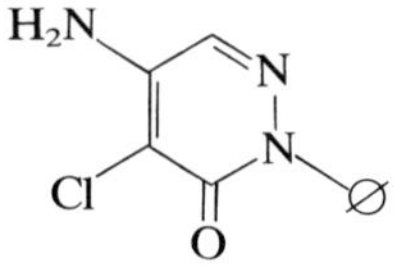 H_2N, N, N, Cl, O, $\varnothing$
Chlormephos	$ClCH_2.S{-}$ EtO– EtO– ≡PS
Chlormequat	$ClCH_2.CH_2.\overset{+}{N}(Me)_3$
Chloroacetic acid	$Cl.CH_2.CO_2H$

Chlorobenzilate	(4 Cl)∅–C(OH, CO.OEt)–∅(4 Cl)
Chloromethiuron	(2 Me, 4 Cl)∅–NH.CS.NMe$_2$
Chloroneb	1,4-dichloro, 2,5-dimethoxy benzene
Chloropicrin	Cl$_3$C.NO$_2$
Chloropropylate	(4 Cl)∅–C(OH, CO.O^{i}Pr)–∅(4 Cl)
Chlorothalonil	1,3-dicyano, 2,4,5,6-tetrachloro benzene
Chloroxuron	(4 Cl)∅–O–∅(4–NH.CO.NMe$_2$)
Chlorphoxim	(2 Cl)∅.C(CN)=N–O– EtO– EtO– ≡PS
Chlorpropham	(3 Cl)∅.NH.CO.O^{i}Pr
Chlorpyrifos	Cl, N, O–, Cl, Cl (pyridine ring) EtO– EtO– ≡PS
Chlorpyrifos-methyl	as above with Me in place of Et
Chlorthal-dimethyl	CO.OMe, Cl, Cl, Cl, Cl, CO.OMe (benzene ring)
Chlorthiamid	(2,6 di Cl)∅.CS.NH$_2$
Chlorthiophos	(di Cl, MeS)∅– EtO– EtO– ≡PS
Chlortoluron	(3 Cl, 4 Me)∅–NH.CO.NMe$_2$
Chlofop isobutyl	(4 Cl)∅–O–∅(4 ^{i}BuO.CO.CHMe.O–)
Crimidine	NMe$_2$, N, Cl, N, Me (pyrimidine ring)
Crotoxyphos	∅.CHMe.O.CO.CH=CMe.O– MeO– ≡PO MeO–
Crufomate	(2 Cl, 4 ^{t}Bu)∅–O– MeO– MeNH– ≡PO
Cyanazine	Cl– EtNH– NC.CMe$_2$.NH– ≡sT
Cyanofenphos	(4 CN)∅–O– ∅– EtO– ≡PS

Cyanophos	(4 CN)∅–O– MeO– MeO– ≡PS
Cycloate	∅.NEt.CO.SEt
Cycloheximide	O, HN, O, Me, Me, CH_2, C, H, HO, H
Cycluron	C_7H_{14} CH–NH–CO.NMe_2
Cyhexatin	$\left(\bigcirc\!-\right)_3$.Sn.OH
Cypermethrin	Me Me, $Cl_2C{=}CH$–(cyclopropane)–CO.O.CH(CN)–∅(3∅O)
2,4-D	(2,4 di Cl)∅CH_2CO_2H
Dalapon	Me.CCl_2.CO_2H
Daminozide	Me_2N.NH.CO.CH_2CH_2.CO_2H
Dazomet	S, S, MeN, NMe
2,4-DB	(2,4 di Cl)∅$CH_2CH_2CH_2$.CO_2H
DDT	(4 Cl)∅–CH(CCl_3)–∅(4 Cl)
Dehydroacetic acid	Me, O, O, CO.Me, O
Demeton-O	Et.S.CH_2CH_2O– EtO– EtO– ≡PS
Demeton-*S*	Et.S.CH_2CH_2S– EtO– EtO– ≡PO
Demeton-*S*-methyl	as above with MeO in place of EtO (twice)
Demeton-*S*-methyl sulphone	Et.SO_2.CH_2CH_2S– MeO– MeO– ≡PO
Desmedipham	(3 EtO.CO.NH–)∅–O.CO.NH∅

Desmetryne	MeS– MeNH– ^{i}PrNH– ≡sT
Diallate	$^{i}Pr_2N.CO.S.CH_2.CCl{=}CHCl$
Diazinon	
1,2-Dibromo-3-chloropropane	$BrCH_2.CHBr.CH_2Cl$
Dicamba	(2 OMe 3,6 di Cl)∅CO_2H
Dichlobenil	(2,6 di Cl)∅–CN
Dichlofenthion	(2,4 di Cl)∅–O– EtO– EtO– ≡PS
Dichlofluanid	$Me_2N.SO_2$– FCl_2C–S– =N∅
p-Dichlorobenzene	(4 Cl)∅Cl
Dichlorophen	(2 OH, 4 Cl)∅–CH_2–∅(2 OH, 4 Cl)
3,6-Dichloropicolinic acid	
1,2-Dichloropropane	$ClCH_2.CHCl.CH_3$
1,3-Dichloropropene	$ClCH{=}CH{-}CH_2Cl$
Dichlorprop	(2,4 di Cl)∅–O–CHMe–CO_2H
Dichlorvos	CCl_2=CH–O– MeO– MeO– ≡PO
Diclofop-methyl	(2,4 di Cl)∅–O–∅(4-O–CHMe–CO_2Me)
Dicloran	(2,6 di Cl, 4 NO_2)∅–NH_2
Dicofol	(4 Cl)∅–C(OH, CCl_3)–∅(4 Cl)
Dieldrin	

Dienochlor	Cl Cl Cl Cl / Cl / Cl / Cl Cl Cl Cl
Diethatyl-ethyl	$ClCH_2.CO–$ $EtO.CO.CH_2–$ ≡N∅(2,6 di Et)
Diethyltoluamide	(3 Me)∅–CO.NEt_2
Difenoxuron	(4 MeO–)∅–O–∅(4 NMe_2.CO.NH–)
Difenzoquat	Me–N–$\overset{+}{N}$–Me / ∅ ∅
Diflubenzuron	(4 Cl)∅–NH.CO.NH.CO–∅(2,6 di F)
Dimefox	$Me_2N–$ $Me_2N–$ F– ≡PO
Dimethachlor	$ClCH_2.CO–$ $MeO.CH_2CH_2–$ =N∅(2,6 di Me)
Dimethametryn	MeS– EtNH– ^{i}Pr.CHMe.NH– ≡sT
Dimethirimol	Bu / Me OH / N N / NMe_2
Dimethoate	$MeNH.CO.CH_2.S–$ MeO– MeO– ≡PS
Dimethyl arsinic acid	Me– Me– HO– ≡AsO
Dimethyl phthalate	(2 Me.O.CO)∅CO.OMe
Dimetilan	Me / Me_2N.CO–N N –O.CO.NMe_2
Dinitramine	(2,4 di NO_2, 3 Et_2N, 6 F_3C)∅NH_2
Dinoseb (DNBP)	(2,4 di NO_2, 6 ^{s}Bu)∅OH
Dioxacarb	(2 Me.NH.CO.O)∅– O / O
Diphenamid	∅–CH(CO.NMe_2)–∅
Dipropetryn	EtS– ^{i}PrNH– ^{i}PrNH– ≡sT

Diquat

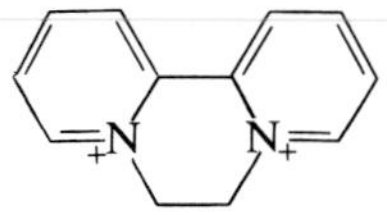

Disulfoton $EtSCH_2CH_2S-$ EtO– EtO– ≡PS

Ditalimfos

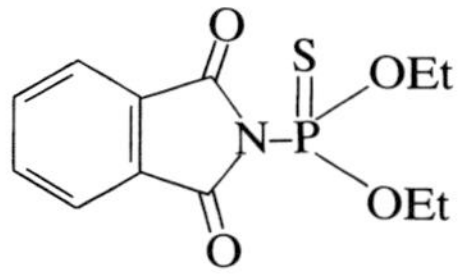

Diuron (3,4 di Cl)∅NH.CO.NMe_2

DNOC (2,4 di NO_2, 6 Me)∅OH

Dodemorph

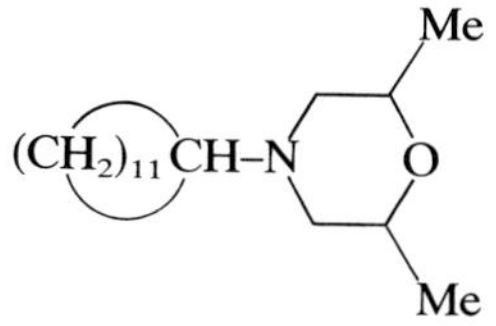

Dodine $nC_{12}H_{25}NH.C(\overset{+}{N}H_2).NH_2$

Drazoxolon

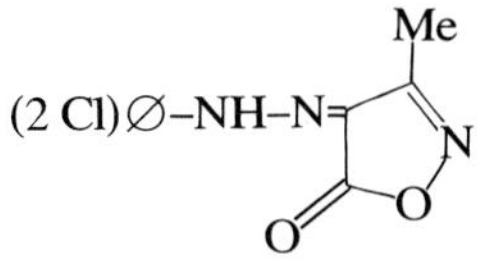

Eglinazine-ethyl Cl– EtNH– $EtO.CO.CH_2.NH-$ ≡sT

Endosulfan

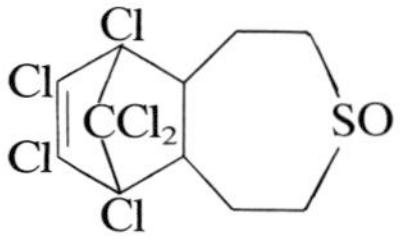

Endothal

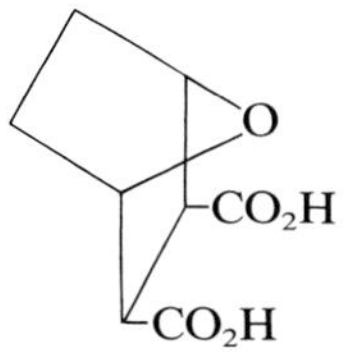

Compound	Formula
Endrin	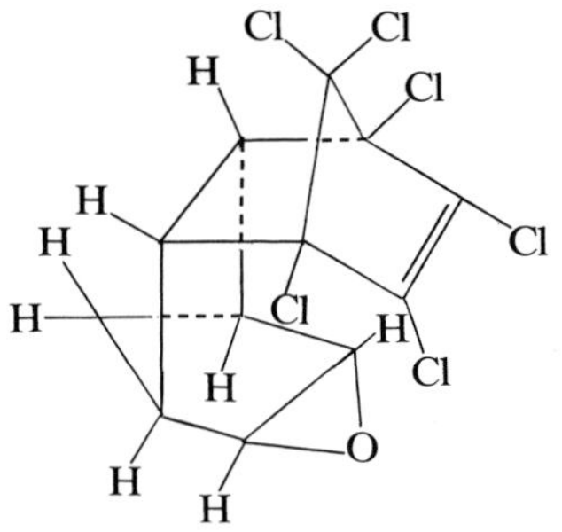
EPN	$(4\,NO_2)\varnothing$–O– $\varnothing$– EtO– ≡PS
EPTC	$Pr_2N.CO.SEt$
Etacelasil	$(MeOCH_2CH_2O)_3SiCH_2CH_2Cl$
Ethalfluralin	$(2,6$ di $NO_2,\ 4\,F_3C)\varnothing$–NEt–$CH_2$–CMe=$CH_2$
Ethiofencarb	$(2\,EtSCH_2)\varnothing$–O.CO.NHMe
Ethirimol	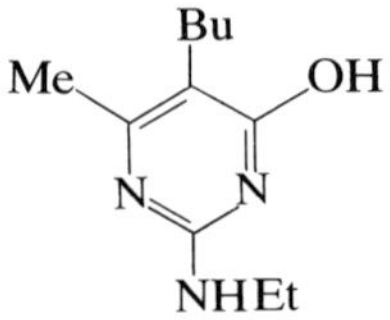
Ethofumesate	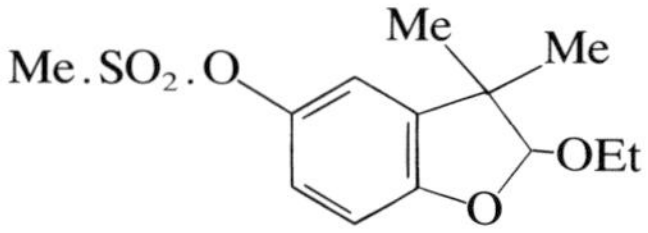
Ethoprophos	PrS– PrS– EtO– ≡PO
Ethohexadiol	Et.CH(CH_2OH).CH(OH).Pr
Ethylene dibromide (EDB)	$BrCH_2.CH_2Br$
Ethylene dichloride	$ClCH_2.CH_2Cl$
Etridiazole	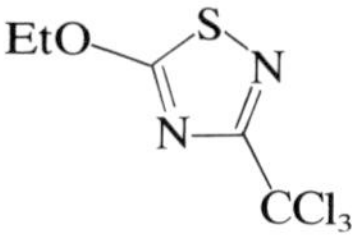
Etrimfos	

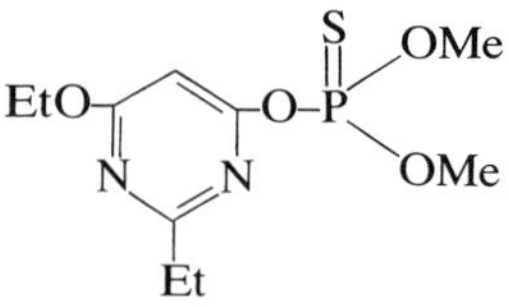

Fenaminosulf — $(4\,Me_2N)\varnothing$–N=N–SO_3^-

Fenamiphos — (3 Me, 4 MeS)∅– EtO– ${}^iPrNH$– ≡PO

Fenarimol — OH / C / Cl / N N

Fenbutinoxide — ∅.CMe_2.CH_2–Sn–O–Sn–CH_2.CMe_2.∅

Fenchlorphos — (2,4,5 tri Cl)∅–O– MeO– MeO– ≡PS

Fenfuram — O / Me / CO.NH.∅

Fenitrothion — (3 Me, 4 NO_2)∅–O– MeO– MeO– ≡PS

Fenoprop — (2,4,5 tri Cl)∅–O–CHMe–CO_2H

Fensulfothion — (4 Me.SO–)∅–O– EtO– EtO– ≡PS

Fenthion — (3 Me, 4 MeS)∅–O– MeO– MeO– ≡PS

Fentin acetate — $\varnothing_3$.Sn.O.CO.Me

Fenuron — ∅NH.CO.NMe_2

Fenvalerate — (4 Cl)∅.CH($CHMe_2$).CO.O.CH(CN).∅(3∅O)

Flamprop-isopropyl — ∅.CO– ${}^iPr$.O.CO.CH(Me)– =N∅(3 Cl, 4 F)

Flamprop-methyl — above with Me in place of ${}^iPr$

Fluometuron — (3 F_3C)∅–NH.CO.NMe_2

Fluoroacetamide — FCH_2.CO.NH_2

Fluorodifen — (4 NO_2)∅–O–∅(2 NO_2, 4 F_3C)

Fluotrimazole — (3 F_3C)∅.$C\varnothing_2$–N (N, N)

Flurecol-butyl — HO CO.OBu

Fluridone

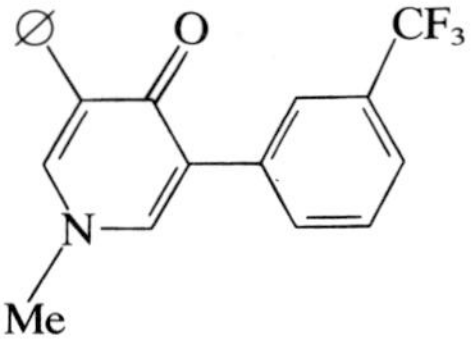

Folpet — N–S–CCl_3

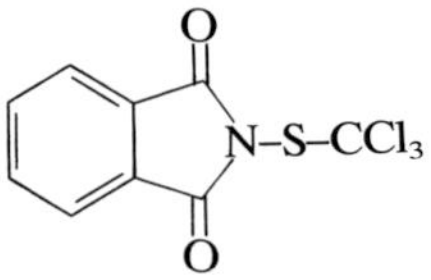

Formetanate — (3 MeNH.CO.O)∅–N=CH.NMe_2

Formothion — H.CO.NMe.CO.CH_2S– MeO– MeO– ≡PS

Fosthietan — =N– EtO– EtO– ≡PO

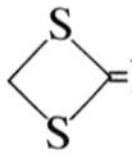

Fuberidazole

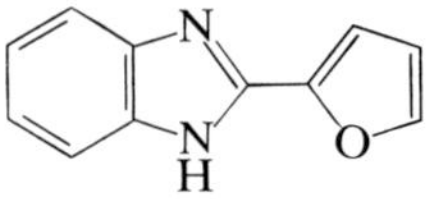

Furalaxyl 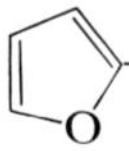 — –CO– MeO.CO.CH(Me)– =N∅(2,6diMe)

Gamma HCH — (γ-BHC, Lindane)

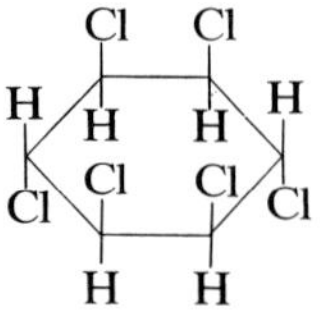

Glyphosate — $HO_2C.CH_2.NH.CH_2.PO_3H^-$

Heptachlor

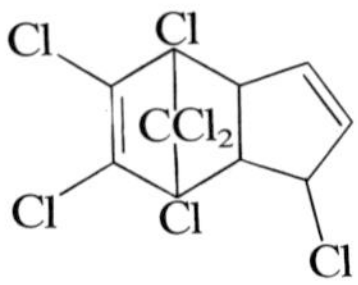

Heptenophos 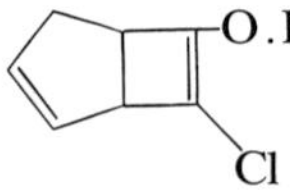 — O.PO.$(OMe)_2$

Hexazinone

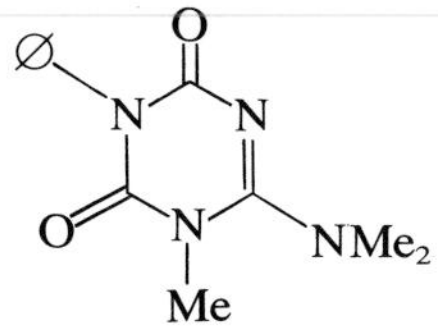

$\varnothing$–N(ring: C(=O), N, C(NMe$_2$), N(Me), C(=O))

Imazalil

(imidazol-1-yl)N–CH_2– CH_2=CH–CH_2–O– =HC–$\varnothing$(2,4diCl)

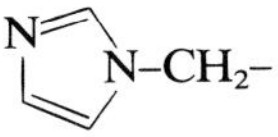

Indol-3-butyric acid

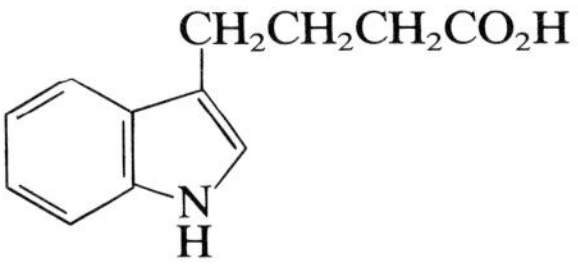

$CH_2CH_2CH_2CO_2H$ (indole, N–H)

Iodofenphos — (2,5 di Cl, 4 I)$\varnothing$–O– MeO– MeO– ≡PS

Ioxynil — (3,5 di I, 4 OH)$\varnothing$–CN

Iprodone

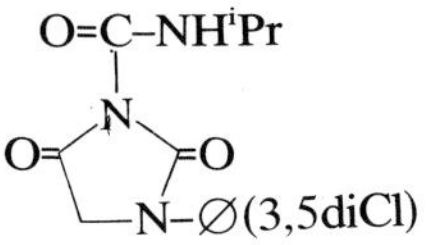

O=C–NH^{i}Pr; O=, N, =O; N–$\varnothing$(3,5diCl)

Isocarbamid

HN N–CO.NH.CH_2.^{i}Pr; O

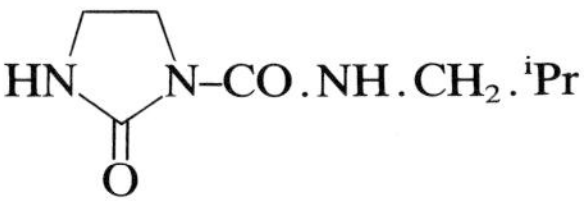

Isofenphos — (2 ^{i}PrO.CO)$\varnothing$–O– EtO– ^{i}PrNH– ≡PS

Isomethiozin

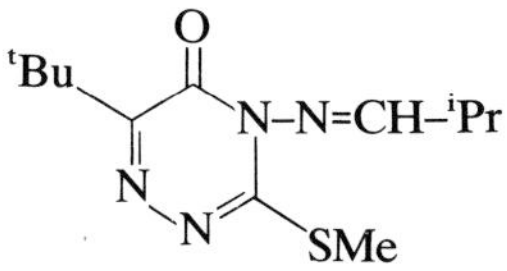

^{t}Bu; O; N–N=CH–^{i}Pr; N; N; SMe

Isopropalin — (2,6 di NO_2, 4 ^{i}Pr)$\varnothing$–NPr_2

Isoproturon — (4 ^{i}Pr)$\varnothing$–NH.CO.NMe_2

Lenacil

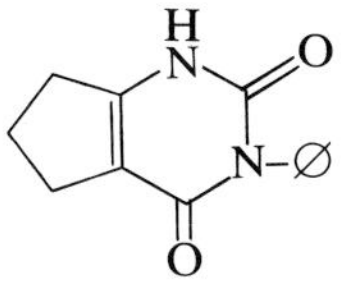

H; N; O; N–$\varnothing$; O

Leptophos — (2,5 di Cl, 4 Br)$\varnothing$–O– MeO– $\varnothing$– ≡PS

Linuron	(3,4 di Cl)∅–NH.CO.NMe(OMe)
Malathion	EtO.CO.CH(–CH$_2$.CO.OEt)– MeO– MeO– ≡PS
Maleic hydrazide	O=⟨ring: CH=CH, N–N(H)⟩–OH
MCPA	(2 Me, 4 Cl)∅.O.CH$_2$.CO$_2$H
MCPB	(2 Me, 4 Cl)∅.O.CH$_2$.CH$_2$.CH$_2$.CO$_2$H
Mecarbam	EtO.CO.NMe.CO.CH$_2$S– EtO– EtO– ≡PS
Mecoprop	(2 Me, 4 Cl)∅–O–CHMe–CO$_2$H
Mefluidide	(2,4 di Me, 5 MeCO.NH–)∅NH.SO$_2$.CF$_3$
Menazon	(4,6-di H$_2$N-1,3,5-triazin-2-yl)–CH$_2$S– MeO– MeO– ≡PS
Mephosfolan	(4-Me-1,3-dithiolan-2-ylidene)=N– EtO– EtO– ≡PO
Metalaxyl	MeO.CH$_2$.CO– MeOCO.CHMe– =N∅(2,6 di Me)
Metaldehyde	cyclic (MeCH–O–)$_4$: MeCH, O, MeCH, O, MeCH, O, MeCH, O
Metamitron	∅–(triazinone ring: C=O, N–NH$_2$, C–Me, N, N)
Methabenzthiazuron	(benzothiazol-2-yl)–NMe.CO.NHMe

Methamidophos	MeO– MeS– H_2N– ≡PO
Metham sodium	MeNH.CS_2^- Na^+
Methazole	(3,4diCl)∅ N–O O= N =O Me
Methidathion	Me O N N–CH_2–S– MeO– MeO– ≡PS S O
Methomyl	MeS.CMe=N–O–CO.NHMe
Methoprotryne	MeS– ^{i}PrNH– $MeOCH_2CH_2CH_2NH$– ≡sT
Methoxychlor	(4 MeO)∅–CH(CCl_3)–∅(4 MeO)
Methyl isothiocyanate	Me–N = C=S
Metobromuron	(4 Br)∅–NH.CO.NMe(OMe)
Metolachlor	$ClCH_2$.CO– MeO.CH_2.CHMe– =N∅(2 Me, 6 Et)
Metoxuron	(3 Cl, 4 MeO)∅–NH.CO.NMe_2
Metribuzin	O ^{t}Bu– N–NH_2 N –SMe N
Mevinphos	MeO.CO.CH=CMe–O– MeO– MeO– ≡PO
Molinate	$(CH_2)_6$ N–CO.SEt
Monalide	(4 Cl)∅–NH.CO.CMe_2Pr
Monocrotophos	NHMe.CO.CH=CMe–O– MeO– MeO– ≡PO
Monolinuron	(4 Cl)∅–NH.CO.NMe(OMe)
Monuron	(4 Cl)∅–NH.CO.NMe_2
Naled	Cl_2CBr.CHBr.O– MeO– MeO– ≡PO

Napthalene

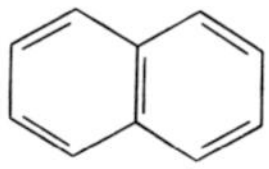

1-Naphthyl acetic acid — $-CH_2CO_2H$

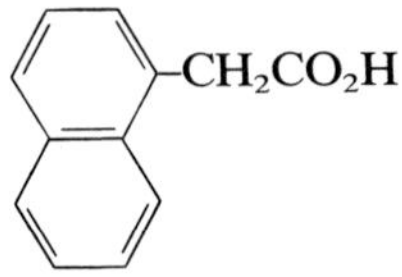

2-Naphthoxy acetic acid — $-OCH_2CO_2H$

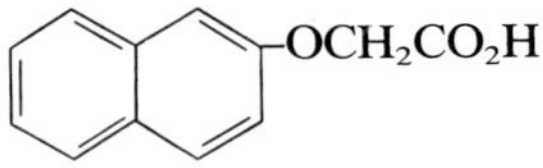

Napromide — $-O.CHMe.CO.NEt_2$

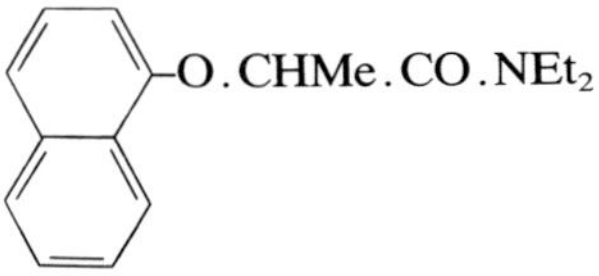

Naptalam (NPA) — $-NH.CO-$; CO_2H

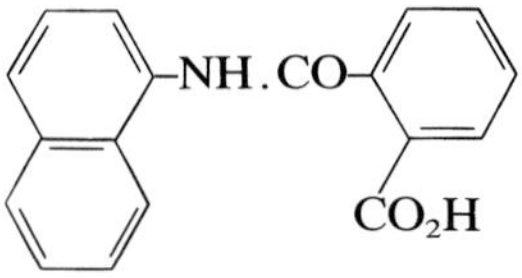

Neburon — (3,4-di Cl)∅.NH.CO.NBu(Me)

Niclosamide — (2 Cl, 4 NO_2)∅–NH.CO–∅(2 OH, 5 Cl)

Nicotine — N ; N ; Me

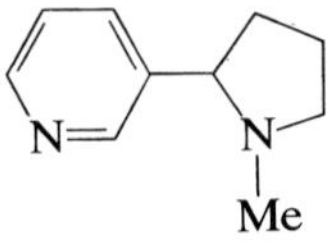

Nitralin — (2,6 di NO_2, 4 $MeSO_2$)∅–NPr_2

Nitrapyrin — Cl ; N ; CCl_3

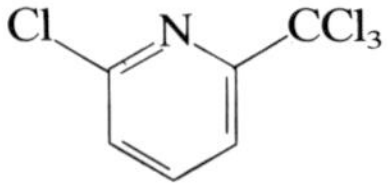

Nitrilicarb — $NC.CH_2.CH_2.CMe_2.CH{=}N.O.CO.NHMe$

Nitrofen — (2,4 di Cl)∅–O–∅(4 NO_2)

Norflurazon — N ; (3 F_3C)∅–N ; –NHMe ; O ; Cl

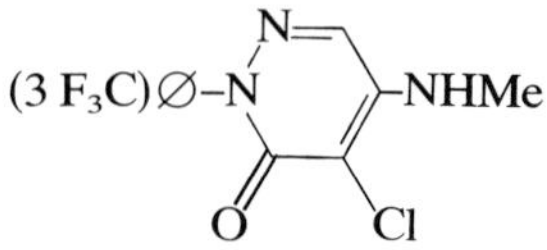

NRDC 161 — $Br_2C{=}CH$; CO.O ; ∅(3 ∅O) ; C ; NC

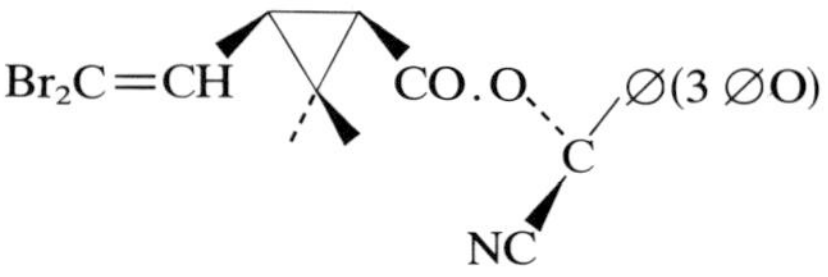

Nuarimol	(2 Cl)∅–C(OH)–∅(4 F) with pyrimidin-5-yl substituent on C
Omethoate	MeNH.CO.CH_2–S– MeO– MeO– ≡PO
Oryzalin	(3,5 di NO_2. 4 Pr_2N)∅–SO_2–NH_2
Oxadiazon	(2,4diCl,5^{i}PrO)∅N (1,3,4-oxadiazol-2(3H)-one ring, 5-^{t}Bu)
Oxamyl	Me_2N.CO.C(SMe)=N–O.CO.NHMe
Oxycarboxin	(dihydro-oxathiin dioxide ring: O, S(=O)(=O); Me; CO.NH∅)
Oxydemeton-methyl	Et.SO.CH_2CH_2S– MeO– MeO– ≡PO
Paraquat	Me–$\overset{+}{N}$(pyridinium)–(pyridinium)$\overset{+}{N}$–Me
Parathion	(4 NO_2)∅–O– EtO– EtO– ≡PS
Parathion-methyl	above with MeO– in place of EtO (twice)
Pebulate	Bu.NEt.CO.SPr
Pendimethalin	(2,6 di NO_2, 3,4 di Me)∅–NH.CH$(Et)_2$
Pentachlorphenol	(2,3,4,5,6 penta Cl)∅OH
Pentanochlor	(3 Cl, 4 Me)∅–NH.CO.CHMe.CH_2CH_2Me
Perfluidone	∅–SO_2–∅(3 Me, 4 NH.SO_2.CF_3)
Permethrin	Cl_2C=CH–(cyclopropane, Me, Me)–CO.O.CH_2.∅(3 ∅O)
Phenmedipham	(3,MeO.CO.NH–)∅–O.CO.NH–∅(3 Me)

Phenothrin	$Me_2C{=}CH-$ (cyclopropane) $-CO.O.CH_2-\varnothing(3\,O\varnothing)$; Me, Me
Phenthoate	$EtO.CO.CH(\varnothing).S-$ MeO– MeO– ≡PS
Phenyl mercury acetate	$Me.CO.O.Hg.\varnothing$
Phenyl mercury nitrate	$\varnothing HgNO_3$ $\varnothing HgOH$
2 Phenyl phenol	$(2\varnothing)\varnothing OH$
Phorate	$Et.S.CH_2.S-$ EtO– EtO– ≡PS
Phosalone	(benzoxazolone) $N-CH_2S-$, =O, O EtO– EtO– ≡PS
Phosfolan	(dithiolane) S, S, =N– EtO– EtO– ≡PO
Phosmet	(phthalimide) O, O, $N-CH_2-S-$ MeO– MeO– ≡PS
Phosphamidon	$Et_2N.CO.CCl{=}CMe.O-$ MeO– MeO– ≡PO
Phoxim	$\varnothing-C(CN){=}N-O-$ EtO– EtO– ≡PS
Picloram	(pyridine) Cl, N, CO_2H, Cl, Cl, NH_2
Pindone	(indandione) O, O, $-CO.{}^tBu$
Piperophos	(2-methylpiperidine) $N-CO.CH_2.S-$ PrO– PrO– ≡PS; Me

Piproctanyl	(piperidinium ring, N^+) CH_2=CH–CH_2– N^+ –$CH_2.CH_2.$CHMe. | $CH_2CH_2CH_2.$CHMe
Pirimicarb	(pyrimidine ring) Me, Me, –O.CO.NMe_2, N, N, NMe_2
Pirimiphos–ethyl	(pyrimidine ring) Me, O–, N, N, NEt_2 EtO– EtO– ≡PS
Pirimiphos-methyl	as above with MeO in place of EtO (twice)
Prochloraz	(imidazole ring) N–CO.NPr.CH_2.CH_2.O–∅(2,4,6triCl)
Procymidone	(3,5diCl)∅–N (bicyclic imide, O, O, Me, Me)
Profenofos	(2 Cl, 4 Br)∅–O– PrS– EtO– ≡PO
Promecarb	(3 Me, 5 ^{i}Pr)∅.O.CO.NHMe
Prometone	MeO– ^{i}PrNH– ^{i}PrNH– ≡sT
Prometryne	MeS– ^{i}PrNH– ^{i}PrNH ≡sT
Propachlor	ClCH$_2$.CO– ^{i}Pr =N∅
Propanil	(3,4 di Cl)∅.NH.CO.Et
Propazine	Cl– ^{i}PrNH– ^{i}PrNH– ≡sT
Propetamphos	^{i}Pr.O.CO.CH=CMe–O– MeO– EtNH– ≡PS
Propham	∅.NH.CO.O^{i}Pr
Propoxur	(2 ^{i}PrO)∅.O.CO.NHMe
Propyzamide	(3,5 di Cl)∅.CO.NH.CMe$_2$.C≡CH
Prothiophos	(2,4 di Cl)∅.O– PrS– EtO– ≡PS

Prothoate — ^{i}Pr.NH.CO.CH_2.S– EtO– EtO– ≡PS

Pyracarbolid — CO.NH.∅ ; O ; Me

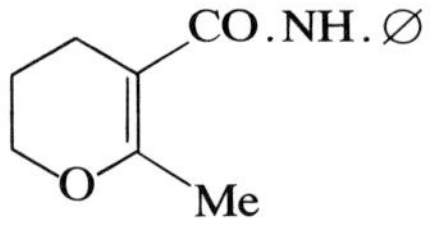

Pyrazophos — EtO.CO ; N–N ; –O– EtO– EtO– ≡PS ; Me ; N

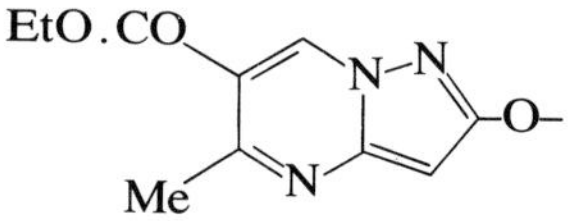

Quinalphos — N ; –O– EtO– EtO– ≡PS ; N

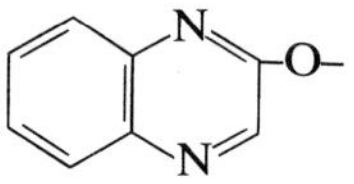

Quinomethionate — Me ; N ; S ; =O ; N ; S

Quinonamid — O ; NH.CO.$CHCl_2$; Cl ; O

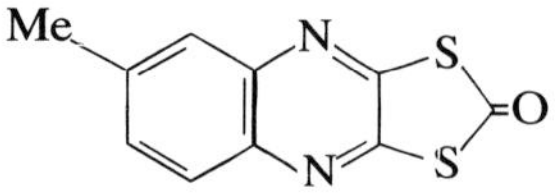

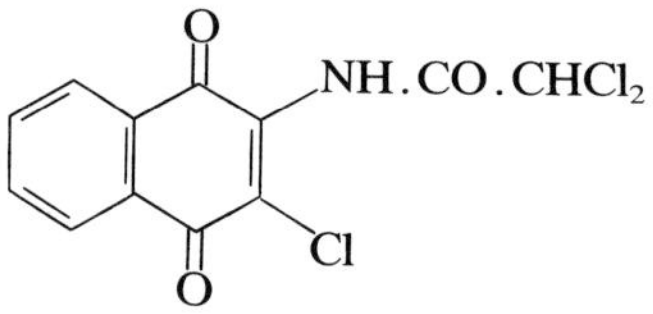

Quintozene — (2,3,4,5,6, penta Cl)∅NO_2

Resmethrin — Me_2C=CH– ; –CO.O.CH_2– ; Me ; Me ; O ; CH_2∅

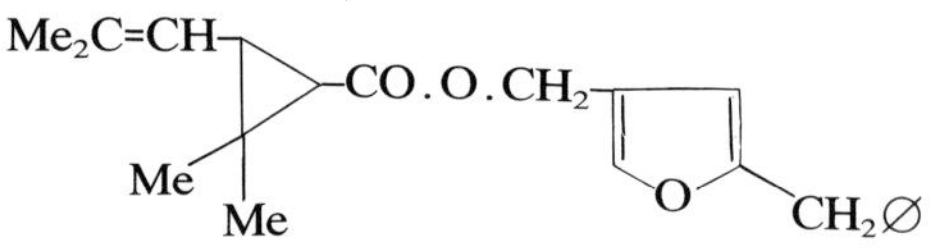

Rotenone — MeO ; MeO ; O ; O ; O ; O ; C=CH_2 ; Me

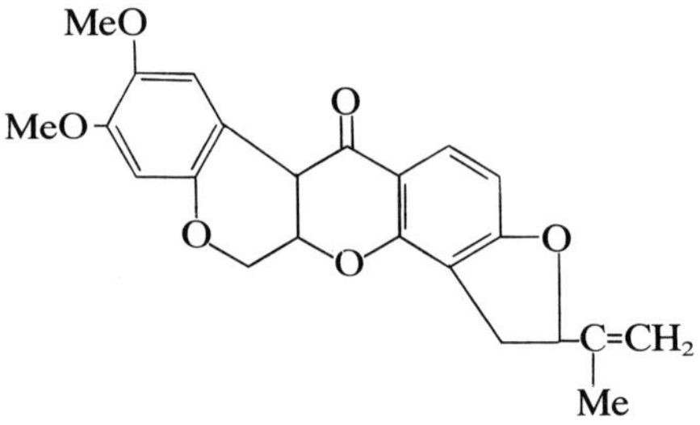

Schradan — $(Me_2N)_2$.PO.O.PO.$(NMe_2)_2$

Secbumeton — MeO– EtNH– ^{s}BuNH– ≡sT

Sesamex	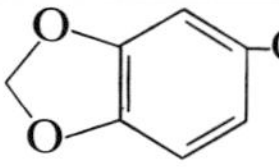 –O.CHMe.O.CH_2CH_2.O.CH_2CH_2.OEt
Siduron	(2 Me)∅.NH.CO.NH.∅
Simazine	Cl– EtNH– EtNH– ≡sT
Simetryne	MeS– EtNH– EtNH– ≡sT
Strychnine	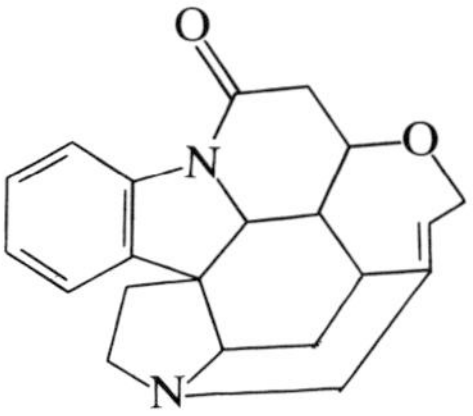
Sulfallate	Et_2N.CS.S.CH_2.CCl=CH_2
Sulfotep	$(EtO)_2$:PS.O.PS:$(OEt)_2$
Sulfoxide	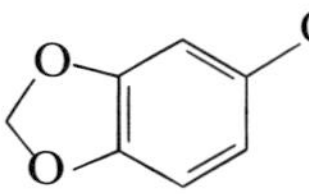 CH_2.CHMe.SO.$(CH_2)_7$.Me
Sulfuryl fluoride	F.SO_2.F
2,4,5-T	(2,4,5 tri Cl)∅.O.CH_2CO_2H
2,3,6-TBA	(2,3,6 tri Cl)∅.CO_2H
TCA	Cl_3C.CO_2H
Tecnazene	(2,3,5,6 tetra Cl)∅.$2NO_2$
Tebuthiuron	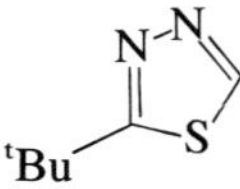 –NMe.CO.NHMe
TEPP	$(EtO)_2$:PO.O.PO:$(OEt)_2$
Terbacil	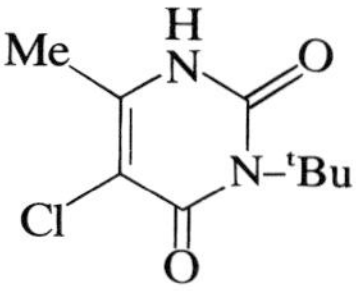
Terbufos	Me_3C.S.CH_2.S– EtO– EtO– ≡PS
Terbumeton	MeO– EtNH– ^tBuNH– ≡sT

Terbuthylazine	Cl– EtNH– ^{t}BuNH– ≡sT
Terbutryne	MeS– EtNH– ^{t}BuNH– ≡sT
Tetrachlorvinphos	(2,4,5 tri Cl)∅.C(=CHCl).O– MeO– MeO– ≡PO
Tetradifon	(2,4,5 tri Cl)∅–SO_2–∅(4 Cl)
Tetramethrin	Me_2C=CH– (cyclopropane, Me, Me)–CO.O.CH_2.N (tetrahydrophthalimide, O, O)
Tetrasul	(2,4,5 tri Cl)∅–S–∅(4 Cl)
Thiabendazole	(benzimidazole N, N–H)–(thiazole N, S)
Thiazafluron	F_3C–(thiadiazole N–N, S)–NMe.CO.NHMe
Thiobencarb	(4 Cl)∅–CH_2.S.CO.NEt_2
Thiocyclam	(trithiane S, S, S)–NMe_2
Thiofanox	MeS.CH_2.C(^{t}Bu)=N–O.CO.NHMe
Thiometon	EtS.CH_2.CH_2.S– MeO– MeO– ≡PS
Thionazin	(pyrazine N, N)–O– EtO– EtO– ≡PS
Thiram	Me_2N.CS.S.S.CS.NMe_2
Tolylfluanid	Me_2N.SO_2– Cl_2CF.S– =N∅(4 Me)
Triadimefon	(triazole N, N)N–CH(CO.^{t}Bu)–O∅(4 Cl)
Triallate	(^{i}Pr)$_2$N.CO.S.CH_2.CCl=CCl_2

Triamiphos	∅, N, N–, N, NH_2 Me_2N– Me_2N– ≡PO
Triazophos	∅, N, N, –O, N EtO– EtO– ≡PS
Trichloronate	(2,4,5 tri Cl)∅–O– EtO– Et– ≡PS
Trichlorphon	Cl_3C.CHOH– MeO– MeO– ≡PO
Triclopyr	Cl, N, Cl–, –O.CH_2.CO_2H, Cl
Tricyclazole	Me, N, N, N, S
Tridemorph	Me, $nC_{13}H_{27}$–N, O, Me
Trietazine	Cl– EtNH– Et_2N– ≡sT
Trifenmorph	$∅_3C$–N, O
Trifluralin	(2,6 di NO_2, 4 CF_3)∅.NPr_2
Triforine	CCl_3, CH–N, N–CH, CCl_3, OHC–HN, NH.CHO
Undecane-2-one	Me.$(CH_2)_8$.CO.Me
Vamidothion	MeNH.CO.CHMe.S.CH_2.CH_2S– MeO– MeO– ≡PO

Vernolate $Pr_2N.CO.SPr$

Vinclozolin

O
O
(3,5diCl)∅–N Me
O $CH=CH_2$

References

Adam, N. K. [1948]. *Disc. Faraday Soc.* **3,** 5.
Adam, N. K. and Elliott, G. E. P. [1962]. *J. Chem. Soc.* p. 2206.
Albert, A. and Phillips, J. N. [1956]. *J. Chem. Soc.* p. 1294.
Ahmed, H. and Gardiner, B. G. [1968]. *Nature* **214,** 1338.
Alders, L. [1954]. *Applied Sci. Res.* A4, 171.
Alexander, A. E. and Johnson, P. [1949]. "Colloid Science". University Press, Oxford.
Allen, M. [1970]. *Pestic. Sci.* **1,** 152.
Al-Rawi, A. A. H. [1964]. Studies on the movement of nutrient ions in soil. Ph.D. Thesis, University of Wales.
Amer. Chem. Soc., Advances in Chemistry Series [1970]. Vol. 114.
Amsden, R. C. [1970]. *Brit. Crop. Prot. Council Monogr. 2*, p. 124.
Anderson, D. K. and Babb, A. L. [1961]. *J. Phys. Chem.* **65,** 1281.
Anderson, W. P. and Reilly, E. J. [1968]. *J. exp. Bot.* **19,** 648.
Angelescu, E. and Popescu, D. M. [1930]. *Koll. Zeit.* **51,** 247.
Anon [1970]. *Chem. Eng. News* **48,** 12.
Archer, B. L., Cockbain, E. G. and McSweeney, G. P. [1963]. *Biochem. J.* **89,** 565.
Archer, B. L., Cockbain, E. G., Cornforth, J. W., Cornforth, Rita H. and Popjak, G. [1965]. *Proc. Roy. Soc. B* **163,** 519.
Archer, R. J. and LaMer, V. K. [1955]. *J. Phys. Chem.* **59,** 200.
Arens, K. [1934]. *Jahrb. wiss. Bot.* **80,** 248.
Arle, H. F. [1967]. *Proc. West. Weed Cont. Conf.* **21**.
Arnett, E. M., Kover, W. B. and Carter, J. V. [1969]. *J. Amer. Chem. Soc.* **91,** 4028.
Arnold, A. J. [1979]. *Proc. 1979 Brit. Crop Prot. Conf.* **1,** 289.
Aronson, J. M. [1965]. "The Cell Wall in the Fungi" (ed. Ainsworth, G. C. and Sussman, A. S), Vol. 1, Chapter III. Academic Press, New York and London.
Ashare, E., Brooks, T. W. and Swenson, D. W. [1975]. *In Proc. 1975 Int. Controlled Release Pestic. Symp.* (ed. Harris, F. W.), p. 42. State University, Dayton, Ohio.
Ashford, R. and Holroyd, J. [1976]. *Brit. Crop Prot. Council Monogr. 18,* "Granular Pesticides", p. 67.
Ashton, F. M. and Sheets, T. J. [1959]. *Weeds* **7,** 88.
Aubertin, G. M. [1971]. *USDA For. Serv. Res. Paper,* NE 192.
Audus, L. J. [1949]. *Plant and Soil* **2,** 31.
Audus, L. J. [1964]. *In* "The Physiology and Biochemistry of Herbicides" (ed. Audus, L. J.), Chapter 5. Academic Press, London and New York.

Badiei, A. A., Basler, E. and Santelman, P. W. [1966]. *Weeds* **14,** 302.
Bagley, E. and Long, F. A. [1955]. *J. Amer. Chem. Soc.* **77,** 2172.
Baker, E. A. and Bukovac, M. J. [1971]. *Ann. appl. Biol.* **67,** 243.
Baker, E. A., Batt, R. F., Roberts, M. F. and Martin, J. T. [1962]. *Rep. agric. hort. Res. Stn. Univ. Bristol* for 1961, p. 114.
Baker, E. G. [1956]. Amer. Chem. Soc., Div. Petroleum Chem. Symp., Dallas, p. 5.
Baker, E. G. [1959]. *Science* **129,** 871.
Baker, G., Bitting, L. E., Lambert, P. A., McClintock, W. L. and Hogan, W. D. [1975]. *Hyacinth Contr. J.* **13,** 21.
Baker, J. B. [1960]. *Weeds* **8,** 39.
Baker, R. W. and Lonsdale, H. K. [1975]. *In Proc. 1975 Int. Contr. Release Pestic. Symp.* (ed. Harris, F. W.), p. 9. State University of Dayton, Ohio.
Baldry, D. A. T. [1963]. *Bull. Ent. Res.* **54,** 497.
Baldwin, B. C. [1963]. *Nature* **198,** 892.
Baldwin, J. P. [1975]. *J. Soil Sci.* **26,** 195.
Baldwin, J. P., Tinker, P. B. and Nye, P. H. [1972]. *Plant and Soil* **36,** 693.
Baldwin, J. P., Nye, P. H. and Tinker, P. B. [1974]. *Plant and Soil* **38,** 621.
Bals, E. J. and Merritt, C. R. [1975]. *Proc. 8th Brit. Insectic. Fungic. Conf.* **1,** 153.
Balson, E. W. [1947]. *Trans. Faraday Soc.* **43,** 48.
Banden, J. D. [1969]. *Crops and Soils* **21,** 15.
Bangham, D. H. and Saweris, Z. [1938]. *Trans. Faraday Soc.* **34,** 554.
Banker, G. S., Gore, A. Y. and Swarbrick, J. [1966]. *J. Pharm. Pharmacol.* **18,** 457.
Bankoff, S. G. [1958]. *Am. Inst. Chem. Eng. J.* **4,** 24.
Banting, J. D. [1967]. *Weed Res.* **7,** 302.
Barkas, W. W. [1948]. *Disc. Faraday Soc.* **3,** 223.
Barlin, G. B. [1974]. *J. Chem. Soc.* (Perkin II), 1199.
Barlow, F. and Hadaway, A. B. [1952]. *Bull. Ent. Res.* **42,** 769.
Barlow, F. and Hadaway, A. B. [1953]. *Bull. Ent. Res.* **43,** 91.
Barlow, F. and Hadaway, A. B. [1956]. *Nature,* **178,** 1299.
Barlow, F. and Hadaway, A. B. [1958a]. *Bull. Ent. Res.* **49,** 315.
Barlow, F. and Hadaway, A. B. [1958b]. *Bull. Ent. Res.* **49,** 333.
Barlow, F. and Hadaway, A. B. [1963]. *Bull. Ent. Res.* **54,** 329.
Barlow, F. and Hadaway, A. B. [1968]. *Soc. Chem. Ind. Monogr. No. 29,* 3.
Barlow, F. and Hadaway, A. B. [1974]. *Br. Crop Prot. Council Monogr. No. 11,* 84.
Barnes, G. T. and LaMer, V. K. [1962]. *In* "Retardation of Evaporation by Monolayers" (ed. LaMer, V. K.), p. 9. Academic Press, London and New York.
Barnett, A. P., Hauser, E. W., White, A. W. and Holladay, J. H. [1967]. *Weeds* **15,** 133.
Baron, H. M. le [1970]. *Residue Rev.* **32,** 311.
Barrentine, W. L. and Warren, G. F. [1971]. *Weed Sci.* **19,** 31, 37.
Barrer, R. M. [1941]. "Diffusion in and through Solids". Cambridge University Press, Cambridge.
Barrer, R. M. [1968]. *In* "Diffusion in Polymers" (eds Crank, J. and Park, G. S.). Academic Press, London and New York.
Barrett, P. R. F. [1978]. *Pestic. Sci.* **9,** 425.
Barrie, J. A. [1968]. *In* "Diffusion in Polymers" (eds Crank, J. and Park, G. S.), p. 293. Academic Press, London and New York.

Barrie, J. A. and Platt, B. [1963]. *J. Polymer Sci.* **4,** 303.
Barrier, G. E. and Loomis, W. E. [1957]. *Plant Physiol.* **32,** 225.
Bartnicki-Garcia, S. [1968]. *Ann. Rev. Microbiol.* **22,** 87.
Bascom, W. D., Cottington, R. L. and Singleterry, C. R. [1964]. Advances in Chem. Series No. 43, p. 355. Amer. Chem. Soc.
Baskin, A. D. and Walker, E. A. [1953A]. *Ag. Chemicals* **8,** (8), 46.
Baskin, A. D. and Walker, E. A. [1953B]. *Weeds* **2,** 280.
Baur, J. R., Bovey, R. W., Baker, R. D. and Riley, Imogen [1971]. *Weed Sci.* **19,** 138.
Beament, J. W. L. [1961]. *Biol. Rev. Camb. Philos. Soc.* **36,** 281.
Beament, J. W. L. [1964]. *Adv. Insect. Physiol.* **2,** 67.
Bear, J. [1969]. *In* "Flow through porous Media" (ed. de Wiest, R. J. M.), p. 109. Academic Press, New York and London.
Bearman, R. J. [1961]. *J. Phys. Chem.* **65**.
Beasley, M. L. and Collins, R. L. [1970]. *Science* **169,** 769.
Becher, P. and Becher, D. [1969]. Advances in Chem. Series, No. 86, p. 15. Amer. Chem. Soc.
Bennett, J. P. and Bathburn, R. E. [1972]. Paper 737, U.S. Geol. Survey. Washington, D.C.
Bent, K. J. [1967]. *Ann. appl. Biol.* **60,** 251.
Bent, K. J. [1970]. *Ann. appl. Biol.* **66,** 103.
Berck, B. [1964]. *Wld. Rev. Pest Control* **3,** 156.
Berck, B. [1974]. *J. agric. Fd Chem.* **22,** 977.
Berck, B. and Soloman, J. [1962]. *J. agric. Fd Chem.* **10,** 163.
Bernett, Marianne K. and Zisman, W. A. [1959]. *J. Phys. Chem.* **63,** 1241.
Bernett, Marianne K. and Zisman, W. A. [1970]. *J. Phys. Chem.* **74,** 2309.
Beynon, K. I. [1973]. *Proc. 7th Brit. Insectic. Fungic. Conf.* Vol. 3, 791.
Beynon, K. I. and Wright, A. N. [1972]. *Residue Reviews* **43,** 23.
Biddulph, O. and Cory, R. [1957]. *Plant Physiol. (Lanc)* **32,** 608.
Biddulph, O. and Cory, R. [1960]. *Plant Physiol. (Lanc)* **35,** 689.
Biddulph, O. and Cory, R. [1965]. *Plant Physiol.* **40,** 119.
Bidlack, D. L. and Anderson, D. K. [1964]. *J. Phys. Chem.* **68,** 3790.
Bielak, E. B. and Mardles, E. W. J. [1954]. *J. Coll. Sci.* **9,** 233.
Bierl, Barbara A., Beroza, M. and Collier, C. W. [1972]. *J. econ. Ent.* **65,** 659.
Biggar, J. W. and Nielsen, D. R. [1962]. *Soil Sci. Soc. Amer. Proc.* **26,** 125.
Biggar, J. W. and Nielsen, D. R. [1967]. *In* "Irrigation of Agricultural Lands" (ed. Hagen, R. M.), *Agronomy* **11,** 254.
Bikerman, J. J. [1956]. *J. Coll. Sci.* **11,** 299.
Birk, L. A. and Roadhouse, F. E. B. [1962]. *Can. J. Plant Sci.* **44,** 21.
Birshtein, T. M. [1969]. *In* "Water in Biological Systems", Consultants Bureau, New York.
Black, A. L., Chiu, Y-C., Fahmy, M. A. H. and Fukuto, T. R. [1973]. *J. agric. Fd Chem.* **21,** 747.
Blackman, G. E. [1958]. *J. exp. Bot.* **9,** 175.
Blackman, G. E. and Sargent, J. A. [1959]. *J. exp. Bot.* **10,** 480.
Blank, I. H. [1964]. *J. invest. Derm.* **43,** 415.
Blokker, P. C. [1957]. *Proc. 2nd Int. Symp. Surf. Activity*, Butterworth (Lond), **1,** 503.
Bode, L. E., Day, C. L., Gebhardt, M. R. and Goering, C. E. [1973a]. *Weed Sci.* **21,** 480.

Bode, I. E., Day, C. L., Gebhardt, M. R. and Goering, C. E. [1973b]. *Weed Sci.* **21,** 485.
Boer, J. H. de [1953]. "The Dynamic Character of Adsorption", p. 90. Oxford University Press, Oxford.
Bouma, J. and Denning, J. L. [1974]. *Soil Sci. Soc. Amer. Proc.* **36,** 124–127.
Bovey, R. W. and Davis, F. S. [1967]. *Weed Res.* **7,** 281.
Bovey, R. W., Ketchersid, M. L. and Merkle, M. G. [1970]. *Weed Sci.* **18,** 447.
Bowcott, J. E. and Schulman, J. H. [1955]. *Zeit für Elektroch.* **50,** 283.
Bowman, M. C., Schechter, M. S. and Carter, R. L. [1965]. *J. agric. Fd Chem.* **13,** 360.
Bowmer, Kathleen H. [1975]. *J. Irrig. Drain. Div., Am. Soc. Civ. Eng.*, p. 230.
Bowmer, Kathleen H. and Adeney, J. A. [1978]. *Pestic. Sci.* **9,** 354.
Bowmer, Kathleen H. and Sainty, G. R. [1977]. *J. Aq. Plant Management* **15,** 40.
Bowmer, Kathleen, H., Lang, A. R. G., Higgins, M. L., Pillay, A. R. and Tchan, Y. T. [1974]. *Weed Res.* **14,** 325.
Boyce, C. B. C. and Milborrow, B. V. [1965]. *Nature* **208,** 537.
Boyles, M., Mason, J. and Santelmann, P. W. [1978]. *Proc. S. Weed Cont. Conf.*
Bracha, P. and O'Brien, R. D. [1966]. *J. econ. Ent.* **59,** 1255.
Bradbury, F. R., Nield, P. and Newman, J. F. [1953]. *Nature* **172,** 1052.
Bradley, E. F. [1968]. *Quart. J. Roy. Met. Soc.* **94,** 361.
Bradley, E. F. and Finnigan, J. J. [1973]. *Proc. 1st Aust. Conf. on heat and mass transfer*, Reviews 57.
Brannock, L. D., Montgomery, M. and Freed, V. H. [1967]. *Bull. Environ. Contam., Toxicol.* **2,** 178.
Brezeale, E. L. and McGeorge, W. T. [1973]. *Soil Sci.* **75,** 293.
Brian, R. C. [1966]. *Weed Res.* **6,** 292.
Brian, R. C. [1967a]. *Ann. appl. Biol.* **59,** 91.
Brian, R. C. [1967b] *Ann. appl. Biol.* **60,** 77.
Brian, R. C. [1968]. *Soc. Chem. Ind. Monogr. No. 29,* p. 316.
Brian, R. C. [1969]. *Ann. appl. Biol.* **63,** 117.
Brian, R. C. [1972]. *Pesticide Sci.* **3,** 121.
Bridges, R. G. [1956]. *J. Sci. Fd Agric.* **7,** 305.
Briggs, G. G. and Robertson, R. N. [1957]. *Ann. Rev. Pl. Physiol.* **8,** 11.
Briggs, G. G. [1969]. *Nature* **223,** 1288.
Briggs, G. G. [1973]. *Proc. 7th Brit. Insectic. Fungic. Cong.* **1,** 83.
British Crop Protection Council 1974 Monograph No. 11. "Pesticide Application by ULV Methods".
Brønsted, J. N. [1930]. *Z. physik. Chem. Bodenstein Festband* 257.
Brønsted, N. B. and Koefoed, J. [1946]. *Kgl. Danske Videnskab, Selskab Mat-fys Medd.* **22,** No. 17.
Brook, P. J. [1957]. *N.Z. J. Sci. Technol. A.* **38,** 506.
Brooker, P. J. and Ellison, M. [1974]. *Chemy Ind.* p. 785.
Brooks, D. H. [1970]. *Outlook on Agriculture* **6,** 122.
Brooks, F. A. [1947]. *Agric. Eng.* **28,** 233.
Brouwer, R. [1965]. *Soc. exp. Biol. Symp.* **19,** 131.
Brown, D. and Woodcock, D. [1975]. *Pestic. Sci.* **6,** 371.
Brown, D. A., Fulton, B. W. and Phillips, R. E. [1964]. *Soil Sci. Soc. Amer. Proc.* **28,** 628.

Brown, R. M. and Thomson, J. H. [1974]. *Brit. Crop Prot. Council Monogr. No. 11*, p. 232.

Brown, W. B., Reynolds, E. M., East, M. F. and Thomas, C. E. [1961]. *Milling* **137,** 401, 432.

Brown, W. R., Jenkins, R. B. and Park, G. S. [1973]. *J. Polymer Sci.* **41,** 45.

Bruinsma, J. [1967]. *Acta Bot. Neerl.* **16,** 73.

Brunauer, S., Emmett, P. H. and Teller, E. [1940]. *J. Amer. Chem. Soc.* **62,** 1723.

Brunskill, R. T. [1956]. *Proc. 3rd Brit. Weed Contr. Conf.* **2,** 593.

Buckingham, E. [1904]. *USDA Dept. Soils, Bull. No. 25.*

Buerger, A. A. [1966]. *J. theor. Biol.* **11,** 131.

Buerger, A. A. [1967]. *J. theor. Biol.* **14,** 66.

Buerger, A. A. and O'Brien, R. D. [1965]. *J. Cell Comp. Physiol.* **66,** 227.

Bukovac, M. J. [1975]. *Proc. Int. Hortic. Congr. 19th, 1974,* **3,** 273.

Bukovac, M. J. [1976]. *In* "Physiology and Biochemistry of Herbicides" (ed. Audus, L. J.). Academic Press, London and New York.

Bukovac, M. J. and Norris, R. F. [1966]. *VI Symposio Int. Agrochim.*, Varenna, Italy, p. 296.

Bukovac, M. J., Sargent, J. A., Powell, R. G. and Blackman, G. E. [1971]. *J. exp. Bot.* **22,** 598.

Burnett, G. F. [1956]. *Nature* **177,** 663.

Burnside, O. C. [1964]. *Proc. 21st N. Cent. Weed Cont. Conf.* p. 140.

Burns Brown, W. and Heuser, S. G. [1953]. *J. Sci. Fd Agric.* **4,** 378.

Burr, R. J. and Warren, G. F. [1971]. *Weed Sci.* **19,** 701.

Burr, R. J., Lee, W. O. and Appleby, A. P. [1972]. *Weed Sci.* **20,** 180.

Burt, P. E., Lord, K. A., Forrest, J. M. and Goodchild, R. E. [1971]. *Ent. Exp. Appl.* **14,** 255.

Busvine, J. R. [1968]. *Soc. Chem. Ind. Monogr. No. 29.*, 18.

Butler, J. A. V., Thomson, D. W. and Maclennan, W. H. [1933]. *J. Chem. Soc.* 674.

Buzágh, A. von [1930]. *Koll Zeit.* **51,** 105, 230; **52,** 46; **53,** 294.

Buzágh, A. and Wolfram, E. [1958]. *Koll. Zeit.* **157,** 50.

Byass, J. B. [1969]. *Chemy Ind.* 1502.

Byass, J. B., Lake, J. R. and Frost, A. R. [1976]. *A.R.C. Res. Rev.* **2,** 45.

Byers, C. H. and King, C. J. [1967]. *Amer. Inst. Chem. Eng. J.* **13,** 628.

Call, F. [1957a]. *J. Sci. Fd Agric.* **8,** 143.

Call, F. [1957b]. *J. Sci. Fd Agric.* **8,** 591.

Call, F. [1957c]. *J. Sci. Fd Agric.* **8,** 630.

Calvert, R., Terce, M. and LeRenard, L. [1975]. *Weed Res.* **15,** 387.

Canny, M. J. [1971]. *Ann. Rev. Plant Physiol.* **22,** 237.

Canny, M. J. [1973]. "Phloem Translocation". Cambridge University Press, London.

Canny, M. J. and Askham, M. J. [1967]. *Ann. Bot.* **31,** 409.

Canny, M. J. and Phillips, O. M. [1963]. *Ann. Bot.* **27,** 379.

Cardarelli, N. [1979]. "Advances in Pesticide Science" (4th Int. Congr.) (ed. Geissbühler, H.), Vol. 3, p. 744. Pergamon Press, Oxford.

Carey, M. C. and Small, D. M. [1978]. *J. clinical Inv.* **61,** 998.

Caro, J. H. [1971]. *J. agric. Fd Chem.* **19,** 78.

Caro, J. H. and Taylor, A. W. [1971]. *J. agric. Fd Chem.* **19,** 379.

Carslaw, H. S. and Jaeger, J. C. [1947]. "Conduction of Heat in Solids". Oxford University Press, Oxford (2nd ed. 1959).
Carter, M. C. [1969]. *In* "Degradation of Herbicides" (eds Kearney, P. C. and Kaufman, D. D.), p. 187. Marcel Dekker, New York.
Cassie, A. B. D. and Baxter, S. [1944]. *Trans. Faraday Soc.* **40,** 546.
Castelfranco, P. and Deutsch, D. B. [1962]. *Weeds* **10,** 244.
Cathey, M. H. and Steffens, G. L. [1968]. *Soc. Chem. Ind. Monogr. 31,* p. 224.
Chamberlain, A. C. [1955]. *Phil. Mag.* **44,** 1145.
Chamberlain, A. C. [1966a]. *Proc. Roy. Soc. A* **290,** 236.
Chamberlain, A. C. [1966b]. *Proc. Roy. Soc. A* **296,** 45.
Chamberlin, R. and Harrison. A. C. [1972]. *Polymer Age,* **3,** No. 9.
Chambers, T. C. and Possingham, J. V. [1963]. *Aust. J. biol. Sci.* **16,** 818.
Chan, A. F., Evans, D. F. and Cussler, E. L. [1976]. *A.I.Ch.E.J.* **22,** 1006.
Chang, F. Y., Stephenson, G. R. and Bandeen, J. D. [1973]. *Weed Sci.* **21,** 292.
Chang, T. M. S. [1964]. *Science* **146,** 524.
Chang, T. M. S. [1972]. "Artificial Cells". C. C. Thomas, Springfield, Illinois.
Chapman, S. and Cowling, T. G. [1952]. "Mathematical Theory of Non-uniform Gases", p. 245. Cambridge University Press, Cambridge.
Chapman, T., Gabbott, P. A. and Osgerby, J. M. [1970]. *Pestic. Sci.* **1,** 56.
Chem, T. M., Seaman, D. E. and Ashton, F. M. [1968] *Weed Sci.* **16,** 28.
Chepil, W. S. and Woodruff, N. P. [1963]. *Adv. Agron.* **15,** 211.
Cherrett, J. M. and Lewis, T. [1973]. *In* "Biology in Pest and Disease Control" (eds Price-Jones, D. and Solomon, M.) p. 130. Blackwell Scientific Publications, Oxford.
Childs, E. C. [1969]. "An Introduction to the Physical Basis of Soil Water Phenomena". Wiley, London.
Childs, E. C. and Collis-George, N. [1950]. *Proc. Roy. Soc. A* **201,** 392.
Christiansen, J. and Arrhenius, S. [1918]. *Medd. K. Vetenskapsakad, Nobel-Inst.* **4,** No. 2.
Christy, A. L. [1976]. *In* "Transport and Transfer Processes in Plants" (eds Wardlaw, I. F. and Passioura, J. B.). Academic Press, London and New York.
Christy, A. L. and Ferrier, J. M. [1973]. *Plant Physiol.* **52,** 531.
Churchill, M. A., Elmore, H. L. and Buckingham, R. A. [1962]. *J. San. Eng. Div. Am. Soc. Civ. Eng.* **88,** SA41.
Clark, M. W. and King, C. J. [1970]. *Amer. Inst. Chem. Eng. J.* **16,** 64.
Clemons, G. P. and Sisler, H. D. [1970]. *Pest. Biochem. Physiol.* **1,** 32.
Clifford, D. R. and Hislop, E. C. [1975]. *Pestic. Sci.* **6,** 409.
Clor, M. A., Crafts, A. S. and Yamaguchi, S. [1963]. *Plant Physiol.* **38,** 501.
Coats, K. H. and Smith, B. D. [1964]. *Soc. Pet. Eng. J.* **4,** 73.
Cobblentz, W. W. [1949]. *Bull. Amer. Met. Soc.* **30,** 204.
Coffee, R. A. [1979]. *Proc. Brit. Crop Prot. Conf.* **3,** 777.
Coldman, M. F., Kalinovsky, T. and Poulsen, B. J. [1971]. *Br. J. Derm.* **85,** 457.
Collander, R. [1949]. *Acta Chem. Scand.* **3,** 717.
Collander, R. [1950]. *Acta Chem. Scand.* **4,** 1085.
Collander, R., [1951]. *Acta Chem. Scand.* **5,** 774.
Collier, C. F., Graham-Bryce, I. J., Knight, B. A. G. and Coutts, J. [1979]. *Pestic. Sci* **10,** 50.
Conibear, Dorthy I. and Furmidge, C. G. L. [1965]. *J. Sci. Fd Agric.* **16,** 144.
Cookson, R. F. and Cheeseman, G. W. H. [1972]. *J. Chem. Soc. (Perkin II),* 392.
Cooper, W. F. and Nuttall, W. H. [1915]. *J. agric. Sci.* **7,** 219.

Cotton, R. T. [1953]. "Pests of Stored Grain and Grain Products". Burgess Publishing, Minneapolis.
Courshee, R. [1959]. *J. Ag. Eng. Res.* **4,** 229.
Courshee, R. J. [1960]. *Int. Ag. Aviation Conf.* The Hague, Netherl., p. 70.
Courshee, R. J. [1967]. *In* "Fungicides" (ed. Torgeson, D. C.), p. 244. Academic Press, New York and London.
Courshee, R. J. and Ireson, M. J. [1961]. *J. agric. Eng. Res.* **6,** 175.
Coutant, C. C. [1964]. *Science* **146,** 420.
Couzens, D. C. F. and Trevena, D. H. [1969]. *Nature* 222, 473.
Cowan, I. R. [1965]. *J. appl. Ecol.* **2,** 221.
Cox, T. I. [1974]. *Weed Res.* **14,** 379.
Crafts, A. S. [1956]. *Hilgardia* **26,** 287.
Crafts, A. S. [1961]. "Translocation in Plants". Holt, Rinehart and Winston, New York.
Crafts, A. S. [1964]. *In* "Physiology and Biochemistry of Herbicides" (ed. Audus, L. J.), p. 75. Academic Press, London and New York.
Crafts, A. S. [1967]. *Hilgardia* **37,** 625.
Crafts, A. S. and Crisp, C. E. [1971]. "Phloem Transport in Plants", p. 1. Freeman, San Franscisco.
Crafts, A. S. and Foy, C. L. [1962]. *Residue Rev.* **1,** 112.
Crafts, A. S. and Reiber, H. G. [1946]. *Hilgardia* **16,** 487.
Crafts, A. S., Currier, H. B. and Drever, H. R. [1958] *Hilgardia* **27,** 723.
Crank, J. [1956]. "The Mathematics of Diffusion". Oxford University Press, Oxford.
Crank, J. and Henry, M. E. [1949]. *Proc. Phys. Soc. B* **62,** 257.
Crisp, C. E. [1965]. The biopolymer cutin. Ph.D. thesis. University of California, Davis, California.
Crisp, C. E. [1971]. *Proc. 2nd Int. Congr. Pestic. Chem., Israel* (IUPAC) (ed. Tahori, A. S.), p. 211.
Crisp, D. J. [1975]. *Chemy Ind.* 187.
Crisp, D. J. and Meadows, P. S. [1962]. *Proc. Roy. Soc. B.* **156,** 500.
Crosby, D. G. [1972]. *In* "Degradation of Synthetic Organic Molecules in the Biosphere", p. 260. National Academy of Sciences, Washington.
Crosby, D. G. [1976]. *In* "Degradation of Herbicides" (eds Kearney, P. C. and Kaufmann, D. D.), (2nd ed.) p. 835. Marcel Dekker, New York.
Crosby, D. G. and Moilanen, K. W. [1974]. *Archiv. Env. Contam. Toxicol.* **2,** 62.
Crowdy, S. H. [1959]. *In* "Plant Pathology and Progress" (ed. Holton, C. S.), p. 231. University of Wisconsin Press. Wisconsin.
Crowdy, S. H. (1972). *In* "Systemic Fungicides" (ed. Marsh, R. W.), p. 92. Longman, London.
Crowdy, S. H. [1973]. *Proc. 7th Br. Insectic. Fungic. Conf.* **3,** 831.
Crowdy, S. H. and Rudd-Jones, D. [1956]. *J. exp. Bot.* **7,** 335.
Crowdy, S. H., Rudd-Jones, D. and Witt, A. V. [1958]. *J. exp. Bot.* **9,** 206.
Crowdy, S. H., Grove, J. F., Hemming, H. G. and Robinson, Kathleen C. [1956]. *J. exp. Bot.* **7,** 42.
Cunningham, R. T. and Steiner, L. F. [1972]. *J. econ. Ent.* **65,** 505.
Cunningham, R. T., Brann Jr., J. L. and Fleming, G. A. [1962]. *J. econ. Ent.* **55,** 192.
Currie, J. A. [1960a]. *Brit. J. appl. Physics* **11,** 314.
Currie, J. A. [1960b]. *Brit. J. appl. Physics,* **11,** 318.

Currie, J. A. [1961]. *Brit. J. appl. Phys.* **12,** 275.
Currie, J. A. [1965]. *J. Soil Sci.* **16,** 279.
Currier, H. B., Pickering, E. R. and Foy, C. L. [1964]. *Weeds* **12,** 301.
Cussans, G. W. and Taylor, W. A. [1976]. *Proc. Brit. Crop Prot. Conf. Weeds*, p. 885.
Cutting, C. L. and Jones, D. C. [1955]. *J. Chem. Soc.* p. 4067.
Dagley, S. [1972]. *In* "Degradation of Synthetic Organic Molecules in the Biosphere", p. 166. Proc. of conf. San Francisco, National Academy of Sciences, Washington.
Dainty, J. [1963]. *In* "Advances in Botanical Research" (ed. Preston, R. D.), Vol. I., p. 279. Academic Press, London and New York.
Dalbro, S. [1956]. *Proc. 14th Int. Hort. Congr.* 770.
Dalton, R. L., Evans, A. W. and Rhodes, R. C. [1966]. *Weeds* **14,** 31.
Danckwerts, P. V., Kennedy, A. M. and Roberts, D. [1963]. *Chem. Eng. Sci.* **18,** 63.
Darlington, W. A. and Barry, J. B. [1965]. *J. agric. Fd Chem.* **13,** 76.
Darlington, W. A. and Cirulis, N. [1963]. *Plant Physiol. Lanc.* **38,** 442.
Davidson, J. M. and Chang, R. K. [1972]. *Soil Sci. Soc. Amer. Proc.* **36,** 257.
Davisdon, J. M., Rieck, C. E. and Santelmann, P. W. [1968]. *Soil Sci. Soc. Amer. Proc.* **32,** 629.
Darcy, H. [1956]. "Les Fontaines Publiques de la Ville Dijon". Dalmont, Paris.
Davies, C. N. [1952]. *Inst. Mechn. Eng. Proc.* **1B,** 199.
Davies, C. N. [1966]. *Proc. Roy. Soc. A* **289,** 235.
Davies, J. T. and Rideal, E. K. [1961]. "Interfacial Phenomena". Academic Press, London and New York.
Davies, P. J. and Seaman, D. E. [1968]. *Weed Sci.* **16,** 293.
Davies, P. J., Drennan, D. S. H., Fryer, J. D. and Holly, K. [1967]. *Weed Res.* **7,** 220.
Davis, D. E. [1956]. *Weeds* **4,** 227.
Davis, D. E., Gramlich, J. V. and Funderburk, H. H. [1965]. *Weeds* **13,** 252.
Davson, H. and Danielli, J. F. [1952]. "The Permeability of Natural Membranes". Cambridge University Press, Cambridge.
Dawson, J. H. (1979). *Weed. Sci.* **27,** 274.
Day, B. E., Jordan, L. S. and Russell, R. C. [1963]. *Res. Progr. Report West. Weed Control Conf.* p. 80.
Day, P. R. and Forsyth, W. M. [1957]. *Soil Sci. Soc. Amer. Proc.* **21,** 477.
Daykin, P. N. [1967]. *Canad. Ent.* **99,** 303.
Daykin, P. N., Kellogg, F. E. and Wright, R. H. [1965]. *Canad. Ent.* **97,** 239.
Daynes, H. [1920]. *Proc. Roy. Soc. A* **97,** 286.
Deacon, E. L. (1973). *Proc. 1st Aust. Conf. on heat and mass transfer* (Reviews), p. 1.
Deans, H. H. [1963]. *Soc. Pet. Eng. J.* **3,** 49.
Dearden, J. C. and Townend, M. A. [1977]. *In* "Herbicides and Fungicides, Factors Affecting their Activity" (ed. McFarlane, N. R.). Chemical Society Special Publication No. 29.
Debye, P. and Hückel, E. [1925]. *Physik. Z.* **24,** 185, 305.
Deming, J. M. [1963]. *Weeds* **11,** 91.
Dept. of Env. (U.K.) [1973]. *Notes on Water Pollution*, No 61.
Derjaguin, B. V. and Landau, L. [1941]. *Acta Physiochim. USSR* **14,** 633.
Derjaguin, B. V., Kusakov, M. and Lebedeva, L. [1939]. *Dokl. Akad, Nauk. USSR* **23,** 670.

Dethier, V. G. [1957]. *Surv. Biol. Prog.* **3,** 149.
Dethier, V. G., Browne, L. B. and Smith, C. N. [1960]. *J. econ. Ent.* **53,** 134.
Dettre, R. H. and Johnson, R. E. [1967]. *Soc. Chem. Ind. Monogr. No. 25,* p. 144.
Devonshire, A. L. and Needham, P. H. [1974]. *Pestic. Sci.* **5,** 161.
Dewey, O. R., Gregory, P. and Pfeiffer, R. K. [1956]. *Proc. 3rd Brit. Weed Contr. Conf.* **1,** 315.
Dewey, O. R., Hartley, G. S. and MacLauchlan, J. W. G. [1962]. *Proc. Roy. Soc. B.* **155,** 532.
Diamond, J. M. and Wright, E. M. [1969]. *Proc. Roy. Soc. B* **172,** 273.
Dickens, R. and Hiltbold, A. E. [1967]. *Weeds* **15,** 299.
Dijk, J. W. and Kaess, G. [1947]. *Wochbl. Papier Fabrik* **75,** 73.
Dolzman, P. [1965]. *Planta* **64,** 76.
Donalley, W. F. and Ries, S. K. [1964]. *Science* **145,** 497.
Donnan, F. G. [1911]. *Z. Elektroch* **17,** 572; *J. Chem. Soc.* **99,** 1554.
Douglas, H. W., Collins, A. E. and Parkinson, D. [1959]. *Biochim. biophys. Acta.* **33,** 535.
Dreschel, P., Hoard, J. L. and Long, F. A. [1953]. *J. Polymer Sci.* **10,** 241.
Dubey, H. D. and Freeman, J. F. [1965]. *Weeds* **13,** 360.
Dubey, H. D., Sigafus, R. E. and Freeman, J. F. [1966]. *Agron. J.* **58,** 228.
Dudman, W. F. and Grncarevic, M. [1962]. *J. Sci. Food. Agric.* **13,** 221.
Duffy, S. L. [1972]. *Weed Res.* **12,** 169.
Dugger, W. M., Taylor, O. C., Cardiff, E. and Thompson, C. R. [1962]. *Plant Physiol.* **37,** 487.
Dybing, C. D. and Currier, H. B. [1961]. *Plant Physiol.* **36,** 169.
Ebeling, W. [1964]. *In* "The Physiology of Insects" (ed. Rockstein, M.), Vol. 3, p. 507. Academic Press, London and New York.
Ebeling, W. and Wagner, R. E. [1965]. *J. econ. Ent.* **58,** 240.
Edwards, Clive A. [1966]. *Residue Reviews* **13,** 83.
Edwards, Chas. A. and Ripper, W. E. [1953]. *Proc. 1st Brit. Weed Cont. Conf.*, p. 348.
Edwards, Clive A., Thompson, A. R., Beynon, K. I. and Edwards, M. J. [1970]. *Pestic. Sci.* **1,** 169.
Edwards, M. J., Beynon, K. I., Edwards, Clive A. and Thompson, A. R. [1971]. *Pestic. Sci.* **2,** 1.
Ehlers, W., Letey, J., Spencer, W. F. and Farmer, W. J. [1969a]. *Soil Sci. Soc. Amer. Proc.* **33,** 501.
Ehlers, W., Farmer, W. J., Spencer, W. F. and Letey, J. [1969b]. *Soil Sci. Soc. Amer. Proc.* **33,** 505.
Einstein, A. [1905]. *Ann. der Physik,* **27,** 10.
Einstein, A. [1908]. *Z. Elektrochemie* **14,** 235.
Eisner, H. S., Quince, B. W. and Slack, C. [1950]. *Faraday, Soc. Disc.* **30,** 86.
Eley, D. D. and Pepper, D. C. [1946]. *Trans. Faraday Soc.* **42,** 697.
Elle, G. O. [1951]. *Weeds* **1,** 141.
Elliott, M., Ford, M. G. and Janes, N. F. [1970]. *Pestic. Sci.* **1,** 220.
Elliott, M. and Janes, N. F. [1978]. *Chem. Soc. Rev.* **7,** 473.
Elliott, M., Janes, N. F. and Potter, C. [1978]. *Ann. Rev. Entomol.* **23,** 443.
Elliott, M., Farnham, A. W., Janes, N. F., Needham, P. H. and Pulman, D. A. [1974]. *In* "Mechanisms of Pesticide Action". A.C.S. Symposium Series, No. 2, 80.
Ellis, P. R. and Hardman, J. A. [1975]. *Ann. appl. Biol.* **79,** 253.

Ellison, A. H. and Zisman, W. A. [1959]. *J. Phys. Chem.* **63,** 1121.
Ellsworth, J. E. and Harris, D. A. [1973]. *Proc. 7th Brit. Insectic. and Fungic. Conf.* 349.
El-Nakeeb, M. A. and Lampen, J. O. [1965]. *J. gen. Microbiol.* **39,** 285.
El-Nahel, A. K. M. [1954]. *J. Sci. Fd Agric.* **5,** 205, 369.
Elrick, D. E. and French, L. K.[1966]. *Soil Sci. Soc. Amer. Proc.* **30,** 153.
Elwell, H. M. [1968]. *Weed Sci.* **16,** 131.
Emmett, R. W. and Parbery, D. G. [1975]. *Ann. Rev. Phytopath.* **13,** 147.
Englin, B. A., Plate, A. F., Tugolukov, V. M. and Pryanishnikova, M. A. [1965]. From *C.A.* **63,** 14608f.
Ennis Jr., W. B. and Boyd, F. T. [1946]. *Bot. Gaz.* **107,** 552.
Ennis, Jr., W. B., Williamson, R. E. and Dorschner, K. P. [1952]. *J. Ass. Reg. Weed Control. Conf.* **1,** 274.
Epstein, E. [1955]. *Plant Physiol. (Lanc)* **30,** 529.
Epstein, E. and Grant, W. J. [1968]. *Soil Sci. Soc. Amer. Proc.* **32,** 423.
Ercegovich, C. D. and Frear, D. E. H. [1964]. *J. agric. Fd Chem.* **12,** 26.
Erickson, L. C. [1965] *Weeds* **13,** 100.
Eschrich, W. [1975]. *Encycl. Plant Physiol.*, New Series, **1,** 39. Springer-Verlag.
Etheridge, P. and Phillips, F. T. [1976]. *Bull. ent. Res.* **66,** 569.
Evans, A. C. and Martin, H. [1935]. *J. Pomol.* **13,** 261.
Evans, E. [1968]. "Plant Diseases and their Chemical Control". Blackwell, Oxford.
Evans, R. A. and Eckert, R. E. [1965]. *Weeds* **13,** 150.
Everitt, C. T. and Haydon, D. A. [1969]. *J. theor. Biol.* **22,** 9.
Evetts, L. L., Rieck, W. L., Carlson, D. and Burnside, O. C. [1976]. *Proc. N. Cent. Weed Control Conf.* **31,** 69.
Fahmy, M. A. H., Fukuto, T. R., Myers, R. O. and March, R. B. [1970]. *J. agric. Fd Chem.* **18,** 793.
Fancher, G. H., Lewis, J. A. and Barnes, K. B. [1933]. *Mining Ind. Exp. Sta., Penstate Bull. No 12.*
Fang, S. C., Thiessen, P. and Freed, V. H. [1961]. *Weeds* **9,** 569.
Farkas, S. R. and Shorey, H. H. [1972]. *Science* **178,** 67.
Farkas, S. R. and Shorey, H. H. [1974]. *In* "Pheromones" (ed. Burch, M. C.). North-Holland Publishing, Amsterdam.
Farkas, S. R., Shorey, H. H. and Gaston, L. K. [1974]. *Environ. Ent.* **3,** 876.
Farmer, W. J. and Jensen, C. R. [1970]. *Soil Sci. Soc. Amer. Proc.* **34,** 28.
Farmer, W. J., Igue, K., Spencer, W. F. and Martin, J. P. [1972]. *Soil Sci. Soc. Amer. Proc.* **36,** 443.
Farrell, T. P. [1978]. *1st Conf. Council Aust. Weed Sci. Socs* p. 179.
Fasting-Jonk, n.v. [1959]. *World's Paper Trade Review* **152,** 843.
Fatica, N. and Katz, D. L. [1949]. *Chem. Eng. Prog.* **45,** 661.
Faulwetter, R. F. [1917a]. *J. agric. Res.* **8,** 457.
Faulwetter, R. F. [1917b] *J. agric. Res.* **10,** 639.
Fawcett, C. H., Ingram, J. M. A. and Wain, R. L. [1954]. *Proc. Roy. Soc. B* **142,** 60.
Feld, W. A., Post, L. K. and Harris, F. W. [1975]. *In* "Proc. 1975 Int. Controlled Release Symp." (ed. Harris, F. W.), p. 113.
Fensom, D. S. [1972]. *Canad. J. Bot.* **50,** 479.
Ferguson, J. [1939]. *Proc. Roy. Soc. B* **127,** 387.
Field, R. J. and Peel, A. J. [1971]. *New Phytol.* **70,** 997.

Finholt, P., Kristiansen, H., Schmidt, O. C. and Wold, K. [1966]. *Medd. Norske Farm. Sellskap* **28,** 31.
Finney, D. J. [1943]. *Ann. appl. Biol.* **30,** 71.
Fisher, R. W. [1952]. *Canad. J. Zool.* **30,** 254.
Flügge, C. [1934]. *Z. vergl. Physiol.* **20,** 463.
Flynn, G. L. [1974]. *In* "Controlled Release of Biologically Active Agents", p. 73. Plenum Press, New York.
Flynn, G. L. and Smith, R. W. [1972]. *J. Pharm. Sci.* **61,** 61.
Foerster, L. A. and Galley, D. J. [1976]. *Pestic. Sci.* **7,** 436.
Fogg, G. E. [1944]. *Nature* **154,** 515.
Fogg, G. E. [1947]. *Proc. Roy. Soc. B* **134,** 503.
Fogg, G. E. [1948]. *Ann. appl. Biol.* **35,** 315.
Ford, R. E. and Furmidge, C. G. L. [1967]. *Soc. Chem. Ind. Monogr. No. 25*, p. 417.
Ford, R. E. and Furmidge, C. G. L. [1969]. *In* "Pesticide Formulation Research" (ed. Gould, R. F.), Adv. in Chem. Series 86, p. 155. Amer. Chem. Soc.
Forde, B. J. [1966]. *Weeds* **14,** 178.
Forgash, A. J., Cook, B. J. and Riley, R. C. [1962]. *J. econ. Ent.* **55,** 544.
Foster, E. E. [1949]. "Rainfall and Run-off". McMillan, New York.
Fowler, M. C. and Robson, T. O. [1974]. *Proc. 4th int. symp. aquatic weeds*, Wien, p. 230.
Fox, H. W. and Zisman, W. A. [1952]. *J. Coll. Sci.* **7,** 428.
Foy, C. L. and Smith, L. W. [1969]. *In* "Pesticide Formulation Research", Adv. in Chem. Series 86, p. 55. Amer. Chem. Soc.
Frank, H. S. [1970]. *Science* **169,** 635.
Frank, H. S. and Evans, Marjorie E. [1945]. *J. Chem. Phys.* **13,** 507.
Franke, W. [1967]. *Ann. Rev. Pl. Physiol.* **18,** 281.
Franks, F. [1966]. *Nature* **210,** 87.
Franks, F., Gent, M. and Johnson, H. H. [1963]. *J. Chem. Soc.* 2716.
Fraser, R. P., Eisenklam, P., Dambrowski, N. and Hasson, D. [1962]. *Am. Inst. Chem. Eng. J.* **8,** 672.
Free, J. B., Needham, P. H., Racey, P. A. and Stevenson, J. H. [1967]. *J. Sci. Fd Agric.* **18,** 133.
Free, S. M. and Wilson, J. W. [1964]. *J. Med. Chem.* **7,** 395.
Freed, V. H. and Witt, J. M. [1969]. *In* "Pesticide Formulation Research" (ed. Gould, R. F.), Adv. in Chem. Series 86, p. 70. Amer. Chem. Soc.
Frick, E. L. [1970]. *Brit. Crop. Prot. Monogr. No. 2*, p. 23.
Frick, E. L. and Burchill, R. T. [1972]. *Pl. Dis. Reptr.* **56,** 770.
Fried, J. J. and Combarnous, M. A. [1971]. *Adv. Hydrosci.* **7,** 169.
Friedrich, K. and Stammbach, K. [1964]. *J. Chromatog.* **16,** 22.
Frisch, H. L. [1956]. *J. Phys. Chem.* **60,** 1177.
Frissel, M. J. [1961]. Versl. Landbouwk. Onderz., No. 67, 3, Wageningen.
Frissel, M. J. and Poelstra, P. [1967]. *Plant and Soil* **26,** 285.
Frissel, M. J., Poelstra, P. and Reiniger, P. [1970]. *Plant and Soil* **33,** 161.
Frössling, N. [1938]. *Beitr. Geophysik.* **52,** 170.
Fuchs, N. A. [1959]. "Evaporation and Droplet Growth in Gaseous Media". Pergamon, Oxford.
Fujita, H. [1968]. *In* "Diffusion in Polymers" (eds Crank, J. and Park, G. S.), p. 75. Academic Press, London and New York.

Fujita, H., Kishimoto, A. and Matsumoto, K. [1960]. *Trans. Farad. Soc.* **56,** 424.
Fujita, T. [1972]. *In* Adv. in Chem. Series, No. 114, Amer. Chem. Soc., p. 1.
Fujita, T., Isawa, J. and Hansch, C. [1964]. *J. Amer. Chem. Soc.* **86,** 5175.
Fuller, E. N. and Schettler, P. E. [1966]. *Ind. Eng. Chem.* **58,** 18.
Furmidge, C. G. L. [1959]. *J. Sci. Fd Agric.* **8,** 419.
Furmidge, C. G. L. [1962a]. *J. Coll. Sci.* **17,** 309.
Furmidge, C. G. L. [1962b]. *J. Sci. Fd Agric.* **13,** 127.
Furmidge, C. G. L. [1962c]. *Chemy Ind.* 1917.
Furmidge, C. G. L. [1964]. *J. Sci. Fd Agric.* **15,** 542.
Furmidge, C. G. L. [1965]. *J. Sci. Fd Agric.* **16,** 134.
Furmidge, C. G. L. and Osgerby, J. M. [1967]. *J. Sci. Fd Agric.* **18,** 269.
Furmidge, C. G. L., Hill, A. C. and Osgerby, J. M. [1966]. *J. Sci. Fd Agric.* **17,** 518.
Furmidge, C. G. L., Hill, A. C. and Osgerby, J. M. [1968]. *J. Sci. Fd Agric.* **19,** 91.
Gabbott, P. A. and Larman, V. N. [1968]. *Soc. Chem. Ind. Monogr. No. 29*, 269.
Galley, D. J. and Foerster, L. A. [1976a]. *Pestic. Sci.* **7,** 301.
Galley, D. J. and Foerster, L. A. [1976b]. *Pestic. Sci.* **7,** 549.
Gameson, A. L. H. [1957]. *J. Inst. Water Eng.* **11,** 477.
Gardner, W. R. [1958]. *Soil Sci.* **85,** 228.
Gardner, W. R. [1960]. *Soil Sci.* **89,** 63.
Gardner, W. R. and Brooks, R. H. [1957]. *Soil Sci.* **83,** 295.
Gary-Bobo, C. M. and Weber, Heather W. [1969]. *J. Phys. Chem.* **73,** 1155.
Gasser, R. (1950). *Mitt.-Schweiz. Entomol. Ges.* **23,** 257.
Gast, R. G. [1962]. *J. Colloid Sci.* **17,** 492.
Gates, D. M. [1968]. *Ann. Rev. Plant Physiol.* **19,** 211.
Geissbühler, H., Haselbach, C., Aebi, H. and Ebner, L. [1963a]. *Weed Res.* **3,** 181.
Geissbühler, H., Haselbach, C. and Aebi, H. [1963b]. *Weed Res.* **3,** 140.
Gentner, W. A. [1964]. *Weeds* **12,** 239.
Genuchten, M. Th. van, and Wierenga, P. J. [1976]. *Soil Sci. Soc. Amer. Proc.* **40,** 473.
Genuchten, M. Th. van, Davidson, J. M. and Wierenga, P. J. [1974]. *Soil Sci. Soc. Amer. Proc.* **38,** 29.
Gerber, H. R., Ziegler, P. and Dubach, P. [1970]. *Proc. 10th Brit. Weed Conf.* **1,** 118.
Gerolt, P. [1963]. *Nature* **197,** 721.
Gerolt, P. [1969]. *J. Insect. Physiol.* **15,** 563.
Gerolt, P. [1970]. *Pestic. Sci.* **1,** 209.
Gerolt, P. [1972]. *Pestic. Sci.* **3,** 43.
Gerolt, P. [1975a]. *Pestic. Sci.* **6,** 561.
Gerolt, P. [1975b]. *Pestic. Sci.* **6,** 233.
Getzin, L. W. and Chapman, R. K. [1959]. *J. econ. Ent.* **52,** 1160.
Gibich, J. and Pederson, J. R. [1963]. *Cereal Sci. Today* **8,** 345.
Gillett, J. W., Harr, J. R., Lindstrom, F. T., Mount, D. A., St. Clair, A. D. and Weber, L. J. [1972]. *Residue Rev.* **44,** 115.
Ginsburg, H. and Ginzburg, B. Z. [1971]. *J. exp. Bot.* **22,** 337.
Glendinning, Dorothy, MacDonald, J. A. and Grainger, J. [1963]. *Trans. Brit. Mycol. Soc.* **46,** 595.

Glueckauf, E. [1955a]. *Trans. Faraday Soc.* **51,** 34.
Glueckauf, E. [1955b]. *In* "Ion Exchange and its Applications", Soc. Chem. Ind., London, 1955.
Glueckauf, E. [1955c]. *Trans. Faraday Soc.* **51,** 1540.
Golumbic, C., Orchin, M. and Weller, S. [1949]. *J. Am. Chem. Soc.* **71,** 2624.
Goodgame, T. H. and Sherwood, T. K. [1954]. *Chem. Eng. Sci.* **3,** 37.
Goodknight, R. C., Klikoff, A. and Fatt, I. [1960]. *J. Phys. Chem.* **64,** 1162.
Goodman, D. S. [1958]. *J. Amer. Chem. Soc.* **80,** 3887.
Goodman, R. N. and Addy, S. K. [1963]. *Phytopath. Z.* **46,** 1.
Goodwin, J. T. and Somerville, G. R. [1974]. *In* "Microencapsulation" (ed. Vandegaar, J. E.), p. 155. Plenum Press, New York.
Goodwin, W., Salmon, E. S. and Ware, W. M. [1929]. *J. agric. Sci.* **19,** 185.
Gopal, E. S. R. [1963]. *In* "Rheology of Emulsions" (ed. Sherman, P.), p. 27. Pergamon, Oxford.
Goring, C. A. I. [1967]. *Ann. Rev. Phytopathol.* **5,** 285.
Gottlieb, D. and Kumar, K. [1970]. *Phytopathology* **60,** 1451.
Gottschlich, C. F. [1963]. *Am. Inst. Chem. Eng. J.* **9,** 88.
Gough, H. C. [1946]. *Bull. Ent. Res.* **37,** 251.
Goulding, R. [1978]. *Symp. on Controlled Drop Appl. Brit. Crop. Prot. Council.*, p. 243.
Gouy, G. [1910]. *J. physique* **9,** 457.
Graham-Bryce, I. J. [1961]. Some Aspects of the Movements of Ions in Soils, p. 81. D.Phil. Thesis, University of Oxford.
Graham-Bryce, I. J. [1963a]. *J. Soil Sci.* **14,** 188.
Graham-Bryce, I. J. [1963b]. *J. Soil Sci.* **14,** 195.
Graham-Bryce, I. J. [1965]. *Technical Rept. No. 48*, IAEA, Vienna, p. 42.
Graham-Bryce, I. J. [1967]. *J. Sci. Fd Agric.* **18,** 72.
Graham-Bryce, I. J. [1968]. *Soc. Chem. Ind. Monogr. No. 29*, p. 251.
Graham-Bryce, I. J. [1969]. *J. Sci, Fd Agric.* **20,** 489.
Graham-Bryce, I. J. [1972]. *Proc. 11th Brit. Weed Cont. Conf.*, Vol. III, p. 1193.
Graham-Bryce, I. J. [1973a]. *In* "Pollution: Engineering and Scientific Solutions" (ed. Barrakette, E. S.), p. 133. Plenum Press, New York.
Graham-Bryce, I. J. [1973b]. *Proc. 7th Brit. Insectic. Fungic. Conf.* **3,** 921.
Graham-Bryce, I. J. [1975]. *C. R. 5e Symp. Lutte intégrée en vergers* OILB/SROP, 315.
Graham-Bryce, I. J. and Briggs, G. G. [1970]. *RIC Reviews* **3,** 87.
Graham-Bryce I. J. and Coutts, J. [1971]. *Proc. 6th Brit. Insectic. Fungic. Conf.* **II,** 419.
Graham-Bryce, I. J. and Etheridge, P. [1970]. *Proc. VIIth Int. Congr. of Plant Prot.*, Paris.
Graham-Bryce, I. J. and Hartley, G. S. [1979]. *In* "Advances in Pesticide Science" (ed. Geissbühler, H.), Part 3, p. 718. Pergamon Press, Oxford and New York.
Graham-Bryce, I. J., Stevenson, J. H. and Etheridge, P. [1972]. *Pestic. Sci.* **3,** 781.
Grainger [1954]. *W. Scot. Agric. Coll. Res. Bull. No. 10.*
Gratwick, M. [1957a]. *Bull. ent. Res.* **48,** 733.
Gratwick, M. [1957b]. *Bull. ent. Res.* **48,** 741.
Gray, R. A. and Weierich, A. J. [1965]. *Weeds* **13,** 141.
Gray, R. A. and Weierich, A. J. [1968]. *Weed Sci.* **16,** 77.

Gray, V. R. [1967]. *Soc. Chem. Ind. Monogr. No. 25*, p. 99.
Green, H. L. and Lane, W. R. [1957]. "Particulate Clouds, Dusts, Smokes and Mists". Spon, London.
Green, R. E. and Corey, J. C. [1971]. *Soil Sci. Soc. Amer. Proc.* **35,** 561.
Greene, D. W. and Bukovac, M. J. [1971]. *J. Amer. Soc. Hort. Sci.* **96,** 240.
Greene, D. W. and Bukovac, M. J. [1972]. *Plant Cell Physiol.* **13,** 321.
Greene-Kelly, R. [1974]. *Geoderma* **11,** 243–257.
Greenland, D. J. [1970]. *Soc. Chem. Ind. Monogr. No. 37*, p. 79.
Gregory, M. D., Christian, Sherrill D. and Affsprung, H. E. [1967]. *J. phys. Chem.* **71,** 2283.
Gregory, P. H. [1973]. "The Microbiology of the Atmosphere" (2nd edn). Chapter V. Leonard Hill, Aylesbury.
Gregory, P. H., Gurhrie, E. J. and Bunce, Maureen E. [1959]. *J. gen. Microbiol.* **20,** 328.
Griffiths, D. C. and Scott, G. C. [1967]. *Proc. 4th Brit. Insec. Fungic. Conf.* **I,** 118.
Griffiths, D. C., Scott, G. C., Maskell, F. E., Mathias, P. L. and Roberts, P. F. [1970]. *Plant Pathol.* **19,** 111.
Griffiths, D. C., Jeffs, K. A. and Scott, G. C. [1974]. *Rep. Rothamsted Exp. Stn.*, 1973.
Groot, S. R. de and Mazur, P., [1962]. "Non-equilibrium Thermodynamics". Interscience, New York.
Grover, R. [1975]. *Weed Sci.* **23,** 529.
Grover, R., Maybank, J. and Yoshida, K. [1972]. *Weed Sci.* **20,** 320.
Grubb Jr., T. C. [1973]. *Science* **180,** 1302.
Gruhn, K. [1971]. *Allg. Papier Rundschau Special IPEX issue*, p. 33.
Gückel, W., Rittig, F. R. and Synnatschke, G. [1974]. *Pestic. Sci.* **5,** 393.
Guenzi, W. D. and Beard, W. E. [1968]. *Soil Sci. Soc. Amer. Proc.* **32,** 552.
Guenzi, W. D. and Beard, W. E. [1970]. *Soil Sci. Soc. Amer. Proc.* **34,** 443.
Guggenheim, E. A. and Adam, N. K. [1933]. *Proc. Roy. Soc. A* **139,** 218.
Hadaway, A. B. and Barlow, F. [1951]. *Bull. ent. Res.* **41,** 603.
Hadaway, A. B. and Barlow, F. [1956]. *Nature* **178,** 1299.
Hadaway, A. B. and Barlow, F. [1963]. *Bull. ent. Res.* **54,** 329.
Hadaway, A. B., Barlow, F. and Turner, C. R. [1970]. *Bull. ent. Res.* **60,** 17.
Hadaway, A. B., Barlow, F. and Flower, L. S. [1976]. Centre for Overseas Pest Research, London: *Misc. Rept.*, No. 22.
Hall, D. M., Matus, A. I., Lamberton, J. A. and Barber, H. N. [1965]. *Aust. J. biol. Sci.* **18,** 323.
Haighton, A. J., Vermaas, L. F. and Hollander, C. den [1971]. *J. Amer. Oil Chem. Soc.* **48,** 7.
Hamaker, J. W. [1972a]. *In* "Organic Chemicals in the Soil Environment" (eds Goring, C. A. I. and Hamaker, J. W.), p. 253. Marcel Dekker, New York.
Hamaker, J. W. [1972b]. *In* "Organic Chemicals in the Soil Environment" (eds Goring, C. A. I. and Hamaker, J. W.), p. 341. Marcel Dekker, New York.
Hamaker, J. W. and Kerlinger, H. O. [1969]. *In* "Pesticide Formulation Research" (ed. Gould, R. F.), Advances in Chem. Series No. 86, p. 39. Amer. Chem. Soc.
Hamaker, J. W. and Thompson, J. M. [1972]. *In* "Organic Chemicals in the Soil Environment" (eds Goring, C. A. I. and Hamaker, J. W.), p. 49. Marcel Dekker, New York.

Hamaker, J. W., Goring, C. A. I. and Youngson, C. R. [1966]. Adv. in Chem. Series, No. 60, p. 23. Amer. Chem. Soc.
Hamlin, W. E., Northam, J. I. and Wagner, L. J. G. [1965]. *J. Pharm. Sci.* **54,** 1651.
Hammett, L. P. [1970]. "Physical Organic Chemistry", p. 347. McGraw Hill, New York.
Hanai, T., Haydon, D. A. and Redwood, W. R. [1966]. *Ann. N.Y. Acad. Sci.* **137,** 731.
Hance, R. J. [1969a]. *Weed Res.* **9,** 108.
Hance, R. J. [1969b]. *J. Sci. Fd Agric.* **17,** 667.
Hance, R. J. [1970]. *Soc. Chem. Ind. Monogr. No. 37,* 92.
Hance, R. J. [1976]. *Pestic. Sci.* **7,** 363.
Hance, R. J. [1981]. "Interactions between Herbicides and the Soil". Academic Press, London and New York. In press.
Hancock, C. K. and Falls, C. P. [1961]. *J. Amer. Chem. Soc.* **83,** 4214.
Hancock, C. K., Myers, E. A. and Yager, B. Y. [1961]. *J. Amer. Chem. Soc.* **83,** 4211.
Hangartner, W. [1967]. *Z. vergl. Physiol.* **57,** 103.
Hanks, J. R. and Bowers, S. A. [1962]. *Soil Sci. Soc. Amer. Proc.* **26,** 530.
Hanks, J. R. and Gardner, H. R. [1965]. *Soil Sci. Soc. Amer. Proc.* **29,** 495.
Hannan, P. J. [1961]. *Appl. Microbiol.* **9,** 113.
Hansch, C. [1972]. Adv. in Chem. Series, No. 114, p. 20. Amer. Chem. Soc.
Hansch, C. and Anderson, Susan M. [1967]. *J. Org. Chem.* **32,** 2583.
Hansch, C. and Fujita, T. [1964]. *J. Amer. Chem. Soc.* **86,** 1616.
Hanus, L., Hrubon, L. and Krejci, Z. [1975]. *Acta Univ. Palacki Olumuc Fac. Med.* **74,** 167.
Hardy, W. B. [1922]. *Proc. Roy. Soc.* A **100,** 573.
Hare, E. F. and Zisman, W. A. [1955]. *J. phys. Chem.* **59,** 335.
Hare, E. F., Shafrin, E. G. and Zisman, W. A. [1954]. *J. phys. Chem.* **58,** 236.
Harper, D. B., Smith, R. V. and Gotto, D. M. [1977]. *Environ. Pollut.* **12,** 223.
Harris, C. I. [1966]. *Weeds* **14,** 6.
Harris, C. I. [1967]. *Weeds* **15,** 214.
Harris, R. L. [1964]. *Int. Congr. Occupational Health 14th,* Madrid, **2,** 356.
Hartley, G. S. [1935]. *Trans. Faraday Soc.* **31,** 31.
Hartley, G. S. [1938]. *J. Chem. Soc.* 1968.
Hartley, G. S. [1948]. *Disc. Faraday Soc.* **3,** 223.
Hartley, G. S. [1949]. *Trans. Faraday Soc.* **45,** 820.
Hartley, G. S. [1951]. *XVth Int. Congr. pure and appl. Chem.*, N.Y., Section XIII.
Hartley, G. S. [1963]. *J. theor. Biol.* **5,** 57.
Hartley, G. S. [1965]. *Soc. Chem. Ind. Monogr. No. 21,* p. 122.
Hartley, G. S. [1967]. *Soc. Chem. Ind. Monogr. No. 25,* p. 438.
Hartley, G. S. [1969]. Adv. in Chem. Series, No. 86, p. 115. Amer. Chem. Soc.
Hartley, G. S. [1973]. *In* "Physiology and Biochemistry of Herbicides" (ed. Audus, L. J.), (2nd ed.), Vol. 2, p. 7. Academic Press, London and New York.
Hartley, G. S. and Brunskill, R. T. [1958]. *In* "Surface Phenomena in Chemistry and Biology" (eds Danielli, J. F., Pankhurst, K. G. A., and Riddiford, A. C.), p. 214. Pergamon, Oxford.
Hartley, G. S. and Crank, J. [1949]. *Trans. Faraday Soc.* **45,** 801.
Hartley, G. S. and Heath, D. F. [1951]. *Nature* **167,** 816.
Hartley, G. S. and Howes, R. [1961]. *1st Brit. Insectic. Fungic. Conf.* p. 533.

Hartley, G. S. and Robinson, C. [1931]. *Proc. Roy. Soc. A* **134,** 20.
Hartley, G. S. and Roe, J. W. [1940]. *Trans. Faraday Soc.* **36,** 101.
Hashimoto, I., Deshpande, K. B. and Thomas, H. C. [1964]. *Ind. Eng. chem. Fundamentals* **3,** 213.
Haskell, R. [1908]. *Phys. Rev.* **27,** 145.
Hassenstein, B. [1959]. *Z. Naturf.* **14b,** 659.
Hawke, J. G. and Alexander, A. E. [1962]. Amer. Chem. Soc. Symp. on "Retardation of Evaporation by Monolayers". p. 67. Academic Press, New York and London.
Hay, J. R. [1976]. *In* "Herbicides, Physiology, Biochemistry and Ecology" (ed. Audus, L. J.), Vol. 1, p. 365. Academic Press, London and New York.
Hay, J. R. and Thimann, K. V. [1956]. *Plant Physiol.* **31,** 446.
Hayes, M. H. B. [1970]. *Residue Rev.* **32,** 131.
Hayes, M. H. B., Stacey, M. and Standley, J. [1968]. *Trans. IXth Int. Congress Soil Sci.* **3,** 247.
Hayes, M. J. and Park, G. S. [1955]. *Trans. Faraday Soc.* **51,** 1134.
Haydon, D. A. [1962]. *Chem. Ind.* p. 1922.
Haydon, D. A. [1970]. *In* "Permeability and Function of Biological Membranes" (ed. Katchalsky, E.). North-Holland, Amsterdam.
Haydon, D. A. Hladky, S. B. [1972]. *Quart. Rev. Biophysics* **5,** 187.
Heard, A. J. [1973]. *Rept. Grassland Res. Inst.* 1972, p. 39.
Heath, D. F. [1956]. *J. chem. Soc.* 3796.
Heath, D. F., Lane, D. W. J. and Park, P. O. [1955]. *Phil. Trans. Roy. Soc. B* **239,** 191.
Helling, C. S. and Turner, B. C. [1968]. *Science* **162,** 562.
Helmholtz, H. [1879]. *Ann. Physik* **7,** 337.
Hemwall, J. B. [1959]. *Soil Sci.* **88,** 184.
Hemwall, J. B. [1960]. *Soil Sci.* **90,** 157.
Hemwall, J. B. [1962]. *Phytopathology* **52,** 1108.
Hermann, K. W. [1962]. *J. phys. Chem.* **66,** 295.
Hermann, K. W., Bushmiller, J. G. and Courchene, W. L. [1966]. *J. phys. Chem.* **70,** 2909.
Heymann, E. and Boye, E. [1932]. *Koll. Zeit.* **59,** 153.
Hilby, J. W. [1962]. *Inst. Chem. Eng., Symp. Series No. 9*, 312.
Hickman, C. J. [1965] *In* "The Fungi" (eds Ainsworth, G. C. and Sussman, A. S.), Vol. 1, Chapter II. Academic Press, London and New York.
Higuchi, T. [1960]. *J. Soc. cosm. Chem.* **11,** 85.
Higuchi, T. [1963]. *J. Pharm. Sci.* **52,** 1145.
Higuchi, T. and Davis, S. S. [1970]. *J. Pharm. Sci.* **59,** 1376.
Higuchi, W. I. and Misra, J. [1962]. *J. Pharm. Sci.* **51,** 459.
Hildebrand, J. H. and Scott, R. L. [1964]. "The Solubility of Non-Electrolytes" (3rd edn). Dover Publ. Inc., New York.
Hildebrand, J. H., Ellefson, E. T. and Beebe, C. W. [1917]. *J. Am. Chem. Soc.* **39,** 2301.
Hill, G. S., McGahen, J. W., Baker, H. M., Finnerty, D. W. and Bingeman, C. W. [1955]. *Agronomy J.* **47,** 93.
Himel, C. M. [1969a]. *J. econ. Ent.* **62,** 919.
Himel, C. M. [1969b]. *Proc. 4th Int. agric. Aviation Congr.* p. 275.
Himel, C. M. and Moore, A. D. [1967]. *Science* **156,** 1250.

Himel, C. M. and Uk, S. [1972]. *J. econ. Ent.* **65,** 990.
Hirst, J. M. [1958]. *Outlook Agr.* **2,** 16.
Hislop, E. C. [1967]. *Ann. appl. Biol.* **60,** 265.
Hislop, E. C. [1969]. *Ann. appl. Biol.* **63,** 71.
Hislop, E. C. and Clifford, D. R. [1976]. *Ann. appl. Biol.* **82,** 557.
Höhn, K. [1954]. *Beitr. Biol. Pflanz* **30,** 159.
Holloway, P. J. [1970]. *Pestic. Sci.* **1,** 156.
Holly, K. [1956]. *Ann. appl. Biol.* **44,** 195.
Honert, T. H. van den [1948]. *Disc. Faraday Soc.* **3,** 146.
Hopkinson, P. A. [1974]. *Brit. Crop Prot. Council Monogr. No. 11.* p. 166.
Hopp, J. and Linder, P. J. [1946]. *Am. J. Bot.* **33,** 598.
Hörmann, W. D. and Eberle, D. O. [1972]. *Weed Res.* **12,** 199.
Horn, D. H. S. and Lamberton, J. A. [1963]. *Aust. Chem. J.* **16,** 475.
Hornsby, A. G. and Davidson, J. M. [1973]. *Soil Sci. Soc. Amer. Proc.* **37,** 823.
Horton, C. W. and Rogers, F. T. [1945]. *J. appl. Phys.* **16,** 367.
Horwitz, L. [1958]. *Plant Physiol. (Lanc.).* **33,** 81.
House, C. R. and Findlay Nele [1966]. *J. exp. Bot.* **17,** 627.
Howes, R. [1961]. Personal communication.
Huelin, F. E. [1959]. *Aust. J. biol. Sci.* **12,** 175.
Huelin, F. E. and Gallop, R. A. [1951]. *Aust. J. scient. Res. B* **4,** 526, 533.
Hughes, E. E. [1968]. *Weeds* **16,** 486.
Hughes, Helen R. and Meeklah, F. A. [1977]. *Proc. 30th N.Z. Weed Pest Contr. Conf.* p. 135.
Huisman, F. and Mysels, K. J. [1969]. *J. phys. Chem.* **73,** 489.
Hull, H. M. [1970]. *Residue Reviews* **31.**
Hull, H. M. and Morton, H. L. [1971]. *Weed Sci.* **19,** 102.
Humphreys, W. J. [1940]. "Physics of the Air". McGraw-Hill, New York.
Hunsley, D. and Burnett, J. H. [1970]. *J. gen. Microbiol.* **62,** 203.
Hurst, H. [1941]. *Nature* **147,** 388.
Hurst, H. [1948a]. *Disc. Faraday Soc.* **3,** 193.
Hurst, H. [1948b]. *Disc. Faraday Soc.* **3,** 224.
Hussain, M., Fukoto, T. R. and Reynolds, H. T. [1974]. *J. agric. Fd Chem.* **22,** 225.
Hyde, R. M. [1975]. *J. med. Chem.* **18,** 231.
Ignoffo, C. M., Berger, R. S., Graham, H. M. and Martin, D. F. [1963]. *Science* **141,** 902.
Igue, K., Farmer, W. J., Spencer, W. F. and Martin, J. P. [1972]. *Soil Sci. Soc. Amer. Proc.* **36,** 447.
Imms, A. D. [1973]. "Insect Natural History", *New Naturalist,* Vol. 8, Collins.
Ingold, C. K. and Mohrhenn, H. G. G. [1935]. *J. chem. Soc.,* p. 949.
Irmann, F. [1965]. *Chemie-Ing. Technik* **37,** 789.
Isawa, J., Fujita, T. and Hansch, C. [1965]. *J. med. Chem.* **8,** 150.
Isensee, A. R., Jones, G. E. and Turner, B. C. [1971]. *Weed Sci.* **19,** 727.
Jackson, J. E. and Weatherley, P. E. [1962]. *J. exp. Bot.* **13,** 128.
Jackson, R. D. [1964]. *Soil Sci. Soc. Amer. Proc.* **28,** 172.
Jackson, R. D. and Klute, A. [1967]. *Soil Sci. Soc. Amer. Proc.* **31,** 122.
Jacobson, M. and Beroza, M. [1965]. *Science* **147,** 748.
Jacobson, M., Beroza, M. and Yamamoto, R. T. [1963]. *Science* **139,** 48.
Jaeger, J. C. [1942]. *Proc. Roy. Soc. Edinb.* **61A,** 223.
Jansen, L. L. [1966]. *Weed Soc. Am. Abstr.* 34.

Jarvis, P. and Thaine, R. [1971]. *Nature* **232,** 236.
Jeffcoat, B. and Harries, W. N. [1973]. *Pestic. Sci.* **4,** 891.
Jeffcoat, B. and Harries, W. N. [1975]. *Pestic. Sci.* **6,** 283.
Jeffcoat, B., Harries, W. N. and Thomas, D. B. [1977]. *Pestic. Sci.* **8,** 1.
Jeffreys, H. [1930]. *Proc. Camb. Phil. Soc.* **26,** 204.
Jeffs, K. A. [1973]. *Proc. 7th Brit. Insectic. Fungic. Conf.* **1,** 341.
Jeffs, K. A. [1976]. *Rothamsted Exp. Station Rept.* for 1976, part 1, 171.
Jeffs, K. A. [1979]. *CIPAC Seed Treatment Monograph.*
Jeffs, K. A. and Griffiths, D. C. [1973]. *Rothamsted Exp. Station Rept.* for 1973, part 1, 179.
Jeffs, K. A. and Tuppen, R. J. [1979]. *CIPAC, Seed Treatment Monograph.*
Jeffs, K. A., Comely, D. R. and Tuppen, R. J. [1972]. *J. agric. Eng. Res.* **17,** 315.
Jenkins, R. B. [1976]. Diffusion in polybutadienes. Ph.D. Thesis, University of Wales.
Jenner, C. F., Saunders, P. F. and Blackman, G. E. [1968]. *J. exp. Bot.* **19,** 333.
Jeschke, D. and Stuart, H. A. [1961]. *Z. Naturf,* **16a,** 37.
Johnson, Mary P. and Bonner, J. [1956]. *Physiol. Plant* **9,** 102.
Johnstone, D. R. [1969a]. *Proc. 4th Int. agric. aviation. Congr.* (Kingston, 1969), 225.
Johnstone, D. R. [1969b], *Trop. Pestic. Res. Unit Rept.* for 1967–68, p. 9.
Johnstone, D. R. [1970]. *J. agric. Eng. Res.* **15,** 188.
Johnstone, D. R. [1971]. *Trop. Pestic. Res. Unit Rept.* for 1969, 1970, p. 11.
Johnstone, D. R. [1973]. *In* "Pesticide Formulations" (ed. Valkenburg, W. van). Marcel Dekker, New York.
Jones, D. W. and Foy, C. L. [1972]. *Weed Sci.* **20,** 21.
Jones, D. C. and Mill, G. S. [1957]. *J. chem. Soc.* p. 213.
Jones, J. R. and Monk, C. B. [1963]. *J. chem. Soc.* p. 2633.
Jones, D. C. and Ottewill, R. H. [1955]. *J. chem. Soc.* p. 4076.
Jordan, L. S., Day, B. E. and Hendrixon, R. T. [1963]. *Weeds* **11,** 198.
Jordan, T. N. [1977]. *Weed Sci.* **25,** 448.
Joris, G. G. and Taylor, H. S. [1948]. *J. Chem. Phys.* **16,** 45.
Jost, W. [1953]. "Diffusion in Solids, Liquids and Gases". Academic Press, London and New York.
Julian, A. C. [1978]. *1st Conf. Council of Aust. Weed Sci. Socs,* p. 327.
Jyung, W. H. and Wittwer, S. H. [1964]. *Am. J. Bot.* **51,** 437.
Kalmus, H. [1942]. *Nature* **150,** 405, 524.
Kalkat, G. S., Davidson, R. H. and Brass, C. L. [1961]. *J. econ. Ent.* **54,** 1186.
Kamimura, S. and Goodman, R. N. [1964a]. *Phytopath. Z.* **51,** 324.
Kamimura, S. and Goodman, R. N. [1964b]. *Phytopathology* **54,** 1467.
Kanellopoulos, A. G. [1974]. *Chemy. Ind.* p. 951.
Karlson, P. and Lüscher, M. [1959]. *Nature* **183,** 55.
Katz, M. [1973]. *In* "Drug Design" (ed. Ariens, E. J.), Vol. IV, Chapter 4. Academic Press, New York and London.
Kay, B. D. and Elrick, D. E. [1967]. *Soil Sci.* **104,** 314.
Kearney, P. C. and Kaufman, D. D. [1969]. "Degradation of Herbicides". Marcel Dekker, New York.
Kearney, P. C. and Kaufman, D. D. [1972]. *Proc. of conf., San. Francisco,* 1971 (Nat. Acad. Sci., Washington), p. 166.
Kellogg, F. E., Frizel, D. E. and Wright, R. H. [1962]. *Canad. Ent.* **94,** 486.

Kemper, W. D. and Quirk, J. P. [1972]. *Soil Sci. Soc. Amer. Proc.* **36,** 426.
Kemper, W. D., Maasland, D. E. L. and Porter, L. K. [1964]. *Soil Sci. Soc. Amer. Proc.* **28,** 164.
Kennedy, J. S. [1939]. *Proc. zool. Soc. A* **109,** 221.
Kennedy, J. S. [1966]. *Symp. No. 3 Roy. Ent. Soc. (London),* p. 97.
Kennedy, J. S. [1976]. *In* "Chemical Control of Insect Behaviour: Theory and Application" (eds McKelvey, J. J., Jr. and Shorey, H. H.). Wiley Interscience, New York.
Kennedy, J. S. and Moorhouse, J. S. [1969]. *Ent. exp. appl.* **12,** 487.
Kerk, G. J. M. van der [1961]. *Soc. Chem. Ind. Monogr. No. 15,* p. 67.
Kershaw, W. E. [1976]. *Brit. Crop Prot. Council Monogr. 18,* p. 123.
Khazanova, N. E. and Kal'sina, M. V. [1961]. *Inzh-Fiz.Zh., Akad Nauk* **4,** 43.
King, P. H. and McCarty, P. L. [1968]. *Soil Sci.* **106,** 248.
King, L. J., Lambrech, J. A. and Finn, T. P. [1950]. *Contrib. Boyce Thompson Inst.* **16,** 191.
Kinzer, G. D. and Gunn, R. [1951]. *J. Meteorol.* **8,** 71.
Kipling, J. J. [1965]. "Adsorption from Solutions of non-Electrolytes", Chapter 10. Academic Press, London and New York.
Kirkwood, R. C., Dalziel, J., Matlib, A. and Somerville, L. [1968]. *Proc. 9th Brit. Weed Cont. Conf.* p. 651.
Kirkwood, R. C., Dalziel, J., Matlib, A. and Somerville, L. [1972]. *Pestic. Sci.* **3,** 307.
Kissel, D. E., Ritchie, J. T. and Burnett, E. [1973]. *Soil Sci. Soc. Amer. Proc.* **37,** 21.
Klinger, H. [1936]. *Arb. physiol. angew. Ent. Berl.* **3,** 49.
Knake, E. L., Appleby, A. P. and Furtick, W. R. [1967]. *Weeds* **15,** 228.
Knight, B. A. G. and Tomlinson, T. E. [1967]. *J. Soil Sci.* **18,** 233.
Knight, B. A. G., Coutts, J. and Tomlinson, T. E. [1970]. *Soc. Chem. Ind. Monogr. No. 37,* p. 54.
Knight-Jones, E. W. [1953]. *J. exp. Biol.* **30,** 584.
Kokes, R. J. and Long, F. A. [1953]. *J. Amer. Chem. Soc.* **75,** 6142.
Kolattukudy, P. E. [1968]. *Science* **159,** 498.
Kolpakov, F. I. [1963]. *Arkh. Patol.* **25**(6), 38.
Kolpakov, F. I. [1970]. *Vest. Dermatol. Venerol.* **44**(7), 16.
Koren, E. and Ashton, F. M. [1973]. *Weed Sci.* **21,** 241.
Korthüm, G., Vogel, W. and Andrussow, K. [1961]. "Dissociation Constants of Organic Acids in Aqueous Solution", Butterworth, London, for IUPAC.
Kramer, P. J. [1940]. *Plant Physiol.* **15,** 63.
Kramer, P. J. [1956]. "Handb. der Pflanzenphysiol" (ed. Ruhland, W.), Vol. 3, p. 126. Springer, Berlin.
Kramer, E. [1975]. *Proc. 5th Int. Symp. Olfaction and Taste,* **5,** 329. Academic Press, London and New York.
Kramer, E. [1976]. *Physiol. Entomol.* **1,** 27.
Kratky, B. A. and Warren, G. F. [1971]. *Weed Sci.* **19,** 79.
Krueger, H. R. and Casida, J. E. [1957]. *J. econ. Ent.* **50,** 356.
Krueger, H. R. and O'Brien, R. D. [1959]. *J. econ. Ent.* **52,** 1063.
Krueger, H. R., O'Brien, R. D. and Dautermann, W. C. [1960]. *J. econ. Ent.* **53,** 25.
Kruyt, H. R. [1952]. "Colloid Science". Elsevier, New York.
Kuiper, P. J. C. [1964]. *Meded. LandbHoogesch. Wageningen* **64.**

Kuiper, P. J. C. [1972]. *Ann. Rev. Plant. Physiol.* **23,** 157.
Kurtin, S. L., Mead, C. A., Mueller, W. A., Kurtin, B. C. and Wolf, E. D. [1970]. *Science* **167,** 1720.
Kydonieus, A. F., Smith, T. K. and Hyman, Sy. [1975]. *In Proc. 1975 Int. Controlled Release Pestic Symp.* (ed. Harris, F. W.), p. 60.
Lachman, L. and Drubulis, A. [1964]. *J. pharm. Sci.* **53,** 639.
Laffort, P. [1963]. *Arch. Sci. Physiol.* **17,** 75.
Lai, T. M. and Mortland, M. M. [1961]. *Soil Sci. Soc. Amer. Proc.* **25,** 353.
Lai, T. M. and Mortland, M. M. [1962]. *Clays Clay minerals* **9,** 229.
Lake, G. R. [1977]. *Pestic. Sci.* **8,** 515.
Lake, G. R. and Taylor, W. A. [1974]. *Weed Res.* **14,** 13.
Lambert, S. M. [1967]. *J. agric. Fd Chem.* **15,** 572.
Lambert, S. M. [1968]. *J. agric. Fd Chem.* **16,** 340.
Lambert, S. M., Porter, P. E. and Schieferstein, R. H. [1965]. *Weeds* **13,** 185.
Landt, E. and Knop, W. [1932]. *Zeit. phys. Chem.* A **162,** 331.
Langmuir, I. [1917]. *J. Amer. Chem. Soc.* **39,** 1883.
Langmuir, I. [1926]. *In* Alexander's "Colloid Chemistry", N.Y., **1,** 525.
Langmuir, I. and Blodgett, K. [1944]. *Report R.L. 225, Gen. Elec. Res. Lab.*
Lapidus, L. and Amundson, N. R. [1952]. *J. phys. Chem.* **56,** 984.
Larsen, S. [1967]. *Adv. Agronomy* **19,** 151.
Lauren, D. R. and Henzell, R. F. [1977]. *Proc. 30th N.Z. Weed and Pest Cont. Conf.* p. 207.
Lavy, T. L. [1968]. *Soil Sci. Soc. Amer. Proc.* **32,** 377.
Lavy, T. L. [1970]. *Weed Sci.* **18,** 53.
Lee, W. O. [1973]. *Weed Sci.* **21,** 537.
Lee, Beatrice and Priestley, J. H. [1923]. *Ann. Bot.* **38,** 525.
Lees, A. D. [1948]. *Disc. Faraday Soc.* **3,** 187.
Legg, B. J. [1975]. *Quart. J. R. Met. Soc.* **101,** 597.
Legg, B. J. and Long, I. F. [1975]. *Quart. J. R. Met. Soc.* **101,** 611.
Legg, B. J. and Monteith, J. [1975]. *In* "Heat and Mass Transfer in the Biosphere" (eds de Vries, D. A. and Afgan, N. H.), Part 1, p. 67. Wiley, New York.
Leistra, M. [1971]. *Pestic. Sci.* **2,** 75.
Leistra, M. [1973]. *Res. Rev.* **49,** 87.
Leistra, M., Smelt, J. H. and Nollen, H. M. [1974]. *Pestic. Sci.* **5,** 409.
Leonard, O. A. and Crafts, A. S. [1956]. *Hilgardia* **26,** 366.
Lester-Smith, E. [1932]. *J. phys. Chem.* **36,** 1401, 1672, 2455.
Letey, J. and Oddson, J. K. [1972]. *In* "Organic Chemicals in the Soil Environment" (eds Goring, C. A. I. and Hamaker, J. W.), p. 399. Marcel Dekker, New York.
Leuthold, U., Brücher, Ch., and Ebert, E. [1978]. *Weed Res.* **18,** 265.
Lewis, C. T. [1962]. *Nature* **193,** 904.
Lewis, C. T. [1965]. *J. Ins. Physiol.* **11,** 683.
Lewis, C. T. and Hughes, J. C. [1957]. *Bull. ent. Res.* **48,** 755.
Lewis, F. J. [1948]. *Disc. Faraday Soc.* **3,** 159.
Lewis, T. and Macaulay, E. D. [1976]. *Ecol. Ent.* **1,** 175.
Lewis, D. G. and Quirk, J. P. [1965]. *Nature* **205,** 765.
Lichtenstein, E. P. and Schultz, K. R. [1961]. *J. econ. Ent.* **54,** 517.
Lindquist, D. A. and Dahm, P. A. [1956]. *J. econ. Ent.* **49,** 579.
Lindström, O. [1958]. *J. agric. Fd Chem.* **6,** 283.

Lindström, O. [1959]. *J. agric. Fd. Chem.* **7,** 562.
Lindstrom, F. T., Boersma, L. and Gardner, H. [1968]. *Soil Sci.* **106,** 107.
Lindstrom, F. T., Boersma, L. and Stockard, D. [1971]. *Soil Sci.* **112,** 291.
Linnett, J. W. [1970]. *Science* **167,** 1719.
Linscott, D. L. and Hagin, R. D. [1968]. *Weed Sci.* **16,** 114.
Linscott, D. L. and Hagin, R. D. [1969]. *Weed Sci.* **17,** 46.
Linscott, J. J., Burnside, O. C. and Lavy, T. L. [1960]. *Weed Sci.* **17,** 170.
Locke, M. [1964]. *In* "The Physiology of Insecta" (ed. Rockstein, M.), Vol. III, p. 379. Academic Press, London and New York.
Long, D. B. [1958]. *Proc. Roy. ent. Soc. Lond.* **33,** 1.
Long, F. A. and Thompson, L. J. [1954]. *J. Polym. Sci.* **14,** 321.
Longsworth, L. G. [1953]. *J. phys. Chem.* **75,** 5705.
Lonsdale, H. K., Cross, B. P., Graber, F. M. and Milstead, C. E. [1971]. *J. Macromol. Sci-Phys. B* **5,** 167.
Lord, K. A., Jeffs, K. A. and Tuppen, R. J. [1971a]. *Pestic. Sci.* **2,** 49.
Lord, K. A., Jeffs, K. A. and Tuppen, R. J. [1971b]. *Proc. 6th Brit. Insectic. Fungic. Conf.* **1,** 9.
Lubatti, D. F. and Harrison, A. [1944]. *J. Soc. chem. Ind.* **63,** 353.
Luckwill, L. C. and Lloyd-Jones, C. P. [1962]. *J. Hort. Sci.* **37,** 190.
Lukens, R. J. [1971]. "Chemistry of Fungicidal Action". Chapman and Hall, London.
Lukens, R. J. and Horsfall, J. G. [1967]. *Phytopathology* **57,** 876.
Lukens, R. J. and Sisler, H. D. [1958]. *Phytopathology* **48,** 235.
Lusis, M. A. and Ratcliff, G. A. [1968]. *Canad. J. Chem. Eng.* **46,** 385.
Lyndsay, Ruth V. and Hartley, G. S. [1963]. *Weed Res.* **3,** 195.
Lyndsay, Ruth V. and Hartley, G. S. [1966]. *Weed Res.* **6,** 221.
McAuliffe, C. [1966]. *J. phys. Chem.* **70,** 1267.
McBain, M. E. L. and Hutchinson, E. [1955]. "Solubilisation and Related Phenomena". Academic Press, London and New York.
McCall, D. W. and Douglass, D. C. [1967]. *J. phys. Chem.* **71,** 987.
McCallan, S. E. A. [1957]. *Plant Prot. Conf.* 1956, p. 77.
McCready, C. C. [1963]. *New Phytol.* **62,** 3.
McCready, C. C. [1966]. *Ann. Rev. Plant Physiol.* **17,** 283.
MacCuaig, R. D. and Watts, W. S. [1963]. *J. econ. Ent.* **56,** 850.
MacCuaig, R. D. and Yeates, M. N. D. B. [1972]. *Anti-Locust Bull. No. 49.*
McDowell, C. M. and Usher, F. L. [1932]. *Proc. Roy. Soc. A* **138,** 133.
McFarlane, N. R. [1976]. *Proc. Brit. Crop Prot. Council Symp.* "Persistence of Insectic. and Herbic.", p. 241.
McGlamery, M. D. and Slife, F. W. [1966]. *Weeds* **14,** 237.
McGowan, J. C. [1954]. *J. appl. Chem.* **4,** 41.
McGowan, J. C., [1965]. *Rec. Trav. chim. Pays-Bas et Belg.* **84,** 99.
McGowan, J. C. [1968]. *Soc. Chem. Ind. Monogr. No. 29,* p. 141.
McLean, J. R. F. and Dixon, A. T. [1972]. *Proc. N.Z. Weed Pest Contr. Conf.* **25,** 51.
McMahon, M. A. and Thomas, G. W. [1974]. *Soil Sci. Soc. Amer. Proc.* **38,** 727.
McWhorter, C. G. and Barrentine, W. L. [1970]. *Weed Sci.* **18,** 500.
Macey, R. [1948]. *J. Ind. Hyg. Toxicol.* **30,** 140.
Maharajh, D. M. and Walkley, J. [1973]. *Canad. J. Chem.* **51,** 944.
Makepeace, R. J. [1978]. *Brit. Crop. Prot. Council Symp. on Controlled Drop* Appl. p. 249.

Mardles, E. J. W. [1951]. *Brit. J. appl. Phys. Suppl. 1*, p. 7.
Marriott, F. H. C. and Nye, P. H. [1968]. *Trans. 9th Congr. Int. Soil Sci. Soc. Adelaide* **I,** 127.
Marsh, P. B. [1945]. *Phytopathology* **35,** 54.
Marshall, T. J. [1959]. *J. Soil Sci.* **10,** 79.
Marth, P. C. and Mitchell, J. W. [1949]. *Bot. Gaz.* **110,** 632.
Marth, P. C., Davis, F. F. and Mitchell, J. W. [1945]. *Bot. Gaz.* **107,** 129.
Marthe, D. E. [1968]. *Phytopathology* **58,** 1464.
Martin, H. and Worthing, C. R. [1974]. "Pesticide Manual" (4th edn), Brit. Crop Prot. Council.
Martin, J. T. [1960]. *J. Sci. Fd Agric.* **11,** 635.
Martin, H. [1965a]. *Z. Vergl. Physiol.* **48,** 481.
Martin, H. [1965b]. *Nature* **208,** 59.
Martin, Hubert [1973]. "The Scientific Principles of Crop Protection" (6th edn). Edward Arnold, London.
Martin, A. J. D. and Synge, R. L. [1941]. *Biochem. J.* **35,** 1358.
Martin, J. T. and Juniper, B. E. [1970]. "The Cuticles of Plants". Edward Arnold, London.
Martin, R. J. L. and Stott, G. L. [1957]. *Austr. J. Agric. Res.* **8,** 444.
Mason, B. J. [1971]. "Physics of Clouds", Oxford, App. B.
Massini, P. [1958]. *Acta Botan. Neerl.* **7,** 524.
Massini, P. [1961]. *Weed Res.* **1,** 142.
Matano, C. [1933]. *Jap. J. Phys.* **8,** 109.
Matell, M. and Lindenfors, S. [1967]. *Acta Chem. Scand.* **11,** 324.
Matic, M. [1956]. *Biochem. J.* **63,** 168.
Matsumura, F. [1972]. *In* "Environmental Pollution by Pesticides" (ed. Edwards, C. A.), p. 494. Plenum Press, London and New York.
Matsumura, F. [1963]. *J. Insect. Physiol.* **9,** 207.
Matsumura, F. and Dauterman, W. C. [1964]. *Nature* **202,** 1356.
Matthews, G. A. [1978]. *Brit. Crop Prot. Council Symp. on Controlled Drop Appl.* p. 213.
Matthews, G. A. and Mowlam, M. D. [1974]. *Brit. Crop Prot. Council Monogr. No. 11*, p. 44.
Mattson, S. [1931]. *Soil Sci.* **32,** 343.
Maxwell, J. Clark [1890]. *Collected Scientific Papers*, Cambridge, **11,** 625.
Mayer, S. W. and Tompkins, E. R. [1947]. *J. Amer. Chem. Soc.* **69,** 2866.
Mayer, R., Letey, J. and Farmer, W. J. [1974]. *Soil Sci. Soc. Amer. Proc.* **38,** 563.
Meares, P. [1958]. *J. Polymer Sci.* **27,** 391, 405.
Meares, P. [1965]. *J. appl. Polymer Sci.* **9,** 917.
Megalhaes, A. C., Ashton, F. M. and Foy, C. L. [1968]. *Weed Sci.* **16,** 240.
Meggitt, W. F., Aldrich, R. J. and Shaw, W. C. [1956]. *Weeds* **4,** 131.
Mehltretter, C. L., Roth, W. B., Weakley, F. B., McGuire, T. A. and Russell, C. R. [1974]. *Weed Sci.* **22,** 415.
Meikle, R. W. [1972]. *In* "Organic Chemicals in the Soil Environment" (eds Goring, C. A. I. and Hamaker, J. W.), p. 145. Marcel Dekker, New York.
Mellor, R. S. and Salisbury, F. B. [1965]. *Plant Physiol.* **40,** 506.
Menges, R. M. [1964]. *Weeds* **12,** 236.
Merritt, C. R. and Taylor, W. A. [1977]. *Weed Res.* **17,** 241.
Merritt, C. R. and Taylor, W. A. [1978]. *Brit. Crop Prot. Council Monogr. No. 22,* p. 59.
Metcalf, R. L. [1964]. *World Rev. Pest Control* **3,** 28.

Metcalf, R. L., Fukuto, T. R. and March, R. B. [1959]. *J. econ. Ent.* **52,** 44.
Metcalf, R. L., Kapoor, I. P. and Hirwe, A. S. [1971]. *Environ. Sci. Technol.* **5,** 709.
Metcalf, R. L., Kapoor, I. P. and Hirwe, A. S. [1972]. *In* "Degradation of synthetic organic Molecules in the Biosphere", Proc. of a conf., San Francisco, 1971, Nat. Acad. of Sci., Washington, p. 244.
Meyer, K. H. [1937]. *Trans. Faraday Soc.* **33,** 1062.
Meyer, M. [1938]. *Protoplasma* **29,** 552.
Michaels, A. S., Bixler, H. J. and Fein, H. L. [1964]. *J. appl. Phys.* **35,** 3165.
Michaels, A. S., Bixler, H. J. and Hodges, R. M. [1965]. *J. Coll. Sci.* **20,** 1034.
Micks, D. W., Gaddy, N. K. and Chambers, G. V. [1972]. *J. econ. Ent.* **65,** 158.
Middleton, M. R. [1973]. *Proc. 7th Brit. Insectic. Fungic. Conf.* **1,** 357.
Miller, L. and Carman, P. C. [1959]. *Trans. Faraday Soc.* **55,** 1831, 1839.
Miller, E. E. and Klute, A. [1967]. *In* "Irrigation of Agricultural Lands" (eds Hagan, R. M., Haise, H. R. and Edminster, T. W.), Chapter 13. Amer. Soc. of Agronomy, Madison, Wisconsin.
Miller, S. D. and Nalewaja, J. D. [1976]. *Weed Sci.* **24,** 134.
Miller, L. P., McCallan, S. E. A. and Weed, R. M. [1954]. *Proc. Radioisotope Conf.* 381.
Millington, R. J. [1959]. *Science* **130,** 100.
Minshall, W. H. [1954], *Canad. J. Bot.* **32,** 795.
Minshall, W. H. [1969]. *Weed Sci.* **17,** 197.
Miskus, R. P., Andrews, T. L. and Look, M. L. [1969]. *J. agric. Fd Chem.* **17,** 842.
Mitchell, J. W., Smale, B. C. and Preston, W. H. [1959]. *J. Agric. Fd Chem.* **7,** 841.
Mitchell, J. W., Linder, P. J. and Robinson, M. B. [1961]. *Bot. Gaz.* **123,** 134.
Moffitt, Susan and Blackman, G. E. [1972]. *J. exp. Bot.* **23,** 128.
Molen, W. H. van der [1956]. *Soil Sci.* **81,** 19.
Monteith, J. L. [1973]. "Principles of Environmental Physics". Edward Arnold, London.
Moody, K., Kust, C. A. and Buchholtz, K. P. [1970]. *Weed Sci.* **18,** 642.
Moon, P. [1940]. *J. Franklin Inst.* **230,** 583.
Moorby, J. and Squire, H. M. [1963]. *Radiation Botany*, **3,** 95.
Moore, W. and Graham, S. A. [1918]. *J. Agric. Res.* **13,** 523.
Morgan, P. W. and Kuclek, S. L. [1959]. *J. Polymer Sci.* **40,** 299.
Morrison, G. H. and Freiser, H. [1966]. "Solvent Extraction". Wiley, New York.
Morrison, H. L., Rogers, F. T. and Horton, C. W. [1949]. *J. appl. Phys.* **20,** 1027.
Morton, H. L. [1966]. *Weeds* **14,** 136.
Mott, C. J. B. and Nye, P. H. [1968]. *Soil Sci.* **105,** 18.
Mueller, P. and Rudin, D. O. [1967]. *Biochem. biophys. Res. Commun.* **26,** 398.
Müller, G. T. A. and Stokes, R. H. [1957]. *Trans. Faraday Soc.* **53,** 642.
Mullin, J. W. and Cook, T. P. [1965]. *J. appl. Chem.* **15,** 145.
Mullins, L. J. [1954]. *Chem. Rev.* **54,** 289.
Münch, E. [1930]. *Ber. deutsche bot. Ges.* **44,** 68.
Munden, A. R. and Palmer, H. J. [1950]. *J. Textile Inst.* **41,** 609P.
Mysels, K. J. [1969]. *In* "Pesticidal Formulations Research" (ed. Valkenburg, J. W. Van). Adv. in Chem. Series, No. 86. Amer. Chem. Soc.
Mysels, K. J., Shinoda, K. and Frankel, S. [1959]. "Soap Films". Pergamon, Oxford.

Nagakawa, T. and Tori, K. [1960]. *Koll. Zeit.* **168,** 132.
Nakayama, F. S. and Jackson, R. D. [1963]. *Soil Sci. Soc. Amer. Proc.* **27,** 255.
Napper, D. H. and Hunter, R. J. [1972]. *Phys. Chem. Series One* **7,** 241.
Naughton, M. A. and Swinstead, J. M. [1955]. *Nature* **175,** 531.
Nelson, Nancy H. and Faust, S. D. [1969]. *Env. Sci. Technol.* **3,** 1186.
Neogi, A. N. and Allan, G. G. [1974]. *In* "Controlled Release of Biogically Active Agents", *Adv. in Exp. Med. Biol.*, Vol. 47. Plenum Press, New York.
Nernst, W. [1888]. *Z. physikal Chem.* **2,** 613.
Neumann, A. W. [1978]. *In* "Wetting, Spreading and Adhesion" (ed. Padday, J. F.) (for S.C.I.), p. 3. Academic Press, London and New York.
Neville, A. C. [1967]. *Adv. Insect Physiol.* **4,** 213.
Newman, E. I. [1969]. "Resistance to Water Flow in Soil and Plant", Vol. I, 1.
Nielsen, D. R. and Biggar, J. W. [1961]. *Soil Sci. Soc. Amer. Proc.* **25,** 1.
Nishimoto, R. K. and Warren, G. F. [1971]. *Weed Sci.* **19,** 156.
Noble-Nesbitt, J. [1970]. *Pestic. Sci.* **1,** 204.
Nordby, A. [1970]. *Brit. Crop Prot. Monogr. No. 2*, p. 21.
Nordby, A. and Skuterud, R. [1974]. *Weed Res.* **14,** 385.
Norris, R. F. [1973]. *Weed Sci.* **21,** 6.
Norris, R. F. [1975]. Personal communication.
Norris, R. F. and Bukovac, M. J. [1968]. *Am. J. Bot.* **55,** 975.
Norris, R. F. and Bukovac, M. J. [1969]. *Physiol. plant.* **22,** 701.
Noyes, A. A. and Whitney, W. R. [1897]. *J. Amer. Chem. Soc.* **19,** 930.
Nye, P. H. [1967]. *Trans. Congr. Int. Soil Sci. Soc.*, Aberdeen, 317.
Nye, P. H. [1968]. *J. Soil Sci.* **19,** 205.
Nye, P. H. and Marriott, F. H. C. [1969]. *Plant and Soil* **30,** 459.
Nye, P. H. and Spiers, J. A. [1964]. *Trans. 8th Int. Congr. Soil Sci.* **3,** 535.
Nye, P. H. and Tinker, P. B. [1977]. "Solute movement in the Soil-root System". Blackwell Scientific Publications, Oxford.
Oberst, F. W. [1961]. *In* "Inhaled Particles and Vapours" (ed. Davies, C. N.), p. 251. Pergamon Press, London.
O'Brien, R. D. and Danelley, C. E. [1965]. *J. agric. Fd Chem.* **13,** 245.
O'Connor, D. J. [1962]. *J. Water Pollut. Cont. Fed.* **34,** 905.
Oddson, J. K., Letey, J. and Weeks, L. V. [1970]. *Soil Sci. Soc. Amer. Proc.* **34,** 412.
O'Donovan, J. T. and Prendeville, G. N. [1975]. *Weed Res.* **15,** 413.
O'Donovan, J. T. and Prendeville, G. N. [1976]. *Weed Res.* **16,** 331.
Ogarev, V. A., Timonina, T. N., Arslanov, V. V. and Trapeznikov, A. A. [1974]. *J. Adhes.* **6,** 637.
Ogata, A. and Banks, R. B. [1961]. Prof. Paper No. 411-A. U.S. Geol. Survey.
Ogle, R. E. and Warren, G. F. [1954]. *Weeds* **3,** 257.
Ogura, Y. [1971]. *In* "Neuropoisons, their Pathophysiological Actions", Vol. 1, p. 139. Plenum, New York and London.
Olinger, L. D. and Kerr, S. H. [1969]. *J. econ. Ent.* **62,** 403.
O'Loughlin, E. M. [1975]. *Techn. Conf. Aust. Inst. Eng. Hydrol. Symp.*, p. 56.
O'Loughlin, E. M. and Bowmer, Kathleen H. [1975]. *J. Hydrol.* **26,** 217.
Olsen, S. R. and Kemper, W. D. [1968]. *Adv. Agronomy* **20,** 91.
Olson, W. P. and O'Brien, R. D. [1963]. *J. Inst. Physiol.* **9,** 777.
Onsager, L. [1931]. *Phys. Rev.* **37,** 405, **38,** 2265.
Oorschot, J. L. P. van [1970]. *Weed Res.* **10,** 230.

Osgerby, J. M. [1970]. *Soc. Chem. Ind. Monogr. No. 37*, p. 63.
Osgerby, J. M. [1974]. *Pestic. Sci.* **5,** 327.
Osgerby, J. M. [1975]. *Pestic. Sci.* **6,** 675.
O'Toole, M. A. [1965]. *Irish J. Agric. Res.* **4,** 231.
Otto, H. W. and Daines, R. H. [1969]. *Science* **163,** 1209.
Overbeek, J. van [1956]. *Ann. Rev. Plant Physiol.* **7,** 355.
Overbeek, J. van and Blondeau, R. [1954]. *Weeds* **3,** 55.
Overton, E. [1901]. "Studien über Narkose". G. Fischer, Jena.
Owens, R. G. and Miller, L. P. [1957]. *Contrib. Boyce Thompson Inst.* **19,** 177.
Painter, R. H. [1951]. "Insect Resistance in Crop Plants". McMillan, New York.
Pallas Jr., J. E. [1960]. *Plant Physiol.* **35,** 575.
Park, G. S. [1950]. *Trans. Faraday Soc.* **46,** 684.
Park, G. S. [1952]. *Trans. Faraday Soc.* **48,** 11.
Park, K. S. and Bruce, W. N. [1968]. *J. econ. Ent.* **61,** 770.
Parker, C. [1966]. *Weeds,* **14,** 117.
Parmele, L. H., Lemon, E. R. and Taylor, A. W. [1972]. *Water, Air Soil Pollution* **1,** 433.
Parochetti, J. V. and Warren, G. F. [1966]. *Weeds* **14,** 281.
Pasquill, F. [1962]. "Atmospheric Diffusion". Van Nostrand, London.
Passioura, J. B. [1963]. *Plant and Soil* **18,** 225.
Passioura, J. B. [1971]. *Soil Sci.* **111,** 339.
Passioura, J. B. [1976]. *In* "Transport and Transfer Processes in Plants" (eds Wardlaw, I. F. and Passioura, J. B.). Academic Press, New York and London.
Passioura, J. B. and Rose, D. A. [1971]. *Soil Sci.* **111,** 345.
Patel, M., Patel, J. M., and Lemberger, A. P. [1964]. *J. Pharm. Sci.* **53,** 286.
Pedley, J. B. [1976]. Kinetics of solute transfer in emulsions. Ph.D. Thesis, London.
Peel, A. J. [1974]. "Transport of Nutrients in Plants". Butterworth, London.
Pence, R. J. and Viray, M. S. [1969]. *J. econ. Ent.* **62,** 622.
Penman, H. L. [1940]. *J. agric. Sci.* **30,** 437.
Penman, H. L. [1948]. *Proc. Roy. Soc. A* **193,** 120.
Penman, H. L. [1976]. Personal Communication.
Penner, D. [1971]. *Weed Sci.* **19,** 571.
Penniston, J. T., Beckett, L., Bently, D. L. and Hansch, C. [1969]. *Mol. Pharmacol.* **5,** 333.
Pereira, J. F., Splitstoesser, W. E. and Hopen, H. J. [1971]. *Weed Sci.* **19,** 622.
Perrin, D. D. [1965]. "Dissociation Constants of Organic Bases", Butterworth for I.U.P.A.C. [Suppl. 1972].
Perrin, J. [1909]. *Ann. di chim. et de phys.* **18,** 5.
Perry, A. S., Pearce, G. W. and Buckner, A. J. [1964]. *J. econ. Ent.* **57,** 867.
Perry, M. W. [1973]. *Weed Res.* **13,** 325.
Perry, M. W. and Greenway, H. [1973]. *Ann. Bot.* **37,** 225.
Peters, R. H. [1968]. *In* "Diffusion in Polymers" (eds Crank, J. and Park, G. S.), p. 357. Academic Press, London and New York.
Peterson, Carol A. and Edgington, Ll. V. [1976]. *Pestic. Sci.* **7,** 483.
Pfeiffer, R. K. Dewey, O. R. and Brunskill, R. T. [1957]. *IV. Int. Cong. Crop Prot. Hamburg I,* 523.
Philip, J. R. [1957]. *Soil Sci.* **83,** 345, 435.
Philip, J. R. [1957]. *Soil Sci.* **84,** 163, 257, 329.

Philip, J. R. [1958]. *Soil Sci.* **85,** 278.
Phillips, F. T. [1964]. *J. Sci. Fd Agric.* **15,** 444.
Phillips, F. T. [1971]. *Plastic. Sci.* **2,** 255.
Phillips, F. T. [1974]. *Chem. Ind.* 193.
Phillips, F. T. and Gillham, E. M. [1968]. *Rep. Rothamsted exp. Stn.* for 1967, 170.
Phillips, F. T., Etheridge, P. and Scott, G. C. [1976]. *Bull. ent. Res.* **66,** 579.
Phillips, R. E. and Brown, D. A. [1966]. *J. soil Sci.* **17,** 200.
Pilar, F. L. [1960]. *J. Polymer Sci.* **45,** 205.
Piper, W. D. and Maxwell, K. E. [1971]. *J. econ. Ent.* **64,** 601.
Pitre, H. N., Pluenneke, R. H., Bhirud, K. M. and Palmer, S. E. [1972]. *J. econ. Ent.* **65,** 1194.
Plimmer, J. R. [1976]. *In* "Herbicides, Chemistry Degradation and Mode of Action" (eds Kearney, P. C. and Kaufman, D. D.), Vol. 2, p. 892. Marcel Dekker, New York.
Poidevin, N. le [1956]. *Phytochem.* **4,** 525.
Polles, S. G. and Vinson, S. B. [1969]. *J. econ. Ent.* **62,** 89.
Polon, J. A. [1973]. *In* "Pesticide Formulations" (ed. Valkenburg, W. van), Chapter 5. Marcel Dekker, New York.
Porter, L. K., Kemper, W. D., Jackson, R. D. and Stewart, B. A. [1960]. *Soil Sci. Soc. Amer. Proc.* **24,** 460.
Possingham, J. V., Chambers, T. C., Radler, F. and Grncarevic, M. [1967]. *Aust. J. biol. Sci.* **20,** 1149.
Potts, K. T. [1961]. *Chem. Rev.* **61,** 87.
Poulsen, B. J. [1973]. *In* "Drug Design" (ed. Ariens, E. J.), Vol. IV, Chapter 5. Academic Press, New York and London.
Prager, S. and Long, F. A. [1951]. *J. Amer. Chem. Soc.* **73,** 4072.
Prasad, R. and Blackman, G. E. [1965]. *J. exp. Bot.* **16,** 545.
Prasad, R., Foy, C. L. and Crafts, A. S. [1967]. *Weeds* **15,** 149.
Prendeville, G. N. and Warren, G. F. [1975]. *Weed Res.* **15,** 162.
Preston, J. M. and Pal, P. [1947]. *J. Soc. Dyers Colour.* **63,** 430.
Price-Jones, D. [1972]. *In* "Biology in Pest and Disease Control" (eds Price-Jones, D. and Solomon, M.), Blackwell Scientific Publications, Oxford.
Priestley, J. H. [1924]. *Bot. Rev.* **9,** 593.
Que-Hee, S. S. and Sutherland, R. G. [1974]. *Weed Sci.* **22,** 313.
Que-Hee, S. S. and Sutherland, R. G. [1975]. *Weed Sci.* **23,** 119.
Quisenberry, V. L. and Phillips, R. E. [1976]. *Soil Sci. Soc. Amer. Proc.* **40,** 484.
Radler, F. and Horn, D. H. S. [1965]. *Aust. J. Chem.* **18,** 1059.
Rahman, A. and Ashford, R. [1970]. *Weed Sci.* **18,** 754.
Rahn, E. M. and Baynard, R. E. [1958]. *Weeds.* **6,** 432.
Ranz, W. E. and Marshall, W. R. [1952]. *Chem. Eng. Progr.* **48,** 141.
Rawicz, F. M., Cato, D. M. and Rutherford, H. A. [1961]. *Amer. Dyestuff Reptr.* **50,** 320.
Rayleigh, Lord [1916]. *Phil. Mag.* **32,** 529.
Rayner, R. W. [1962]. *Ann. appl. Biol.* **50,** 245.
Razdan, R. K. and Zitko, B. A. [1969]. *Tetrahedron Lett.* (56) 4947.
Read, A. D. and Kitchener, J. A. [1967]. *Soc. Chem. Ind. Monogr. No. 25,* p. 300.
Reinhold, Leonora [1954]. *New Phytol.* **53,** 217.
Rhodes, R. C., Belasco, I. J. and Pease, H. L. [1970]. *J. agric. Fd Chem.* **18,** 524.

Rich, S. and Horsfall, J. G. [1952]. *Phytopathology* **42,** 457.
Richards, A. G. [1951]. "The Integument of Arthropods". Minnesota University Press, St. Paul.
Richards, A. G. [1953]. *In* "Insect Physiology" (ed. Roeder), p. 1. Wiley, New York.
Richards, L. A. [1952]. *Int. Symp. on Desert Res.*, Bull. Res. Council of Israel.
Richardson, R. G. [1975]. *Weed Res.* **15,** 33.
Richardson, R. G. [1977]. *Weed Res.* **17,** 259.
Richmond, D. V. and Somers, E. [1962a]. *Ann. appl. Biol.* **50,** 33.
Richmond, D. V. and Somers, E. [1962b]. *Ann. appl. Biol.* **50,** 45.
Richmond, D. V. and Somers, E. [1966]. *Ann. appl. Biol.* **57,** 231.
Riley, D. [1973]. Personal Communication.
Riley, D. and Hawkins, A. F. [1974]. *Proc. 12th Brit. Control. Conf.* **1,** 193–201.
Ripper, W. E. and Scott, J. K. [1957]. *Z. Pflanzenkr. Pflanzenschutz* **64,** 469.
Ripper, W. E., Greenslade, R. M., Heath, J. and Barker, K. [1948]. *Nature* **161,** 484.
Ripper, W. E., Greenslade, R. M. and Hartley, G. S. [1951]. *J. econ. Ent.* **44,** 448.
Risebrough, R. W. [1968]. *Science* **159,** 1233.
Ritchie, J. T., Kissel, D. E. and Burnett, E. [1972]. *Soil Sci. Soc. Amer. Proc.* **36,** 874.
Ritschel, W. A. [1973]. *In* "Drug Design" (ed. Ariens, E. J.), Vol. IV, Chapters 2 and 3. Academic Press, New York and London.
Ritter, W. F., Johnson, H. P. and Lovely, W. G. [1973]. *Weed Sci.* **21,** 381.
Robbins, W. W., Crafts, A. S. and Raynor, R. N. [1952]. "Weed Control" (2nd edn). McGraw-Hill, New York.
Robertson, M. Margaret, Parham, P. H. and Bukovac, M. J. [1971]. *J. agric. Fd Chem.* **19,** 754.
Robertson, M. Margaret, and Kirkwood, R. C. [1970]. *Weed Res.* **10,** 102.
Rohrbaugh, L. M. and Rice, E. L. [1949]. *Bot. Gaz.* **111,** 85.
Rosano, H. L. [1951]. *Mém. services chim. état (Paris)* **36,** 437.
Rose, D. A. [1973]. *J. Soil Sci.* **24,** 284.
Ross, R. G. and Ludwig, R. A. [1957]. *Canad. J. Bot.* **35,** 65.
Rouse, P. E. [1947]. *J. Amer. Chem. Soc.* **69,** 1068.
Rousseau, D. L. and Porto, S. P. S. [1970]. *Science* **167,** 1715.
Rowell, D. L., Martin, M. W. and Nye, P. H. [1967]. *J. Soil Sci.* **18,** 204.
Rubin, J., Steinhardt, R. and Reiniger, P. [1964]. *Soil Sci. Soc. Amer. Proc.* **28,** 1.
Russell, E. W. [1974]. "Soil Conditions and Plant Growth" (10th edn), p. 89. Longmans, London.
Russell, J. D., Cruz, M. I. and White, J. L. [1968]. *J. agric. Fd Chem.* **16,** 21.
Russell, R. Scott and Possingham, J. V. [1961]. *Progr. in Nuclear Energy*, Series B, Biol. Sci. **3,** 2. Pergamon, Oxford.
Sagar, G. R., Parker, C., Powell, R. G. and Sargent, J. A. [1968]. "Weed Control Handbook" (5th edn) (eds Fryer, J. D. and Evans, S. A.), Vol. 1, pp. 53, 55. Blackwell, Oxford.
Saha, J. G. and Stewart, W. W. A. [1967]. *Canad. J. Sci.* **47,** 79.
Sanders, F. E., Tinker, P. B. and Nye, P. H. [1970]. *Plant and Soil* **34,** 453.
Sands, R. and Bachelard, E. P. [1973a]. *New Phytologist* **72,** 69.
Sands, R. and Bachelard, E. P. [1973b]. *New Phytologist* **72,** 87.

Sargent, J. A. [1966]. *Proc. 8th Brit. Weed Contr. Conf.* **3,** 804.
Sargent, J. A. and Blackman, G. E. [1962]. *J. exp. Bot.* **13,** 348.
Sargent, J. A. and Blackman, G. E. [1965]. *J. exp. Bot.* **16,** 24.
Sargent, J. A. and Blackman, G. E. [1969]. *J. exp. Bot.* **20,** 542.
Sargent, J. A. and Blackman, G. E. [1972]. *J. exp. Bot.* **23,** 830.
Sargent, J. A., Powell, R. G. and Blackman, G. E. [1969]. *J. exp. Bot.* **20,** 426.
Saunders, B. C. [1947]. *Nature* **160,** 179.
Saunders, P. F., Jenner, C. F. and Blackman, G. E. [1966]. *J. exp. Bot.* **17,** 241.
Savigny, C. B. de and Ivy, E. E. [1974]. "Microencapsulation" (ed. Vandegaer, J. E.), p. 89. Plenum Press, New York.
Savory, B. M. [1973]. *Int. Sugar. J.* **75,** 195.
Sawicki, G. C. [1978]. *In* "Wetting, Spreading and Adhesion" (ed. J. F. Padday) (for S.C.I.), p. 361. Academic Press, London and New York.
Scatchard, G. [1931]. *Chem. Rev.* **8,** 329.
Schatzberg, P. [1963]. *J. phys. Chem.* **67,** 776.
Scheibel, E. G. [1954]. *Ind. Eng. Chem.* **46,** 2007.
Scheidegger, A. E. [1955]. *J. appl. Phys.* **25,** 994.
Scheuplein, R. J. and Blank, I. H. [1971]. *Physiological Revs.* **51,** 702.
Scheuplein, R. J. and Blank, I. H. [1973]. *J. invest. Derm.* **60,** 286.
Schieferstein, R. H. and Loomis, W. E. [1959]. *Am. J. Bot.* **46,** 625.
Schmidt, C. H. and Dahm, P. A. [1956]. *J. econ. Ent.* **49,** 729.
Schmidt, E. W. [1924]. *Ber. dtsch. bot. Ges.* **42,** 131.
Schofield, R. K. [1935]. *Trans. 3rd Int. Congr. Soil Sci.* (Oxford), **2,** 37.
Schofield, R. K. [1949]. *Trans. Brit. Ceram. Soc.* **48,** 207.
Schofield, R. K. and Dakshinamurti, C. [1948]. *Disc. Faraday Soc.* **3,** 56.
Schofield, R. K. and Graham-Bryce, I. J. [1960]. *Nature* **188,** 1048.
Schönherr, J. and Bukovac, M. J. [1972]. *Plant Physiol.* 49, 813.
Schönherr, J. and Bukovac, M. J. [1973]. *Planta* **109,** 73.
Schotland, R. M. [1960]. *Disc. Faraday Soc.* **30,** 72.
Schreiber, K. and Lacroix, L. H. [1967]. *Canad. J. Plant Sci.* **47,** 455.
Schulman, J. H. and McRoberts, T. S. [1946]. *Trans. Faraday Soc.* **42B,** 165.
Schwartz, A. M. and Minor, F. W. [1959]. *J. Coll. Sci.* **14,** 572.
Schwetz, B. A., Norris, J. M., Sparschu, G. L., Rowe, V. K., Gehring, P. J., Emerson, J. L. and Gerbig, C. G. [1973]. *In* "Chlorodioxins, Origin and Fate" (ed. Blair, E. H.), Adv. in Chem. Series, No. 120, p. 55. Amer. Chem. Soc.
Scott, A. F., Shoemaker, D. P., Tanner, K. N. and Wendel, J. G. [1948]. *J. chem. Phys.* **16,** 495.
Scott, G. C. [1974]. *Ann. appl. Biol.* **77,** 107.
Scott, H. D., Phillips, R. E. and Paetzold, R. F. [1974]. *Soil Sci. Soc. Amer. Proc.* **38,** 558.
Seaman, D. and Warrington, R. P. [1972]. *Pestic. Sci.* **3,** 799.
Sebba, F. and Briscoe, H. V. A. [1940]. *J. chem. Soc.* p. 106, 128.
Sehlin, R. C., Cussler, E. L. and Evans, D. F. [1975]. *Biochim. Biophys. Acta* **388,** 385.
Seymour, K. G. [1969]. *In* "Pesticide Formulation Research" (ed. Valkenburg, J. W. van), Adv. in Chem. Series, No. 86, p. 135. Amer. Chem. Soc.
Shaw, D. J. [1970]. "Introduction to Colloid and Surface Chemistry". Butterworth, London.
Sharma, M. P. and Vanden Born, W. H. [1970]. *Weed Sci.* **18,** 57.

Sheets, T. J. [1961]. *Weeds* **9,** 1.
Shellhorn, S. J. and Hull, H. M. [1971]. *Weed Sci.* **19,** 102.
Shephard, M. C. [1973]. *Proc. 7th Brit. Insectic. Fungic. Conf.* **III,** 841.
Shinoda, K., Yamanaka, T. and Kinoshita, K. [1959]. *J. phys. Chem.* **63,** 648.
Shone, M. G. T. and Wood, Ann V. [1972]. *J. exp. Bot.* **23,** 141.
Shone, M. G. T. and Wood, Ann V. [1973]. *Proc. 7th Brit. Insectic. Fungic. Conf.* **1,** 151.
Shorey, H. H. [1973]. *Ann. Rev. Entomol.* **18,** 349.
Silva Fernandes, A. M. S. [1965]. *Ann. appl. Biol.* **56,** 297, 305.
Simpson, L. L. [1971]. *In* "Neuropoisons, their Pathophysiological Actions", Vol. 1, p. 303. Plenum, New York and London.
Skopp, J. and Warrick, A. W. [1974]. *Soil Sci. Soc. Amer. Proc.* **38,** 545.
Skoss, J. D. [1955]. *Bot. Gaz.* **117,** 55.
Slade, P. [1965]. *Nature* **207,** 515.
Slade, P. and Bell, E. G. [1966]. *Weed Res.* **6,** 267.
Slatyer, R. O. [1960]. *Bot. Rev.* **26,** 331.
Slatyer, R. O. [1967]. "Plant Water Relationships". Academic Press, New York and London.
Slomka, M. B. [1970]. "Facts about No-Pest DDVP Strips". Shell Chem. Co.
Smelt, J. H. and Leistra, M. [1974]. *Pestic. Sci.* **5,** 401.
Smith, H. H. [1946]. *Bot. Gaz.* **107,** 544.
Smith, L. W. and Foy, C. L. [1966a]. *Weeds* **15,** 67.
Smith, L. W. and Foy, C. L. [1966b]. *J. agric. Fd Chem.* **14,** 117.
Smith, J. M. and Sagar, G. R. [1966]. *Weed Res.* **6,** 314.
Smith, R. and Tanford, C. [1973]. *Proc. Nat. Acad. Sci., U.S.A.* **70,** 289.
Smith, D. D. and Wischmeier, W. H. [1962]. *Adv. Agron.* **14,** 109.
Smith, R. E., Friess, E. T. and Morales, M. F. [1955]. *J. phys. Chem.* **59,** 382.
Smith, A. E., Zukel, J. W., Stone, G. M. and Riddell, J. A. [1959]. *J. agric. Fd Chem.* **7,** 341.
Solel, Z. and Edgington, L. V. [1973]. *Phytopath.* **63,** 505.
Somers, E. [1956]. *Ann. Rep.* (1955) Agr. Hort. Res. Stn., Long Ashton, Bristol, p. 111.
Somers, E. [1957]. *J. Sci. Fd Agric.* **8,** 520.
Somers, E. [1958]. *Canad. J. Bot.* **36,** 997.
Somers, E. [1962]. *Science Progress* Vol. L., No. 198, p. 218.
Somers, E. [1963]. *Ann. appl. Biol.* **51,** 425.
Somers, E. [1968]. *In* "Physico-chemical and Biophysical Factors Affecting the Activity of Pesticides", Monogr. No. 29, p. 243. S.C.I., London.
Somers, E. and Fisher, D. J. [1967]. *J. gen. Microbiol.* **48,** 147.
Somers, E. and Pring, R. J. [1966]. *Ann. appl. Biol.* **58,** 457.
Spanner, D. C. [1970]. *J. exp. Bot.* **21,** 325.
Spencer, W. F. and Cliath, M. M. [1969]. *Env. Sci. Tech.* **3,** 670.
Spencer, W. F. and Cliath, M. M. [1970]. *Soil Sci. Soc. Amer. Proc.* **34,** 574.
Spencer, W. F. and Cliath, M. M. [1972]. *J. agric. Fd Chem.* **20,** 645.
Spencer, W. F. and Cliath, M. M. [1973]. *J. environ. Qual.* **2,** 284.
Spencer, W. F. and Cliath, M. M. [1975]. *In* "Environmental Dynamics of Pesticides" (eds Haque, R. S. and Freed, V. H.), p. 61. Plenum, New York.
Spencer, W. F., Cliath, M. M. and Farmer, W. J. [1969]. *Soil Sci. Soc. Amer. Proc.* **33,** 509.

Spencer, W. F., Farmer, W. J. and Cliath, M. M. [1973]. *Residue Rev.* **49,** 1.
Sprangle, P., Meggitt, W. F. and Penner, D. [1975]. *Weed Sci.* **23,** 229.
Stanley, J. S. and Radley, J. A. [1960]. *Proc. 3rd Int. Cong. Surface Activity* **2,** 464.
Starr, R. I. and Johnson, R. E. [1968]. *J. agric. Fd Chem.* **16,** 411.
Staverman, A. J. and Smit, J. A. M. [1975]. "Phys. Chem. Enriching Topics in Coll. and Surf. Sci.", p. 343. Theorex, La Jolla, California.
Steffens, G. L. and Cathey, H. M. [1969]. *J. agric. Fd Chem.* **17,** 312.
Steiner, L. F., Mitchell, W. C., Harris, E. J., Kozuma, T. T. and Fujimoto, M. S. [1965]. *J. econ. Ent.* **58,** 961.
Stern, O. [1924]. *Z. Elektrochem.* **30,** 508.
Stokes, B. M. [1956]. *Nature* **178,** 801.
Stokes, R. H. [1964]. *J. Amer. Chem. Soc.* **86,** 979.
Strang, R. H. and Rogers, R. L. [1971]. *Weed Sci.* **19,** 355.
Stumpf, P. K. [1965]. *In* "Plant Biochemistry" (eds Bonner, J. and Varner, J. E.), p. 322. Academic Press, London and New York.
Sun, Y. P. [1968]. *J. econ. Ent.* **61,** 949.
Sund, K. A. [1956]. *J. agric. Fd Chem.* **4,** 57.
Sutcliffe, J. F. [1962]. "Mineral Salt Absorption in Plants". Pergamon, Oxford.
Sussman, A. S. and Lowry, R. J. [1955]. *J. Bacteriol.* **70,** 675.
Sutton, O. G. [1953]. "Micrometeorology", McGraw-Hill, New York.
Sutton, D. L. and Bingham, S. W. [1969]. *Weed Sci.* **17,** 431.
Synerholm, M. E. and Zimmerman, P. W. [1947]. *Contrib. Boyce Thomson Inst.* **14,** 369.
Szeicz, F. M., Plapp, F. W. and Vinson, S. B. [1973]. *J. econ. Ent.* **66,** 9.
Taft, R. W. [1956]. *In* "Steric Effects in Organic Chemistry" (ed. Newman, M. S.), p. 556. Wiley, New York.
Talbert, R. E. and Fletchall, O. H. [1965]. *Weeds* **13,** 46.
Talbert, R. E., Smith, D. R. and Frans, R. E. [1971]. *Weed Sci.* **19,** 6.
Tames, R. S. and Hance, R. J. [1969]. *Plant and Soil* **30,** 221.
Tanford, C. [1973]. "The Hydrophobic Effect". Wiley, New York.
Tattersfield, F. and Potter, C. [1943]. *Ann. appl. Biol.* **30,** 259.
Tattersfield, F. and Roberts, A. W. R. [1920]. *J. agric. Sci., Camb.* **10,** 199.
Tawashi, R. [1968]. *Science.* **160,** 76.
Taylor, W. A. and Merritt, C. R. [1975]. *8th Brit. Insectic. Fungic. Conf.* p. 161.
Taylor, R. L., Hermann, D. B. and Kemp, A. R. [1936]. *Ind. Eng. Chem.* **28,** 1225.
Taylor, A. W., Glotfelty, D. E., Glass, B. L., Freeman, H. P. and Edwards, W. M. [1976]. *J. agric. Fd Chem.* **24,** 625.
Taylor, W. A., Merritt, C. R. and Drinkwater, J. A. [1976]. *Weed Res.* **16,** 203.
Terjesen, S. G. [1959]. *Dechema Monogr.* **32,** 190.
Terry, D. L. and McCants, C. B. [1970]. *Soil Sci. Soc. Amer. Proc.* **34,** 371.
Thaine, R. [1969]. *Nature* **222,** 873.
Thaine, R., Probine, M. C. and Dyer, P. Y. [1967]. *J. exp. Bot.* **18,** 110.
Thom, A. S. [1968]. *Quart. J. Roy. Met. Soc.* **94,** 44.
Thom, A. S. [1971]. *Quart. J. Roy. Met. Soc.* **97,** 414.
Thom, A. S. [1972]. *Quart. J. Roy. Met. Soc.* **98,** 124.
Thomas, W. D. E. and Potter, L. [1966]. *J. Sci. Fd Agric.* **17,** 65.

Thompson, A. R. [1973]. *In* "Environmental Pollution by Pesticides" (ed. Edwards, C. A.). Plenum, London and New York.
Thompson, F. B. [1969]. *Chemy. Ind.* p. 1495.
Thompson, A. R. Edwards, C. A., Edwards, M. J. and Beynon, K. I. [1970]. *Pestic. Sci.* **1,** 174.
Thornthwaite, C. W., Mather, J. R. and Nakamura, J. K. [1960]. *Science* **131,** 1015.
Thrower, Stella, L., Hallam, N. D. and Thrower, L. B. [1965]. *Ann. appl. Bot.* **55,** 253.
Tinker, P. B. [1970]. *Soc. Chem. Ind. Monogr. No. 37,* p. 120.
Tomlinson, T. E., Knight, B. A. G., Bastow, A. W. and Heaver, A. A. [1968]. *Soc. Chem. Ind. Monogr. No. 29.* p. 317.
Treherne, J. E. [1957]. *J. Insect. Physiol.* **1,** 178.
Trichell, D. W., Morton, H. L. and Merkle, M. G. [1968]. *Weed Sci.* **16,** 447.
Trofinova, M. G. and Mitrofanov, A. M. [1974]. *Biol. Abs.* **57,** 42435.
Tsvigolou, E. C. and Wallace, J. R. [1972]. *U.S. Env. Prot. Agency* R3-72-012.
Tukey Jr., H. B. [1970]. *Ann. Rev. Plant Physiol.* **21,** 305.
Tukey Jr., H. B. and Morgan, J. V. [1963]. *Physiol. Plant.* 16, 557.
Turner, D. J. [1972]. *Pestic. Sci.* **3,** 323.
Turner, G. A. [1958]. *Chem. Eng. Sci.* **7,** 156.
Turner, D. J. and Loader, M. P. C. [1972]. *Proc. 11th Brit. Weed Cont. Conf.* **2,** 654.
Turner, D. J. and Loader, M. P. C. [1974]. *Proc. 12th Brit. Weed Cont. Conf.* **1,** 117.
Turner, D. J. and Loader, M. P. C. [1975]. *Pestic. Sci.* **6,** 1.
Tyree, M. T. and Dainty, J. [1975]. *Encycl. Plant Physiol.* New Series, **I,** 367, Springer Verlag.
Upchurch, R. P. and Pierce, W. C. [1957]. *Weeds* **5,** 321.
Uhlenbroek, J. H. and Bijloo, J. D. [1958]. *Rec. Trav. chim. Pays. Bas.* **77,** 1004.
Uhlenbroek, J. H. and Bijloo, J. D. [1959]. *Rec. Trav. chim. Pays. Bas.* **78,** 382.
Vaidyanathan, L. V., Drew, M. C. and Nye, P. H. [1968]. *J. Soil Sci.* **19,** 94.
Valkenberg, W. van [1967] *In* "Solvent Properties of Surfactant Solutions" (ed. Shinoda, K.), p. 263. Edward Arnold, London.
Vault, D. de, [1943]. *J. Amer. Chem. Soc.* **65,** 532.
Verloop, A. and Nimmo, W. B. [1969]. *Weed Res.* **9,** 357.
Vermeulen, T. and Hiester, N. K. [1952]. *Ind. Eng. Chem.* **44,** 636.
Verwey, E. J. W. and Overbeek, J. T. G. [1948]. "Theory of Stability of Lyophobic Colloids". Elsevier, New York.
Volmer, M. and Adikhari, G. [1925]. *Z. Physik.* **33,** 320.
Volmer, M. and Estermann, J. [1921]. *Z. Physik.* **7,** 13.
Voros, J. [1963]. *Nature* **199,** 1110.
Vostral, H. J., Buchholtz, H. P. and Kust, C. A. [1970]. *Weed Sci.* **18,** 115.
Vries, D. A. de [1952]. *Meded. Landbouwhoogesch* **52,** 1.
Wade, P. [1954]. *J. Sci. Fd Agric.* **5,** 184.
Wain, R. L. [1964]. *In* "The Physiology and Biochemistry of Herbicides" (ed. Audus, L. J.), Chapter 16. Academic Press, London and New York.
Walker, A. [1971]. *Pestic. Sci.* **2,** 56.
Walker, A. [1972]. *Pestic. Sci.* **3,** 139.

Walker, A. [1973]. *Weed Res.* **13,** 407, 416.
Walker, A. and Crawford, D. V. [1970]. *Weed Res.* **10,** 126.
Walker, A. and Featherstone, R. M. [1973]. *J. exp. Bot.* **24,** 450.
Wang, G. H., Robinson, C. V. and Edelman, I. S. [1953]. *J. Amer. Chem. Soc.* **75,** 466.
Ward, A. F. H. [1946]. *Trans. Faraday Soc.* **42,** 399.
Ward, T. M. and Holly, K. [1966]. *J. coll. interf. Sci.* **22,** 221.
Ward, W. J. and Quinn, J. A. [1965]. *Am. Inst. Chem. Eng. J.* **11,** 1005.
Ward, T. M. and Upchurch, R. P. [1965]. *J. agric. Fd Chem.* **13,** 334.
Wardlaw, I. F. and Passioura, J. B. (eds.) [1976]. "Transport and Transfer Processes in Plants". Academic Press, London and New York.
Washburn, E. W. [1921]. *Phys. Rev.* **17,** 273.
Waters, E. [1950]. *J. Soc. Dyers Colour* **66,** 609.
Watkin, E. M. and Sagar, G. R. [1971]. *Weed Res.* **11,** 1.
Wauchope, D. [1976]. *Rec. Trav. chim.* **71,** 497.
Wax, L. M. and Behrens, R. [1965]. *Weeds* **13,** 107.
Way, M. J. [1959]. *Ann. appl. Biol.* **47,** 783.
Weaver, R. J., Minarik, C. E. and Boyd, F. T. [1946]. *Bot. Gaz.* **107,** 540.
Weber, J. B. [1966]. *Amer. Mineralog.* **51,** 1657.
Weber, J. B. [1970]. *Residue Rev.* **32,** 93.
Weij, H. G. van der [1970]. *Brit. Crop Prot. Council Symp. on Pestic. Appln.*, p. 12.
Weiss, P. W. [1962]. *Austral. Weed. Res. Newsletter* **1,** 19.
Weintraub, R. L., Yeatman, J. N., Brown, J. W., Throne, J. A., Skoss, J. D. and Conover, J. R. [1954]. *Proc. 8th N.E. Weed Contr. Conf.*, p. 5.
Wellman, R. H. and McCallan, S. E. A. [1946]. *Contr. Boyce Thompson Inst.* **14,** 151.
Wenzel, R. N. [1936]. *Ind. Eng. Chem.* **28,** 988.
Wershaw, R. L., Goldberg, M. C. and Pinckney, D. J. [1967]. *Water Resource Res.* **3,** 511.
Wescott, E. W. and Sisler, H. D. [1964]. *Phytopathology* **54,** 1261.
Wetselaar, R. [1962]. *Plant and Soil* **16,** 19.
Wheatley, G. A. [1965]. *12th Int. Cong. Ent.* 556.
Wheatley, G. A. and Hardman, J. A. [1962]. *Rep. Nat. Veg. Res. Sta.* Wellesbourne, for 1961, 52.
Wiesmann, R. [1946]. *Verh. Schweiz. Naturf. Ges.* **126,** 166.
Wigglesworth, V. B. [1957]. *Ann. Rev. Ent.* **2,** 37.
Wigglesworth, V. B. [1965]. "The Principles of Insect Physiology" (6th edn). Methuen, London.
Wiklander, L. [1955]. *In* "Chemistry of the Soil" (ed. Bear, F. E.), p. 107. Reinhold, New York.
Wild, A. [1972]. *J. Soil Sci.* **23,** 315.
Wilke, C. R. [1949]. *Chem. Eng. Progr.* **45,** 219.
Wilke, C. R. and Chang, P. [1955]. *Am. Inst. Chem. Eng. J.* **1,** 264.
Williams, J. D. H. [1968]. *Weed Res.* **8,** 327.
Wilson, J. N. [1940]. *J. Amer. Chem. Soc.* **62,** 1583.
Wilson, B. and Nishimoto, R. K. [1975a]. *Weed Sci.* **23,** 289.
Wilson, B. and Nishimoto, R. K. [1975b]. *Weed Sci.* **23,** 197.
Winsor, P. A. [1948]. *Trans. Faraday Soc.* **44,** 451.

Winsor, P. A. [1954]. "Solvent Properties of Amphiphilic Compounds". Butterworth, London.
Winston, P. W. and Beament, J. W. L. [1969]. *J. exp. Biol.* **50,** 541.
Winteringham, F. P. W., Harrison, A., Jones, C., McGirr, J. and Templeton, W. [1956]. *J. Sci. Fd Agric.* **7,** 214.
Wit, C. T. de and Keulen, H. van, [1972]. Monogr. ISBN 90 220 0417 1. Centre for Ag. Publishing and Documentation, Wageningen.
Witherspoon, P. A. and Saraf, D. N. [1965]. *J. phys. Chem.* **69,** 3752.
Wood, A. L. and Davidson, J. M. [1975]. *Soil Sci. Soc. Amer. Proc.* **39,** 820.
Woods, J. D. and Mason, B. J. [1964]. *Quart. J. Roy. Met. Soc.* **90,** 373.
Woodwell, G. W., Craig, P. P. and Johnson, A. [1971]. *Science* **174,** 1101.
Wootten, N. W. and Sawyer, K. F. [1954]. *Bull. ent. Res.* **45,** 177.
Wright, E. M. and Diamond, J. M. [1969]. *Proc. Roy. Soc. B* **172,** 202, 227.
Wright, R. H. [1958]. *Can. Entomol.* **90,** 81.
Wright, R. H. [1966]. *Can. Entomol.* **98,** 1083.
Wright, R. H. [1964]. "The Science of Smell". Allen and Unwin, London.
Wurster, D. A. [1959]. *J. Amer. Pharm. Assoc.*, Scientific Edition.
Wurtz, A. [1961]. *Melliand Text. Ber.* **42,** 913.
Yamada, Y., Wittwer, S. H. and Bukovac, M. J. [1964]. *Plant Physiol. (Lancs).* **39,** 28.
Yamaguchi, S. and Crafts, A. S. [1958]. *Hilgardia* **28,** 161.
Yarwood, C. E. [1950]. *Phytopathology* **40,** 971.
Yates, W. E. and Akesson, N. B. [1973]. *In* "Pesticide Formulations" (ed. Valkenburg, W. van), p. 275. Marcel Dekker, New York.
Youngson, C. R. and Goring, C. A. I. [1962]. *Soil Sci.* **93,** 603.
Youngson, C. R., Baker, R. C. and Goring, C. A. I. [1962]. *J. agric. Fd Chem.* **10,** 21.
Zechmeister, L. and Sease, J. W. [1947]. *J. Am. Chem Soc.* **69,** 273.
Ziegler, H. [1975]. *Encycl. of Plant Physiol.*, New Series, **I,** 92. Springer Verlag.
Zimmermann, M. H. and Milburn, J. A. [1975]. *Encycl. of Plant Physiol.*, New Series, **I**. Springer Verlag.
Zisman, W. A. [1964]. Adv. in Chem. Series, No. 43, p. 1. Amer. Chem. Soc.
Zobel, H. F., Hellman, N. N. and Senti, F. R. [1955]. *J. Amer. Oil Chemists Soc.* **32,** 706.
Zschintzsh, J., O'Brien, R. D. and Smith, E. H. [1965]. *J. econ. Ent.* **58,** 614.
Zweep, W. van der [1965]. *Z. Pflanzenkr. Pflanzenschutz* **3,** 123.

Index

Only pesticides which are nominated in the main text are listed in the index. (Appendix 4 gives a more extensive list (alphabetical by accepted common names) of pesticides for which physico-chemical data are available and Appendix 9 gives their structural formulae). Many other chemicals appear in the text but only those used to illustrate important aspects of behaviour appear in the index.

A

Accumulation of atrazine near roots, 273
  of fungicides by fungi, 686–697
  of solutes at barriers, 602, 622, 630
  of solutes at evaporating surfaces, 203
Acetone
  exchange of between air and water, 198
  permeation of cellulose acetate by, 490
  permeation of plant cuticles by, 575
  solubility in mixtures of water and, 41, 85
Acrolein, 396
Acromyrmex sp., 741
Activation energy, 488
Activity coefficient, 94
Additives, effect on penetration, 646–697
Adhesion of airborne particles, 370
  energy of, 406–409
  of particles,
    effect of liquid, 467, 801
    effect of shape, 802
    effect of size, 468, 717
  of pesticides to seeds, 705, 851
  selective, 803–809
Adhesives in formulations, 809–810
  weathering of, 805–806
Adsorption on active charcoal, 76, 747, 759, 850
  amphipathic, 70–72
  in aquatic use, 392
  Donnan-Mattson, 95–100
  double layer, 87–92
  and partition, 65–70
  reversibility of, 81–84
  of salts, 84–106
  on soil, 238–241
  and solution of vapour, 18–22
Aedes aegypti, 713, 721
Aeration of rivers, 195
Agar block technique, 578, 584
Agropyron repens, 651
Airborne pesticides, collection of, 365–379
Alcohol, *see* Methanol, Ethanol etc.
Aldicarb, chemical formulation, 231
Aldrin (HEOD) *vs* soil insects, 701
  evaporation from free surface, 355
  evaporation from disturbed soil, 222
  solubility in water, 30
Alfalfa straw, effect on DDT persistence, 222
Alginate in formulations for aquatic use, 389
Alternaria tenuis, 690
Ametryne, solubility and SVC, 49
  effect in shoot zone, 760
Amiben, root uptake, 598
Amine oxide wetting agents, 430, 654
Amine stearate formulations, 727
Aminotriazole, anomalous base strength, 103, 894
  transport in plants, 621, 656
  as tracer for drift, 813

Ammonia, exchange of between air and water, 198
corrosion of Zn alloys by, 780
Ammonium salts of active acids, 223, 780
salts, effect on penetration, 654
soaps, as transient emulsifiers, 808
Amphipathy, 44, 71, 428
Anemotaxis, 743, 746
Anemomenotaxis, 745
Angle, *see also* Contact
of spray incidence, significance of, 788–792, 937.
of receiving surface
effect on retention, 444
effect on spread, 420
Anopheles
gambiae, 721
stephensii, 714
Antidotes to herbicides, 850
Anthonomus grandis, 732
Anthracene solubility, 38
Antonow's rule, 405
Aphis spp., 839
tapping of phloem by, 616, 631
fabae, 631
Apis mellifera, 730
Apoplast, symplast and free space, 557–559
Apple fruit cuticle, 548, 568
Applicance, definition, 7
control of, 781–792
Application methods, 765–774
of pesticides to soil, 820–848
Aprocarb, 721
Apricot leaf cuticle, 578
Arrestants, 709, 743
Arrhenius activation energy, 488
Arsenate, humectant effect on, 648
Asparagus beetle, 686
Aspergillus sp., 685
niger, 689
oryzae, 694
Aspirin, rate of solution of, 837
Asymmetry of permeation, 505–513, 594
Atrazine effect of temperature, 644
accumulation near roots, 273
adsorption on humus, 83
SVC, 49
Atta spp., 741
Attractants, 742, 755
Autophobic liquids, 419, 456
Availability, selective, 811
effect on of adsorption and movement, 315–320
Availance, definition, 8
estimation of, 128–138, 255
Avena sp. hypocotyl uptake, 589, 599

B

Baffles, oblique, to assist separation, 258
Baits, 774
carrying of, 741
banana, attraction of flies by, 744
Band spraying, 766
BAP (mixed butyl acid phosphates), 656
Barnacles, 392, 743
Base exchange and adsorption, 95–101
of cutin, 551
Basic salts, 104
Beans, *see* Vicia, Phaseolus, Glycine
Bees, 729, 754
Behaviour-controlling chemicals, 740–756
Benzene, v.p.-T in liquid and solid, 36
water solubility, 28, 394
Benzoic acid, triethanolamine salt, 640
acids, chlorinated, permeation of, 573
see also TBA
Benzoylprop-ethyl, 227, 233
BET (Brünauer, Emmett and Teller) isotherm, 74
γ BHC, 714
evaporation of, 353, 358
Biophase, 59
Bipyridilium herbicides
strong adsorption of, 102, 268
translocation of, 635
Blockage, of filters, 767
of pores, 493, 562
Bloom, 411
Bombyx mori, 746
Bordeaux mixture, 684, 810

Botrytis spp., 685
control by soap, 433
fabae, 682, 739
Botulinum toxin, 686
Boundary layers, 334
effective thickness, 338
laminar, transport across, 338–340
atmospheric, transport across, 341–349
Brenner number, 293
Bread mould, 686
Brown rot, 686
Brønstead principle, 62
Bunsen's law, 24
nButanol, uptake by cuticle, 568

C

Caking of powders, 775
Caliphora sp., 505
Capacity, in permeation cf. electrical, 486
factor, in diffusion, 123, 143
Capillary rise, in soil, 80, 296
in tubes, 415
Capping of soils, 282
Captan, 684, 694
Carbamate insecticides, 716, 720
see also individual names
Carbaryl, 231, 725
Carbendazim, 686, 694
Carbofuran, 231
Carbon dioxide, in atmosphere, 926
acidifying effect, 808
effect on translocation, 637
Carbon disulphide, effect on air density, 164
partition water/air, 246
Carbon tetrachloride, effect on air density, 164
in grain silos, 261
Carbophenothion, 707
Carboxyethyl cellulose, 438
Carrot, root uptake in, 599
Cationic fungicides, 432, 691
wetting agents, 432, 780
Caustic soda, as root killer, 760
CDA (Controlled Drop Application), 452, 734, 814
Cellosolve, 650
Cellulose, 61, 64, 548, 721
permeability of, 486
Cellulose acetate, permeability, 490, 608
as leaf surface model, 553
Cellulose esters of active acids, 865
Cerium, 692
Cetylpyridinium chloride, solubilization by, 41, 45
Charcoal, active
adsorption properties, 76
seed protection by, 759, 850
in testing repellent action, 747
Chironomids, 392
Chitin, 61, 689
Chlorfenvinphos, 707
soil surface run-off, 329
Chloride ions, non-adsorbed solute, 291, 295, 304, 314
Chloroacetamides, *N*-alkyl, 578
Chloroform, special features as solvent, 38, 549, 649
Chloropicrin, cf. methyl bromide, 180
partition water/air, 246
adsorption on grain, 262
Chlorpropham, with oil on leaf, 651
root uptake 598, 760, 761
Chlorotriazines, decay assisted by adsorption, 212
Chlorthal methyl, *see* DCPA
Chlorthiamid, 217
evaporation, 358
Cholesterol, 433
Chromatography, of soil samples, 299
leaching through soil considered as, 302–313
Citronella, oil of, 747
Clapyron-Clausius equation eq (1), p. 15, 271
Cleaning, redistribution by, in insects, 723, 724
Coalescence of drops, 444–451
Coccinelids, 729
Cockroach, American, 752
Coleoptyle node, 758, 761
Coleus blumei, 639
Collection of particles from air, 365–370
of vapour, 374–379

Compatibility of formulations, 780
Concentration, in rivers, 395–399
  of saturated vapour, *see* SVC
  see also Pulse forms, Gradients
Concertina pores, diffusion through, 120
  wetting of, 459
Condensed phases, 14
Conduction of electricity cf. diffusion, 113, 486
  of heat cf. diffusion, 167, 361
Contact of granules and soil water, 840
Contact angle, 409
  on rough surfaces, 410–414, 460
  advancing and receding, 443, 567
  apparent, 422
  see also Autophobic liquids
Controlled release, advantages, 857–860
  physical methods, 860–864
  chemical methods, 864–866
Convection in porous beds, 258
  within drops, 807
Convolvulus cneorum, 456
  arvensis, 640
Cooling, adiabatic, 162
Copepods, 392
Copper, as root poison, 760
  fungicides, uptake, 692
    coverage, 442, 736
    sedimentation, 463
Coprosma repens leaf cuticle, 550
Corncob granules, 772
Corrosion, 779
Costelytra zealandica, 719
Cotton, 757, 760
Cotton stainers, 723
Crawling, pick-up during, 725–727
Coverage, by sprays with limited spread, 439–442, 932
  by microgranules, 820
Cross-linking of macro molecules, 60–65, 864
Crystallinity of fats, reduction of permeability by, 561
Crystallization in sprays, effect on deposits, 467
  in polymers
    effect on solubility, 64
    effect on permeability, 500
Cucumber, 651
Cucurbits, phloem transport in, 617
Cuelure, 755
Cuphea platycentra, 639
Curtain, anti-drift, 813
Cuticle
  of higher plants, 547–552
    swelling of, 566–568
    permeation by water, 563–566
    permeation by other substances, 569–579
    see also individual species
  of insects, 659–662
    penetration of, 662–679
Cutin, 550–553
Cyanatrin, 401
Cyclohexane, uptake by cuticle, 568
Cycloheximide, 686, 694
Cynodon dactylon, 640
Cypermethrin, 719
Cyperus rotundus, 634, 651
Cytoplasmic membrane, see Plasmalemma

D

2,4–D
  entry into leaves, 589, 640, 647, 651
  esters, volatile, 357, 363, 374
  esters of poly alcohols, 392, 865
  permeation of tomato fruit cuticle, 577
  translocation, 613, 621, 635
  run-off from soil, 330
Dalapon, uptake, 641, 646
Darcy's law, 276, 277
2,4-DB, progenitor of 2,4-D, 233, 859
  washing out, 595
DCPA, 760
DDD, 714
DDT
  biodegradable analogues, 221
  degradation, 222
  diffusion in polymers, 489, 491
  distribution in soil, 242
  oil formulation, 776
  pick-up from deposits, 713–720
  transport in dust, 327
  toxicity to housefly, 686

use in water, 392, 394
Deactivation of acid centres, 777
Debye-Hückel theory, 90
Decamethrin, 686, 719
Decanoate (Caprate), methyl, 468
Decay, chemical, effect on transfer, 168–181, 184–187, 397
Deliquescence of dusts, 768
Demeton, 220
Density, of water and air, effect of pesticides and temperature, 163–168
see also Population
Deposits, evaporation of non-uniform, 353–358
see also Pick-up, Residues
Depth protection, 315, 319, 757
Detection of gradients, 749–754
Diallate, 761
Diallyl dichloro acetamide, 850
Diazinon, 270
1,2-Dibromo-3-chloropropane, 246
Dicamba, 330
Dicarboxylic acids, in cutin, 551
as hydrophilic substituents, 226
Diclobenil
distribution in soil, 242
evaporation, 358
exchange between water and air, 197
formed from chlorthiamid, 217
Dichlofluanid, 682
Dichlone, 682
dichlorpropene, 242, 246, 263
Dichlorvos
density vapour, 164
evaporation, 358
formed from trichlorphon, 219, 225
slow release strips, 862
Dieldrin (HEOD)
diffusion in soil, 271
evaporation, 355
pick-up from deposits, 714–720
run-off from soil, 329
seed treatment, 701, 705, 707
solubility in water, 30
Diffusion, *see also* Permeation, Decay
accelerated, 617
eddy, 341–349, *see also* Transfer coefficients
equations, 137–146
into limited volume 141, 823–834
molecular, coefficients in gas, liquid and solid, 871–881
across interfaces, 147–152
along surfaces, 265, 426
of matter and temperature, 167
near surfaces, 192–198
in porous media, 119–126, 245–276
perpendicular to flow, 160
with or against flow, 156–160
Dimefox, 229, 633
Dimethirimol, 683
Dimethoate, in soil, 242, 267, 269, 274
formulation, 777
Dimethylphthalate (DMP), 747
Dimethyl sulphoxide (DMSO), 515, 649
Dinocap, 455
Dinonylphenol ethylene oxide condensate, 390
Dinoseb (DNBP), 635, 651, 758, 776
Diquat, 102, 389, 401, 636
Diphenamid, 320, 760
Diphenylacetic acid and amides, 577
Dipoles, 17
Disintegration of granules, 838
Dispersion and localization of spray, 765
Dispersion, hydrodynamic, 289
Displacement of liquid in pores, 839
Distillation and diffusive evaporation, 199–201
Distribution of particles in air, 371–374
Disulfoton, in soils, 220, 242, 270
evaporation, 358
Diuron
decay, 213
in dry ditches, 392
root uptake, 599, 607
shoot uptake, 760
DNOC
bacterial decay, 211
uptake, 642
wind blow, 803
Dodecyl dimethyl ammonium ion, 432
Dodecyl alcohol. 468

Dodine, 432, 688
Donnan-Mattson adsorption, 95–100
Dose,
  definition, 7, 781
  measures of, 6–9
  received by individuals, variation of, 3–5
    relation to availance and applicance, 132–136, 676–678
    effect of chemical decay on, 170–177
Double-layer, ionic, 88–91
Drainage, of continuous film, 433–438
  from incomplete wetting, 442–451
Drazoxolon, 682
Drift, factors affecting, 379–385
  control of, 812–820
Drops, population density, 932
  floating, 403–405
  sliding, 426, 442–444
  spreading, 415–427
  see also Reflection
Drosophila melanogaster, 748
Dust, formulation, 768
  drift of, 819
  see also Collection
Dynamic catch, *see* Impaction
Dysdercus fasciatus, 723

E

Echinochloa crus-galli, 761
Ectodesmata, 554
Eddies, 194, *see also* Diffusion
Eelworm, potato, 263
Efficiency of transfer in reactive systems, 169, 535
Effective height of theoretical plate (EHTP), 311
Electro-osmosis mechanism in phloem, 619
Electrostatic attraction, 798–800
  dispensers, 800
Elodea canadensis, 636
Emulsifiable concentrates (e.c.), 767
Emulsion glues, 809
Emulsions,
  invert, 817
  unstable, 468, 808
Encapsulation, 860, 868, 870
Energy
  of adhesion and cohesion, 406–409
  kinetic, of impact, 441
Enrichment, of microorganism population, 210
Entry of water into soil, 281–284
EPTC, 264, 363, 598, 758, 760, 850
Evaporation
  from porous material, 202
  of spray drops, 358–363
  from solutions, 192–199
  from soil surface, 846–849
  see also Vapour, Deposits, Distillation

F

Fall-out, 382
Feronia madida, 723
Fick's law, 113
  application in porous media, 124
Film, wetting, 419
Filters, 767
Flocculation, 767
Flamprop, ethyl and isopropyl, 234
Flow
  chemical decay during, 184–187
  in phloem, 616–619, 632
  simultaneous diffusion and, 156–161
  near surfaces, 333–337
  transport by, 152–156, 288–313
  of water in soil, 276–288
  see also Convection; Dispersion,
Fluometuron, 304, 309
Fluorescent particles, 733
Fluoroacetic acid, 232
Foam, 391, 773
Folpet, 684
Forces
  close-range interparticle, 92–95
  competing, in adsorption, 67
  coulombic, 87–92
  intermolecular, 16
    contribution to solubility, 37
Formulation
  chemical, 214–235
  for controlled release, 860–869
  limiting factors in, 774–781
  for spray, 766
Formulations, 764–774

Fragaria leaf cuticle, 548
Free path, 13
Freeze drying, 614
Freundlich isotherm, 75, 122, 306
Fruit flies, 744
Fumigants, 242, 244–263, 768
  effect of vapour density of, 257–263
Fungi, penetration into, 679–697
Fungicides
  accumulation in fungi, 686–697
  redistribution by rain, 739

G

Gaussian error curve, 139
Gelatin, 869
Generation of active compound, 182–184, 214–221
Geometrical factors in coverage, leaf-shading, 451–453
  mm scale, 453–461
Gibbs-Duhem equation, 23
Glass
  evaporation from, 587
  pick-up from, 714, 721
Glossina palpalis, 713
Glycerol, 648
Glycine (amino acetic acid), 893,
Glycine max (Soyabean), 598, 634, 758
Glycol, 607
Glyodin, 432
Glyphosate, 105, 640, 656
Gouy double-layer, 89
Gradients of concentration, diffusion down, 111–203
  sensory response to, 746–754
  equilibrium, 874
Granules
  disintegration of, 835
  evaporation from, 358, 848
  formulation of, 770–773
  release from, 821–849, 860–869
  use in rivers, 391
  micro, 393, 772
Grape cuticle, 549
Grapes, drying of, 562
Grasses
  as collectors of particles, 373
  oil formulations on, 652
Grass grub, 719
Gravity, significance for spread of drops, 420–422
  see also Settlement, Sedimentation
Grease, 660
Group factors in partition, 54
Growth
  of plants in relation to uptake, 757–762
  of roots, rate of, 943
Guard cells, 559

H

Half-life, 209
Haloxidine, 219, 297
Hansch approach, 520–522
Harkins spreading coefficient, 405
HCN, 255
Heat,
  exchange, analogy with xylem/phloem, 627
  latent, of vaporization, 17
    of fusion, 35
  specific, 167
Hedera helix leaf cuticle, 505, 563, 564, 634
Heliothis zea, 732
Helmholtz double-layer, 88
Henry's law, 24
HEOD, *see* Dieldrin
Heptachlor, solubility in water, 30
  evaporation from soil, 222
Hexadecanol, effect on air density, 164
  retarding water evaporation, 503
HHDN, *see* Aldrin
Honeydew, 632, 648
Housefly, 686
Human, 686
Humectants, 647
Humidity, effect
  on adsorption, 68, 321, 720
  on pick-up, 720–722
  on uptake by leaves, 639–642
  on uptake by roots, 642–646
  see also Temperature, wet-bulb
Hydraulic conductivity, 279–281
Hydrofugal reactions, 226
Hydrogen, diffusion in soils, 250
Hydrogen bonds, 17, 55
  and adsorption, 68, 108

Hydrolysis
limiting formulations, 778
of macromolecular esters of active acids, 394, 865
Hydrophilicity, 106
Hydrophobic bond, 107

I

Ilex vomitoria, 596
2-Imidazoline,-2-*n*-alkyl, 686, 687, 694
Impaction, 367–371
Infiltration capacity, 281
Insects
interception of sprays by flying, 730–735
and pathogens attacking aerial parts, 710–735
pick-up by crawling, 725–727
pick-up by walking, 713–720
sites of uptake on, 722–725
uptake into, from vapour, 676
Instant coffee, 768
Interception of spray in relation to angle, 937
Inversion, atmospheric, 112, 163, 383
Interface
resistance to diffusion in, 147
surfactant effect on, 199
see also Surface replacement
Iodine, as model for vapour collection, 344, 379
Ionization constant, 86, 887, 895
Ipomea lacunosa, 648
hederifolia, 651
Iso paraffins, 651
Isopropanol, 649

K

Kaolin, 715
Keratin, 65
Kieselguhr, 777
Klinotaxis, 746
Kozeny–Karman equation, 280

L

Lag (time) in permeation, 143, 486
Lagarosiphon major, 389
Lamium amplexicaule leaf cuticle, 548
Langmuir isotherm, 19, 73
Latex, 556
Layer, unstirred, 189
Layering in fluids, 164
Leaching, 293, 295, 320
from discrete sources, 835–837
Lead (Thorium B) as model for vapour collection, 344, 379
Leaf inclination, 789, 937
Lecithin, 433
Lemna minor, 646
Leptohylemyia coarctata, 700
Lettuce, 599, 631
Light, effect on uptake, 633–637
Lime sulphur, 682, 780
Lindane (γBHC)
in soil
distribution, 242
movement, 304
effect of moisture, 322, 325
action on insects, 701, 707
effect on air density
Lines of force, diffusion equivalent, 846
Linseed, 759, 822
Linuron
distribution in soil, 242
as example of dipole interaction, 51
leaf uptake with isoparaffins, 651
root uptake, 598, 607
temperature effect on, 644
zone of effect after leaching, 320
Lipoid phase, choice of standard, 57
Liquids mutual solubility of, 23–28
as adhesives, 801
Local concentrations, relation to activity coefficient, 94
see also Double-layer
Locust swarms, 732
Loss (by evaporation) from surface application, 846–849
Lycopodium spores, as model dust, 373

M

Malathion, 230, 719
Maleic hydrazide, 621
Macromolecular substances
diffusion in, 500–502, 881

as coatings, 869
solubility of and in, 60–65
Mancozeb, 682
Maneb, 682
Mannitol, 607
Marginal regeneration, 437
Markers, for spray swath, 773
MCPA, as weak acid, 86
humidity and solubility of salts, 641, 776
localized root action, 759
MCPB, 233
Meerschaum, 777
Melon fly, 755
Melting point, 35
a major factor in solubility, 37
and solubility limits in formulation, 775
Mercury diffusion of vapour, 873
diffusion in surface of solid, 426
evaporation, 353
toxicity, 686
Metering, 781–784
Metham sodium, 182, 216
Methamidophos, 227
Methanol, swelling of cuticle by, 568
permeation of cuticle by, 575
Methoxychlor, 714
Methoxyphenoxy acetic acid, 633
Methyl bromide, vapour density, 164
escape from soil, 180
partition water/air, 246
Methyl cellulose, 438
Methylene blue, as model for cation uptake, 690
Methylisothiocyanate, 181, 216, 246
Methyl orange, permeation by, 566
Metribuzin, 266
Metrosideros excelsa leaf cuticle, 548, 568
Micelles, critical concentration for (CMC), 437
Michaelis–Menten equation, 208
Microencapsulation, 728, 868
Microgranules, 772, 824
Mildew,
apple, 455
cucumber, 683
Milk powder, 768
Mineral extenders, 776
Molasses
crystal inhibition by, 465
as humectant, 648
Molinate, 761
Morning glory, 651
Morphology of higher plants, 547–561
Mosaic surfaces, 413
Mosquito larvae, 389
Movement, of pesticides to soil surface, 320–325
Mud blocks, availability of insecticides on surface, 714
Mud, bottom, in rivers, retention in, 391
Mulching, 282
Musca domestica 721, 732
see also House fly

N

Naphthalene
evaporation, 352
as fungicide, 681
solubility, 37
Naphthalene acetic acid, 637
Nerium oleander leaf cuticle, 548, 568
Neurospora crassa, 681, 690, 694
sitophila, 689
tetrasperma, 690
Nicotine, 712
Nitralin, 760
Nitrate ion, non-adsorbed solute, 295
Nitropyrazoles, alkyl, 687
Noise level limiting gradient perception, 750
Notostira erratica, 723
NPA (Naptalam) leaching, 314
Nozzles
fan and swirl, 784
solid cone, 791
spacing of, 786
Nutrient additives to systemic sprays, 654–657

O

Octadecanol, retarding water evaporation, 503
Oats, 646

wild in cultivated, 850
see also Avena
Octyloxyethanol, volatile wetting agent, 438, 651
Oil
citronella, 747
enhancing herbicidal action, 651
olive, 687
solutions, pick-up by insects, 715
Oleanolic acid, 549, 563
Oleyl alcohol (octadecenol), 684, 688
Order of reaction, 208
Organic matter of soils,
adsorption on, 238–240
models for, 66, 70
Organomercury fungicides, 681
Orientation of long molecules, 501–502
see also Paraffinic monolayers
Osmosis, 482–485, 607, 619
raverse, 608
Osmotic pressure, 619, 645
Oxalate, for cuticle separation
see Dibasic acids
Oxathiins, 686, 693, 694
β-oxidation, 232
Oxygen exchange water/air, 197

P

Paraffinic monolayers, 502–505
Paraquat, strong adsorption of, 102, 301
effect of light on action, 636
Paraoxon, 686
Parathion, 686
Particle size
deposition and drift, 380–383
and impaction, 368
and pick-up, 713
and retention, 820
and sedimentation, 367
shape and adhesion, 468, 718, 802
shape and pick-up, 717
Partition coefficient
definition, 22
log. of additive, 50–54
independent of m.p., 49
optimum values, 522–541
and adsorption, 65–70
between phases, 47–58
between vapour and solution in soil, 241
of salts, 86
Paspalum dilatatum, 761
Pea, 429, 589, 599, 759, 822
Peach fruit cuticle, 553, 566, 568
Pear leaf cuticle, 637
Pectin, 548, 558
Pectinophora gossypiella, 744
Pelleting of seeds, 856
Penetration
of liquid into pores, 458–461, 845
of liquid and vapour under equivalent pressure, 479–482
of liquid in and vapour out, 512
see also Infiltration
in molecular dispersion, used synonymously with permeation, q.v.
Penicillin, 686
Penicillium italicum, 690
atrovenetum, 694
Pentachlor phenol, 681
Peristalsis in phloem, 618, 621
Permeation, *see also* Penetration, Swelling, Uptake
anomalous, 493–502
asymmetric, 505–514
dependence on
concentration, 489
molecule size and shape, 488
temperature, 488
from different solvents, 514–518, 669–672
of incompatible substances, 491–493
into fungi, 679–686
into insects, 659–679
of leaf cuticle by water, 563–569
by other substances, 569–579
measurement, 473–476
mechanisms of, 477–493
modification by chemical change, 223–228
of paraffinic monolayers, 502–505
Permeability, definition, 473
Permeance, definition, 473
Peroxide links in cutin, 551
Persistence modification by chemical change, 215–223

Phaseolus sp.
  autography of phloem, 614
  uptake of 2,4-D into leaves, 589, 635
  depth protection, 758
  leaf cuticle, 548
  leaf uptake of dalapon, 641
Phases of matter, 13
  gaseous and condensed, 12–14
  in soil, distribution of pesticides between, 242
Phenanthrene, solubility, 38
Phenoxyacetic acids, chloro, permeation by, 573, 577
Phenoxyacetic acid, permeation of tomato fruit cuticle, 485
Phenoxybutyric acids, uptake, 233
  permeation of cuticle by, 578
Phenyl mercury chloride, 682
Phloem, 557
  transport in, 616–633
pH of plant tissues, 615
  and adsorption, 98
  and partition of weak acids, 86
Phorate, 220, 631, 633
Phormia teraenovae, 427, 716, 722
Phosphate esters, effect on permeation, 655
Photo decomposition, 205, 384
Phytophagus larvae, 726
Phytophthora infestans, 442, 685, 736
  parasitica, 681
Pieris brassicae, 725
Pick-up, *see* Insects
Picloram
  distribution in soil, 242
  uptake
    effect of DMSO, 649
    effect of ammonium sulphate, 656
    effect of temperature, 646
Pisum, *see* Pea
Plasmalemma (cytoplasmic membrane), 558, 559, 679, 683, 693
Plasmopara viticola, 683
Plaster surface, 714
Plywood surface, 714, 720
Poaching of soils, 326
Poiseuille's equation, 154
Pollinators, 729
Polybutenes, 705, 854
Polyethylene
  adsorption by, 377
  permeation of, 486, 503, 565
Polyethylene glycol, 648, 650
Polymers
  as coating, 863
  interfacial, 869
  see also Macromolecular substances
Poly(vinyl acetate), 61, 495, 512
Population density
  of drops, 726, 932
  of granules, 823
  of weeds, 762
Population surveys, 754
Pores, *see also* Concertina
  blind, 123
  crumb, and voids, 248, 292
  divergent, 497
  displacement of liquid from, 839
  gel-filled wedge, 496
  liquid and vapour flow in, 479–482
  and matrix diffusion, 485–488
  micro, in continuous film, 477
  shape, effect on penetration, 120, 248, 497
  and swelling, 487, 493–495
Potential of water, governing movement, 153
Potting composts, 294
Predators, 729
Prill, 797
Pronamide (Propyzamide), 761
Propanol, permeation of cuticle by, 575
Propoxur, 231
Propylene glycol, 648
Prosopis juliflora, 638
Prunus armeniaca leaf cuticle, 578
Pteridium aquilinum, 564
Puccinea recondita, 683
Pulse, of concentration, forms of, 138, 318, 530, 538
  complete integral, *see* Availance
  finite integral, *see* Exposure
Pyrazon, 644
Pyrethroids, synthetic, 215, 711, 719, 870
Pyrimidine fungicides, 602

Pythium sp., 685

Q

Quercus robur, 564
  suber, 555
  marilandica, 640
  virginiana, 596

R

Radiation, solar, 926
  and evaporation
    from soil, 346
    from drops, 361
Radio autographs
  of plants, 613, 623, 625
  of soil sections, 841
Raffinose, 607
Rain
  effect on deposits, 803–806
  rates of fall, 929
  see also Washing
Raoult's law
Rat, 686
Rate
  of reaction, 206–211
    during diffusion and flow, 168–187
  of spread of liquids between surfaces, 415–416
    on surfaces, 417–427
  of solution, *see also* Evaporation
    acceleration of, 837–840
    anomalous, of Li salts, 187, 838
    control of, by diffusion, 187–190
    of granules, 824–831
Rate theories of chromatography, 303–310
Reception, reduction of, 818
Redistribution of seed dressing, 702
  from leaves, 736–740
Reflection
  of drops from leaves, 370, 414, 441, 443, 793–796
  of drops from water surface, 794
  principle in space-limited diffusion, 141, 529, 752
Release, from source in soil, 821–823
  see also Controlled Release
Repellents, 742, 747, 822
Residues of spray, forms of, 461–467
  see also Deposits
Resistance
  in interfaces, 147
  to mass transfer (diffusion) 340, 473 847
  of monolayers, 502–505
  of species to pesticides, vi, 179, 724
Respiration inhibitors, 597
Restriction of pesticides in water
  in space, 388–394
  in time, 394–401
Retention, *see also* Reflection, Run-off
  in cuticles, 735
  factor for non-wetting liquids, 446
  of solid particles, 370, 796–799
  of spray, definitions, 443
  of wetting liquid on inclined surfaces, 434
Reversibility
  of adsorption, 81–83
  of chemical processes, 77
  of physical processes, 78–81
  of uptake, 587–595
Reynold's number, 155, 277, 334
Rhagonycha fulva, 723
Rhizoctonia solanae, 686, 694
Root or shoot uptake from soil, 760–761
Roots; 555
  entry into, 596–600
    environmental effects on, 642–646
    significance of local damage, 759, 760
    inhibition of adventitious, 760
    translocation from, 600–606
Rotostat seed dresser, 852
Roughness of surfaces
  effect on contact angle, 410–415
  effect on drainage, 434
  effect on retention, 414
Rubus procerus, 356, 587, 638
Run-off
  from leaves, 447–451
  from soil surface, 326–331

S

Saccharomyces cerevisiae, 686
  fragilis, 685

pastorianus, 685, 694
Salt additives to systemic sprays, 654–657
Saltation, 326
Salvinia spp., 391, 429, 455, 456, 554
Satellite drops, 816
Saturated Vapour Concentration (SVC), 14–18, 49, 883
Schizophyllum commune, 681
Schmidt number, 339
Sedimentation
equilibrium (molecular), 874
of small particles in turbulent air, 372
rate in air (terminal velocity), 368–369
in liquids, 463
of smokes, 770
in stored formulations, 771
within drops, 463, 807
Seed treatment, 849–857
action of, 701–703
evaluation, 853, 854
interaction with biological factors, 704–710
methods, 851–853
Selectivity between insect species, 727–730
ecological, 228, 729
modification by chemical formulation, 228–234
physiological, 728
Selective adhesion, 806–809
availability, 811
entry into transport systems, 631–633
barriers in root, 606, 610
Sensitivity of olfaction, 752
Sepiolite, 777
Setaria viridis, 761
Settlement, gravitational, 366–372
Setting, of insects, pick-up during, 722
Sevin (Carbaryl), 720
Shattercane, 850
Sherwood number, 339
Shoot activity of soil herbicides, 761
Sign strengths, 750
Silver, 686, 692
Simazine, 49, 645, 761
distribution in soil, 242
depletion near root, 273
evaporation, 353
Skin, human
asymmetric permeation by cations, 506
permeation by alcohols from water and oil, 516
Smokes, 769
Smoke generators, 769
Smoke plumes, 744
Snails, aquatic, 395
Soaps
as fungicides, 433, 685
as unstable emulsifiers, 808
Social insects, 740
Sodium chloride, restricted permeability in cellulose acetate, 609
Soil, *see also* Adsorption, Pores, Roots, Seed treatment, Structure
application of pesticides to, 820–849
behaviour in, and other porous materials, 236–331
Soil inhabiting pests and pathogens, 699–710
Solanum nigrum, 850
Solid-cone jet, 452, 791
Soluble liquid (s.l.) = solution concentrate, 766
Solubility
ideal, of crystalline solid, 37
measurement of very low, 28–32
in mixed solvents, 39–47
mutual, of liquids, 23–28
product, 85
relationships in macromolecular substances, 44, 60–65
of salts, 84
super-additive, 43
and volatility of crystalline solids, 34–39
Solubilization, 45–47, 840
Solutions
ideal, 25
super-ideal, 32–34
Solvation of ions, 84
Solvent power, 44
as measure of non-crystalline fraction in fats, 561

Solvents, auxiliary, effect on permeation, 649–652
Sorghum sp. 651, 850
vulgare, 761
Sources and sinks
in diffusion theory, 144
in physiology, 622
see also Diffusion
Soybean (*Glycine*), temperature effect, 644
Spinning-disk sprayers, 452, 783, 813, 814
seed treaters, 852
Splash
dispersal of fungus spores by, 737–740
redistribution of fungicides by, 739
Spodoptera litoralis, 338
Spot density (population), 725, 932
Spreading, *see also* Rate
geometrical factors in, 454, 459
liquid on liquid, 403–408
limited, on solids, 408–410
Spreading coefficient (Harkins), 405
Spruce budworm, 733
Square of distance principle in crescent diffusion or flow, 114, 115
Stability (chemical) limits in formulation, 777
Stanton number, 348
Stearic acid 727, 841
Stern double layer, 89
Stickers, deposit, 809, 818
Stomach poisons, 711
Stomata
entry through, 559
penetration of liquid into, 459
trichomes and grooves, 553
Stratification, turbulence and eddy diffusion, 161–168
see also Fumigant density, Layering
Stratum corneum, permeation of, 493, 565, 664
Structure, crumb, in soil, importance for movement, 289–294
destruction by grinding, 299
molecular, in water, 53, 107
Suberin, 555, 679
Suction and hydraulic conductivity, 281
Sulphur, 682, 686
Sulphones and sulphoxides from thioethers, 220
Surface
forces operating in spread, 415–420
replacement between fluids, 193–198
tension, 404
critical, 427
effect of changing, 422–426
and energies of adhesion and cohesion, 406–409, 453
treatment of, on water, 389, 391
Surfactants as fungicides, 689
effect on interface movement, 194
effect on penetration, 652–654
molecular mechanism of, 71, 428
against mosquito larvae, 390
Suspension concentrates (flowables), 767
Sway of spray booms, 790
Swelling
effect on permeance
via pores, 487, 494, 496–500, 581
via matrix, 489–491
of plant cuticle, 566–569
Symplast, apoplast and free space, 557–559
Systemic insecticides and fungicides, 545, 611–616, 711
see also chemical names

T

2,4,5-T
DMSO as additive, 649
esters, evaporation of, 356, 363, 374, 587
surface run-off, 330
Taft constant, 521
Tamarix pentandra, 641
TBA (2,3,6-trichlorobenzoic acid), 218
Tears in wine, 422
Temperature
effect on uptake by leaves, 637–639
effect on uptake by roots, 642–646
variation with
of diffusion coefficient, 882
of permeability, 488, 638
of vapour pressure, 15, 883
wet bulb, 361
Terthienyl, 225

Tetrachlorodibenzodioxin (TCDD), 686
Tetradotoxin, 686
Thiabendazole, 690, 694
Thickeners, 815–817
Thiocyanate, ammonium, 656
Thiram, 682
Threshold concentration
for smell, 752
for toxic response, 536
Tillandsia sp., 456
Tobacco budworm, 726
Tomato fruit cuticle, 410, 485, 550, 568, 573, 582
Tortuosity of pores, 120
Toxicity,
physical, 58–60
relative, among chemicals and species, 686
Trails, scent, 740, 746
Transfer
coefficients, for momentum, heat and mass, 347
efficiency, for reactive pesticides, 169
dependence on site of reaction, 172–177
within canopies, 341–349
Translocation
in apoplast properties required for, 611
stream concentration factor (TSCF), 601
modification of, by chemical change, 223–228
from roots to shoot, 600–606
system in plants, 556
Transport,
active and passive, 619–621
by flow
in rivers, 395–398
in soils, 288–313
in xylem, 610
mechanism of phloem, 616–618
in symplast, 612–633
systems, selective entry into, 631–633
Traube's rule, 50
Triazine herbicides, *see also* individual chemicals
depth effects, 758, 760
ionic adsorption on clays, 98
partition water/air, 49
cylohexane/water, 56
root uptake, 599
Trichloroacetic acid, 235
2,3,6-trichlorobenzoic acid, 218
2,3,6-trichlorobenzyl compounds, 233
Trichlorphon, 218, 219, 225
Trichoplusia ni, 748, 755
Triethanolamine salts of hormone acids, 642
Trifluralin, 272, 325, 644, 758
Trimethylamine, 808
Trouton's rule, 16
Turbulence, 161, 194, 335, 341
see also Diffusion, Eddy
Two-pack formulations, 810, 817

U

ULV (Ultra Low Volume application), 719, 727, 734, 814
Uniformity of applicance
across swath, 784–790
along swath, 790
of penetrated spray, 791, 937
Uptake
of fungicides by fungi, 684–686
into plants, measurement of, 582–587
reversibility of, 587–595
into roots and xylem, 596–612
from vapour into insects, 676–679
see also Vapour
Uranyl ion, 691
Urea
as carbon source in autography, 623
3,4-dichlorophenyl, 213
permeation of cuticle by, 573
Urea herbicides, *see* individual chemicals
Urtica sp., 455
Ustilago maydis, 686, 694

V

Vaporization, reduction of, 818
Vapour
action (transfer), 165, 676, 682
drift, 384

saturated, 14–18, 49, 883
transport and uptake from, 374–379
Velocity, terminal in free fall, 368–369
Venturia inaequalis, 432, 683
Verbascum thapsus, 455
Vespula vulgaris, 732
Vicia faba, 578, 631, 637, 729, 739, 758, 839
Vine leaves, 563

W

Walking, pick-up by insects during, 722–725
Wallboard, 715
Walnut shell, 772
Washing
off surfaces, 585–587
out of plants, 587–595
out by rain, 595
Water
application to, 386–401
atmosphere content and SVC, 926
effect on air density, 163, 164
Wax
insect, 660
leaf, cuticular, 561–563
epi-cuticular, 549
bio-synthesis, 225, 551
as granule coating, 863
Waxed paper, permeability of, 566
Weathering of deposits, 803–806
Wettable powder (w.p.), 767
Wetting agents mechanism, 427–430
types of compound, 430–433, 780
Wheat bulb fly, 700
Wick evaporation, 325
applicators, 766
Wind-blow, from crops, 803
of herbicidal soil, 331
Wireworms, 701

X

Xylem, 556
models, 604
transport in, 610
reverse flow in, 636
Xylene
exchange between air and water, 197
evaporation rate, 352

Y

Young's equation, 409
Yeast, 686

Z

Zebrina
pendula, 623
purpusii, 559
Zone of control, 181